S0-FJU-282

MOLECULAR CYTOLOGY

Volume 2
Cell Interactions

MOLECULAR CYTOLOGY

Volume 2
Cell Interactions

JEAN BRACHET

Laboratoire de Cytologie et Embryologie Moléculaires
Département de Biologie Moléculaire
Université Libre de Bruxelles
Brussels, Belgium

1985

ACADEMIC PRESS, INC.

Harcourt Brace Jovanovich, Publishers
Orlando   San Diego   New York   Austin
London   Montreal   Sydney   Tokyo   Toronto

COPYRIGHT © 1985 BY ACADEMIC PRESS, INC.
ALL RIGHTS RESERVED.
NO PART OF THIS PUBLICATION MAY BE REPRODUCED OR
TRANSMITTED IN ANY FORM OR BY ANY MEANS, ELECTRONIC
OR MECHANICAL, INCLUDING PHOTOCOPY, RECORDING, OR
ANY INFORMATION STORAGE AND RETRIEVAL SYSTEM, WITHOUT
PERMISSION IN WRITING FROM THE PUBLISHER.

ACADEMIC PRESS, INC.
Orlando, Florida 32887

United Kingdom Edition published by
ACADEMIC PRESS INC. (LONDON) LTD.
24–28 Oval Road, London NW1 7DX

Library of Congress Cataloging in Publication Data

Brachet, J. (Jean), Date
    Molecular cytology.

    Includes bibliographical references and indexes.
    1. Cytology.   2. Molecular biology.   I. Title.
QH581.2.B73   1985   574.87   84-24418
ISBN 0–12–123371–5 (v. 2. : alk. paper)
ISBN 0–12–123373–1 (v. 2. : paperback)

PRINTED IN THE UNITED STATES OF AMERICA

85 86 87 88   9 8 7 6 5 4 3 2 1

To my dear wife, Françoise,
the faithful companion of the
good and evil days.

I build a castle of comfort and indulgence
for him, and stand sentinel always to
keep little vulgar cares out.
G. B. Shaw, *Candida*, Act III

CONTENTS

PREFACE

When I was asked to write a second edition of "Biochemical Cytology" (published by Academic Press in 1957), I wondered where my own copy of this old book might be. When it became useless for teaching purposes, it had to leave the desk for some distant bookshelf. I finally found it and put it back on the desk to be subjected to autopsy. My verdict was that only two things retained some value: the Preface, because the author's general philosophy has remained almost unchanged, and the general backbone (the book's skeleton), because cells have remained cells. However, the picture of the cells has become increasingly complex; we know much more about them and understand them better than we did 25 years ago.

Today "Biochemical Cytology" is of little more than historical interest; our ignorance of major facts a quarter of a century ago makes the reading of this old book almost painful. Reading it now does illustrate fashions in science: many facts and hypotheses which were hotly disputed in "Biochemical Cytology" do not arouse the slightest interest today. Because cell biology has undergone a complete revolution, thanks to the explosive progress of knowledge in the field of molecular biology, the book evolved into an entirely new one—hence the title change to "Molecular Cytology." Nonetheless, the general philosophies of both the author and the book have remained unchanged; the Preface of "Biochemical Cytology" remains valid for "Molecular Cytology."

We have attempted in this book to present an integrated version of what is currently known about the morphology and the biochemistry of the cell. There are many excellent cytology and biochemistry textbooks. What remains to be done is the difficult and important task of linking these two sciences more closely together, now that they have so much in common. This is what we tried to do, with the hope that the book will prove useful to advanced students and research workers.

It has been assumed that the reader already knows the fundamentals of descriptive cytology, biochemistry, embryology, genetics, and molecular biology. Our goal will have been reached if the reader enjoys the attempt we have made to show that structure and metabolism are so closely linked together that they cannot be separated.

Special emphasis has been laid on the problems which are most familiar to the author. This will perhaps excuse the apparent imbalance of the book. If too much is said on embryos and too little on cancer cells, it is because the author has spent much of his life working with embryos and has so far not touched a cancer cell.

Emphasis has been given to the more dynamic aspects of cytology, not to detailed description. More is said about nucleocytoplasmic interactions in unicellular organisms and eggs than about the pure description of cytoplasmic and nuclear constituents.

Hypotheses and personal opinions have not been forgotten, for hypotheses, provided that they can be tested experimentally, may become more important than dry facts. Ideas are as vital for scientists as engines for cars or airplanes. Nowadays, some scientists forget that thinking may sometimes be more useful than performing an experiment.

The book has been written directly in English, and the author may not have expressed the ideas and facts as precisely as he would have wished. But what has been lost in subtlety has perhaps been gained in directness and clarity.

The need for a book dealing with biochemical or molecular cytology is obviously much less acute today than it was 25 years ago. At that time "Biochemical Cytology" was not accepted easily by either biochemists or cytologists. I remember vividly a very distinguished professor of anatomy and histology accusing me publicly of having produced a dreadful bastard; my answer was that hybridization can lead to improvement of crops. Today the battle is over, and there are several excellent books dealing with the molecular aspects of cell biology. Some of them, for instance those of Dyson (1978),* DeRobertis and DeRobertis (1981),† and Alberts et al. (1983),‡ are textbooks for students; these texts are remarkably complete, clearly written, and illustrated. I used them frequently for the preparation of these volumes, since a scientist remains a student all his life.

This book, like "Biochemical Cytology," is intended for advanced students and for research workers, but it is not an encyclopedia: to aim at completeness is an impossible task in view of the tremendous growth of the scientific literature during the past years. Already the monumental six-volume treatise "The Cell," which was edited 20 years ago by the late Alfred Mirsky and myself, is obsolete and incomplete. The voluminous literature and the specialization of prospective authors have so far precluded a second edition. A second edition would be a giant treatise and, thanks to computer-assisted literature searches, the reader might be crushed under the weight of documentation. If we believe the Preface of Anatole France's "L'Ile des Pingouins" ("Penguin Island"), such a situation has arisen: The author, who is writing a book on the history of the penguins, seeks information from the greatest art critic in the world. His office is filled with files from top to bottom. Finally, A. France, after climbing on the top of a scale, finds the file dealing with penguin art and lets it drop. All the files escape from their

---

*Dyson, R. D. (1978). "Cell Biology: A Molecular Approach." Allyn & Bacon, Newton, Massachusetts.

†DeRobertis, E., and DeRobertis, E. M., Jr. (1981). "Essentials of Cell and Molecular Biology." Holt, New York.

‡Alberts, B., et al. (1983). "Molecular Biology of the Cell." Garland, New York.

boxes, and the critic dies under their weight. His last words are "Que d'Art!" Since many more papers have been published on cell biology than on art in Penguin Island, the utilization of a complete documentation system would have been exceedingly dangerous for the author and boring for the readers.

No computer has been used for the preparation of the present book except an old rusty one, my brain, where lipofuscins and melanins should be expected to accumulate. My major task has been selection, a process which always brings out criticism because it is arbitrary. Most of the selected references are those of recent papers (where the previous literature is summarized) and of review articles published in easily available journals. Despite their number and quality, papers published in specialized symposia have been seldom quoted.

An author should take responsibility for the choice of the topics, the way he deals with them, and the correctness of the references. He should, as much as possible, avoid factual errors. However, the great English biochemist Frederick Gowland Hopkins told me, many years ago, that all textbooks, including his own, are full of errors. Reading through "Biochemical Cytology," which was once praised as "remarkably free of factual errors," I was ashamed to find the book full of errors. They are in general due to the progress of science, which will always move ahead. Errors will accumulate in the present book and, in agreement with Orgel's "error catastrophe theory" of aging (see Chapter 3, Volume 2), they will ultimately lead to its death. An author is also expected to avoid repetition. But there is repetition (perhaps too frequently) in this book, because I believe that important facts should be told more than once and presented again, in a slightly different way, when they are examined from another angle—at least this is a lesson I learned after more than 40 years of teaching and scientific direction of a laboratory.

In short, the aim of the present book is to present a critical and synthetic view, not an encyclopedic description, of living dynamic cells to young scientists. Whether or not this aim has been reached will be decided by the readers themselves.

It is a pleasure and a duty for me to thank all those who kindly agreed to read the manuscript, in particular Professor Werner W. Franke (Heidelberg), who carefully revised the first three chapters. Thanks go also to my colleagues from the University of Brussels, Professors A. Burny, H. Chantrenne, A. Ficq, and P. Van Gansen, who corrected many errors that had escaped my notice. The heavy burden of selecting and preparing the illustrations went to Professors P. Van Gansen and H. Alexandre and Mr. D. Franckx and that of checking the correctness and completeness of the references to Mrs. A. Pays. Last, but not least, my warmest thanks are due to Mrs. J. Baltus, who had the long and unpleasant job of typing the manuscript, and to the publishers for encouragement and patience.

JEAN BRACHET

CONTENTS OF VOLUME 1

The Cell Cycle

NUCLEOCYTOPLASMIC INTERACTIONS IN SOMATIC CELLS AND UNICELLULAR ORGANISMS: GENE TRANSFER IN SOMATIC CELLS

I. GENERAL BACKGROUND

Many of the results summarized in Chapters 3, 4, and 5 of Volume 1 were obtained from fractionated homogenates that are particularly suitable for biochemical studies. However, these studies must be controlled in order to reduce the introduction of artifacts, such as loss or adsorption of enzymes during isolation and fractionation. Careful EM observations are needed to prove that the isolated fractions are identical to the preexisting intracellular organelles: rupture of nuclei or mitochondria, aggregation of cytoplasmic particles, or degradation of macromolecules by hydrolytic enzymes during differential centrifugation of homogenates are dangerous pitfalls. Very careful work is therefore needed to be certain that the factions obtained in this way are cytologically homogeneous.

The limitations of these methods become more apparent when, as in this chapter, interest shifts from mere description of the chemical composition of the various cell organelles to a more dynamic approach: what is the nature of the interactions that occur between the various cell constituents, and between the nucleus and the cytoplasmic organelles, in particular? It is reasonable to combine the various fractions obtained by differential centrifugation of homogenates. For example, Vishniac and Ochoa (1952) studied the biochemical events that occur when chloroplasts are mixed with mitochondria isolated from animal tissues. In an effort to understand the biochemical role of the cell nucleus, Potter *et al.* (1951) and Johnson and Ackermann (1953) have studied the effects of adding nuclei on oxidative phosphorylations in mitochondria. However, such experiments have little value when the nature of the interactions between nuclei and mitochondria in the intact living cell becomes our objective. There is a very serious reason for doubting, in the case of nuclei, at least, the value of experiments performed on mixed fractions recovered by differential centrifugation of homogenates. If we consider as a test of survival for isolated nuclei the capacity of dividing when they are reintroduced into adequate cytoplasm, there is no doubt that, as a rule, they will be quickly inactivated by contact with the outside medium. The experiments of Comandon and de Fonbrune (1939) on the transplantation of a nucleus into an anucleate ameba fragment showed that brief contact of the nucleus with the outside medium resulted in the "death" and elimination of the transplanted nucleus. The same is true of the embryonic nuclei transplanted by Briggs and King (1953) into anucleate, unfertilized frog eggs (see Chapter 2). However, progress has been made in this field, as in so many others. A method that allows the isolation of "living" somatic nuclei has been devised by Gurdon (1976). If nuclei, such as those described above, are injected into *Xenopus* oocytes or eggs, they can, as we shall see in Chapter 2, either synthesize various RNA species or replicate their DNA.

In order to better understand the biochemical role of the cell nucleus in the living cell, experimental approaches other than homogenization and centrifugation are needed. One approach is found in molecular genetics, but this is such a huge field that an entire book would be needed to provide all of the details. It must suffice here to discuss briefly the exciting gene transfer experiments that are now being done in several laboratories. This chapter will discuss only the interactions that take place between the cytoplasm and the nucleus as a whole, and not the expression of a specific gene. This question will be reviewed when we discuss cell differentiation in Chapter 3.

Two main approaches have been used to study nucleocytoplasmic interactions in somatic cells and unicellular organisms. One is to work on intact cells and to use autoradiography and other cytochemical techniques as the major tools; the other is to compare the biological and biochemical properties of nucleate and anucleate halves obtained by merotomy of cells, protozoa, or eggs.

The classic experiments of merotomy were done more than 85 years ago by Verworn (1892), Balbiani (1888), Klebs (1889), Townsend (1897), and many others on eggs, protozoa, and animal and vegetal cells. They led to an important conclusion, which was emphasized by Mazia (1952) in a review article: "there is not a single case where an activity has not continued in an enucleated cell." However, the life span of an anucleate cytoplasm varies considerably from cell to cell. Anucleate fragments from mammalian cells (cytoplasts) seldom survive for more than a couple of days; on the other hand, anucleate fragments of the giant unicellular green alga *Acetabularia* can survive for as long as 3 months. An anucleate half of an ameba remains alive for about 10 days. In all cases, the life span of anucleate cytoplasm is somewhat shorter than that of its nucleate counterpart, but the difference between the two is never striking, provided they are maintained under the same conditions.

The fact that anucleate fragments of protozoa and eggs survive and retain their biological activities (even ciliary or ameboid motility) for a certain length of time came as a complete surprise to the author when he attended, more than half a century ago, the first lecture of a cytology course given by Pol Gérard. Naively, the author believed that removing the nucleus would be the same as cutting off the head of a man; this led him to study nucleocytoplasmic interactions in amphibian and sea urchin eggs, as well as in unicellular organisms such as *Amoeba proteus* or *Acetabularia mediterranea*. Similar work currently being done in many laboratories has provided quite satisfactory answers to many of the questions raised by the survival of the anucleate cytoplasm. This work has also put to rest many of the ideas that had been proposed to explain the inferiority of anucleate cytoplasm to nucleate fragments of cells. For instance, Loeb (1899) had proposed that anucleate cytoplasm dies because the cell nucleus is the main center of energy production. This idea was based on inadequate cytochemical evidence about the intracellular localization of respiratory enzymes, and was disproved by work done on the respiration of nucleate and anucleate fragments of eggs and unicellular organisms (and by the already described experiments on the distribution of the respiratory enzymes in centrifuged homogenates). While the question of whether isolated nuclei are capable of restricted energy production remains debatable, there is no doubt that energy production by the nucleus—if it takes place at all—is negligible in comparison to that which occurs in the cytoplasm.

Based on the facts known in 1925, E. B. Wilson concluded that "the nucleus might be a storehouse of enzymes, or of substances that activate the cytoplasmic enzymes, and that these substances may be concerned with synthesis as well as destructive processes." In other words, the nucleus would be a site for enzyme and coenzyme synthesis or accumulation. Stated in such a general way, this idea is no longer tenable although we may still consider the nucleus a storehouse of some important enzymes synthesized in the cytoplasm. This is the case for the polymerases involved in both DNA and RNA synthesis, and probably for the enzymes that control the metabolism of the dinucleotide coenzyme (NAD), the synthesis of which takes place mainly in the nuclei of both liver cells (Hogeboom and Schneider, 1952) and starfish oocytes (Baltus, 1954). In addition, synthesis of poly(ADP) ribose at the expense of NAD and poly(ADP) ribosylation of chromosomal proteins are also nuclear functions. Thus, E. B. Wilson's suggestion remains valid for enzymes involved in dinucleotide and polynucleotide synthesis but can no longer be accepted for the hydrolytic enzymes that are accumulated in the lysosomes. The same criticism holds for Caspersson's (1950) theory, which proposed that the cell nucleus is the main center of protein synthesis. We know (see Chapters 3 and 4, Volume 1) that most, if not all, of the proteins present in the nucleus are synthesized in the cytoplasm; some of them migrate into the nucleus and eventually accumulate there. The question remains as to whether isolated nuclei are capable of *in vitro* protein synthesis. Nevertheless, it is clear that neither energy production nor protein synthesis can be the major biochemical activity of the nucleus. This negative conclusion was reinforced by the biochemical analysis of merotomy experiments; they demonstrated that the nucleus is specialized in nucleic acid synthesis, as we shall now see.

II. WORK ON INTACT CELLS

Autoradiography is a very valuable method for the study of macromolecule synthesis and has the great advantage that it can be used with intact cells. The principle and the methodology are simple (Ficq, 1959): radioactive precursors, usually tritium labeled, are added to living cells, and their incorporation into macromolecules is observed photographically. The classic precursors are [^{3}H]leucine for the study of protein synthesis, [^{3}H]uridine for RNA synthesis, and [^{3}H]thymidine for DNA synthesis. The technique can be used at the EM level in order to visualize, with excellent definition, the localization of newly synthesized proteins or nucleic acids. Autoradiography also has the unique advantage of showing which cells in a heterogeneous cell population are engaged in macromolecule synthesis. However, autoradiography also has its limitations. It does not allow easy measurement of the uptake of soluble precursors. Thus, if as a result of enucleation the uptake of the precursor is increased or decreased, erroneous conclusions might be drawn. In order to obtain quantitative data, therefore, information about the size of the precursor pool and the chemical composition of the newly synthesized macromolecules should be obtained by other methods.

In the early experiments described in "Biochemical Cytology" (Brachet, 1957), treatment of the cells with the appropriate precursors was a lengthy process, and no adequate inhibitors of DNA, RNA, and protein synthesis were available. Our conclusion at that time, therefore, was that "very active RNA metabolism occurred in the nucleus and that protein metabolism, on the other hand, was not necessarily more active in the nucleus than in the cytoplasm." This conclusion was substantially correct, although it was based on weak and sometimes contradictory evidence. The situation became much clearer when many kinds of cells were submitted to short pulses (1 hr or less) with the tritiated precursors. It was found that thymidine and uridine were quickly incorporated into the nuclei, while labeling of proteins began in the cytoplasm. If the pulse was followed by a chase (usually a treatment with the nonradioactive precursor), kinetic analysis showed that thymidine incorporation into chromatin remained unchanged and that it was limited to nuclear DNA. There was no increase in radioactivity unless DNA replication occurred during a long chase. Labeled nuclear RNA, however, moved quickly from the nucleus into the cytoplasm. Conversely, proteins that had been synthesized on cytoplasmic polyribosomes moved the nuclei. Thus, as a rule, RNAs flowed from the nucleus to the cytoplasm, while proteins moved in the opposite direction. Autoradiography, besides leading to this important conclusion, has largely contributed to the demonstration of a number of facts that have already been discussed in this book, for example, the existence of an S phase during the cell cycle of DNA semiconservative replication in eukaryotes and of sister chromatid exchanges. The great lability of the nuclear RNAs compared to the long half-life of the cytoplasmic ribosomal RNAs and the stability of the nuclear DNA molecules were discovered utilizing a combined approach of autoradiography and biochemical methods. The use of inhibitors of macromolecule synthesis (HU, aphidicolin, actinomycin D, α-amanitin, cordycepin, puromycine, cycloheximide, etc.) combined with auto-radiography also played an important role in increasing our understanding of nucleic acids and protein synthesis at the cellular level. Although all of this work is now mainly of historical interest, it is important to recall here that progress in cell biology still requires a combination of biochemical and cytochemical data. Safe conclusions can never be drawn unless the results are obtained with diverse techniques, such as differential centrifugation of homogenates and autoradio-graphy.

Another cytochemical approach that is now of mainly historical interest but that, like autoradiography, has retained its usefulness is the binding to chromatin, in intact cells, of substances that bind selectively to DNA. For example, the fluorescent dye acridine orange (Ringertz and Bolund, 1969; Ringertz et al., 1971) or tritiated actinomycin D (Brachet and Hulin, 1970) have been used successfully for the study of genetic activity in intact cells. Acridine orange gives a bright green fluorescence with DNA and a red fluorescence with RNA. The work of Ringertz and Bolund (1969) has shown that there is a close correlation between the amount of dye bound by chromatin and its genetic activity. The same is true for the binding of [3H]-labeled actinomycin D, which binds prefer-entially to guanylic acid residues of DNA and can be detected by autoradiogra-phy. If chromatin is in a repressed, inactive state, its DNA is covered by chro-mosomal proteins and there will be little or no binding of acridine orange or actinomycin D. If lymphocytes are activated by treatment with lectins (which leads to increased RNA synthesis before the induction of DNA replication), binding of both actinomycin D and acridine orange markedly increases (Ringertz and Bolund, 1969). On the other hand, there is a decrease in [3H]-labeled actinomycin D binding when embryonic cells differentiate and become spe-cialized in the synthesis of a major protein such as hemoglobin (Brachet and Hulin, 1970). Actinomycin D binding can even be studied at the ultrastructural level (Steinert and Van Gansen, 1971).

Although the techniques discussed in this section may seem obsolete to many readers—particularly since they obviously cannot compete with more specific methods, such as in situ hybridization or immunocytochemistry—they deserve mention for two reasons: (1) they played an important role in increasing our understanding of nucleocytoplasmic interactions and (2) they remain a useful first approach (like Feulgen or Unna staining) in the study of nucleic acid dis-tribution and metabolism in intact cells.

III. ENUCLEATION EXPERIMENTS

In the following section, we shall deal with somatic cells in culture (where enucleation can be obtained by treatment with cytochalasin B), with reticulocytes (where loss of the nucleus is a natural process), and with merotomy experiments on protozoa and the giant unicellular alga *Acetabularia*. Merotomy experiments on sea urchin and amphibian eggs will be discussed in the next chapter.

A. KARYOPLASTS AND CYTOPLASTS

As mentioned in the preceding volume, disruption of the microfilament (MF) cytoskeleton by treatment with cytochalasin B induces, in many cells, the extrusion of the nucleus together with a small amount of cytoplasm (Fig. 1). Low-speed centrifugation helps separate the cytochalasin B–treated cells into small karyoplasts and large cytoplasts, which represent 80–90% of the initial cell volume (Ladda and Estensen, 1970); Poste and Reeve, 1971). Karyoplasts and cytoplasts can be isolated, on a large scale, by taking advantage of their difference in size and density (centrifugation methods) or of the fact that cytoplasts adhere to the substratum better than karyoplasts (Follett, 1974; Wigler and Weinstein, 1974, 1975). When cells undergoing mitosis are treated with cytochalasin B, so-called mitoplasts which are cytoplasmic fragments of dividing cells (Sunkara *et al.,* 1977), can be obtained. Albrecht-Buehler (1980) has described a method, involving cytochalasin B treatment and pipetting, for the preparation of microplasts that represent only 2% of the initial volume of a fibroblast.

Anucleate cytoplasts can be fused to karyoplasts or whole cells of the same or a different species by treatment with uv-inactivated Sendai virus or with polyethylene glycol. Such "reconstituted" cells and the *cybrids* obtained by fusion of a cell with the cytoplast of a different species can survive and divide (Ladda and Estensen, 1970; Veomett *et al.,* 1974; Ege *et al.,* 1974a,b). It has even been possible to fuse cytoplasts from different species, called "heteroplasmons" by Wright and Hayflick (1975). We shall leave cell hybrids and cybrids for a later section of this chapter (Section IV) and limit ourselves here to a comparison between karyoplasts and cytoplasts.

Karyoplasts are formed by the cell nucleus surrounded by a thin, ribosome-containing cytoplasmic layer (Wise and Prescott, 1973). According to Shay *et al.* (1974), they lack centrioles and microtubules (MTs). Karyoplasts remain spherical and die after 72 hr. In the case of myoblasts, karyoplasts do not divide or

---

FIG. 1.   Cytoplasts and karyoplasts. (a) Several enucleated Chinese hamster cells (cytoplasts) and a single nucleated cell. (b) An electron micrograph of an L-cell karyoplast surrounded by cytoplasmic fragments. [(a) Prescott and Kirkpatrick, 1978; (b) Lucas (1977).]

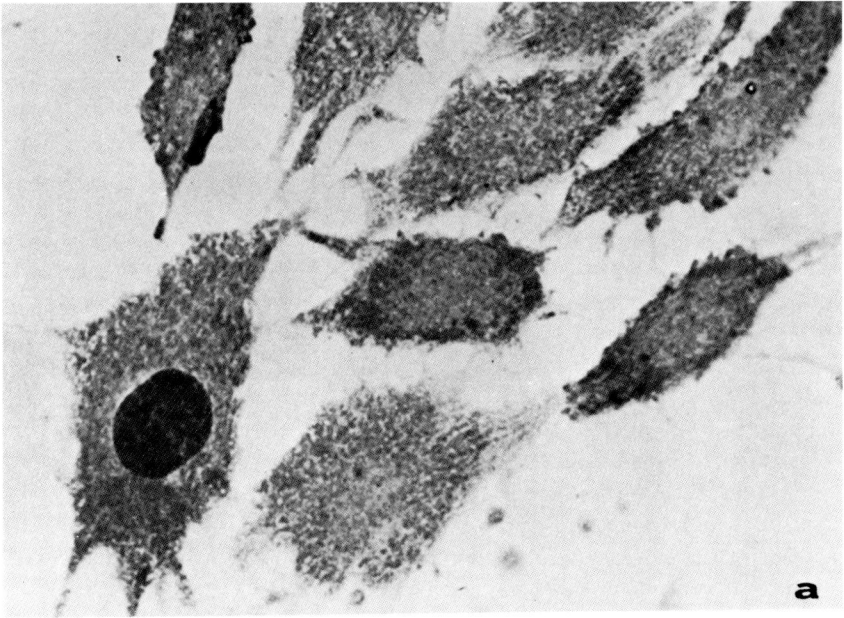

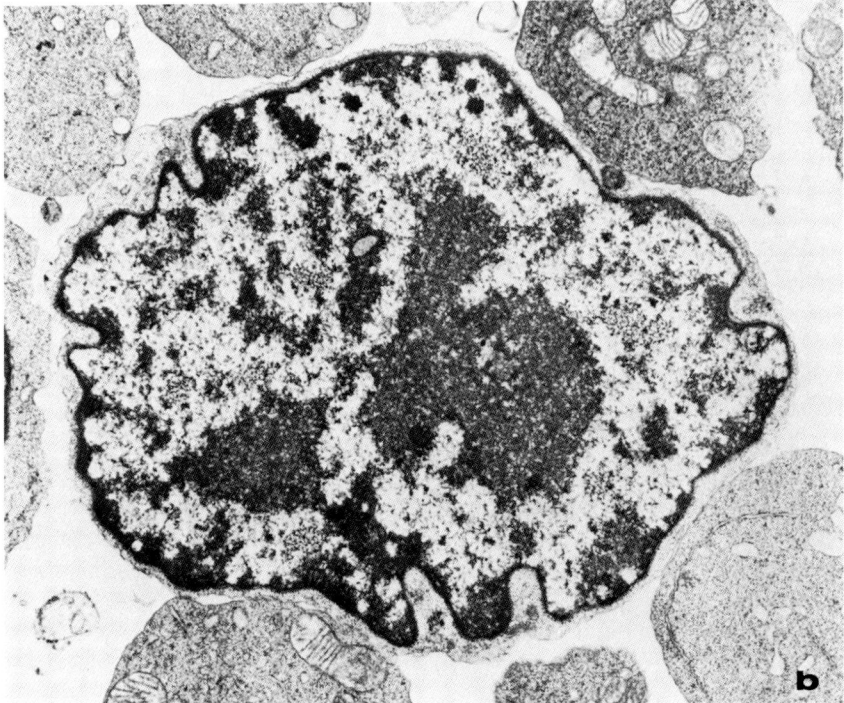

regenerate, dying within 36 hr (Ege *et al.*, 1974a). According to a more recent quantitative study by Zorn *et al.* (1979), karyoplasts represent 10% of the cell volume, possessing 11% of the mitochondrial and only 3% of the Golgi volumes. Zorn *et al.* (1979) have confirmed that karyoplasts lack centrioles, but that they appear, presumably *de novo,* in the karyoplasts that have attached to the substratum. This occurs in only 10% of the population, but exerts a favorable effect on the life span.

It has been possible to select karyoplasts capable of growing in a culture medium; the morphology and biochemistry of these regenerating karyoplasts, which ultimately become complete cells, have been studied by White *et al.* (1983). They found that the pattern of protein synthesis changes in a characteristic, almost stage-specific way as a function of time. The synthesis of polypeptides associated with the nucleus and MTs, as well as that of soluble proteins, is closely regulated, while MF and intermediate filament (IF) proteins are synthesized at a constant rate. MFs and IFs assemble spontaneously during karyoplast regeneration; even MTs form in the absence of visible centriole-associated organizing centers. The last events to take place during karyoplast regeneration are DNA synthesis (which does not occur until the karyoplasts have regenerated almost all of their cytoplasm) and *de novo* centriole formation.

More is known about cytoplasts. In 1956, Crocker *et al.* reported that enucleated HeLa cells remained apparently normal for 40 hr, after which shrinkage and loss of motility were observed. More recent work on cytoplasts obtained by the cytochalasin B technique has shown that, while the small karyoplasts usually round up, the locomotion of the cytoplasts—despite the fact that their survival is shorter (about 48 hr) than that of the karyoplasts—remains normal for many hours. These findings led Shay *et al.* (1974, 1975) to conclude that maintenance of shape and motility are cytoplasmic functions. Similar findings were made on epithelial cells by Goldman *et al.* (1973), who noted that cytoplasts look essentially normal from the viewpoints of attachment to the substratum, spreading, pinocytosis, locomotion, and contact inhibition. In anucleate leukocytes phagocytosis is normal, but digestion is slow and chemotaxis is reduced (Roos *et al.*, 1983). That enucleation has no rapid effect on cytoplasmic activities has been confirmed by the observations of Keller and Bessis (1975) on anucleate fragments of leukocytes, which are capable of normal locomotion, chemotaxis, and phagocytosis. In addition, binding and endocytosis of ConA in cytoplasts occur normally (Wise, 1974, 1976). It seems clear, therefore, that the main properties of the cell surface and of the cytoskeleton remain unaffected for some time by enucleation. In agreement with this conclusion, electron microscopy has shown that cytoplasts retain apparently normal MTs (Brown *et al.*, 1980) and IFs (Laurila *et al.*, 1981). However, removal of the nucleus is followed by the breakdown of the Golgi complexes (Shay *et al.*, 1974; Wise and Larsen, 1976). As we have just seen, this does not impede the binding of lectins such as ConA to the cell surface. Euteneuer and Schliwa (1984) studied the locomotion of cytoplasts from fish epidermal keratinocytes. Despite the absence of micro-tubules and centrioles, these cytoplasts are capable, as whole cells, of oriented locomotion.

The reactions of cytoplasts to agents that induce morphological changes in whole cells are also perfectly normal. Among such agents are dibutyryl-cAMP and protaglandin E_1, which affect cell shape in both intact cells and cytoplasts by an MT assembly–dependent process (Schröder and Hsie, 1973). Neuroblastoma cells as well as neuroblastoma cytoplasts respond to the addition of cAMP by neurite formation (Miller and Ruddle, 1974). Finally, viral DNA (obtained from an avian sarcoma virus) replicates in cytoplasts made from fibroblasts (Varmus *et al.*, 1974).

Interestingly, mitoplasts obtained by cytochalasin B treatment of dividing cells behave like mitotic cells; they remain spherical and do not attach to the sub-stratum (Sunkara *et al.*, 1977). Even the tiny microplasts studied by Albrecht-Buehler (1980) retain some of the activities of fibroblasts for some time. These very small pieces of cytoplasm, which die within 8 hr, can form filopodia and membrane ruffles, but are incapable of coordinated locomotion.

One might be tempted to conclude from this discussion that the only function of the nucleus is to allow cell division and reproduction. This is such an essential function that life would soon disappear if all cells lost their chromosomal DNA. However, it would be inappropriate to claim that cytoplasts are equivalent to whole cells in all respects. For instance, Otteskog *et al.* (1981) have shown that cytochalasin B can induce capping of antibodies in transformed (malignant) cells, but not in normal ones. Although cytochalasin B binds to cytoplasts formed from transformed cells, it does not induce capping of antibodies. The conclusion is that either the nucleus or a nucleus-associated organelle controls the mobility of plasma membrane molecules, at least in transformed cells.

It is in their *biochemical properties* that cytoplasts are, in general, inferior to whole cells. In cytoplasts there is no synthesis of nuclear DNA, and RNA synthesis stops very quickly. Although protein synthesis continues, it proceeds at a lower rate than that found in whole cells and its rate decreases continuously (Poste, 1972; Follett, 1974; Shay *et al.*, 1974; Bruno and Lucas, 1983); the rate of both RNA and protein synthesis is very low in mitoplasts (Sunkara *et al.*, 1977). Cholesterol synthesis is not affected for at least 6 hr by removal of the nucleus (Cavenee *et al.*, 1981), but both synthesis and turnover of the coenzyme NAD are modified. According to Rechsteiner and Catanzarite (1974), the en-zyme responsible for NAD synthesis (NAD pyrophosphorylase) is present only in the karyoplasts and is thus localized predominantly or entirely in the nucleus. There is no rapid destruction or turnover of NAD in the cytoplasts, whereas turnover of NAD is faster in karyoplasts because the nucleus contains most or all of the poly(ADP) ribose phosphorylase activity; the substrate of this enzyme is NAD (Rechsteiner *et al.*, 1976). Cytoplasts are also deficient in polyamine synthesis. They synthesize putrescine at a normal rate but are unable to convert it into the spermidine, leading to the conclusion that in fibroblasts ornithine decarboxylase (ODC) is probably ubiquitous in the cell, while S-adenosylmethionine-decarboxylase (SAM) is apparently not present in the cytoplasm (McCormick, 1977). In contrast to complete cells, cytoplasts are unable to control amino acid transport properly when the amino acid concentration of the medium is changed experimentally (Hume *et al.*, 1975). It is well known that hepatoma cells respond to the addition of glucocorticoids by the synthesis of tyrosine aminotransferase (TAT). This induced enzyme synthesis is controlled by nuclear genes, since glucocorticoids do not induce TAT activity in the cytoplasts of hepatoma cells (Ivarie *et al.*, 1977). In these cells, enucleation slows the degradation of another enzyme, glutamine synthetase, suggesting the existence of a nuclear control on lysosomal activity (Freidkopf-Cassel and Kulka, 1981). That such a control exists is not very surprising since, as we have seen, removal of the nucleus is quickly followed by disorganization of the Golgi bodies.

If cytoplasts are inferior to complete cells in many respects, the same might be true for the much less studied karyoplasts. Although RNA and proteins are synthesized, this occurs for a few hours only (Ege *et al.*, 1974b; Shay *et al.*, 1974). Karyoplasts are unable to regenerate and divide unless they succeed in attaching to the substratum. The poor macromolecular synthesis in karyoplasts is probably the result of their small size and very abnormal nucleocytoplasmic ratio. A similar condition prevails in the polar bodies expelled during oocyte maturation, which might be responsible for their still unexplained rapid degradation.

Finally, it would be a mistake to believe that the controls exerted by the nucleus on the cytoplasm are always positive. England *et al.* (1978) observed a temporary increase in mitochondrial protein synthesis in cytoplasts. Other examples of this negative nuclear control of mitochondrial activity will be discussed when we deal with eggs (Chapter 2).

B. RETICULOCYTES

Nature itself performs an enucleation experiment during mammalian erythropoiesis, when nuclei and mitochondria are cast out of differentiating cells. The resulting anucleate reticulocytes contain a basophilic "reticulofilamentous" substance. Electron microscopy has revealed the presence of many polyribosomes in this substance whose main function is the synthesis of a major specific protein, hemoglobin. Synthesis of this protein, coded by the nuclear genes for α- and β-globins, continues for a few days after the nucleus has been expelled. This is possible because the α- and β-globin mRNAs remain functional in the absence of the corresponding nuclear genes. During erythrocyte maturation (Fig. 2), these mRNAs break down and the polyribosomes disintegrate into individual monoribosomes. Finally, the latter also undergo complete degradation, leaving an adult red blood cell (or erythrocyte), which is little more than a bag surrounded by a membrane filled with hemoglobin. Mammalian red blood cells are easily

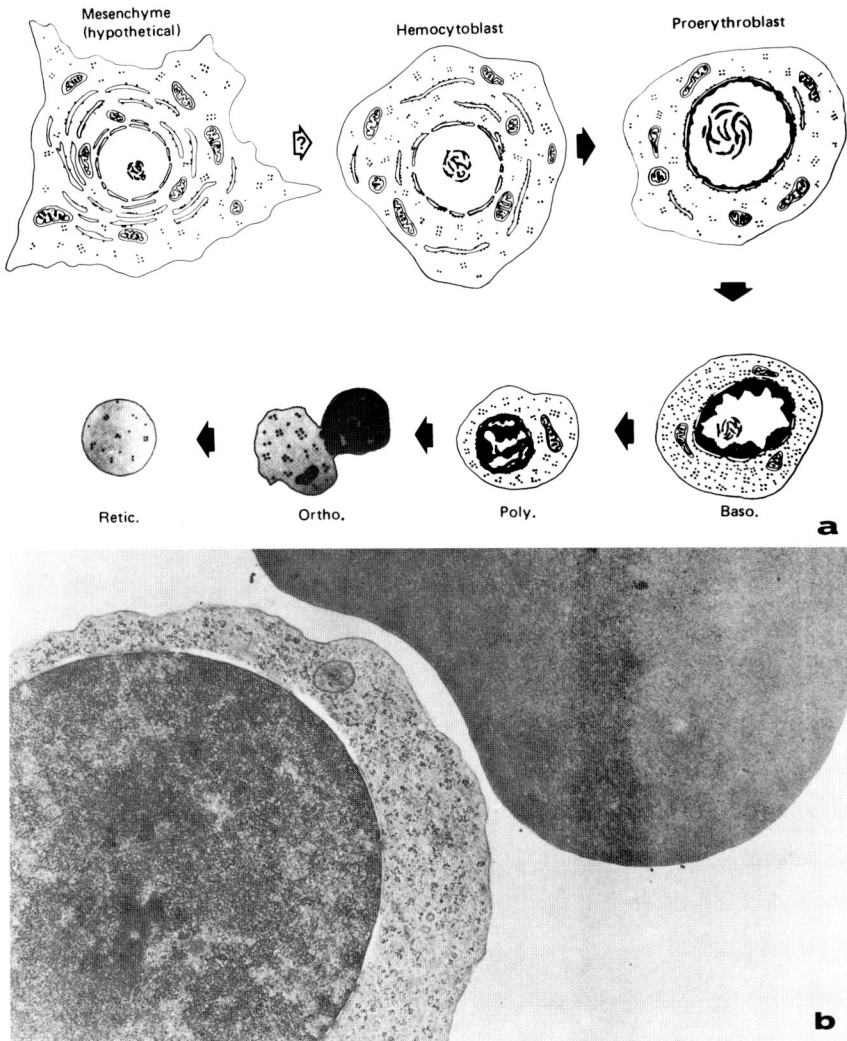

FIG. 2. Erythropoietic cells. (a) Differentiation of erythropoietic cells in the embryonic liver and in all adult erythropoietic tissues. Baso., basophilic erythroblast; poly., polychromatophilic erythroblast; ortho., orthochromic erythroblast; retic., reticulocyte. (b) Electron micrograph of a rat erythroblast (left) and an erythrocyte. [(a) Rifkind (1972); (b) Porter and Bonneville (1968).]

available in large quantities and are a favorite material for those working on cell permeability, membrane-bound enzymes (in particular, Na^+,K^+–ATPase, the so-called sodium pump), and the spectrin-actin contractile system. Treatment with hypotonic media results in hemolysis; the red blood cells swell, and hemoglobin leaks out. The remaining red cell ghosts can, if one wishes, be resealed and used as vectors for the introduction of large molecules (DNA, for instance) into recipient cells by cell fusion. Since the mitochondria, as well as the nuclei, disappear during erythropoiesis, it is not surprising that Warburg (1909) was able to show that reticulocytes respire much more than mature red blood cells.

It is very doubtful that the striking changes that take place during mammalian erythropoiesis directly depend upon the complete disappearance of the nucleus, since very similar changes occur in the nucleate red blood cells of birds and amphibians. Here too, the mitochondria ultimately disappear and the hemoglobin content increases at the stage when polyribosomes become abundant. However, the nuclei of bird erythrocytes are genetically inactive. This is due, in part, to the presence of histone H5, and to the facts that their chromatin is in a highly condensed form (Fig. 3) and their DNA is hardly transcribed. A similar situation probably prevails in the erythrocytes of fishes, amphibians, and reptiles, but we know little about them. Thus, nonmammalian erythrocytes are nucleated, but their nucleus is in such a strongly repressed condition that its existence can almost be disregarded.

That mammalian reticulocytes (thus, immature red blood cells that have lost their nucleus but retained their polyribosomes) are capable of protein (including hemoglobin) synthesis despite the lack of a nucleus has been known for many years (Borsook et al., 1952; Koritz and Chantrenne, 1954; Nizet and Lambert,

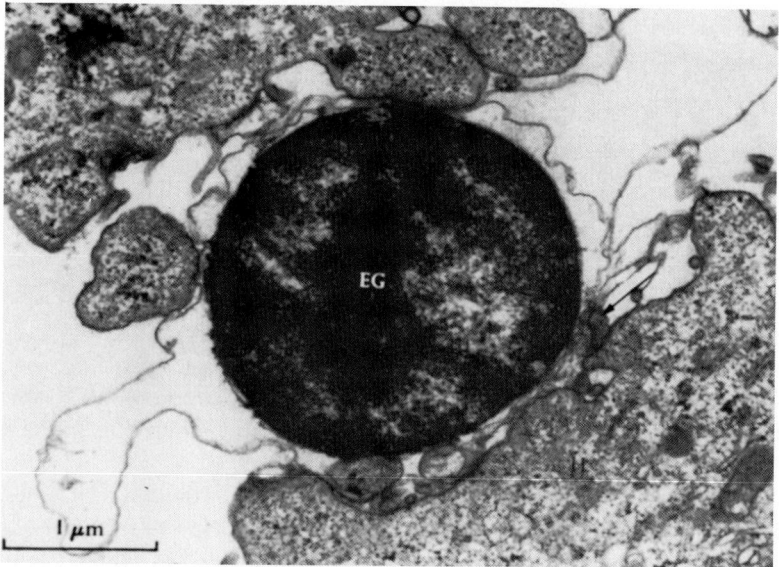

FIG. 3.   An erythrocyte ghost (EG) adherent to an HeLa cell (H). The arrow shows a virus particle wedged between the two cell membranes. [Harris (1974).]

1953; Gavosto and Rechenmann, 1954). During the maturation of the erythrocytes, a loss of RNA and a decrease in protein synthesis go hand in hand.

Innumerable papers have been devoted to reticulocytes. The fact that they specialize in hemoglobin synthesis allowed Chantrenne and his colleagues (1967) to isolate, for the first time, a specific mRNA from a eukaryotic source, namely, globin mRNA, which is a mixture of the two mRNAs coding, respectively, for the α- and β-chains of globin associated with several proteins. Since the time of this pioneer work, an enormous amount of research has been done (Chapter 4, Volume 1) and is still being done on the organization and regulation of the globin genes. An important point in the present context is that the isolation of globin mRNA from reticulocytes has made it possible to answer an essential question: is the number of copies of the hemoglobin genes the same in all cells? Molecular hybridization experiments using labeled cDNA as a probe have clearly answered that question: the number of globin genes is small (two to five copies per haploid genome), and is the same in all cells, whether or not they will synthesize hemoglobin. There is thus no selective amplification of these genes in hematopoietic cells, the only ones that produce hemoglobin under physiological conditions. It follows that the probable reason why erythropoietic cells synthesize hemoglobin to the exclusion of other cells is that they are the only ones in which the globin genes are derepressed (Bishop *et al.,* 1972). This problem will be discussed in more detail when we deal with cell differentiation in Chapter 3. It suffices to point out here that reticulocytes and erythrocytes, whether nucleated or not, are highly specialized cells, adapted through evolution to synthesize hemoglobin and to survive for weeks in the bloodstream. They can hardly be compared to cytoplasts that have been artificially severed from a cell (a fibroblast, for instance) by experimental manipulations such as cytochalasin B treatment and centrifugation. Such manipulations presumably alter the coordination of metabolic processes in the cells more significantly than the slow physiological changes that occur during normal erythropoiesis.

At the time when a cytoplast is severed from the whole cell it is synthesizing hundreds of proteins in a coordinated way. How it copes with such an intricate situation when the regulatory signals emitted by the nucleus suddenly fail to be received is not known. A study by Bruno and Lucas (1983) has shown that cytoplasts produce the same proteins as normal cells over a 12-hr period, but at a reduced rate. After that time, the pattern of protein synthesis becomes aberrant; certain polypeptides are even produced in excess. Enucleation clearly disrupts the mechanisms that control mRNA translation; it probably affects mRNA stability in the cytoplasm. Clearly, the situation is simpler for a reticulocyte in which extrusion of the nucleus is a slow physiological process and a single protein, hemoglobin, is almost the only product of the synthetic machinery. It is, therefore, not surprising that the life span of a normally differentiating red blood cell greatly exceeds that of a cytoplast which has not recovered very well from "chemical enucleation."

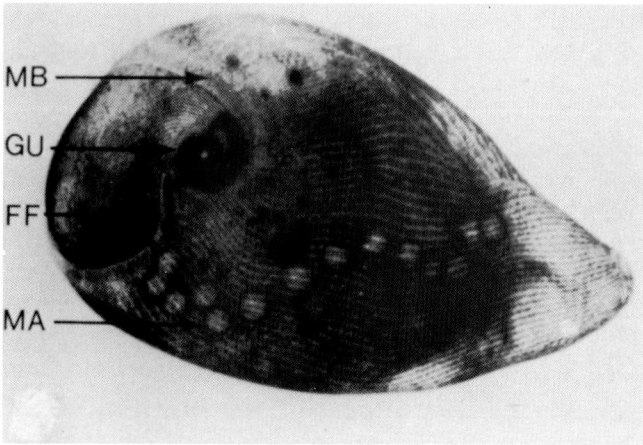

Fig. 4. *Stentor,* flattened in a rotocompressor. MA, chain macronucleus; MB, membranellar band; GU, gullet; FF, frontal field. [de Terra (1981).]

C. Merotomy and Nuclear Transplantations in Protozoa

Microsurgery has been used for many years to cut large protozoa such as *Amoeba proteus* or the ciliate *Stentor ceruleus* into two halves. The equipment needed for merotomy is a fine glass thread; exceptional ability is not required to perform the operation. Nuclear transplantations between amebas of the same or different strains require more skill and a more elaborate microinstrumentation. The results obtained by a number of workers with this type of technique have been described in detail in a special issue of the *International Review of Cytology* [Supplement **9** (1979)]. Only a broad outline of the major results will be given here.

A favorite material for the study of regeneration in protozoa has long been *Stentor* (Fig. 4), because many interesting problems of morphogenesis can be studied in this organism. Tartar (1953) has studied the effects of nuclear trans-plantations and graft combinations between two species of *Stentor*. His general conclusion was that completely successful regeneration occurs when there is a preponderance of the nucleus of one species in a preponderance of its own specific cytoplasm. The nucleus apparently supplies something essential to cell differentiation; this substance of nuclear origin is not stored and is apparently used up. If one may venture a guess in the complete absence of experimental data, it is that Tartar's (1953) substance is a mixture of mRNAs with a relatively short life span in foreign cytoplasm. Tartar's (1953) results, in agreement with the earlier observations of Balamuth (1940), indicated that the continuous pres-ence of the nucleus is required for growth and differentiation in *Stentor*. In 1956, Tartar concluded from new experiments that it is only the development of the

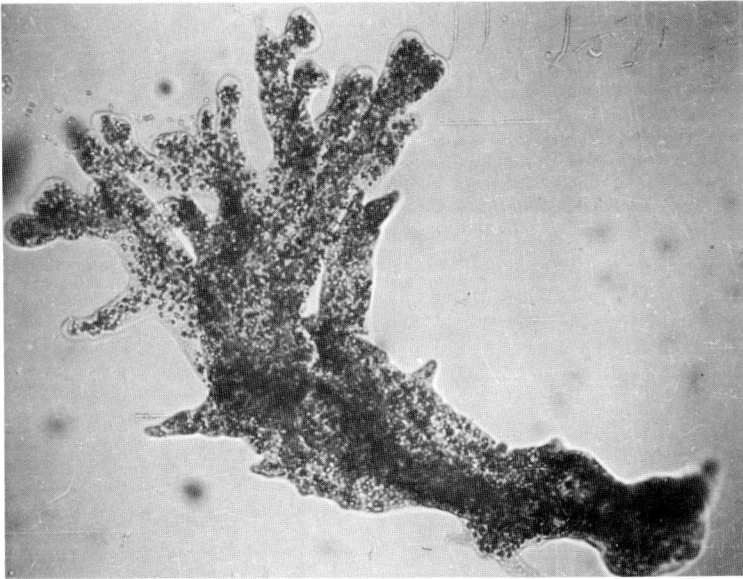

FIG. 5. *Amoeba proteus,* seen in phase contrast. [P. Van Gansen.]

oral primordium that requires the continuous presence of the nucleus; anucleate fragments of *Stentor* survive as long as starving nucleate pieces and retain ciliary activity and contractility of the myonemata. The chemical nature of the myonemata has not been studied by immunocytochemistry, so far as we know, but one can assume that their main constituents are actomyosin fibers. Grafting of a nucleus from a different species into an anucleate fragment may permit the formation of an oral primordium but does not allow the continued life of the cell (Tartar, 1956). Before leaving *Stentor,* we should mention the interesting work of de Terra (1974). She surgically removed part of the cortex of *Stentor* and, following [³H]thymidine incorporation by autoradiography, concluded that DNA replication and cell division were under cortical control. This conclusion also seems valid for eggs (Brachet and Hubert, 1972) and (as we have seen in Chapter 5, Volume 1) for tissue culture cells.

Much work has been done on amebas, especially *Amoeba proteus* (Fig. 5), despite the fact that, in contrast to *Stentor,* they show little differentiation other than a rigid cortex and a more fluid endoplasm. In *A. proteus,* removal of the nucleus by merotomy is quickly followed by loss of motility; even after a few minutes, pseudopod formation is sporadic and the anucleate halves become spherical (Fig. 9). They are unable to feed on living prey, in contrast to normal amebas or nucleate halves. If both anucleate halves are kept fasting, they may survive for as long as 2 weeks, whereas nucleate ones die within 3 weeks.

Delicate experiments by Comandon and de Fonbrune (1939) have shown that motility (i.e., pseudopod formation) resumes dramatically when an ameba nucleus is reintroduced into a cytoplasmic fragment that had been severed from the nucleus 2 or 3 days previously. If, however, the anucleate half is derived from an ameba that had been operated on 6 days earlier, the graft of a nucleus no longer has a favorable effect. Apparently, irreversible changes occur sooner or later in nonnucleated cytoplasm.

Is there a species specificity in the reactivation of locomotion activity after transplantation of a nucleus into an anucleate ameba half? In nuclear transfer experiments, Lorch and Danielli (1950) and Danielli (1955) exchanged nuclei between two distinct, but closely related, species of amebas and produced "hybrids" in organisms where no sexual reproduction was known to occur. They found that the nucleus of either species was capable of restoring locomotion activity in anucleate cytoplasm. Flickinger (1973) has confirmed that grafting a heterologous nucleus (i.e., a nucleus belonging to another species of amebas) is followed by a rapid and spectacular reappearance of motility. However, he found that after such heterologous transplantations of nuclei, the anucleate cytoplasm survived for only 10–12 days. The molecular bases of these facts remain unknown. It is known, however, that amebas contain large amounts of contractile proteins (actin and myosin) in their cortex and that these proteins are directly involved in locomotion. We also know that the addition of ATP to living amebas induces their contraction; we have found a higher ecto-ATPase activity in anucleate than in nucleate halves (Sells *et al.,* 1961), but this does not explain why amebas should contract as soon as their nucleus is removed and relax when a nucleus (even from a different species) is reintroduced. The results of unpublished experiments by our colleague R. Tencer add to the mystery. She found that in amebas treated with cytochalasin B, the nucleus and the surrounding cytoplasm make a small projection. If this projection is severed with a glass thread and the ameba thus enucleated, locomotion remains perfectly normal. Thus, a cytochalasin B-treated and surgically enucleated ameba (with little loss of cytoplasm) behaves exactly like the cytoplasts discussed in the preceding section. Only a thorough study of the MF organization in nucleate and anucleate fragments of *A. proteus,* obtained by different methods of enucleation, will shed some light on this mystery. Its solution is important, since the entire issue of the existence or nonexistence of nuclear control of cell adhesion and locomotion is at stake. This solution might well turn out to be trivial, localized changes such as changes in pH or free Ca^{2+}.

It is possible to go further than nuclear transplantation in the acrobatic field of ameba microsurgery. Jeon *et al.* (1970), using microdissection, isolated and reassembled the nucleus, endoplasm, and contractile cortex of an ameba. A normal ameba capable not only of locomotion, but also of division after proper feeding, was obtained.

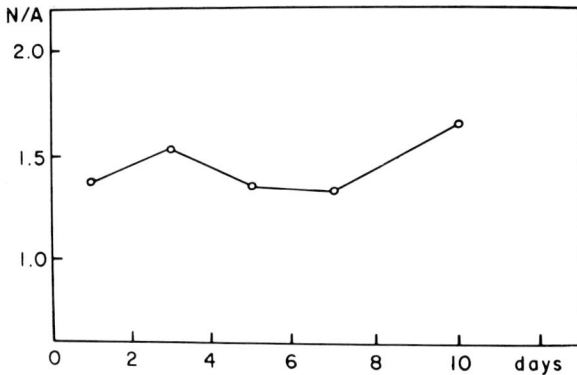

FIG. 6.   Ratio between nucleate and anucleate fragments (N:A) for the oxygen consumption of amebas. Changes with time. [Brachet (1955).]

Enucleation alters the ultrastructure of the cytoplasm. The most conspicuous changes are the disappearance of the Golgi bodies, followed by swelling of the mitochondria, and a decrease in the number of ribosomes. Reintroduction of a nucleus is followed by rapid reappearance of the Golgi bodies (Flickinger, 1973, 1974). Removal of the nucleus also leads to the loss of the cell coat (glycocalix), according to Jeon (1975). This is due to the fact that enucleation decreases the glycosylation of protein in the Golgi bodies and the transport of glycoproteins toward the cell surface. The membrane glycoproteins reappear within 1 hr after a nucleus has been grafted (Flickinger, 1981).

The biochemical changes induced in *A. proteus* by removal of the nucleus have been carefully studied by the author of this book and were described in detail in ''Biochemical Cytology'' (Brachet, 1957). These studies are no longer particularly interesting because we now know that most, if not all, *A. proteus* strains harbor endosymbionts that retain their own synthetic capacities in an anucleate cytoplasm. In particular, the experiments on the incorporation of labeled precursors in RNA and proteins from nucleate and anucleate fragments have little significance, since the endosymbionts synthesize their own DNA, RNA, and proteins. We shall thus limit ourselves here to a brief description of the main biochemical changes that take place after merotomy.

Removal of the nucleus has little effect on the respiratory rate, for 1 week at least (Fig. 6); by that time, some of the anucleate fragments undergo cytolysis and the differences between the two halves become more conspicuous (Brachet, 1955). The uptake of radioactive phosphate is quickly decreased in anucleate halves (Mazia and Hirshfield, 1950); however, their ATP content is, if anything, somewhat higher than in their nucleate counterparts (Brachet, 1955). This increase probably reflects an imperfect utilization of ATP in the nonnucleated

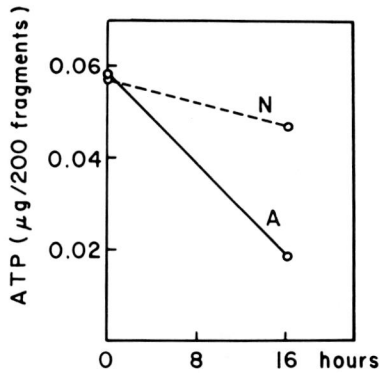

FIG. 7.   Drop in the ATP content of nucleate (N) and anucleate (A) ameba halves under anaerobiosis. [Brachet (1955).]

cytoplasm, where synthetic processes are limited. These results have led to the conclusion that, in contradiction to Loeb's hypothesis, the nucleus does not exert tight control on energy production. However, as shown in Fig. 7, the ATP content falls more rapidly in anucleate than in nucleate halves under anaerobic conditions (Brachet, 1955); anaerobic energy-producing reactions (glycolysis, presumably) are thus less efficient in the absence of the nucleus. This finding is in keeping with the fact that starving anucleate halves hardly utilize their glycogen reserves (Fig. 8); utilization of glycogen ceases completely after 3 days

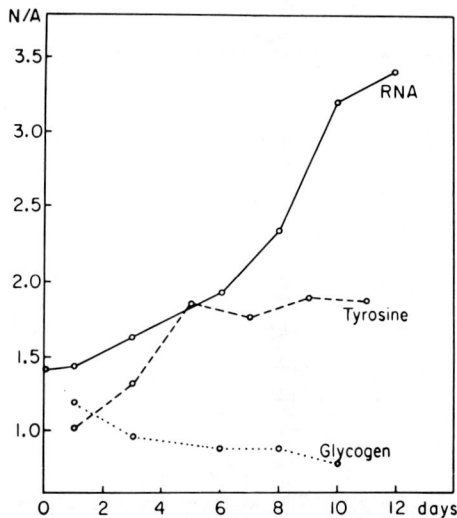

FIG. 8.   Changes in the RNA, protein (tyrosine), and glycogen content in nucleate and anucleate *Amoeba* halves (N:A ratios as in Fig. 6). [Brachet (1955).]

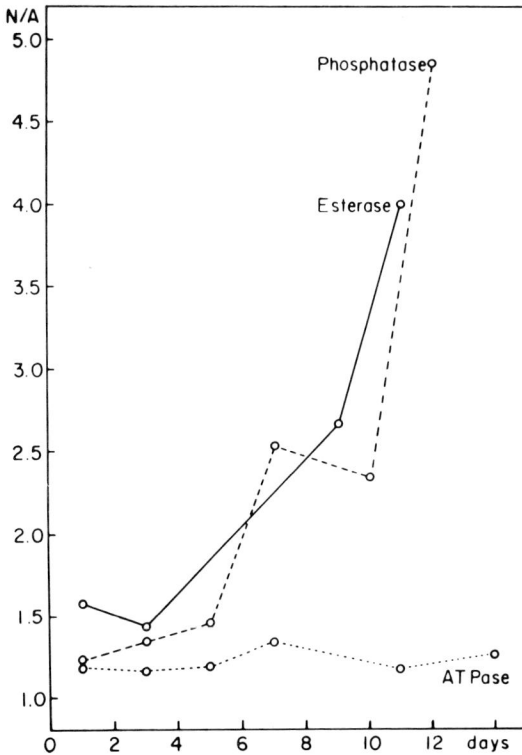

FIG. 9. Changes in the N:A ratio for adenosinetriphosphatase (ATPase), esterase, and phosphatase.

in anucleate fragments, while that of nucleate fragments decreases more slowly (Brachet, 1955). In *Stentor* also, utilization of the carbohydrate reserves is considerably (50%) inhibited in anucleate fragments (Tartar, 1956). The lack of glycogenolysis could be due to a decreased NAD content (Baltus, 1956), since its synthesis is usually a function of the nucleus. However, another possibility should be seriously considered, namely, that the activity of the lysosomal enzymes might somehow depend on factors released by the nucleus. The work of Holter (1955) on the intracellular localization of hydrolases (acid phosphatase, protease, amylase, etc.) in centrifuged amebas has played an important role in demonstrating that de Duve's lysosomes actually exist in intact cells. Interestingly, acid phosphatase and esterase activities practically disappear from the anucleate cytoplasm after a few days (Urbani, 1952; Brachet, 1955), as shown in Fig. 9. If we remember that, in both cytoplasts and anucleate halves of *A. proteus*, the Golgi bodies are the first cell organelles to break down, and that there are very close links between the Golgi apparatus and lysosomes (Chapter 3, Volume 1), the hypothesis that lysosomes do not function properly in anucleate cytoplasm becomes very plausible. The two (among many others) explanations suggested here (lack of NAD synthesis and inactivation of lysosomal enzymes) to explain the lack of utilization of glycogen (and fats) in starving annucleate fragments are not mutually exclusive. The total protein content in anucleate fragments decreases markedly (almost 50%) between the third and fifth days after removal of the nucleus (Fig. 8). It seems that, in these fragments, increased utilization of proteins coincides with the disappearance of the glycogen reserves; when the anucleate half becomes unable to utilize its normal reserve stores, it becomes the site of a kind of autophagy. In either case, nuclear control is not immediate. A lag period of about 3 days is necessary before measurable changes in carbohydrate and protein metabolism can be detected. As already shown by the work of Urbani (1952) and our own experiments (1955), all proteins are not under a single general control. Thus, the activity of some enzymes (protease, enolase, ATPase, amylase) remains practically unchanged 10 days following removal of the nucleus, while acid phosphatase and esterase activities almost disappear. This could be due to differences in the intracellular localization of the various enzymes; compartmentalization might affect the rate of protein degradation. For instance, if lysosomal enzymes are no longer synthesized in anucleate ameba fragments, they might conceivably undergo more rapid degradation than soluble enzymes. Another regulatory factor might be differences in the stability of the mRNAs coding for the various enzymes; methods for testing this possibility are now available. None of the enzymes we studied in 1955 ever showed a predominant nuclear localization, a finding that disposed of Wilson's (1925) theory of nuclear synthesis or storage of enzymes. However, it has been reported that the enzyme that removes sialic acid from cell surface glycoproteins (cytidine-5'-monophosphosialic synthetase) is accumulated in the nucleate halves (Kean and Bruner, 1971). This suggests a role for the nucleus in the composition of the cell surface, which agrees with the previously mentioned results of Jeon (1975) and Flickinger (1981) on protein glycosylation and glycoprotein transport in anucleate halves. We did not study the distribution in nucleate and anucleate halves of enzymes such as the DNA and RNA polymerases or ADPribose polymerase because the existence of these enzymes was unknown in 1955 and the present of endosymbionts in the cytoplasm of *A. proteus* discouraged many investigators, including this author, from working with amebas in the years that followed.

A finding of historical interest was the discovery (Prescott and Mazia, 1954; Brachet, 1955) that the total RNA content quickly and strongly decreased in anucleate halves of *A. proteus*. This was shown by both cytochemical (Fig. 10) and biochemical (Fig. 8) techniques. Thus, while nucleate halves maintain a constant RNA content, even after 12 days of fasting, there is a steady and marked decrease in the RNA content in the anucleate cytoplasm, which drops by 60% within 10 days. These experiments led to the conclusions (which will no longer

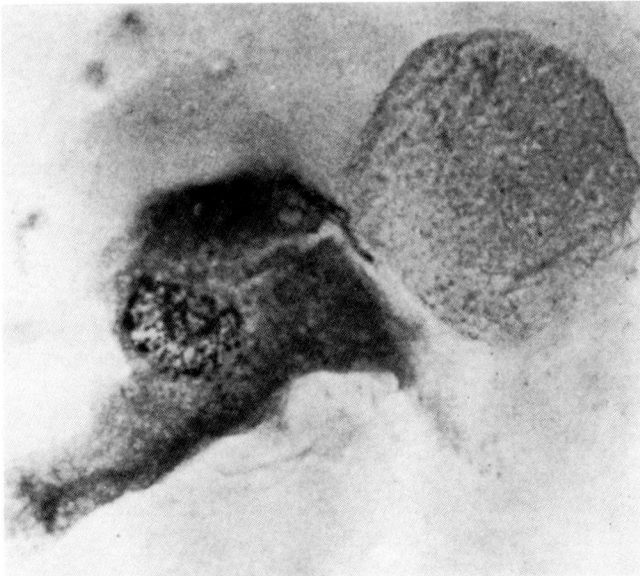

FIG. 10. Drop in basophilia in an anucleate fragment (right) of *Amoeba*. [Brachet (1955).]

be surprising) that cytoplasmic RNAs, particularly ribosomal RNAs, are under close nuclear control. All the data were compatible with the idea that all cytoplasmic RNAs originate from nuclear RNAs, an idea that would have been correct if amebas had been devoid of symbionts and of mitochondria that are capable of independent RNA synthesis.

Our experiments on the incorporation of precursors for the synthesis of macromolecules are no longer significant in view of the presence of endosymbionts in amebas. Much more interesting and important are the nuclear transplantation experiments by Goldstein (1974a,b), in which the nucleus of a radioactive ameba, labeled with either an RNA or a protein precursor, is injected into a normal, "cold" ameba. The movement of the labeled material from the "hot" nucleus to the "cold" cytoplasm and eventually back to the nucleus can be followed using autoradiography or biochemical techniques.

These experiments have clearly demonstrated the existence of cytonucleoproteins, i.e., a class of proteins that migrate back and forth between the cytoplasm and the nucleus (Prescott and Goldstein, 1969). These proteins vary in molecular size and in rate of turnover. The proteins that move very quickly from the nucleus to the cytoplasm and back have a low M_r (about 2300), representing as much as 17% of all soluble proteins and 3% of total proteins (Jelinek and Goldstein, 1973). According to a biochemical study by Goldstein and Ko on *A. proteus* (1981), of 130 proteins analyzed, less than 50% were exclusively nuclear and, among the others, 95% were evenly distributed between the nucleus and the cytoplasm; only the remaining 5% were found exclusively in the cytoplasm, where they might be bound to cell organelles. It can be concluded that the nuclear membrane is freely permeable in both directions, and that accumulation in the nucleus of certain proteins would result from their binding to high-affinity nuclear sites. A paper by Mills and Bell (1981) on the same subject concludes that two distinct classes of protein migrate from the nucleus to the cytoplasm in amebas. Proteins with an isoelectric point higher than 7 would play a role in the transfer of the nuclear RNAs to the cytoplasm; their movement toward the cytoplasm is inhibited by actinomycin D. On the other hand, the "shuttle" proteins, discovered by Prescott and Goldstein in 1969, have an isoelectric point lower than 7; their movements are not affected by actinomycin D, and their function might be the activation of nuclear genes.

No less important are the experiments of Goldstein and his co-workers on the shuttle RNAs. Their story began with the now classic experiments of Goldstein and Plaut (1955), in which ^{32}P-labeled nuclei were transplanted into normal unlabeled amebas or into anucleate halves of the same species. Autoradiography disclosed that the labeled nuclear RNA moved from the nucleus to the cytoplasm and that the cytoplasm did not supply RNA to the nucleus. However, experiments by Goldstein et al. (1973) left little doubt that certain RNA molecules shuttled back and forth between the nucleus and the cytoplasm. According to Goldstein and Trescott (1970), RNA species with sedimentation constants of 4–6, 19, and even 30 S easily moved from the cytoplasm into the nucleus. This migration of a 30 S RNA into the nucleus was surprising, since the same authors found that certain 4 S RNA species never left the nucleus. Indeed, in another paper, Goldstein and Ko (1975) concluded that there were four different shuttle RNAs with sedimentation constants between 5 and 8 S. Whether they are identical to the snRNAs discussed in Chapter 4, Volume 1 is not yet known. Needless to say, the physiological role of both the cytonucleoproteins and the shuttle RNAs remains unknown. Goldstein's hypothesis that they might play a role in gene regulation by binding to specific chromatin segments is attractive and plausible, but has not yet been substantiated experimentally.

Before we leave amebas for algae, a few points should be made. The first observation, which might have something to do with the shuttle RNAs, is that nuclear RNA leaves the nucleus at the beginning of mitosis and reenters the nuclei of the daughter amebas after cell division (Yudin and Neyfakh, 1973; Goldstein, 1974a,b). This conclusion is based on autoradiography studies; the nuclear and cytoplasmic RNAs have not been further identified. What seems to be clear is that amebas possess stable RNA species in their nuclei. The second point is that amebas have proved useful in studying the control of cell division. Nuclear transplantation experiments analyzed by autoradiography after [^{3}H]thymidine incorporation have shown that, in amebas, neither the continuation nor the termination of DNA synthesis is controlled by the cytoplasm (Ord, 1971).

However, similar transplantation experiments, but with heterologous nuclei, led
Rao and Chatterjee (1974) to conclude that chromosome replication, in amebas
as in other cells, is under cytoplasmic control. It is likely that, as in *Stentor* and
cells in culture, the cytoplasm, after receiving signals from the outer layers of the
ameba, controls the initiation, but not the length, of the DNA replication period.

D. NUCLEOCYTOPLASMIC INTERACTIONS IN *ACETABULARIA*

The giant unicellular green alga *Acetabularia* has been a fascinating subject
for biologists ever since the discovery by J. Hämmerling (1934, 1953) that
anucleate fragments not only survive for months but even regenerate species-
specific "caps" (umbrellas).

Acetabularia has been the subject of many review articles, of several symposia, and of a book by Puiseux-Dao (1970). The more recent reviews are by
Brachet (1976, 1981), Schweiger (1977), and Schweiger and Berger (1979),
which deal mostly with biochemical aspects; the ultrastructure and cytochemistry
of the alga have been discussed by Werz (1974) and by Spring *et al.* (1974).

Figure 11 summarizes the life cycle of *Acetabularia*. After fusion of two

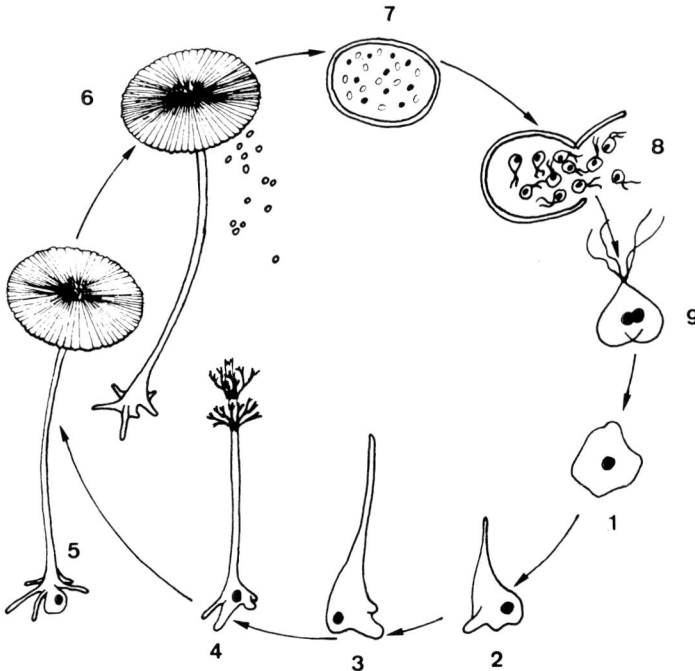

FIG. 11. Life cycle of *A. mediterranea.* 1, zygote; 2–4, growing of the young alga; 4, first
verticilla; 5, young cap; 6, adult cap, liberation of cysts; 7, resting cyst with many nuclei (white) and
plastids (black); 8, germinating cyst (liberation of gametes); 9, conjugation.

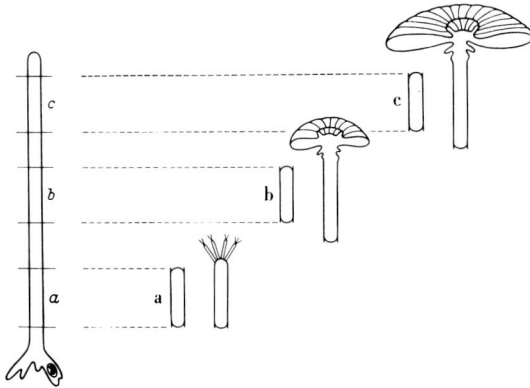

Fɪɢ. 12. Diagram of the polar distribution of morphogenetic substances in *Acetabularia*. [Brachet and Lang (1965).]

swimming haploid gametes, the zygote forms a chloroplast-containing stalk that grows in length. When it has attained its full size, the stalk differentiates a cap at its apical end; this cap (or umbrella) has species-specific characteristics. When the cap has reached its full size, the large, single vegetative nucleus (located at the basal end—the rhizoids—of the alga) breaks down; secondary nuclei actively divide, migrate into the stalk, and colonize the cap, which subdivides into cysts. The thick envelope surrounding the cysts breaks down at the time of germination, and flagellated gametes are released into the sea. The entire life cycle takes about 4–6 months under laboratory conditions (in which cyst germination can be induced by brief exposure to distilled water) and 1 year in nature. The length of the stalk (3–10 cm or more) depends, as shown by Beth (1953b, 1955), upon the amount of light received by the algae. Intense illumination produces algae that have a short stalk and a large cap; insufficient light results in the formation of very long algae with small caps. This phenomenon also occurs in nature. Algae collected from the shallow waters of the Mediterranean Sea are short and have large caps, while those obtained by deep diving have long stalks and much smaller caps.

The main results obtained by Hämmerling (1934) are summarized in Figs. 12 and 13. Anucleate fragments are capable of regeneration, and "morphogenetic substances" are distributed along a gradient that decreases from the apical to the basal end. Fragment *c* of Fig. 12 can form a full-sized cap, while the basal fragment *a* forms only a few sterile whorls at best. However, the source of these morphogenetic substances is the nucleus. This has been shown conclusively by the elegant experiments of Hämmerling (1943, 1946) and his colleagues (Beth, 1943; Maschlanka, 1946) on interspecific grafts (obtained by grafting a fragment containing the nucleus of species A onto an anucleate stalk of species B). As

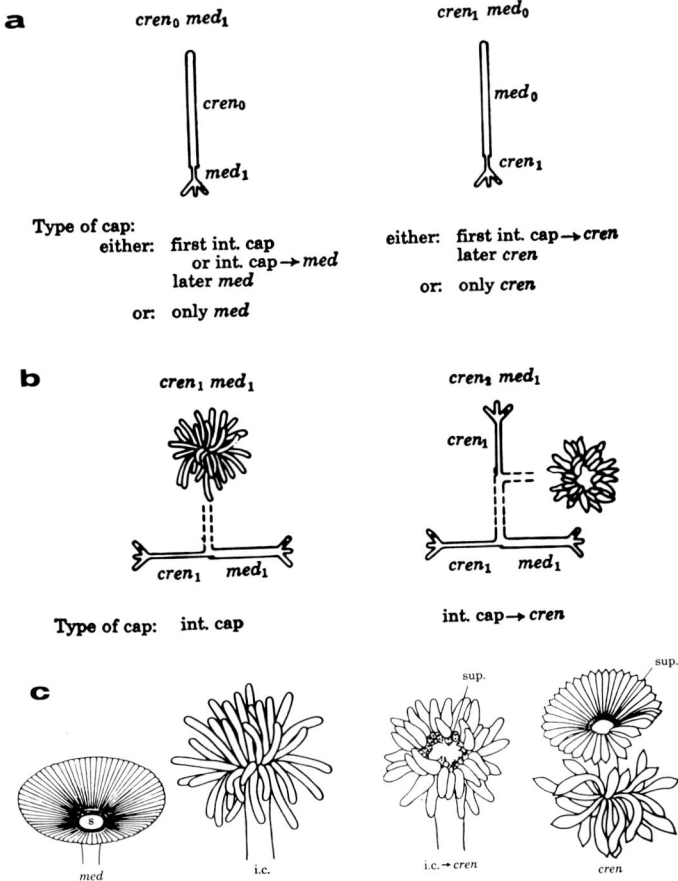

a $cren_0\ med_1$ $cren_1\ med_0$

$cren_0$ med_0

med_1 $cren_1$

Type of cap:
   either: first int. cap
            or int. cap → *med*
            later *med*
   or: only *med*

either: first int. cap → *cren*
          later *cren*
or: only *cren*

b $cren_1\ med_1$ $cren_1\ med_1$

$cren_1$

$cren_1\ \ med_1$ $cren_1\ \ med_1$

Type of cap: int. cap

int. cap → *cren*

c

sup. sup.

med i.c. i.c. → *cren* *cren*

FIG. 13. Interspecific grafts between *A. mediterranea* and *A. crenulata*. (a) Grafts between a rhizoid belonging to one species and a stalk belonging to another. Formation of caps belonging to the type of the nucleated fragment. (b) Intermediate caps formed in binucleated and trinucleated grafts of *A. mediterranea* × *A. crenulata*. (c) Types of caps: med, *A. mediterranea;* i.c., intermediate cap; cren, *A. crenulata;* sup, corona superior. [Hämmerling (1953).]

shown in Fig. 13a, this hybrid first forms a cap corresponding to species B, since the apex of stalk B contains morphogenetic substances of its own B type. Later, a second cap forms, which, in general, is intermediate in morphology between the caps of species A and B. This demonstrates that the A nucleus has produced species-specific morphogenetic substances. Figures 13b and 13c summarize the results obtained by Hämmerling and his school on grafts between *A. mediterranea* (*med*) and *A. crenulata* (*cren*); binucleate grafts form "intermediate" caps and trinucleate grafts containing two *cren* nuclei and one *med* nucleus give, as expected, caps that more closely resemble *cren*. If an anucleate *cren* stalk is grafted onto a *med* rhizoid, intermediate caps of various types are obtained; if this intermediate cap is removed, the new cap that forms will always be a typical *med* cap. Hämmerling (1953) concluded from these studies that the nucleus-controlled morphogenetic substances displayed species specificity. Thus, in un-inucleate grafts, the anucleate cytoplasm contained a store of morphogenetic substances of its own species. If this store was large enough, an "intergrade" cap would be formed as a result of competition between the preexisting morphogenetic substances stored in the anucleate piece and those produced by the grafted nucleus. The morphogenetic substances produced under the influence of the nucleus are "products of gene action, which stand between gene and character." It is necessary to point out here that long survival of anucleate fragments of plant cells is not unusual. In 1908, Van Wysselingh insisted upon the importance of the biological activities that persisted in nonnucleated fragments of *Spirogyra*. Not only did these fragments survive for several weeks, but photosynthesis, plasmid formation, cytoplasmic streaming, and even an increase in cell length were said to continue. Synthesis of chlorophyll and starch also occurs in anucleate fragments of plasmolyzed *Elodea* cells (Yoshida, 1956). What is exceptional in the case of *Acetabularia* is not the long survival of anucleate fragments, but their amazing capacity to produce species-specific caps in the absence of the nuclear genome.

It is not known for certain how morphogenetic substances produced by the nucleus (in the rhizoid) accumulate at the apex of the stalk and distribute themselves along an apicobasal gradient. All that is known is that centrifugation of anucleate fragments does not modify the polarity of the stalks (Sandakhiev *et al.*, 1972), but does displace chloroplasts and other cell organelles; the distribution of the morphogenetic substances is not affected. These experiments support the hypothesis (Bonotto *et al.*, 1971) that, in *Acetabularia*, the plasma membrane might contain specific receptors for the morphogenetic substances; the concentration of these receptors would decrease along an apicobasal gradient. In favor of such a hypothesis is the fact that organomercurials, which react with sulfhydryl groups without penetrating into cells, inhibit growth and cap formation in both nucleate and anucleate fragments of the alga. This suggests that proteins containing —SH groups are present in the cell membrane and involved in morphogenesis (Brachet, 1975). It is also known, from the cytochemical work of Werz (reviewed by Werz, 1974), that the tip of the alga contains specific glycoproteins that might be the hypothetical cell surface receptors. We also found that this tip, in contrast to the rest of the cell wall, gives a strong metachromatic staining, characteristic of acid polysaccharides, when algae are vitally stained with toluidine blue (J. Brachet, unpublished observations). Cytoplasmic streaming (which is inhibited by cytochalasin B and thus involves actin MFs) easily explains the movement of chloroplasts, secondary nuclei, and presumably morphogenetic substances toward the apex of the alga. Thus, there is a strict correlation beteen inhibition of morphogenesis and inhibition of streaming (Puiseux-Dao *et al.*, 1980). However, the molecular bases of the gradient distribution of the morphogenetic substances still remain obscure.

In fact, we do not know which factors control polarity in *Acetabularia*, since they operate at a very early stage of development, i.e., when a stalk begins to form soon after germination; the study of these early stages has so far been neglected. However, it is known that, in larger algae, longitudinal growth is correlated with an apicobasal gradient in electrical potential; in addition, spontaneous recurrent action potentials propagate from the apex to the base. Neither the growth in length nor the bioelectrical characteristics are affected by the addition of excess Ca^{2+}, Mg^{2+}, or the bivalent ion ionophore A23127; however, this ionophore suppresses the formation of sterile whorls and caps (Goodwin and Pateromichalekis, 1979).

Before proceeding to the biochemical studies that could identify the chemical nature of morphogenetic substances, a few other points of biological interest should be touched upon. An important finding of Hämmerling (1939, 1953) has been the demonstration that division of the large vegetative nucleus depends on cytoplasmic factors. Thus, if an already large cap is removed by sectioning just prior to division of the vegetative nucleus, nuclear division is postponed until a new cap is formed. If the operation is repeated, mitosis can be delayed indefinitely.Conversely, if a young nucleate fragment (rhizoid) is grafted onto an alga containing a large cap, nuclear division may begin as early as 2 weeks after grafting, instead of at the normal time of about 2 months. That the cytoplasm indeed controls the morphology and physiology of the giant vegetative nucleus has been confirmed by the ultrastructural studies of Berger and Schweiger (1975a) on the transplantation of an old nucleus into a young cytoplasm. This experiment induced rejuvenation of the nucleus and was characterized, among other things, by an increase in the development of the nucleoli, which normally begin to shrink at the time of cap formation (Fig. 14). Thus, aging of the nucleus in a full-grown alga can be prevented by experimental modification of the surrounding cytoplasm; rejuvenation can take place within 10 days. The converse experiment leads to premature aging of the young nucleus in a shorter time.

Another important fact is that the vegetative nucleus of *Acetabularia* does not exert only positive effects on the cytoplasm. Hämmerling (1934, 1953) and Beth (1953a) observed that anucleate fragments of *Acetabularia* form caps more rapidly than complete algae. Cap formation is initially speeded up if the stalk is severed from the rhizoid shortly before the formation of the cap. It seems clear, therefore, that the nucleus exerts negative as well as positive controls on the cytoplasm, and that the interactions between the two are more subtle than was often believed. These interactions take place in both directions, and it is therefore not surprising that the ultrastructure of the perinuclear space is particularly com-

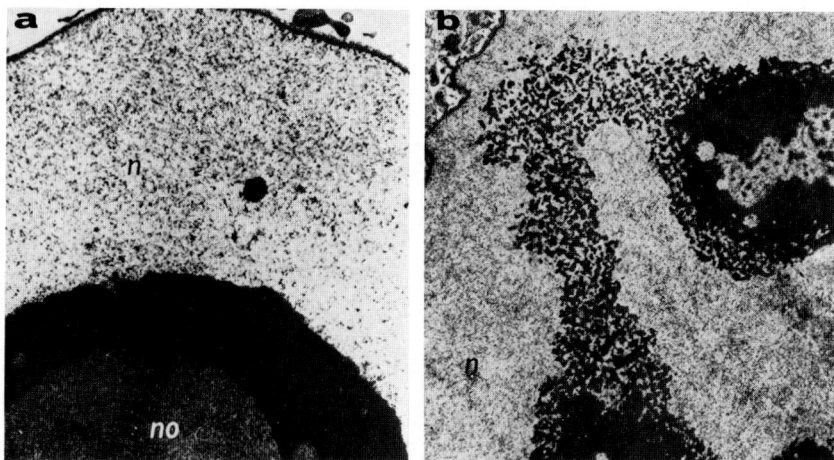

Fig. 14.   Implantation of an old nucleus into a young cytoplasm of *A. mediterranea*. (a) Old nucleus 3 days after implantation in a young cytoplasm. no, nucleolus. (b) Nucleus of an old rhizoid that had been combined with a young stalk 10 days previously. The structures of the nucleolar material and the perinuclear cytoplasm are characteristic of a young nucleus. [Berger and Schweiger (1975a).]

plex in *Acetabularia*. As shown in Fig. 15, a complicated perinuclear "labyrinthum," which has been studied in detail by Bouloukhère (1965, 1970), Werz, (1974), and particularly Franke *et al.* (1976), surrounds the large vegetative nucleus, which will be described later.

Concerning the biochemistry of *Acetabularia*, studies reported in more detail in "Biochemical Cytology" (Brachet, 1957) showed that enucleation has no or little effect on energy production. As shown in Fig. 16, oxygen consumption increases with time in the anucleate as well as the nucleate fragments; however, in the nucleate halves, where growth and regeneration are more active after 4 weeks, the increase is greater than in the anucleate stalks. In addition, photosynthetic activity remains entirely normal in anucleate pieces of *Acetabularia* for at least 4 weeks (Brachet *et al.*, 1955; Craig and Gibor, 1970). This is not surprising, since Shephard and Bidwell (1973) demonstrated that isolated *Acetabularia* chloroplasts synthesize the pigments involved in photosynthesis *in vitro*. The uptake of ^{32}P has been studied by Hämmerling and Stich (1954), Stich (1955), and Brachet *et al.* (1955). No differences were found between nucleate and anucleate fragments that had been separated for less than 4 weeks, except that there was a very strong accumulation of radioactivity in the nucleoli. Radioactive phosphate is mainly incorporated into polyphosphates (metaphosphates). This incorporation is dependent on photosynthetic and oxidative phosphorylations; darkness, cyanide, and dinitrophenol inhibit the process rapidly (Stich,

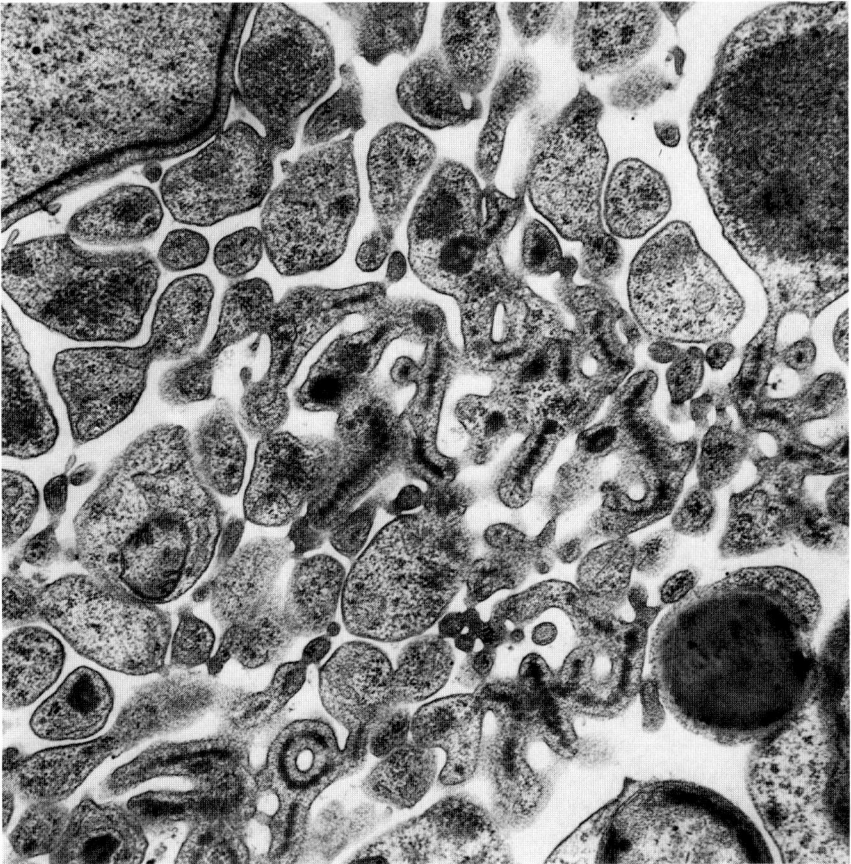

FIG. 15. Labyrinthic form of the perinuclear cytoplasm in *A. mediterranea*. [Courtesy of Dr. M. Boloukhère.]

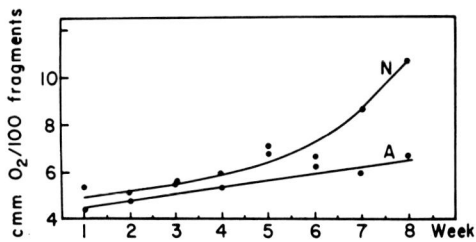

FIG. 16. Oxygen consumption of nucleate (N) and anucleate (A) halves of *Acetabularia*. [Brachet *et al.* (1955).]

$$R^+ \quad\quad R^- \quad\quad\quad N^+S^- \quad N^-S^+$$

R = 2.4   R = 1.0   R = 2.1   R = 1.45

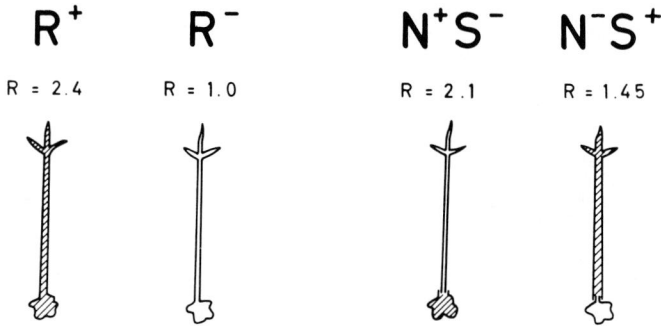

FIG. 17.   Nuclear control of circadian rhythms in *Acetabularia*. Grafts between rhythmic (R^+) and harythmic (R^-) variants. The hybrids display photosynthetic rhythmicity according to the nature of the rhizoid. R, ratio between O_2 produced in the middle of the dark phase and the light phase; N^+, N^-, rhizoids of R^+ and R^-, respectively; S^+, S^-, stalks of R^+ and R^-, respectively.

1955). Polyphosphates, which are frequent constituents of unicellular organisms and which can be detected by their metachromasia (i.e., red staining with toluidine blue), are probably a reserve of energy-rich phosphate bonds. They accumulate in anucleate halves of *Acetabularia* when regeneration ceases, suggesting that they might be utilized for synthetic processes (Stich, 1955).

It has been known for many years that anucleate fragments of *Acetabularia* retain a photosynthetic circadian rhythm for several weeks. However, it is the nucleus that "sets the clock," as was shown by experiments in which nucleate and anucleate halves of algae, which were at opposite phases of the cycle, were combined (Schweiger *et al.*, 1964). The same conclusion has been drawn from experiments (Fig. 17) in which fragments of algae that had lost their rhythm of photosynthetic capacity were combined with anucleate ones that had retained it, and vice versa; the presence or absence of the rhythm is thus nucleus dependent (Vanden Driessche, 1967). Experiments with inhibitors, particularly actinomycin D, indicate that RNA synthesis is involved in the rhythm of photosynthetic capacity. Actinomycin D abolishes the circadian rhythm in nucleate but not anucleate halves (Vanden Driessche, 1966; Mergenhagen and Schweiger, 1975). In contrast, the inhibitors of chloroplastic and mitochondrial RNA and protein synthesis, rifampicin and chloramphenicol, have no effect on the circadian rhythm of photosynthesis. However, the classic inhibitors of cytoplasmic protein synthesis (which act on the cytoplasmic 80 S ribosomes), puromycin and cycloheximide, suppress this rhythm in both nucleate and anucleate halves. Thus, an apparently typical chloroplastic function such as the circadian rhythm of photosynthesis is regulated by protein synthesis on cytoplasmic (80 S) ribosomes (Mergenhagen and Schweiger, 1975; Karakashian and Schweiger, 1976). It should be added that intensive work by Vanden Driessche (reviewed by her in 1973) has shown that many circadian rhythms are superimposed upon the rhythm

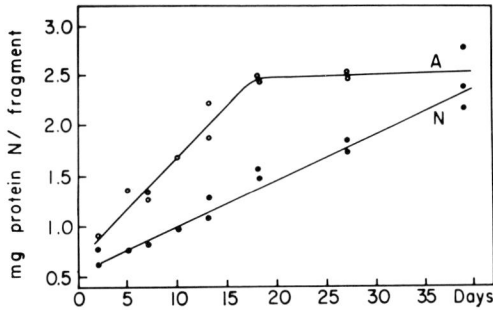

FIG. 18.   Protein synthesis in nucleate (N) and anucleate (A) halves of *Acetabularia* under optimal conditions. [Brachet *et al.* (1955).]

of photosynthetic capacity. They deal with the size and ultrastructure of the chloroplasts, their ATP and polysaccharide contents, the Hill reaction in photosynthesis, etc. In all cases that have been analyzed from that particular viewpoint, anucleate fragments appear to retain these rhythms for a long time, but their phase is modulated by factors of nuclear origin, presumably mRNAs.

When we became interested in *Acetabularia* and its mysterious morphogenetic substances, the first questions to arise were: are anucleate halves of *Acetabularia* capable of protein synthesis? If so, can they synthesize specific proteins such as enzymes? As shown in Fig. 18, the answer to the first question is yes (Brachet *et al.*, 1955); if the anucleate halves form a high proportion of caps, as was the case in the experiment shown in Fig. 18, net protein synthesis definitely occurs more rapidly in the anucleate halves than in the nucleate rhizoids. This finding confirms the existence of negative controls of the nucleus on the cytoplasm. The presence of the nucleus is thus not necessary for protein synthesis, but it is required for prolonged protein synthesis, since net protein synthesis stops completely after 2–3 weeks, that is, when the growth of the cap has ceased in anucleate halves. Studies on the incorporation of $^{14}CO_2$ into the proteins of nucleate and anucleate halves showed that this process becomes less active in the anucleate halves after 3 weeks, but persisted in these anucleate fragments for at least 7 weeks. This indicated that protein turnover continues for a considerable time in the absence of the nucleus (Brachet *et al.*, 1955). In the same paper, we tried to estimate indirectly the "life span" of the morphogenetic substances in anucleate halves; since regeneration does not take place in the absence of light, anucleate fragments were kept in the dark for 1, 2, 3 . . . 6 weeks and then illuminated. It was found that the same percentage of caps was obtained with stalks that had been kept in the dark for 2 weeks as in the controls. As shown in Fig. 19, the morphogenetic potencies of the anucleate fragments decreased after 3 weeks in the dark and vanished after 4 weeks in the absence of light. We shall come back to this question of the stability of morphogenetic substances in

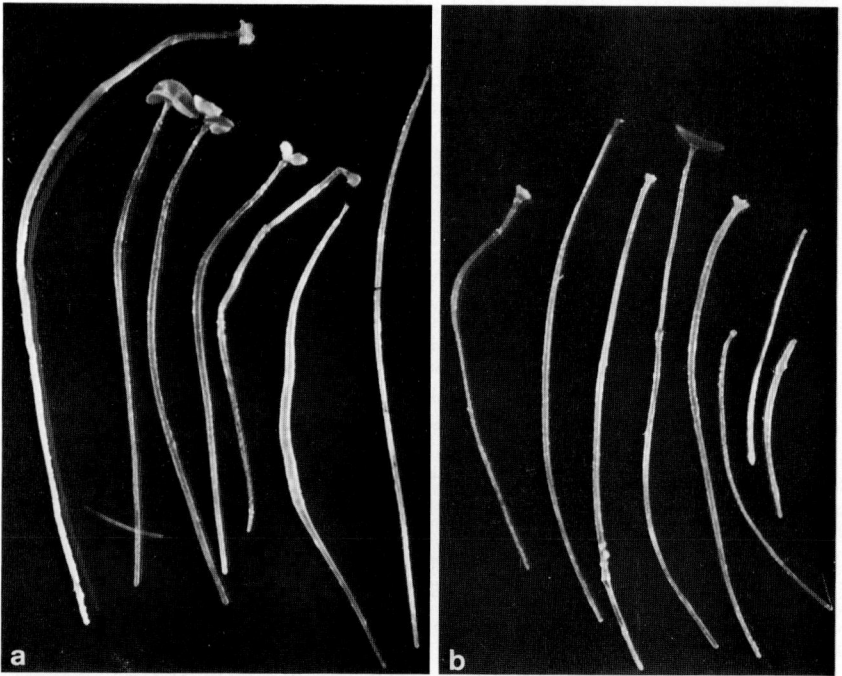

FIG. 19.   Effect of light on cap regeneration in *Acetabularia*. (a) Regeneration (cap formation) of anucleate stalks of *Acetabularia* (controls). (b) Decreased regeneration potencies of anucleate parts of *Acetabularia* that were kept in the dark for 3 weeks. [Brachet *et al.* (1955).]

anucleate fragments. For the time being, we may conclude that they disappear or are inactivated at about the time when net protein synthesis ceases.

Among the newly synthesized proteins in anucleate halves are a number of enzymes, as was first shown by Baltus (1959) for aldolase. The question of enzyme synthesis in *Acetabularia* has been discussed in a detailed review article by the author (Brachet, 1968). The main point is that many enzymes with unrelated functions increase greatly in activity at the time of cap formation, whether or not the nucleus is present. This is the case, for instance, for phosphatases; for UDPGpyrophosphorylase, which is involved in new cell wall synthesis; and for thymidylate kinase, which is required for DNA synthesis. The regulation of enzyme synthesis in the absence of the nucleus will be discussed again later.

The next question is whether or not anucleate fragments of *Acetabularia* are capable of nucleic acid synthesis. After an initial period in which both opposite and confusing results were obtained, the unexpected conclusion was that both RNA and DNA increase as much as two- to threefold when anucleate fragments of *Acetabularia* form caps. The curves obtained for the two nucleic acids were almost identical to that shown in Fig. 18 for protein synthesis (for further discussion of this early work, see our 1968 review article).

Around 1960, net DNA synthesis in an anucleate fragment of *Acetabularia* was a baffling mystery. The only possible explanation was that *Acetabularia* chloroplasts contained DNA and that this chloroplastic DNA could replicate in the absence of the nucleus. Previous work on spinach leaves suggested that chloroplasts contained DNA, but that belief was open to criticism since contamination with nuclear DNA always remained a possibility. Work on chloroplasts isolated from anucleate fragments of *Acetabularia* unequivocally demonstrated the existence of chloroplastic DNA (Baltus and Brachet, 1962). Soon afterward, D. Shephard made careful counts of the number of chloroplasts present in regenerating nucleate and anucleate fragments of *Acetabularia* and concluded (Shephard, 1965) that both number and size of the chloroplasts in anucleate fragments of *Acetabularia* increase; however, this increase is still larger in the nucleate halves of the alga.

In the 1960s, these findings were both surprising and exciting. However, the discovery that chloroplasts isolated from anucleate halves of *Acetabularia* can, as we have seen in Chapter 3, Volume 1, incorporate amino acids into their own proteins (Goffeau and Brachet, 1965) and that they are even capable of RNA synthesis raised many questions. It appeared that chloroplasts in *Acetabularia,* like the endosymbionts of *A. proteus,* were a serious nuisance to those who tried to understand nucleocytoplasmic interactions in molecular terms and to identify the mysterious morphogenetic substances. In particular, the apparent autonomy of the chloroplasts from the nucleus raised an important question: is the synthesis of chloroplastic macromolecules responsible for cap formation? This question has been tackled in several laboratories using specific inhibitors (reviewed by Brachet, 1968), and the answer is no. Inhibitors of chloroplastic DNA (ethidium bromide), RNA (rifampicin), and protein (chloramphenicol) synthesis do not affect the initiation of cap formation; they can, however, slow down the growth of caps that had been initiated in their presence. In contrast, inhibitors of nuclear RNA synthesis (actinomycin D and cordycepin) have different effects on morphogenesis in nucleate and anucleate halves. As shown in Fig. 20, they arrest regeneration in nucleate halves and have no effect, with the exception of a slowdown of growth, on anucleate halves. RNase treatment, which is expected to destroy the preexisting RNAs, prevents morphogenesis in both kinds of fragments. These experiments suggested that, as shown diagrammatically in Fig. 21, the integrity of preexisting RNAs is required for morphogenesis in anucleate halves; in their nucleate counterparts, continuous nuclear RNA synthesis is necessary for growth and morphogenesis. Finally, numerous experiments with puromycin and cycloheximide clearly demonstrated that cytoplasmic protein synthesis is required for growth and regeneration in both nucleate and anucleate halves (Fig. 22).

FIG. 20. Morphogenesis in *Acetabularia*. (a) Cap formation in the absence of the nucleus. (b) Effects of actinomycin on cap formation in anucleate halves. (c) Effects of puromycin on nucleate halves. [Brachet (1968).]

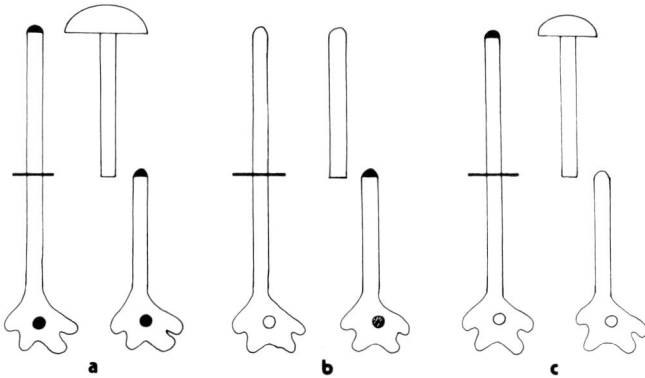

FIG. 21. Diagram of mRNA distribution in *Acetabularia* (black areas). (a) Normal algae, whole and cut into two halves. (b) Algae treated with RNase after cutting. (c) Algae treated with actinomycin after cutting.

All of these experiments suggested that "specific DNA molecules" (or parts of molecules), corresponding to each gene, would act as a template for RNA. There would thus be as many specific RNA molecules as there were genes. Finally, each specific RNA molecule would act as a template for a "specific protein" (Brachet, 1960). Soon after these sentences inspired by our work on *Acetabularia* were written, the concept of messenger RNAs was developed by Jacob and Monod (1961), who were working on *Escherichia coli*. Today it is

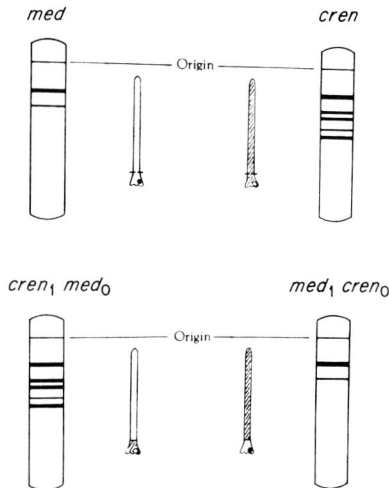

FIG. 22. Transplantation experiments between *A. mediterranea* (*med*) and *A. crenulata* (*cren*). 1 denotes the presence and 0 the absence of the nucleus of that species. [Reprinted with permission from *Nature* **216**, 555. Copyright © 1967 Macmillan Journals Limited.]

generally accepted that Hämmerling's morphogenetic substances are a family of mRNAs whose synthesis is directed by nuclear genes and that display unusual longevity in anucleate cytoplasm. In principle, the net synthesis of DNA, RNA, and proteins in the absence of the nucleus might be entirely due to the multiplication of chloroplasts in the anucleate halves observed by Shephard (1965). If such is the case, the chloroplasts would be fully autonomous organelles, uninfluenced by the nuclear genome. However, elegant interspecific grafting experiments by H. G. Schweiger and his co-workers have shown that this extreme view is untenable and that the nucleus exerts control on the chemical composition of the chloroplasts. Chloroplast enzymes such as lactic and malic dehydrogenases can be distinguished, in various *Acetabularia* species, by the electrophoretic pattern of their isozymes. As shown in Fig. 22, when a nucleate half of a given species is grafted on an anucleate stalk of another species, the isozyme pattern becomes that of the nuclear type after a few days (Schweiger *et al.*, 1967). Similar results were obtained for insoluble chloroplast proteins by Apel and Schweiger (1972, 1973) and by Kloppstech and Schweiger (1973). However, in the case of these chloroplastic proteins, replacement of the initial species-specific proteins by those that were synthesized under the control of the heterologous grafted nucleus took longer (6 weeks) than the replacement of the two dehydrogenases (8 days). This difference might mean that the chloroplastic genome played a more direct role in the synthesis of its membrane and ribosomal proteins than in the synthesis of its dehydrogenases.

Analysis of protein synthesis inhibition in isolated chloroplasts by chloramphenicol and cycloheximide led Apel and Schweiger (1973) to the following conclusion: two out of the three major proteins forming the chloroplast membranes are synthesized by both 70 S (chloroplastic) and 80 S (cytoplasmic) ribosomes, while the third is synthesized by the 80 S cytoplasmic ribosomes only. The main conclusion is that chloroplasts are not fully autonomous of the nucleus and that, in complete algae, the chloroplastic and nuclear genomes must cooperate in order to specify the protein composition of the chloroplasts. Since we have seen that morphogenesis is inhibited by puromycin and cycloheximide, but not by chloramphenicol, it follows that protein synthesis carried out by cytoplasmic polyribosomes and stable mRNAs of nuclear origin is much more important for morphogenesis than chloroplast protein synthesis. This view was shared by Brändle and Zetsche (1973), who concluded from their studies that chloroplasts are dependent on the nucleus, but less than mitochondria, and that the arrest of growth in anucleate *Acetabularia* halves is due to the arrest of cell wall and protein formation. It is probably at this stage that cytoplasmic polyribosomes can no longer function. They are presumably degraded when the preexisting store of mRNAs of nuclear origin has been utilized or is degraded and exhausted.

We have already seen that many enzymes, but not all, increase in activity when caps are formed, even in anucleate halves where transcriptional control at the level of the nuclear genes is no longer possible. Unfortunately, we know little about the posttranscriptional control mechanisms that operate when enzymatic activity increases severalfold at the time of cap formation in the absence of the nucleus. The presence of chloroplasts in the algae does not simplify the problem, since anucleate fragments retain their chloroplastic genome, which might code for enzymes. As we have seen, some of the chloroplastic enzymes (lactic and malic dehydrogenases) are synthesized on cytoplasmic polyribosomes, while others are synthesized on 70 S chloroplastic ribosomes. This is the case for enzymes involved in thymidine phosphorylation, an event that is closely linked to DNA synthesis. Thymidine kinase activity increases considerably when the cap reaches its full size and the stalk becomes yellowish. In this stage in nucleate algae, the large vegetative nucleus breaks down and gives rise to daughter nuclei, which are the sites of repeated cycles of DNA replication. It has been estimated that, in the cap, the multinucleate cysts contain as many as 10^8 nuclei. However, the same increase occurs in thymidine phosphorylation (Bannwarth and Schweiger, 1975) when anucleate halves form caps. Thus, this occurs under conditions in which multiplication of the secondary nuclei does not take place. This is due to an increase in thymidine kinase that is synthesized on 70 S chloroplastic ribosomes, since its increase in activity is inhibited by rifampicin and chloramphenicol, but not by cycloheximide (Bannwarth et al., 1977). The same situation holds for another enzyme involved in DNA replication, deoxycytine monophosphate (dCAMP) deaminase (Bannwarth et al., 1982; Bannwarth and Schweiger, 1983). Enzymatic activity increases at the time of cap and cyst formation; removal of the nucleus, ligation of the stalk above the rhizoid, and actinomycin D treatment all speed up enzyme synthesis. Experiments with suitable inhibitors of protein synthesis show that dCAMP deaminase is synthesized in the chloroplasts. Bannwarth and Schweiger (1983) concluded that the nucleus suppresses cap-specific information in the cytoplasm and that the hypothetical nuclear suppressor has a short life. Another enzyme involved in DNA synthesis, thymidylate kinase, increases in activity when nucleate or anucleate caps form. Since this increase is inhibited by puromycin and cycloheximide, but not by rifampicin and chloramphenicol, it must be concluded that the enzyme is synthesized by the nuclear genome and translated on 80 S cytoplasmic ribosomes; its mRNA is stable for more than 20 days in anucleate fragments of the alga (De Groot and Schweiger, 1983). Another important enzyme for DNA synthesis, ribonucleotide reductase, strongly increases in activity when caps form in both halves; synthesized on chloroplastic ribosomes, its synthesis is faster in anucleate than in nucleate fragments (De Groot and Schweiger, 1985).

The fact that synthesis of these three enzymes is sensitive to rifampicin indicates that they result from the transcription of chloroplastic DNA. Another interesting and puzzling case has been reported by Grieninger and Zetsche (1972). They found that two enzymes involved in the synthesis of UDPglucose and UDPgalactose, and thus in cell wall formation, were regulated independently. Phosphoglucomutase activity increases when caps differentiate, but there is no change in the activity of phosphoglucoisomerase. This suggests that only the

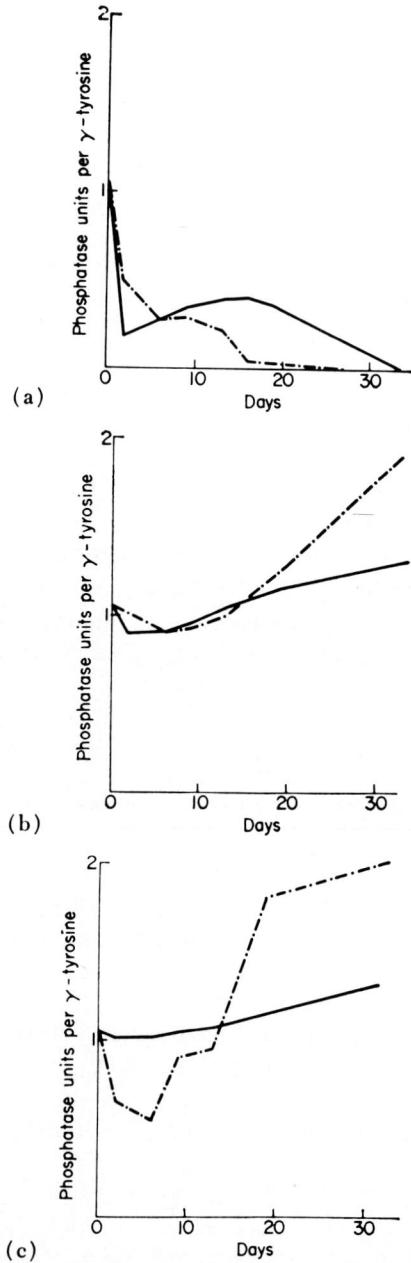

Fig. 23.   Regulation of acid phosphatase activity in *A. mediterranea*. (a) Anucleate fragments; (b) nucleate fragments; (c) complete algae. Solid line, complete medium; dashed line, phosphate-deficient medium. [Brachet (1968).]

former enzyme (and its messenger) is distributed along the apicobasal gradient. In experiments on the regulation of phosphatase activity described in our 1968 review article, we found that the idea of a controlling role of the nucleus should not be hastily rejected. As shown in Fig. 23, when nucleate or anucleate fragments of *Acetabularia* are cultivated in a phosphate-free medium, there is an initial decrease in phosphatase activity. This is probably due to a transitory increase in the inorganic phosphate concentration resulting from polyphosphate breakdown. Afterward, phosphatase activity increases only in the nucleate halves, suggesting that, as in bacteria, derepression of nuclear genes is required for substrate-induced synthesis of phosphatase.

We shall now review briefly what is known about chloroplastic and nuclear DNAs in *Acetabularia*. Chloroplastic DNA has been well characterized by physicochemical methods and electron microscopy (Green *et al.*, 1967; Green, 1973; Lüttke and Bonotto, 1981; Green *et al.*, 1977; Santulli *et al.*, 1983). As mentioned in Chapter 3, Volume 1, it is made up of long, fibrous molecules (up to 400 μm) and small circles (5 μm), in contrast to the 40-μm circles found in chloroplasts from Euglena and from higher plants. On the average, *Acetabularia* chloroplasts possess about 10 times more DNA than spinach chloroplasts; this DNA displays a greater kinetic complexity than any other chloroplast. Thus *Acetabularia* is not only the largest unicellular organism in the world; it also possesses the largest chloroplastic DNA molecules in existence.

Chloroplastic DNA, as we already know, can replicate autonomously. Its synthesis is inhibited by HU and, more specifically, by ethidium bromide (Heilporn and Limbosch, 1971a,b). The fact that these inhibitors do not inhibit morphogenesis has led to the conclusion that chloroplastic DNA replication is not needed for the initiation of cap formation; however, as mentioned, it might play a role in the growth of already initiated caps. Finally, it has been reported that many chloroplasts in *Acetabularia* do not contain enough DNA to stain with the highly sensitive DNA fluorescent stain 4, 6-diamidino-2-phenylindole, (DAPI); interestingly, the DNA-containing chloroplasts are distributed along the classic apicobasal gradient (Lüttke, 1981; Lüttke and Bonotto, 1980, 1981, 1982). In isolated chloroplasts and, of course, in intact algae, chloroplastic DNA is actively transcribed in all types of RNAs, including mRNAs; however, these chloroplastic mRNAs have no poly(A) tail and do not bind to cytoplasmic ribosomes.

It is now time to discuss the large (up to 300 μm in diameter) vegetative nucleus of the alga. In *A. mediterranea,* it contains a single, ribbon-shaped, very basophilic nucleolus (Fig. 24a,c) in which incorporation of labeled RNA precursors occurs very rapidly (Fig. 24b). Chromatin could not be detected by Feulgen staining in this species, probably due to dilution of its DNA in the huge nucleus. Technical improvements allowed Spring *et al.* (1974, 1975, 1976, 1978) to demonstrate that this long held view is not always correct. Figure 25 clearly shows that the nucleus of *A. cliftonii* (a related species) contains several nucleoli;

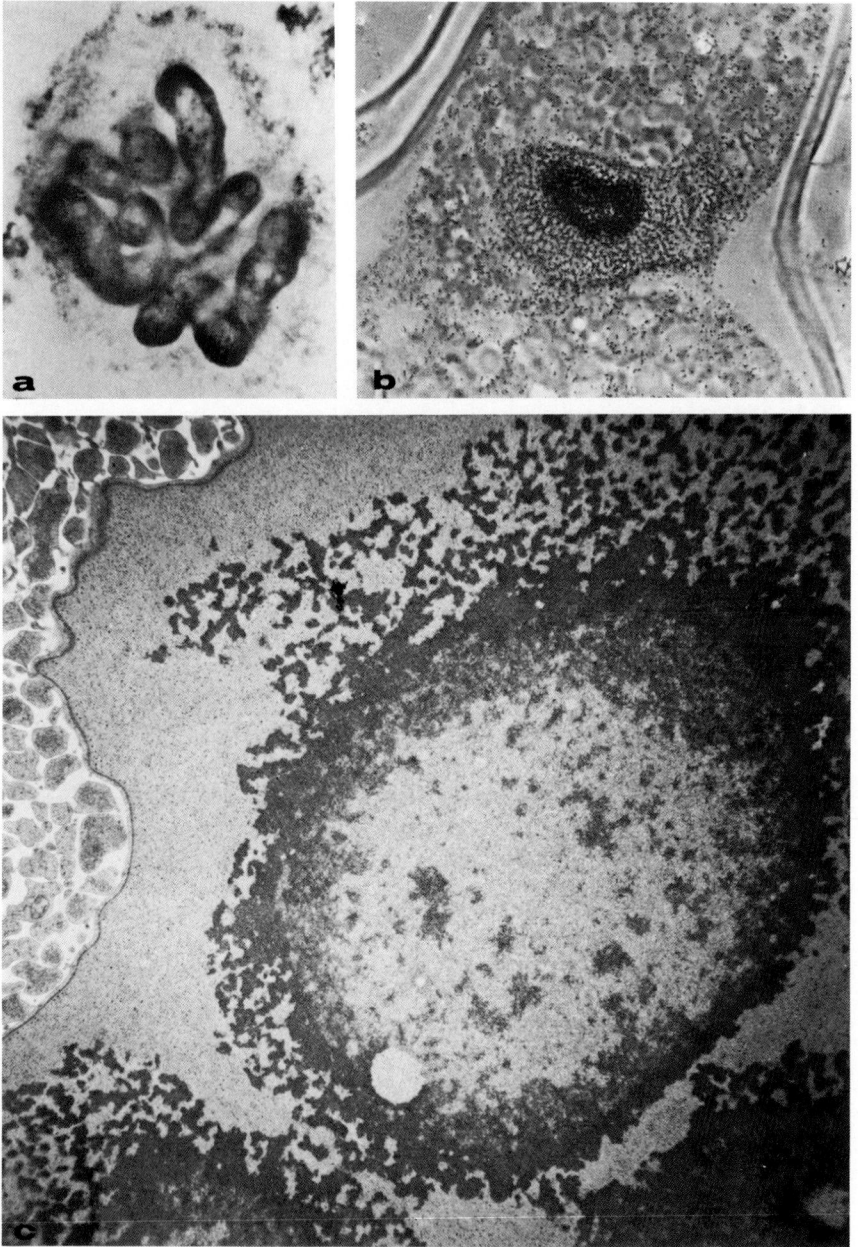

FIG. 24.   Nucleolus of *A. mediterranea*. (a) Ribbon-like structure seen with the light microscope after Unna straining. (b) Autoradiograph of the rhizoid ([³H]uridine pulse). The nucleolus is strongly labeled. (c) Ultrastructure of the nucleolus. [(a) Brachet (1965). (b, c) Courtesy of Dr. M. Boloukhère]

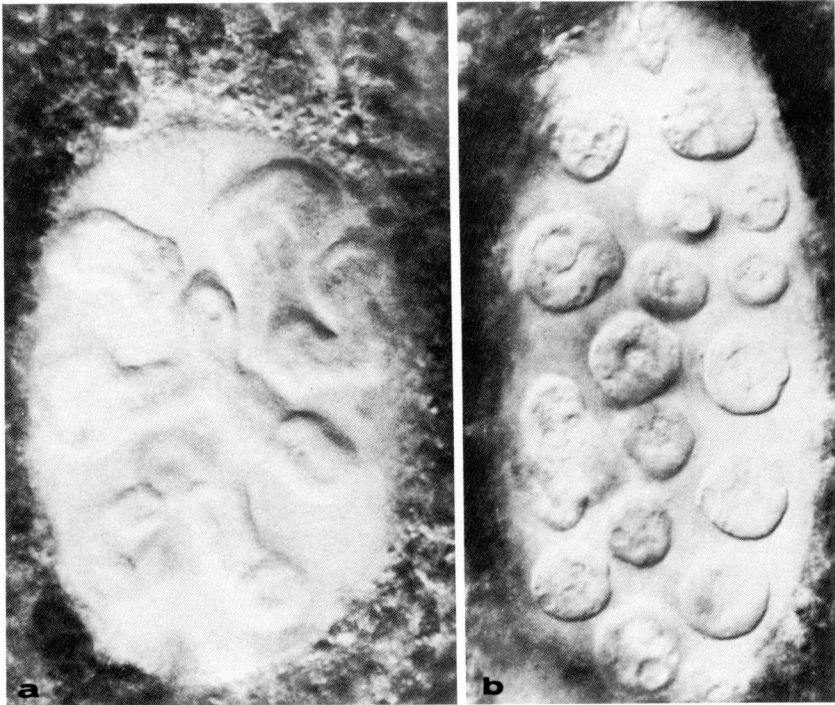

FIG. 25. Living nuclei of two species of *Acetabularia* photographed *in situ* with interference optics. (a) Sausage-shaped nucleolar aggregates in *A. mediterranea*. (b) Separate nucleolar units in *A. cliftonii*. [Spring (1978).]

much more importantly, the vegetative *Acetabularia* nucleus of all species possesses lampbrush chromosomes, that are basically similar to the classic lampbrush chromosomes of the amphibian oocytes (see Chapter 5, Volume 1 and Chapter 2, this volume). The *Acetabularia* chromosomes were not seen earlier because they are in a very extended lampbrush state. Loops that are 20 μm long extend from condensed chromomeres with a diameter of 1–2 μm (Spring *et al.*, 1975). According to Spring *et al.* (1978), the gametes and the zygote of *Acetabularia* contain 0.9 and 1.8 pg of DNA/nucleus, respectively, whereas the large vegetative nucleus contains 2.6 pg of DNA. This nucleus, despite its huge size is diploid (not polyploid). This was already concluded by previous investigators on the basis of its "Feulgen negativity." The small increase in DNA content between the zygote and the full-sized vegetative nucleus is due, as we shall see, to amplification of extrachromosomal (nucleolar) genes. When the alga has reached its full length and forms a cap, the vegetative nucleus begins to shrink; during the growth of the cap, its diameter decreases from 140 to 40 μm. This regression of the vegetative nucleus is followed by the formation of a giant (possibly the largest in the world) intranuclear spindle, which is shown in Fig. 26

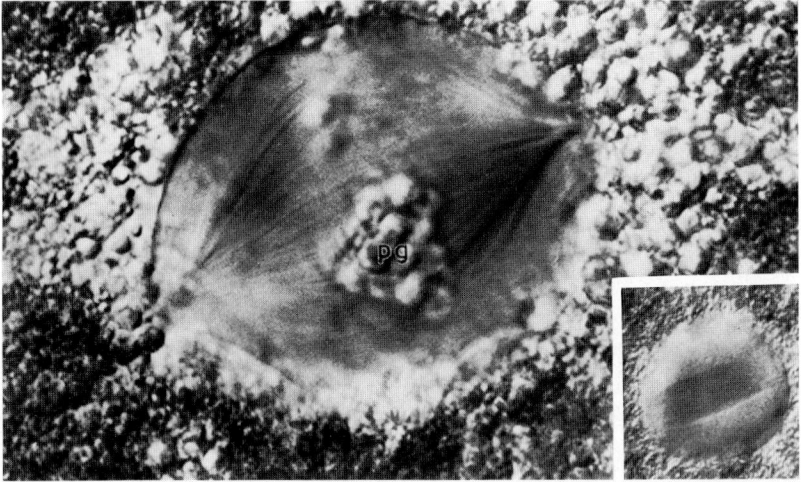

FIG. 26. Division of the primary nucleus of *Acetabularia in situ*. Typical intranusclear spindle in a dividing nucleus of *A. cliftonii*. The inset shows a somewhat different type of spindle (Nomarski interference contrast optics). pg, polyphosphate granules. [Koop (1979).]

(Koop *et al.*, 1979). The molecular mechanisms of spindle formation (assembly of MTs, effects of colchicine, etc.) have not been studied so far, nor do we know whether *Acetabularia* builds up a maturation-promoting factor (MPF) (see Chapter 5, Volume 1 and Chapter 2, this volume) responsible for chromosome condensation and nuclear membrane breakdown, like amphibian oocytes and HeLa cells. What is clear, however, is that, as already discussed, breakdown of the vegetative nucleus and induction of cell division are under cytoplasmic control. As pointed out by Spring *et al.* (1975) and Koop *et al.* (1979), the *Acetabularia* giant vegetative nucleus, like the germinal vesicle of oocytes, is probably in meiotic prophase; thus, it is likely that its huge spindle is a meiotic spindle. Its chromosomes give rise, as in parthenogenetic eggs, to haploid nuclei, which undergo repeated divisions and ultimately give rise to the gamete nuclei.

The ultrastructure of these late stages of development in *Acetabularia* has been reinvestigated by Berger and Schweiger (1975a,c,d) and Franke *et al.* (1976). The latter have pointed out that, at the time of cyst formation, the overall cell surface increases several thousand times, due to membrane invagination and intervention of the Golgi bodies. This increase may be compared to the large increase in the cell surface that occurs during egg cleavage. At first glance, an *Acetabularia* and an egg have nothing in common, except that they are both living creatures. Yet, as we have just seen, their initial processes of development are fundamentally the same. A long period of growth with a large nucleus in meiotic prophase is followed by meiosis and cleavage. Afterward, the sequence of the major events (gamete formation and fertilization) differs in the alga and the egg, but lead to the same ultimate goal: reproduction of the organism. Comparison of *Acetabularia* with sea urchin or amphibian eggs makes one believe more than ever that, in the living world, unity and diversity coexist. While unity is particularly striking for the molecular biologist, diversity is the realm of the ecologist.

Similarities exist between the *Acetabularia* vegetative nucleus and the oocyte's germinal vesicle, one of which is an amplification of the ribosomal genes. Although it is not quite as impressive as in *Xenopus* oocytes (which possess 2×10^6 copies of the 28 and 18 S rRNA genes compared to 10^3 in somatic cells), amplification of the ribosomal cistrons in *Acetabularia* has been firmly established by Spring *et al.* (1974), Trendelenburg *et al.* (1974), and Berger and Schweiger (1975b). Electron microscopy of spread nucleoli has shown that, as in amphibian oocytes, fibril-covered matrix units alternate with untranscribed spacers, giving the cistrons the well-known Christmas tree configuration (Fig. 27). According to the more recent work of Spring *et al.* (1978), the *Acetabularia* nucleus contains 32 nucleoli, each of which would have 110–150 ribosomal genes. There are thus about 4000 copies of the 28 and 18 S rRNA genes in the *Acetabularia* vegetative nucleus (an earlier estimate gave a higher value of about 13,000 ribosomal genes). The number of gene copies, again as in the *Xenopus* oocyte nucleus, does not increase markedly during the growth of the vegetative nucleus. Amplification thus remains moderate and apparently takes place soon after zygote formation. As in amphibian oocytes, the genes are separated by spacers of variable length; however, these spacers are, on the average, much smaller in *Acetabularia* than in *Xenopus*. The size of the rRNA precursor in *Acetabularia* is only 2.3×10^6 daltons compared to about 4.5×10^6 daltons in mammalian cells; however, the final products (the 28 and 18 S rRNAs) have, in both cases, the same M_r of 1.3 and 0.65×10^6, respectively. It follows that the processing of the rRNA precursor is much more "economical" in *Acetabularia* than in our own cells; only a 300,000-dalton piece of RNA is "wasted" in *Acetabularia*, compared to more than 2 million daltons in mammalian cells. The rate of rRNA precursor synthesis is 4×10^7 nucleotides per second per nucleus in intact algae; after removal of part of the stalk, the rate of rRNA synthesis increases 20 times within 3 days and reaches a maximum 5–7 days after section. Ribosomal RNAs are very stable in the cytoplasm of the alga (half-life of 80 days) and are transferred from the rhizoid to the apex at a rate of 2–4 mm/day (Kloppstech and Schweiger, 1975a; Schweiger, 1977; Schweiger and Berger, 1979). According to Naumova *et al.* (1976), the initial transport of rRNA from the nucleus to the surrounding cytoplasm is much more rapid than its migration from the rhizoid to the apex of the alga, which follows. As one might expect, the *Acetabularia* vegetative nucleus possesses the enzymatic machinery required for the synthesis of both rRNAs and mRNAs. As shown by Brändle and Zetsche (1977), isolated nuclei contain the two classic RNA polymerases, I and

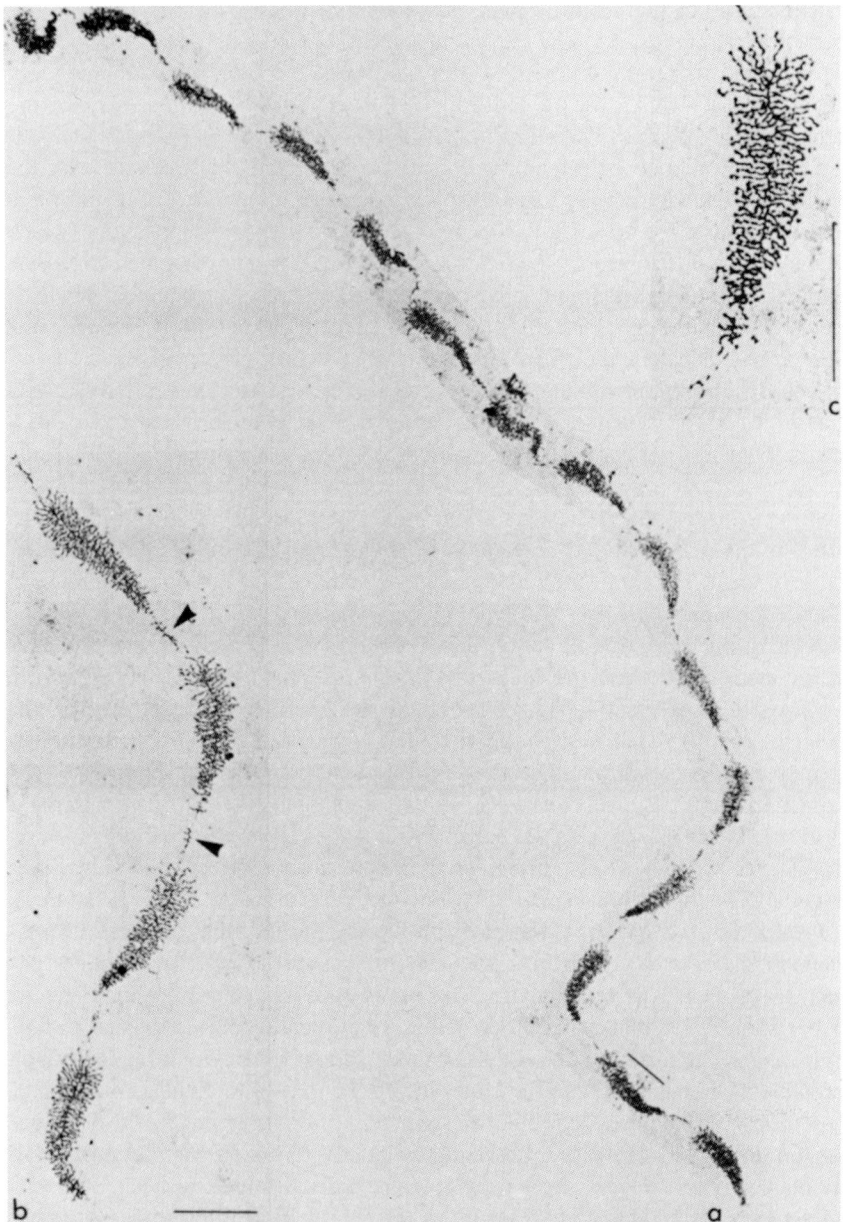

FIG. 27. Structure of nucleolar organizer in *Acetabularia*. (a) Well-spread nucleolar material showing the regular pattern of matrix units and spacer segments. (b) Occasional groups of small fibrils associate with some of the space regions (arrowheads). (c) At higher magnification, the dense packing of dense fibrils within the matrix units is shown. Scales indicate 1 μm. [Trendelenburg *et al.* (1971).]

II, and can synthesize several RNAs *in vitro*. The greatest synthetic activity is located in the nucleoli, where it is, as expected, α-amanitin insensitive.

Due to the work of Kloppstech and Schweiger (1975b), we know something about the mRNAs that are synthesized on the loops of the lampbrush chromosomes and that finally accumulate at the tip of the alga. This is possible because as already mentioned, chloroplastic mRNAs synthesized on chloroplastic DNA, have no 3'-poly(A) terminal sequence. It is therefore possible to "catch" the cytoplasmic poly(A$^+$) RNAs on a poly(U) or poly(dT) column. This allowed Kloppstech and Schweiger (1975b) to determine that polyadenylated RNAs are synthesized only in nucleate halves, while total RNA (which is mainly chloroplastic) is synthesized by both nucleate and anucleate fragments of the alga. As expected for mRNAs, the *Acetabularia* poly(A$^+$) RNAs are polydisperse, having an M_r ranging from 0.5 to 3 × 10^6. They migrate from the nucleus to the apex of the stalk independently of the rRNAs at a speed of 5 mm/day (the growth rate of the stalk is only 1–3 mm/day), which is somewhat faster than the rate of the rRNAs. While synthesis of the rRNAs increases 20-fold during regeneration, the increase in the rate of poly(A$^+$) RNA synthesis under the same conditions is only 2- to 3-fold. This led Schweiger (1977) to conclude that transcription of chromosomal DNA plays only a limited role in the control of genetic activity and morphogenesis in *Acetabularia*. One should point out, however, that the present data refer only to poly(A$^+$) RNAs. That some, if not all, of the poly(A$^+$) RNAs are authentic mRNAs has been shown by Shoeman *et al.* (1983). They obtained the *in vitro* translation of 77 different soluble proteins. In this study, they compared the soluble proteins of three different *Acetabularia* species and showed that only 10% of them are common to all three species. The main conclusion was that, at a given stage of development, the translation of some proteins is not regulated, while that of others may be turned off or increased. It should be added that it is not known whether (poly(A$^-$) mRNAs are produced by the *Acetabularia* lampbrush chromosomes. The half-life of the poly(A$^+$) RNA that are synthesized in the nucleate halves is about 10 days. They are thus much less stable than the rRNAs. It seems likely that the half-life of cytoplasmic poly(A$^-$) mRNAs, if they exist at all, would be still shorter.

Neuhaus *et al.* (1984) have succeeded in isolating the large vegetative *Acetabularia* nucleus and fusing it with chloroplast-containing cytoplasts obtained by fragmenting the stalk of the alga. They injected a DNA-containing virus (SV40) into these reconstituted cells and observed the appearance of a specific viral antigen (the T-antigen) in the *Acetabularia* nucleus. T-antigen appears earlier in the nucleus than the products of other microinjected foreign genes. The techniques used by Neuhaus *et al.* (1984) are very delicate and are not likely to supersede the now almost classical method of microinjection into the germinal vesicle of a *Xenopus* oocyte, but it is important to have the opportunity to compare transcriptional events in plant and animal nuclei after injection of pure genes.

If, as proposed by Schweiger (1977), transcription of chromosomal DNA plays only a limited role in morphogenesis, we are left with only one hypothesis about the chemical nature of the morphogenetic substances: they would be a store of mRNAs accumulated in a stable form along an apicobasal gradient. Since the stability of the stored mRNAs might be due, as in the sea urchin eggs that will be discussed in the next chapter, to binding by proteins that protect RNA against intracellular RNases, one might imagine that mRNA activation could result from protease activity. These enzymes would release mRNAs in translatable form from inactive ribonucleoprotein complexes. In order to test this hypothesis, we treated (Brachet, 1975) nucleate and anucleate fragments of *Acetabularia* with inhibitors of protease activity [which, as we have seen, arrest cell proliferation, according to Schnaebli and Burger (1972)]. These inhibitors completely prevented cap formation in both kinds of fragments, in agreement with our hypothesis. The protease inhibitors might have been acting in a different way. The EM studies of Werz (1974) indicate that lytic changes take place at the tip of the algae at the time of cap formation. If these changes are due to proteases and other hydrolases, arrest of morphogenesis by protease inhibitors would be easily understandable. All that can be concluded from our experiments, however, is that limited proteolysis seems to be a prerequisite for morphogenesis in *Acetabularia*.

More *recently* (Brachet *et al.*, 1978, and unpublished results), we became interested in the effects of two inhibitors of polyamine synthesis [α-methylornithine (α-MeO) and difluoro-α-methylornithine (DFMO)] on morphogenesis in *Acetabularia*. Polyamines, as we have seen in Chapter 5, Volume 1, are believed to play an important role in the control of cell proliferation and cell differentiation (reviewed by Heby, 1981). There is disagreement about their molecular site of action—DNA or RNA synthesis? It was interesting to see whether α-MeO and DFMO, two inhibitors of ODC, the key enzyme in polyamine synthesis, would affect a system (*Acetabularia*) in which differentiation takes place in the absence of cell division. If added at a late stage of the life cycle, after the breakdown of the primary nucleus, would they prevent the multiplication of the secondary nuclei and cyst formation? One reason for using *Acetabularia* as a test system lies in the fact that RNA or DNA synthesis is separated in the course of time in the life cycle of the alga. The results of our experiments are shown in Fig. 28. Growth and regeneration in the two halves are inhibited by treatment with either α-MeO or DFMO. Following treatment of complete algae after the vegetative nucleus has broken down, the number of cysts is greatly reduced and their morphology is abnormal (Fig. 29). Thus, the inhibitors did not act specifically on events correlated with either RNA or DNA synthesis. They affected both of them in a stage-dependent way.

In these experiments, reversibility of the inhibition of morphogenesis was tested. The effects of all inhibitors on *Acetabularia* (including α-MeO and DFMO) have always been found to be reversible in the case of nucleate frag-

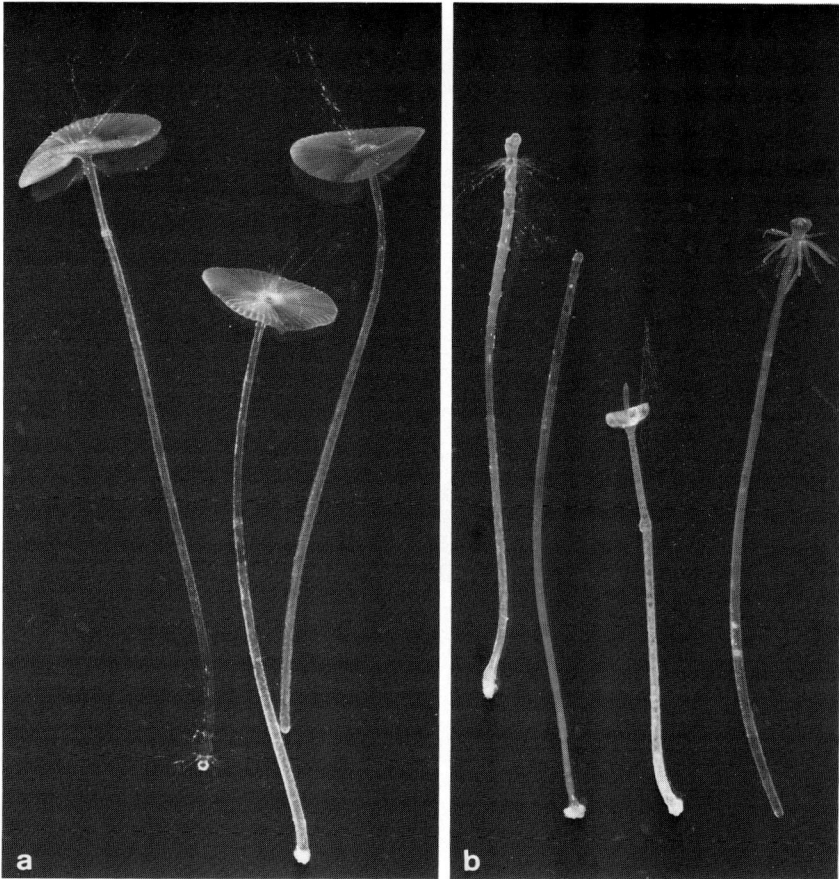

FIG. 28. Effects of α-methylornithine (α-MeO) on growth and cap formation in *Acetabularia*. (a) Controls: both nucleate (left and middle) and anucleate algae (right) formed normal caps. (b) After 1 month of α-MeO (10 mM) treatment, growth and regeneration were inhibited in the two halves. [Brachet and Bouloukhère, original.]

ments. The growth of the stalk, the formation of sterile whorls, and, finally, the formation of caps are resumed after a lag, depending on the length of treatment with the inhibitors. In anucleate halves, however, the effects of the inhibitors eventually become irreversible. Kinetic analysis of reversibility after α-MeO or DFMO treatment of anucleate halves (Fig. 30), showed that arrest of morphogenesis becomes irreversible if the length of the treatment exceeds 2–3 weeks. This is exactly what we had found (Brachet *et al.*, 1955) when we studied the effects of periods of darkness on cap formation in anucleate halves. These results suggest that the half-life of the morphogenetic substances is about 10

FIG. 29.   Effects of α-MeO on cyst formation in *Acetabularia*. Left: control cap. Right: the number of cysts is greatly reduced and their shape is abnormal when the treatment is applied on whole algae where the vegetative nucleus has broken down. [Brachet and Bouloukhère, original.]

days, which is identical to the half-life of the poly(A$^+$) RNAs studied by Schweiger (1977).

Present evidence is strong for the mRNA nature of the morphogenetic substances, but how a family of mRNAs and their translation products can produce a structure as complex as an *Acetabularia* species-specific cap remains a mystery. This problem will perhaps be solved by biophysicists (particularly x-ray crystallographers) interested in the structure and self-assembly of macromolecules. It should also be pointed out that the evidence for the mRNA nature of the morphogenetic substances remains indirect. Attempts by Alexeev *et al.* (1974) to induce the formation of caps after injection of ribonucleoprotein particles into basal anucleate fragments were not really successful. At best, a few sterile whorls and very rudimentary caps were obtained. This failure is not surprising, since Kloppstech and Schweiger (1975b) showed that the injected 40 and 120 S ribonucleoprotein can be synthesized in anucleate fragments and are, therefore, not of nuclear origin. Microinjection experiments are difficult to perform in *Acetabularia* because the cytoplasm is reduced to a thin layer compressed between the rigid cell wall and the turgescent central vacuoles. In addition, one cannot exclude entirely the already mentioned possibility (Bonotto *et al.*, 1971) that morphogenetic substances bind to specific membrane receptors in order to

FIG. 30.   Reversibility of α-MeO treatment on cap formation in *Acetabularia*. After 1 month of α-MeO treatment, the algae were returned to normal seawater. After 1 week of recovery, the nucleate fragments (two algae, left) formed normal caps, which the anucleate fragments (right) were still inhibited. [Brachet and Boloukhère, original.]

produce their effects and that these receptors might not be present in basal fragments of the stalk. Nevertheless, it is hoped that continued microinjection experiments will ultimately yield positive results. Only a bold, direct approach will lead to a full understanding of morphogenesis in the absence of the nucleus in *Acetabularia*.

IV. SOMATIC CELLS: HYBRIDS AND CYBRIDS

As already seen in Chapter 5, Volume 1, cell fusion (which can be greatly facilitated by treatment of the cells with cell surface–modifying agents such as uv-inactivated Sendai virus or polyethylene glycol) is a powerful tool for analyzing the factors that control DNA replication and cell division. As we have seen, cell fusion (books by Harris, 1970; Ringertz and Savage, 1976) gives rise to a heterokaryon in which two nuclei of different origins live in a common cytoplasm; if the two nuclei fuse together, a synkaryon is formed that can eventually give rise to a stable hybrid cell line. Chapter 3 shows that somatic cell hybridization has yielded significant results with respect to three important biological problems: cell differentiation, malignant transformation of normal cells, and cell aging. Here we shall deal mainly with the controls exerted by the cytoplasm on chromatin transcription. Cybrids (i.e., hybrids between a cell and a cytoplast) are more difficult to obtain because of the greater fragility of the cytoplasts, but they have yielded very interesting results, particularly concerning the role played by the mitochondria in heredity. Finally somatic cell hybridization is a very powerful tool for geneticists interested in mapping chromosomes.

Treatment with polyethylene glycol allows fusion between cells of very distant taxonomic origins, even between animal and plant cells, although the latter should be cleared of their rigid cell walls by treatment with appropriate enzymes before fusion can be successfully achieved. Such plant cells deprived by enzymatic digestion of their cell walls are called "protoplasts." Examples of successful fusion of animal cells with plant protoplasts have been reported by Ahkong *et al.* (1975) and Jones *et al.* (1976), who formed somatic cell hybrids between yeast protoplasts and chicken erythrocytes (Ahkong *et al.*, 1975) and between tobacco protoplasts and human HeLa cells (Jones *et al.*, 1976). As one can see, there is really no taxonomic barrier against somatic hybridization between eukaryotic cells. This is, of course, not the case for sexual hybridization, in which many mechanisms prevent the production of hybrids between distant species.

One should draw a clear distinction between the results obtained with heterokaryons (in which two nuclei of different origins lie in a common cytoplasm) and hybrid cell lines (in which, following repeated division of the synkaryon, chromosome losses often occur). The rule, which suffers from many exceptions, seems to be that inactive genes are reactivated in heterokaryons; in contrast, markers of phenotypic differentiation are often lost in hybrid cell lines (extinction of the differentiated phenotype). We shall begin with a survey of the work done on heterokaryons.

That nuclear activity is controlled by the surrounding cytoplasm has been shown by the famous experiments of Henry Harris, who fused chicken erythrocytes with human cells; they have been the subject of a book (Harris, 1970). Briefly, the condensed nucleus of the bird erythrocyte, which is completely inactive in both DNA and RNA synthesis, is reactivated after fusion (Fig. 31). It swells, re-forms a nucleolus, and begins to synthesize RNA. If fusion took place with a cell that synthesizes DNA—a malignant HeLa cell, for instance—DNA replication would follow in the chick erythrocyte nucleus. The initiation of both RNA and DNA synthesis in the reactivated erythrocyte nucleus is completely controlled by cytoplasmic factors present in the active cell (fibroblast or HeLa cell). Swelling of the chick nucleus, as shown by Ringertz and Bolund (1969) in elegant immunological experiments, is due to uptake of human cytoplasmic proteins. Human antigens of cytoplasmic origin first accumulate in the chick erythrocyte nucleus (Ege *et al.*, 1971). The reactivated chick nucleus then synthesizes chick proteins, and some of them move into the human (HeLa) nucleus (Ringertz *et al.*, 1971). This demonstrates that the chick erythrocyte nucleus has actually been reactivated by cell fusion. It has synthesized its own mRNAs, which, in turn, have been correctly translated. Curiously enough, the synthesis of chick surface antigens by the heterokaryons is suppressed by localized uv irradiation of the two chick nucleoli (Déak *et al.*, 1972; Déak and Defendi, 1975). This suggests a still poorly understood role of the nucleolus in the synthesis, processing, or transport of mRNAs. In recent experiments, Nyman *et al.* (1984) fused chick erythrocytes with rat myoblasts and observed that the reactivated erythrocyte nucleus acquires antigens typical of mammalian nuclear sap and nuclear envelope. It is clear that the two nuclei exchange macromolecules of regulatory importance via the common cytoplasm. In other experiments, Woodcock *et al.* (1984) fused chicken erythrocytes with anucleate cytoplasts from mouse fibroblasts and concluded that reactivation of the erythrocyte nucleus does not require DNA or RNA synthesis. Their results suggested that the cytoplasm contains a store of nucleus-specific proteins.

Interesting results have also been obtained with other combinations of heterokaryons. For example, Wright (1984a) found that in heterokaryons between chick myocytes and rat brain cells, the genes coding for muscle proteins are activated in the brain nuclei. On the other hand, in heterokaryons between differentiated myocard cells and malignant cells deriving from an epidermoid carcinoma (KB cells), contractility stops 2–4 hr after fusion (Goshima *et al.*, 1984). In heterokaryons between chick myocytes and mouse adrenal cells, both muscle functions and steroid synthesis continue (Wright, 1984b). Finally, Den Boer *et al.* (1984) fused embryocarcinoma cells (see Chapters 2 and 3, Volume

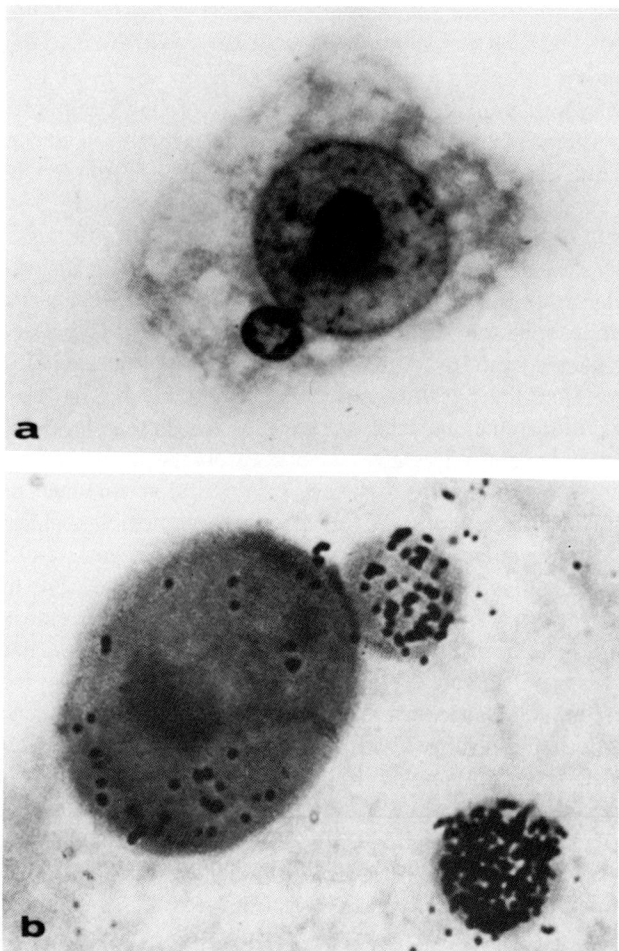

FIG. 31.   Reactivation of an erythrocyte after fusion with an active cell. (a) Heterokaryon containing one HeLa nucleus and one hen erythrocyte nucleus with still condensed chromatin. (b) Autoradiograph of a heterokaryon containing one mouse cell nucleus inactivated by a microbeam of uv light and two hen erythrocyte nuclei actively labeled with a tritiated RNA precursor. [Harris (1974).]

2) with melanoma cells or cytoplasts. In the heterokaryons, embryonic antigens characteristic of embryocarcinoma cells disappeared within 2 days; in the cybrids, disappearance was only transitory. Cytoplasmic trans-acting factors present in melanoma cells thus exert a negative control on the expression of the embryonic antigens. What will happen when two different cells are fused together and form a heterokaryon is unpredictable.

As expected, Appels *et al.* (1974) and Goto and Ringertz (1974) found that the chick erythrocyte's nuclear membrane does not allow free passage of all human proteins in chick and human hybrids; only some of the cytoplasmic human proteins move into the chick nucleus. Among the proteins that migrate into the chick erythrocyte nucleus in rat-chick heterokaryons is RNA polymerase I. In such somatic hybrids, the chick ribosomal genes are transcribed by a mammalian enzyme (Scheer *et al.*, 1983). RNA polymerase II is another one of the proteins that are selectively taken up during reactivation of chick erythrocyte nuclei; this should allow transcription of a variety of genes, and allow the synthesis of chicken globin and a number of constitutive chicken proteins (Zuckerman *et al.*, 1982). We know that, in chick erythrocyte nuclei, transcription of DNA sequences is prevented by the presence of a specific histone called H5. Linder *et al.* (1982) have studied the movements of H5 in heterokaryons between chicken erythroblasts and mammalian cells. They found that H5 can be detected in the mammalian nucleus, but only 9 days after fusion takes place. Summing up the results of these experiments, and of others still unpublished at the time of writing, Ringertz (1983) concluded that, in heterokaryons, RNA polymerases I and II, as well as various nuclear membrane, nucleolus, and nuclear sap antigens of mammalian origin, move into the erythrocyte nucleus; in contrast, histone H5 leaves it. Clearly, cytoplasmic signals modulate the pattern of gene expression. This conclusion is also valid for lysosomal enzymes. In a chicken-human heterokaryon, the reactivated chick erythrocyte nucleus directs the synthesis of a number of lysosomal enzymes within 4 days. It also reactivates the synthesis of N-acetylglucosamine-1-phosphate transferase; this enzyme, which is located in the ER–Golgi region, is required for the phosphorylation of the lysosomal enzymes (Van der Veer *et al.*, 1982).

Surprising, at first, was the finding by Davis and Harris (1975) that hemoglobin synthesis stops within 60 hr when a primitive erythroid cell removed from a 5-day chick embryo is fused with a mouse fibroblast. This is only one of the many examples of "extinction" of the differentiated phenotype, which will be examined in more detail when we deal in Chapter 3 with cell differentiation. More recently, Linder *et al.* (1981) have extended these findings to heterokaryons between chick erythrocytes and rat myoblasts. These hybrids do not synthesize hemoglobin; they produce globin mRNAs, but are apparently unable to synthesize heme. Interestingly, Linder *et al.* (1981) found that the globin mRNAs present in these hybrids coded only for adult-type globin. There was no formation of embryonic globin mRNAs. Since only the adult globin genes were reactivated by fusion, there was no "reprogramming." The embryonic globin genes were not activated before the adult genes by the rat cytoplasm.

Other somatic hybrids are interesting to cell biologists. For instance, Laurila *et al.* (1982) observed the coexpression of vimentin and keratin IFs in heterokaryons between fibroblasts and amnios epithelial cells (the former normally express only vimentin and the latter keratin). In the heterokaryons, both types of IFs are incorporated into the cytoskeleton. Blau *et al.* (1983) studied heterokaryons between human amniocytes and mouse muscle cells. They found that these somatic hybrids synthesize human muscle proteins. In this case, in contrast to the embryonic and adult globin genes, reprogramming took place. Human genes, which were silent in the amniocytes, were somehow activated by the heterokaryon cytoplasm. Silent muscle genes are activated in fibroblasts by fusion with muscle cells (Chiu and Blau, 1985). As one can see, the biochemical consequences of somatic hybridization are, for the time being, unpredictable. More surprises will certainly come in the next years.

Interest in the organization of chromatin and nucleoli in somatic cell hybrids has been growing and is likely to increase. For instance, Sperling *et al.* (1980) found that the repeat length of chromatin (thus, the nucleosome organization) is, in interspecies somatic cell hybrids, intermediate between the values observed for the two parental species. According to Sperling and Weiss (1980), this repeat length varies with cell differentiation, and not with generation time or chromosome number. In the past, there was a good deal of discussion about the control of rRNA synthesis by the nucleolar organizers in interspecific hybrids. Relatively recent contributions to this field are those of Wejksnora and Warner (1981) and Lipszyc *et al.* (1981). According to Wejksnora and Warner, in hybrids between mouse and hamster cells, the mouse ribosomal genes are dominant for the synthesis of the rRNA precursor, but its processing and the synthesis of the ribosomal proteins are species independent. Lipszyc *et al.* (1981), using cytochemical methods (silver staining, autoradiographic detection of [^{3}H]uridine incorporation), studied human–mouse somatic hybrids; they found no dominance for nucleolar RNA synthesis for 2 days. Dominance did not become apparent before chromosomes were lost during the replication of the synkaryons. In rat–mouse hybrids in which no chromosome loss was taking place, the two parental rRNAs were co-expressed and the two mitochondrial DNAs were retained (Hayashi *et al.*, 1982). Attempts have recently been made to understand why human ribosomal genes are repressed in human–mouse somatic hybrids. For Onishi *et al.* (1984), nucleolar dominance results from loss or inactivation of the human ribosomal genes. Inactivation would result from the loss of a specific factor required to recognize the promoter of the human ribosomal genes. This factor is missing in the hybrid cell line since extracts of the hybrids do not initiate the transcription of the human genes. Miesfeld *et al.* (1984) came to similar conclusions: mouse nucleolar dominance, after loss of a few human chromosomes in man–mouse cell hybrids, seems to be due to the loss of specific human rDNA transcription factors. Addition of these factors to extracts from the hybrids rescues human rDNA transcription.

This chromosome loss in human–mouse somatic cell hybrids, which as already been mentioned, allowed geneticists to map hundreds of genes in both humans and mice very precisely by following their expression in hybrid cell lines. The exact localization of a new gene on a given chromosome is determined every month; there is no doubt that human genetics has made great progress in the last few years (reviewed by D'Eustachio and Ruddle, 1983). In general, when synkaryons between human and mouse genomes divide repeatedly, human chromosomes are preferentially lost in the hybrid cell lines. This loss is not random in the human myeloma–mouse myeloma cell line analyzed by Cieplinski *et al.* (1983). Certain chromosomes are conserved longer than others, but it is not known whether this is a general and constant phenomenon. However, this is not an absolute rule. Croce (1976) observed a selective loss of mouse chromosomes in one of five human–mouse hybrid cell lines. Karyogram analysis with banding techniques allows the precise identification of the chromosomes that have been retained (finally, a single chromosome of one of the two species may be present). Studies on the enzyme complement of hybrids that have retained a single human (or mouse) chromosome allow the mapping of the corresponding genes on that particular chromosome. Another approach is to detect the disappearance of proteins after the loss of a given chromosome. For instance, Scoggin *et al.* (1981) reported that the loss of human chromosome 11 is followed by the disappearance of eight polypeptide spots in two-dimensional electrophoresograms. Another valuable approach for mapping the mouse or human genome is fusion of a microcell with a complete cell. Microcells can be obtained by treating cells with colcemid and then with cytochalasin B. The chromosomes, which have been scattered after the colcemid treatment, form micronuclei after the cytochalasin step and arise from a single chromosome (Fournier, 1981). Chromosome-mediated gene transfer has even enabled Fournier and Ruddle (1977) to build up trispecific (human, mouse, hamster) cell hybrids by fusion of microcells and a complete cell and to show that integration of the donor genomes takes place on many chromosomal loci of the receptor genome.

However, it should be pointed out that the kind of work we have just discussed is likely to be superseded by "restriction analysis." As D'Eustachio and Ruddle (1983) said, somatic cell genetics has moved from the phenotype to the genome itself because of recombinant DNA and blotting techniques that allow the visualization of single genes. This kind of analysis allows the study of multigene families and pseudogenes (see Chapter 4, Volume 1) dispersed on different chromosomes. Several hundreds of human genes have been mapped by combination of somatic cell genetics and recombinant DNA technology [reviewed by Kao (1983) and Tunnacliffe (1984)]. Briefly, the technique for mapping human or mouse chromosomes is the following: a number of somatic hybrid cell lines are cultured, and karyotypes of the hybrids, which have lost different chromosomes, are made; radioactive cDNA probes for enzymes that have already been located on given chromosomes are prepared; total DNA is isolated from the hybrid cultures and cleaved with restriction enzymes; the fragments are separated by electrophoresis and hybridized with the cDNA probes specific for the different chromosomes; they will segregate with the markers of one or the other of the chromosomes. *In situ* hybridization techniques, which have greatly improved, provide an additional means for mapping individual genes in metaphase chromosomes stained with a banding technique.

Hybrids and cybrids can tell us a great deal about the interactions between nuclear and mitochondrial genomes. The first papers on the subject dealt with human–mouse somatic hybrids. Attardi and Attardi (1972) showed that there is no synthesis of human mitochondrial DNA in such hybrids. In agreement with this finding, Jeffreys and Craig (1974) reported that only mouse mitochondrial proteins can be detected in human–mouse somatic hybrids. This dominance would be due to the preferential loss of human chromosomes in the hybrids, and this remains a valid explanation of the results, since we have seen that there is no loss of mitochondrial DNAs when there is no chromosome loss (Hayashi *et al.*, 1982).

Cytoplasts became a more popular tool when Ege and Ringertz (1975) found that "reconstituted cells," resulting from fusion between a karyoplast and a cytoplast, are viable even if the nucleus and the cytoplasm belong to different animal species. Furthermore, chick erythrocyte nuclei can be reactivated by fusion of the erythrocyte with a cytoplast isolated from a metabolically active cell, which provides final proof that the reactivating factor is cytoplasmic. Even in such an artificial system, selectivity in the uptake of cytoplasmic proteins into the nucleus can be detected (reviewed by Appels and Ringertz, 1975).

Cybrids have already told us a good deal about the role played by cytoplasmic factors in cell morphology and chemical composition. For instance, Lafond *et al.* (1983) found that fusion of a fibroblast cytoplast with a chick erythrocyte results in the generation of a nuclear matrix that was not visible in the condensed erythrocyte nucleus. Hightower *et al.* (1983) fused human fibroblast cytoplasts with mouse fibroblast karyoplasts and noticed that the cybrids took the shape characteristic of the nucleus-donor cell; however, there were no changes in the organization of the MFs that apparently do not determine cell shape. The same authors found that the cytoplasm possesses all of the factors required for the initiation and support of DNA synthesis, but not for entry in mitosis. This suggests that the cytoplasm of the human fibroblasts used for these fusion experiments contained no MPF (see Chapters 5, Volume 1 and Chapter 2, this volume) or that this factor decays rapidly in cybrids. Finally, Hightower *et al.* (1983) reported that protein synthesis in cybrids is directed by the nucleus 3–6 hr after fusion; after 48 hr, the protein synthesis patterns of cybrids and normal cells become indistinguishable. This is in good agreement with the finding that the enzyme dopa-oxidase (required for melanin synthesis) is not expressed in heterokaryons between fibroblasts and melanoma cells, while it is expressed in corresponding nondividing cybrids (Schwartz and Brumbaugh, 1982). Finally, fusion of cytoplasts from epithelial cells with fibroblasts suppresses the formation of the extracellular fibronectin matrix characteristic of fibroblasts (Laurila and Stenman, 1982).

Wallace *et al.* (1975) have shown that cytoplasts from chloramphenicol-resistant cells can transfer chloramphenicol resistance to normal HeLa cells. This resistance, as is well known from studies on the genetics of yeast mitochondria, is a purely mitochondrial genetic trait, since it results from a mutation in mitochondrial DNA. This conclusion has been confirmed and extended to human mitochondria by Wallace (1981), who studied the restriction endonuclease cleavage pattern of mitochondrial DNA in hybrids and cybrids, and came to the conclusion that chloramphenicol resistance lies in the mitochondrial genome, even in humans. The complex events that may take place when mitochondria from two different species face each other in a common cytoplasm was made apparent in a paper by Hayashi *et al.* (1981). They fused mouse cells deficient in thymidine kinase (TK^-) with rat cells deficient in hypoxanthine guanine phosphoribose transferase ($HGPRT^-$) and cultured the hybrids in a classic selective medium (HAT: hypoxanthine, aminopterin, thymidine);[1] the mitochondria of the selected cell hybrids were isolated and their DNA analyzed by restriction endonuclease digestion. The patterns revealed the presence of some hybrid mouse–rat mitochondrial DNA molecules (Wallace, 1981; Oliver and Wallace, 1982). These findings strongly support the idea, first suggested by Dujon *et al.* (1974) after extensive studies on yeast mitochondrial genetics, that yeast cells possess two distinct populations of mitochondria and that DNA can be exchanged between them. Apparently, such recombination events can take place even between mitochondrial DNAs of two different species, such as mouse and rat.

Of course, all of the effects exerted by the cytoplasm in cybrids are not due to the mitochondria present in the cytoplasts. For instance, Clark and Shay (1982a) formed cybrids between adrenal cell nuclei and cytoplasm from fibroblasts; the cybrids were separated with a cell sorter after vital staining of the mitochondria with the fluorescent dye Rhodamine 123. It was found that the cybrids, in contrast to intact adrenal cells, were unable to synthesize steroids in response to adrenocorticotropic hormone (ACTH). The conclusions of these studies were that the cytoplasm of the fibroblasts contained a nonmitochondrial inhibitor of steroidogenesis and that this inhibitor had a long life.

Present evidence is too scanty to allow us to draw a precise picture of the interactions between the nuclear and mitochondrial genomes. Progress in this field can be expected, since Clark and Shay (1982b) succeeded in introducing mitochondria into chloramphenicol-sensitive recipient cells by endocytosis; resistance to chloramphenicol was conferred to the cells by mitochondria-mediated transfer. For now, it seems that the situation is not very different from what we know about the interactions between the chloroplasts and nucleus in *Acetabularia*. Mitochondria enjoy some autonomy toward the nucleus due to their DNA, but this autonomy is very limited since, as we already know, the great majority

---

[1]*TK^-* and *HGPRT^-* cells do not grow on HAT because they lack the "salvage pathway" that is necessary since aminopterin inhibits the normal pathway of DNA synthesis.

of the mitochondrial proteins are coded for by nuclear genes. Whether this conclusion is an argument for the symbiotic origin of mitochondria is an unresolved question.

V. GENE TRANSFER IN SOMATIC CELLS

The famous experiments of Avery *et al.* (1944) on bacterial transformation by the addition of DNA to cultures of *Pneumococcus* will forever remain a landmark in the history of molecular biology. They proved for the first time that DNA is, indeed, the genetic material. These experiments created the possibility of changing, at will, the genetic background of eukaryotic cells and organisms by the addition of DNA or, better, by the introduction of a specific gene into the genome. What was a dream only a few years ago has now become a reality. Mammalian cells can be transformed by the introduction of pure genes isolated from prokaryotes or eukaryotes (reviewed by Wigler *et al.*, 1979; Scangos and Ruddle, 1981; Ruddle, 1981; Gordon and Ruddle, 1985).

It is only fair to recall here that the presently very active research on gene transfer in eukaryotic cells began with the bold efforts of our former student and co-worker, Lucien Ledoux, in the 1960s. He added DNA of various origins to animal or plant cells and attempted to demonstrate, using then available methods, that part of the foreign DNA had been integrated into the chromatin of the recipient cells. At the time, these attempts met with widespread skepticism; this had also happened to Avery *et al.* when they published the results of their experiments in 1944. The weak point in the work of Ledoux was the lack of a genetic test to prove that a defective gene had been selectively replaced by an active one after DNA addition to an appropriate mutant. This was realized by Ledoux *et al.* (1971) when they attempted to correct mutations in the plant *Aspidistra* by the addition of DNA isolated from wild-type plants.

Currently, several methods are available for the introduction of genetic material into cells (transformation, tranfection).[2] Success in genetic manipulation at the cellular level requires not only the uptake of the added DNA but also its integration into the host genome. This is, of course, a prerequisite for the maintenance of the added genetic material during successive cell divisions, and thus for obtaining a heritable change. The transferred genes persisting in the nucleus as part of large DNA molecules are called "transgenomes" (or "transforming material"); the cells that have received the transforming material are called "transformants." Unfortunately, the percentage of stably transformed cells in cultures is very small (between 1×10^5 and 1×10^6) when the most popular technique (DNA-mediated gene transfer) is used. Selection of the transformants is an absolute necessity for further analysis. This is why so much of the work done so far has focused on the thymidine kinase (*TK*) gene, since selection of

---

[2]Reviewed by Celis (1984) and de Jonge and Bootsma (1984).

TK⁻ cells in HAT medium is easy to perform. The only method that allows a very high (50–100%) percentage of transformations is direct injection of the transforming gene into the cell nucleus (Capecchi, 1980), which requires enormous skill and patience, since injection into the cytoplasm is useless. For this reason, it is doubtful that Capecchi's elegant approach will become very popular, since most cell biologists and almost all biochemists like to work with large populations of cells.

Historically, the first approach was to introduce intact metaphase chromosomes into recipient cells (chromosome-mediated gene transfer). This approach was followed by the addition, as in Avery's initial experiments with bacteria, of unfractionated DNA to cell cultures (DNA-mediated gene transfer). It is now possible to introduce any cloned gene into cells and to obtain transformants. The old dream of modifying, at will, the genetic makeup of a cell has thus been fulfilled. To go further, i.e., to obtain, at will, genetic transformations that can be transferred from one generation to the next, one has to work with eggs. As we shall see in the next chapter, this has recently been successfully achieved in mouse eggs.

Transfer of genetic information through isolated metaphase chromosomes was achieved by McBride and Ozer (1973) and by Degnen *et al.* (1976). Using this technology, Willecke and Ruddle (1975) and Athwal and McBride (1977) transferred the *HGPRT* human gene to mouse cells; however, the frequency of transformants did not exceed 10^{-7}. Chromosome-mediated gene transfer was also used by Fournier *et al.* (1979) to introduce human chromosomes into Chinese hamster ovary cells. An increase in transformant frequency was obtained by Mukherjee *et al.* (1978) when they introduced metaphase chromosomes into liposomes and allowed them to fuse with the plasma membrane of the recipient cells; one cell out of 10,000 could be transformed in this way. Another approach for the introduction of a single chromosome in a cell is to use microcells (also called "minicells") obtained by successive treatments with colchicine and cytochalasin B; microcells possessing a single chromosome swollen into a micronucleus can easily be fused with normal cells (McNeill and Brown, 1980; Fournier, 1981). Microcell clones containing specific mouse chromosomes have been constructed by Fournier and Frelinger (1982). They will be very useful tools for microcell-mediated chromosome transfer.

DNA-mediated gene transfer (reviewed by Scangos and Ruddle, 1981; Ruddle, 1981) is, in principle, a straightforward transformation method. In theory, all one has to do is add purified DNA (or, of available, a cloned gene) to the cell culture. There are, however, problems due to the poor uptake of DNA in the nuclei of the recipient cells. These problems can be partly circumvented by adding a carrier DNA of about 50 kb, precipitated by calcium phosphate to the gene; the coprecipitate is then transferred into the cell (Wigler *et al.,* 1979); Farber *et al.,* 1975). Technical improvements aimed at increasing the yield in transfected cells have been proposed. If the cells are treated with polyethylene glycol (PEG) shortly before the addition of the DNA–calcium phosphate complex, as many as 70% of the cells may be transfected in favorable cell lines; the PEG shock alters the plasma membrane without affecting cell viability (Shen *et al.*, 1982). According to Gorman *et al.* (1983), treatment of recipient mammalian cells with butyrate increases the number of cells in which DNA uptake takes place and enhances the expression of the transfected genes.

In this very popular calcium phosphate–DNA coprecipitation technique, DNA forms a nuclease-resistant complex with calcium phosphate. The complex is taken up by endocytosis in all of the recipient cells but can be detected in the nuclei of only a few cells (Loyter *et al.*, 1982a,b). Penetration into the nucleus is not sufficient for successful transformation of the cell. This will not be achieved unless the foreign DNA is integrated into the host cell genome and finally replicated. According to one paper (Schaefer-Ridder *et al.*, 1982), transfer of the *TK* gene to fibroblasts via liposomes is more efficient than the calcium phosphate–DNA coprecipitation method; as many as 10% of the treated cells stably express the *TK* gene. Still better results (20% of the recipient cells have the transfected DNA in their nuclei after 4 hr) have been obtained by fusing cells with erythrocyte ghosts containing donor DNA (Wiberg *et al.*, 1983).

Using purified DNA, Grof *et al.* (1979) transferred the *HGPRT* gene into cells in culture. With the calcium phosphate–DNA coprecipitation technique, Robins *et al.* (1981) introduced into recipient cells 5–100 copies of the human growth hormone gene. Using *in situ* molecular hybridization, they detected the localization of the transferred gene copies and found that their insertion was not limited to a particular chromosome or to a specific region of a chromosome. In fact, the transgenomes that persisted in the transformants as part of chromosomal macromolecules were at first unstable and tended to be lost. However, it is possible, later on, to obtain cell subpopulations in which the transferred gene is stably maintained. In such stable cell lines, the transgenome is associated with one of the chromosomes of the recipient cell, but this chromosome is not always the same in the different stable cell lines (Scangos and Ruddle, 1981).

A promising and elegant technique for improving the yield of transformants (besides the already mentioned microinjection technique of Capecchi, 1980) has been proposed by Schaffner (1980). Cloned genes can be introduced into mammalian cells by fusing protoplasts of bacteria, which harbor cloned genes, with the recipient cells. Bacterial cell walls are digested with lysozyme, the protoplasts are centrifuged, and their fusion to the mammalian cells is induced with PEG. It is said that the yield of the technique can be as high as 50% for fusion and 1–2% for transformation, and that it is 10–20 times more efficient than transfection with a calcium phosphate–DNA coprecipitate (Rassoulzadegan *et al.*, 1982). A variant of this ingenious technique has been proposed by Slilaty and Aposhian (1983). They transferred genes into recipient cells through empty capsids (envelopes) of polyoma virus (an oncogenic DNA virus). This involves eliminating polyoma DNA, allowing the empty capsids to react with the transfecting DNA, and infecting recipient cells that are susceptible to polyoma virus infection (so-called permissive cells). This method is said to be 50–100 times more efficient than the calcium phosphate–DNA coprecipitation technique.

A popular technique for the transfer of cloned genes is cotransformation. Here the gene is inserted into a vector that can penetrate and replicate in recipient cells. In 1975, Horst *et al.* claimed to have obtained the transfer of the *E. coli* β-galactosidase gene into human cells after addition of a λ phage in which the β-galactosidase gene had been inserted. Currently, the favorite agent for "cotransfection" is the herpes simplex virus (HSV), which can easily infect mammalian cells and replicate in them, since it possesses the base sequences required for initiation of DNA synthesis. Another advantage of HSV is that its genome contains a viral *TK* gene, affording the possibility of selection with the HAT medium if this gene has been successfully introduced into recipient cells.

Using this method of cotransformation, Axel *et al.* (1980) introduced the *TK* viral gene into mouse fibroblasts, erythroleukemic Friend cells, and teratocarcinoma cells. They obtained unstable mutants that switched from TK^+ to TK^- and back again to TK^+. Axel believed this instability to be due to frequent, but reversible, alterations of the DNA molecules. Calos *et al.* (1983) have shown that the frequency of mutations in transfected DNA is exceedingly high (1% as compared to less than 10^{-5} in the donor cell).

Now that we have developed an array of methods allowing successful gene transfer in eukaryotic cells, very interesting results have been obtained. For instance, Kurtz (1981) cotransfected the HSV viral *TK* gene and the two genes for α_2-globulin in hepatoma cells; he induced the expression of these globulin genes by treatment with corticosteroids (dexamethazone). Chen and Nienhuis (1981), using a similar approach, introduced the genes for human δ- and β-globins into mouse fibroblasts. They found that the human genes were inserted in regions of the mouse chromosomes where DNA was unmethylated. Nevertheless, expression of the genes was reduced to fewer than 100 copies of globin mRNA per cell. Corces *et al.* (1981) transformed mouse fibroblasts with *Drosophila* heat shock genes. They observed, in certain cell lines, an accumulation of the transcription products of these genes when the cells were submitted to heat shock. That there is no taxonomic barrier in gene transfer experiments was fully demonstrated by the work of P. Berg (reviewed in 1981). Bacterial genes can be expressed in mammalian cells, as shown by Mulligan and Berg (1981) and Yoder and Ganesan (1981).

Gene transfer experiments have already produced interesting data on DNA replication, DNA repair, and chromatin organization. For instance, Bradshaw (1983) introduced the human *TK* cloned gene into mouse cells and followed thymidine kinase activity during the cell cycle. He found that it strongly increased during the S phase, demonstrating that the foreign human gene is regulated together with the *TK* mouse genes. Rubin *et al.* (1983) transfected with human DNA Chinese hamster ovary (CHO) cells that were deficient in DNA repair after uv irradiation or mitomycin treatment; this resulted in the correction of the defect. These transfection experiments provide evidence for the existence of a human DNA repair gene. Transfected and integrated genes may, under certain experimental conditions, undergo amplification. Kaufman and Sharp (1982) transformed by transfection cells that lacked the enzyme dehydrofolate reductase (DHFR) into cells that expressed this enzyme. It is known that methotrexate, an inhibitor of DHFR, induces amplification of the DHFR genes in resistant cells. Kaufman and Sharp (1982) found that treatment of the transformed cells with methotrexate increased up to 1000 times the number of transforming DNA copies. These experiments and others clearly show that it is possible to "cure" cells that lack an enzyme by introducing into them the appropriate gene by transfection. It is also possible to confer resistance to a drug by transferring a chromosome from a resistant cell. This has been demonstrated, for resistance to HU (an inhibitor of ribonucleotide reductase), by Lewis and Srinivasan (1983). Another possibility is to modify the gene *in vitro* before introducing it into recipient cells; such modification, called "site-directed mutagenesis," can be obtained at will in the laboratory. An example can be found in a paper by Busslinger *et al.* (1983), who methylated *in vitro* specific segments of the human β-globin gene before transfection; they found that the methylation pattern of the 5′-region of the gene plays a direct role in gene expression. Finally, we should mention the curious results obtained by Reeves *et al.* (1983). They introduced the histone H2 gene of *Drosophila* into simian cells and isolated their chromatin. The simian chromatin contained *Drosophila* histones.

This brief account of a developing field shows that we are on the verge of a new and fascinating era. How far gene transfer experiments will lead us remains a matter of speculation. It is clear that one of its applications is to achieve "gene therapy," i.e., substitution of a deficient gene responsible for an inborn disease by a normal one. Attempts to transfer genes in intact animals were made by Cline *et al.* (1980). They reported the induction of resistance to methotrexate by the injection of DNA isolated from methotrexate-resistant cells into the bone marrow of men. These experiments were halted because the necessary permission had not been requested of the appropriate authorities; this interdiction is likely to slow down progress in a promising field of research. However, caution is required because gene transfer in intact organisms—in humans in particular—may prove to be either a blessing or a disaster. Molecular biologists are fallible, and the so-called experiments during World War II are not yet completely forgotten.

VI. SUMMARY AND GENERAL CONCLUSIONS

Figure 32 summarizes, in a very oversimplified way, what has been said in this chapter. All the approaches discussed show that the nucleus is the main—but

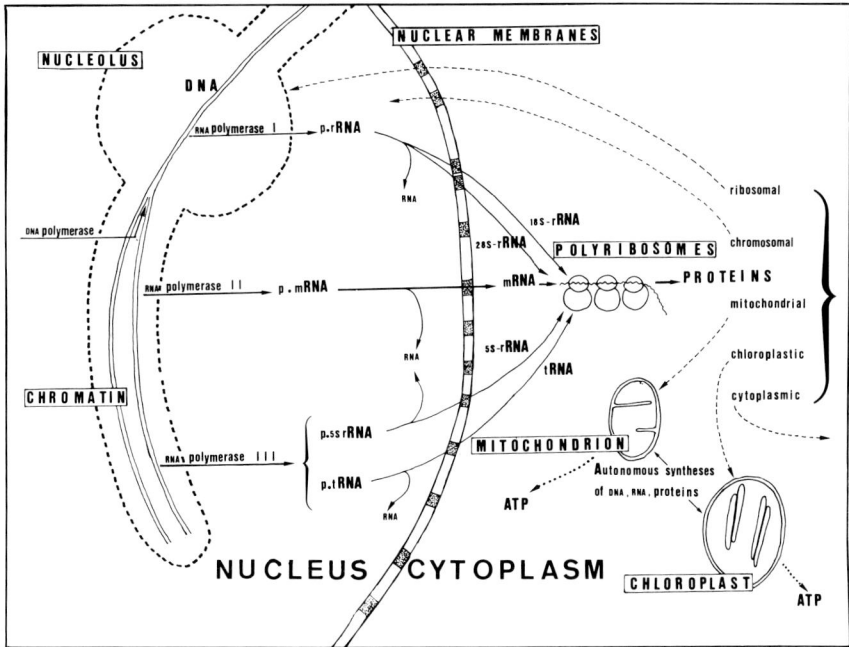

Fig. 32. Diagrammatic representation of nucleocytoplasmic exchanges. [Original drawing by P. Van Gansen.]

not the only—center of the synthesis of all types of nucleic acids. Nuclear DNA, which is replicated during the S phase of the cell cycle, can be transcribed in precursors of mRNAs, tRNAs, and rRNAs. The nucleus contains three different kinds of polymerases that specialize in the synthesis of the precursors for the three main types of RNA. Nucleolar RNA polymerase I directs the synthesis of the 28 and 18 S rRNAs; chromatin contains RNA polymerase II for the synthesis of the pre-mRNAs and RNA polymerase III for that of 5S RNA, tRNAs, and some of the snRNAs. The processing of these precursors, presumably by specific endoribonucleases, is also a function of the nucleus, and the degradation products probably never reach the cytoplasm. When processing is completed, the RNAs that have been synthesized on chromatin and on nucleolar DNAs cross the nuclear membrane. These RNA associate with proteins, which differ for each kind of RNA, and finally assemble to form polyribosomes. In general, once the RNAs are in the cytoplasm, they do not move back into the nucleus. In amebas, however, the existence of "shuttle" RNAs of small molecular weight seems to be established. Unfortunately, it is not known whether they exist in other cells, whether they are related to the snRNAs, and what role they may be playing. Many of the proteins synthesized on the polyribosomes move back into the nucleus. Histones and non-histone chromosomal proteins play an essential role in the molecular organization of chromatin and combine with DNA to exert positive or negative controls on gene transcription.

Prolonged activity of mitochondria and chloroplasts, despite the fact that they contain their own DNA and can synthesize their own rRNAs and tRNAs, as well as a limited number of mRNA species, requires help from the nucleus. This assistance is provided by the intake, into the semiautonomous cell organelles, of proteins synthesized on cytoplasmic polyribosomes; synthesis of these proteins is directed by mRNAs that have been transcribed on nuclear DNA. Chloroplasts— and probably mitochondria as well—are semiautonomous in the sense that they can increase in number in anucleate cytoplasm. In fact, the burst of overall RNA synthesis that follows enucleation in *Acetabularia* is probably due almost entirely to chloroplastic RNA synthesis. This means that the nucleus can exert negative as well as positive controls on cell organelles such as chloroplasts and mitochondria. Chloroplasts and mitochondria play an essential role in cell economy: energy production. In cells all activity ceases when energy production breaks down. If functioning of the mitochondria is halted by the addition of potassium cyanide or dinitrophenol, transcription and translation quickly come to a halt; the same occurs when a green cell is cultivated in the absence of light.

The work on cell hybrids and cybrids has disclosed other important functions of the cytoplasm: control of the choice between DNA replication and DNA transcription, and the responsibility for nuclear membrane breakdown and chromosome condensation at prophase. Unfortunately, we do not yet know the molecular bases of this control. Cytonucleoproteins, which shuttle back and forth between the cytoplasm and the nucleus appear to be likely candidates for this important job.

We can now return to the questions that puzzled us more than 50 years ago: why is anucleate cytoplasm inferior to its nucleate counterpart? Why does the survival of anucleate cytoplasm vary so much according to the biological system chosen (months in *Acetabularia* and hours in a cytoplast from a mammalian fibroblast)? Merotomy, in cells or organisms that possess a single nucleus, is necessarily followed by the immediate cessation of RNA synthesis in the anucleate half, except in mitochondria and chloroplasts, but protein synthesis always continues, at a slowly decreasing rate, as long as the anucleate cytoplasm remains viable. This means that the amount, variety, and stability of the RNAs of nuclear origin (in particular, the mRNAs) that were present in the anucleate cytoplasm at the time of section were sufficient to allow continuation of protein synthesis. Death probably occurs when this residual machinery for protein synthesis becomes incapable of providing sufficient amounts of one or several key proteins that are essential for life. Differences in the stability of the RNAs (especially the mRNAs that are, in general, much more labile than the rRNAs) in different kinds of cells would thus be responsible for the variable longevity of nucleate fragments from different living organisms. Although this seems to be a reasonable explanation, we do not know whether the stability of the mRNA molecules is the same in nucleate and anucleate fragments of unicellular organisms or cells. Investigation of how the stability of mRNAs (and proteins) is controlled, in the presence or absence of the cell nucleus, remains an important task.

REFERENCES

Ahkong, Q. F., Howell, J. I., Lucy, J. A., Safwat, F., Davey, M. R., and Cocking, E. C. (1975). *Nature (London)* **255**, 66.
Albrecht-Buehler, G. (1980). *Proc. Natl. Acad. Sci. U.S.A.* **77**, 6639.
Alexeev, A. B., Betina, M. I., Swolinsky, S. L., Yazikov, A., and Zubarev, T. N. (1974). *Plant Sci. Lett.* **3**, 297.
Apel, K., and Schweiger, H. G. (1972). *Eur. J. Biochem.* **25**, 229.
Apel, K., and Schweiger, H. G. (1973). *Eur. J. Biochem.* **38**, 373.
Appels, R., and Ringertz, N. R. (1975). *Curr. Top. Dev. Biol.* **9**, 137.
Appels, R., Bolund, L., and Ringertz, N. R. (1974). *J. Mol. Biol.* **87**, 339.
Athwal, R. S., and McBride, O. W. (1977). *Proc. Natl. Acad. Sci. U.S.A.* **74**, 2943.
Attardi, B., and Attardi, G. (1972). *Proc. Natl. Acad. Sci. U.S.A.* **69**, 129.
Avery, O. T., McLeod, C. M., and McCarthy, M. (1944). *J. Exp. Med.* **79**, 137.
Axel, R., Jackson, J., Lowy, I., Ostrander, M., Pellicer, A., Pellicer, A., Roberts, J., Robins, D., Sim, G. K., Sweet, R., Wold, B., and Silverstein, S. (1980). *Eur. J. Cell Biol.* **22**, 29.
Balamuth, W. (1940). *Q. Rev. Biol.* **15**, 290.
Balbiani, E. G. (1888). *Rec. Zool. Suisse* **5**, 1.
Baltus, E. (1954). *Biochim. Biophys. Acta* **15**, 263.
Baltus, E. (1956). *Arch. Int. Physiol. Biochim.* **64**, 124.
Baltus, E. (1959). *Biochim. Biophys. Acta* **33**, 337.
Baltus, E., and Brachet, J. (1962). *Biochim. Biophys. Acta* **61**, 157.
Bannwarth, H., and Schweiger, H. G. (1975). *Proc. R. Soc. London, Ser. B* **188**, 203.
Bannwarth, H., and Schweiger, H. G. (1983). *Cell Biol. Int. Rep.* **7**, 859.
Bannwarth, H., Ikehara, N., and Schweiger, H. G. (1977). *Proc. R. Soc. London, Ser. B* **198**, 177.
Bannwarth, H., Ikehara, N., and Schweiger, H. G. (1982). *Eur. J. Cell Biol.* **27**, 200.
Berg, P. (1981). *Science* **213**, 296.
Berger, S., and Schweiger, H. G. (1975a). *J. Cell Sci.* **17**, 517.
Berger, S., and Schweiger, H. G. (1975b). *Protoplasma* **83**, 41.
Berger, S., and Schweiger, H. G. (1975c). *Planta* **127**, 49.
Berger, S., and Schweiger, H. G. (1975d). *J. Cell Sci.* **17**, 529.
Beth, K. (1943). *Z. Indukt. Abstamm. Vererbungsl.* **81**, 252.
Beth, K. (1953a). *Z. Naturforsch., B: Anorg. Chem., Org. Chem., Biochem. Biophys., Biol.* **8B**, 334.
Beth, K. (1953b). *Z. Naturforsch., B: Anorg. Chem., Org. Chem., Biochem., Biophys., Biol.* **8B**, 771.
Beth, K. (1955) *Z. Naturfosch., B: Anorg. Chem., Org. Chem., Biochem., Biophys., Biol.* **10B**, 267.
Bishop, J. O., Pemberton, R., and Baglioni, C. (1972). *Nature (London), New Biol.* **235**, 231.
Blau, H. M., Chiu, C. P., and Webster, C. (1983). *Cell (Cambridge, Mass.)* **32**, 1171.
Bouloukhère, M. (1965). *J. Microsc. (Paris)* **4**, 347.
Bouloukhère, M. (1970). *In* "Biology of Acetabularia" (J. Brachet and S. Bonotto, eds.), p. 145. Academic Press, New York.

Bonotto, S., Puiseux-Dao, S., Kirchmann, M., and Brachet, J. (1971). *C.R. Hebd. Seances Acad. Sci.* **272**, 392.

Borsook, H., Deasy, E. L., Haagen-Smit, A. J., Keighley, O., and Lowy, P. H. (1952). *J. Biol. Chem.* **196**, 669.

Brachet, J. (1955). *Biochim. Biophys. Acta* **18**, 247.

Brachet, J. (1957). "Biochemical Cytology." Academic Press, New York.

Brachet, J. (1960). "The Biological Role of Nucleic Acids," 6th Weizmann Memorial Lecture. Elsevier, Amsterdam.

Brachet, J. (1965). *Endeavour* **25**, 155.

Brachet, J. (1968). *Curr. Top. Dev. Biol.* **3**, 1.

Brachet, J. (1975). *Biochem. Physiol. Pflanz.* **168**, 493.

Brachet, J. (1976). *Encycl. Plant Physiol., New Ser.* **3**, 53.

Brachet, J. (1981). *Forrschr. Zool.* **26**, 15.

Brachet, J., and Hubert, E. (1972). *J. Embryol. Exp. Morphol.* **27**, 121.

Brachet, J., and Hulin, N. (1970). *Exp. Cell Res.* **59**, 486.

Brachet, J., and Lang, A. (1965). In "Handbuch der Pflanzenphysiologie" (W. Ruhland and A. Lang, eds.), Vol. 15, p. 22. Springer-Verlag, Berlin and New York.

Brachet, J., Chantrenne, H., and Vanderhaeghen F. (1955). *Biochim. Biophys. Acta* **18**, 544.

Brachet, J., Mamont, P., Boloukhère, M., Baltus, E., and Hanocq-Quertier, J. (1978). *C.R. Hebd. Séances Acad. Sci.* **287**, 1289.

Bradshaw, H. D., Jr. (1983). *Proc. Natl. Acad. Sci. U.S.A.* **80**, 5588.

Brändle, E., and Zetsche, K. (1973). *Planta* **11**, 209.

Brändle, E., and Zetsche, K. (1977). *Protoplasma* **93**, 43.

Briggs, R., and King, T. J. (1953). *J. Exp. Zool.* **122**, 485.

Brown, R. L., Wible, L. J., and Brinkley, B. R. (1980). *Cell Biol. Int. Rep.* **4**, 453.

Bruno, J., and Lucas, J. J. (1983). *Cell Biol. Int. Rep.* **7**, 651.

Busslinger, M., Hurst, J., and Flavell, R. A. (1983). *Cell* **34**, 197.

Calos, M. P., Lebkowski, J. S., and Botchan, M. R. (1983). *Proc. Natl. Acad. Sci. U.S.A.* **80**, 3015.

Capecchi, M. R. (1980). *Cell (Cambridge, Mass.)* **22**, 479.

Caspersson, T. (1950). "Cell Growth and Cell Function." Norton, New York.

Cavenee, W. K., Chen, H. W., and Kandutsch, A. A. (1981). *J. Biol. Chem.* **256**, 2675.

Celis, J. E. (1984). *Biochem. J.* **223**, 281.

Chantrenne, H., Burny, A., and Marbaix, G. (1967). *Prog. Nucleic Acid Res. Mol. Biol.* **7**, 173.

Chen, M. J., and Nienhuis, A. W. (1981). *J. Biol. Chem.* **256**, 9680.

Chiu, C. P., and Blau, H. M. (1985). *Cell* **40**, 417.

Cieplinski, W., Reardon, P., and Testa, M. A. (1983). *Cytogenet. Cell Genet.* **35**, 93.

Clark, M. A., and Shay, J. W. (1982a). *Proc. Natl. Acad. Sci. U.S.A.* **79**, 1144.

Clark, M. A., and Shay, J. W. (1982b). *Nature (London)* **295**, 605.

Cline, M. J., Stang, H., Mercola, K., Morse, L., Ruprecht, R., Browne, J., and Salser, W. (1980). *Nature (London)* **284**, 422.

Comandon, J., and de Fonbrune, P. (1939). *C.R. Séances Soc. Biol. Ses Fil.* **130**, 740.

Corces, V., Pellicer, A., Axel, R., and Meselson, M. (1981). *Proc. Natl. Acad. Sci. U.S.A.* **78**, 7038.

Craig, I. W., and Gibor, A. (1970). *Biochim. Biophys. Acta* **217**, 488.

Croce, C. M. (1976). *Proc. Natl. Acad. Sci. U.S.A.* **73**, 3248.

Crocker, T. T., Goldstein, L., and Cailleau, R. (1956). *Science* **124**, 935.

Danielli, J. F. (1955). *Exp. Cell Res., Suppl.* **3**, 98.

Davis, T. J., and Harris, H. (1975). *J. Cell Sci.* **18**, 207.

Déak, I. I., and Defendi, V. (1975). *J. Cell Sci.* **17**, 531.

Déak, I. I., Sidebottom, E., and Harris, H. (1972). *J. Cell Sci.* **11**, 379.

Degnen, G. E., Miller, I. L., Eisenstadt, J. M., and Adelberg, E. A. (1976). *Proc. Natl. Acad. Sci. U.S.A.* **73**, 2838.

De Groot, E. J., and Schweiger, H. G. (1983). *J. Cell Sci.* **64**, 27.

De Groot, E. J., and Schweiger, H. G. (1985). *J. Cell Sci.* **78**, 1.

De Jonge, A. J. R., and Bootsma, D. (1984). *Intern. Rev. Cytol.* **92**, 133.

Den Boer, W. C., Van der Kamp, A. W. M. and, Mulder, M. P. (1984). *Exp. Cell Res.* **154**, 25.

de Terra, N. (1974). *Science* **184**, 530.

de Terra, N. (1981). *J. Exp. Zool.* **216**, 368.

D'Eustachio, P., and Ruddle, F. H. (1983). *Science* **220**, 919.

Dujon, B., Slonimski, P. P., and Weill, L. (1974). *Genetics* **78**, 415.

Ege, T., and Ringertz, N. R. (1975). *Exp. Cell Res.* **94**, 469.

Ege, T., Carlsson, S. A., and Ringertz, N. R. (1971). *Exp. Cell Res.* **69**, 472.

Ege, T., Hamberg, H., Krondahl, U., Ericsson, J., and Ringertz, N. R. (1974a). *Exp. Cell Res.* **87**, 365.

Ege, T., Krondahl, U., and Ringertz, N. R. (1974b). *Exp. Cell Res.* **88**, 428.

England, J. M., Costantino, P., and Attardi, G. (1978). *J. Mol. Biol.* **119**, 455.

Eutenever, V., and Schliwa, M. (1984). *Nature* **310**, 58.

Farber, F., Melnick, J. L., and Butel, J. A. (1975). *Biochim. Biophys. Acta* **390**, 298.

Ficq, A. (1959). *In* "The Cell" (J. Brachet and A. E. Mirsky, eds.), Vol. 1, p. 67. Academic Press, New York.

Flickinger, C. J. (1973). *Exp. Cell Res.* **80**, 31.

Flickinger, C. J. (1974). *Exp. Cell Res.* **80**, 31.

Flickinger, C. J. (1981). *J. Cell Sci.* **47**, 55.

Follett, E. A. C. (1974). *Exp. Cell Res.* **84**, 72.

Fournier, R. E. K. (1981). *Proc. Natl. Acad. Sci. U.S.A.* **78**, 6349.

Fournier, R. E. K., and Frelinger, J. A. (1982). *Mol. Cell. Biol.* **2**, 526.

Fournier, R. E. K., and Ruddle, F. H. (1977). *Proc. Natl. Acad. Sci. U.S.A.* **74**, 3937.

Fournier, R. E. K., Juricek, D. K., and Ruddle, F. H. (1979). *Somatic Cell Genet.* **5**, 1061.

Franke, W. W., Kartenbeck, J., and Spring, H. (1976). *J. Cell Biol.* **71**, 196.

Freidkopf-Cassel, A., and Kulka, G. (1981). *FEBS Lett.* **124**, 27.

Gavosto, F., and Rechenmann, R. (1954). *Biochim. Biophys. Acta* **13**, 583.

Goffeau, A., and Brachet, J. (1965). *Biochim. Biophys. Acta* **95**, 302.

Goldman, R. D., Pollack, R., and Hopkins, N. H. (1973). *Proc. Natl. Acad. Sci. U.S.A.* **70**, 750.

Goldstein, L. (1974a). *In* "The Cell Nucleus" (H. Busch, ed.), Vol. 1, p. 387. Academic Press, New York.

Goldstein, L. (1974b). *Exp. Cell Res.* **89**, 421.

Goldstein, L., and Ko, C. (1975). *Exp. Cell Res.* **96**, 297.

Goldstein, L., and Ko, C. (1981). *J. Cell Biol.* **88**, 516.

Goldstein, L., and Plaut, W. (1955). *Proc. Natl. Acad. Sci. U.S.A.* **41**, 874.

Goldstein, L., and Trescott, O. H. (1970). *Proc. Natl. Acad. Sci. U.S.A.* **67**, 1367.

Goldstein, L., Wise, G. E., and Beeson, M. (1973). *Exp. Cell Res.* **76**, 281.

Goodwin, C., and Pateromichelakis, S. (1979). *Planta* **145**, 427.

Gordon, J. W., and Ruddle, F. H. (1985). *Gene* **33**, 121.

Gordon, J. W., Scangos, G. A., Plotkin, D. T., Bardosa, J. A., and Ruddle, F. H. (1980). *Proc. Natl. Acad. Sci. U.S.A.* **77**, 7380.

Gordon, S. (1975). *J. Cell Biol.* **67**, 257.

Gorman, C. M., Howard, B. H., and Reeves, R. (1983). *Nucleic Acids Res.* **11**, 7631.

Goshima, K., Kaneko, H., Wakabayashi, S., Masuda, A., and Matsui, Y. (1984). *Exp. Cell Res.* **151**, 148.

Goto, S., and Ringertz, N. R. (1974). *Exp. Cell Res.* **85,** 173.
Green, B. R. (1973). *J. Cell Biol.* **59,** 123a.
Green, B. R., Heilporn, V., Limbosch, S., Boloukhère, M., and Brachet, J. (1967). *Proc. Natl. Acad. Sci. U.S.A.* **58,** 1351.
Green, B. R., Muir, B. L., and Padmanabhan, V. (1977). *In* "Progress in *Acetabularia* Research" (C. L. F. Woodcock, ed.), pp. 107–122. Academic Press, New York.
Grieninger, G. R., and Zetsche, K. (1972). *Planta* **104,** 329.
Grof, L. H., Jr., Urlaub, G., and Chasin, L. A. (1979). *Somatic Cell Genet.* **5,** 1031.
Gurdon, J. B. (1976). *J. Embryol. Exp. Morphol.* **36,** 523.
Hämmerling, J. (1934). *Wilhelm Roux' Arch. Entwicklungsmech. Org.* **131,** 1.
Hämmerling, J. (1939). *Biol. Zentralbl.* **59,** 158.
Hämmerling, J. (1943). *Z. Indukt. Abstamm.= Verebungsl.* **81,** 114.
Hämmerling, J. (1946). *Z. Naturforsch.* **1,** 337.
Hämmerling, J. (1953). *Int. Rev. Cytol.* **2,** 475.
Hämmerling, J., and Stich, H. (1954). *Z. Natuforsch., B: Anorg. Chem. Org. Chem., Biochem., Biophys., Biol.* **9B,** 149.
Harris, H. (1970). "Cell Fusion, the Dunham Lecture." Oxford Univ. Press (Clarendon), London and New York.
Harris, H. (1974). "Nucleus and Cytoplasm." Oxford Univ. Press (Clarendon), London and New York.
Hayashi, J. I., Gotoh, O., Tagashira, Y., Tosu, M., and Sekiguchi, T. (1981). *Exp. Cell Res.* **131,** 458.
Hayashi, J. I., Tagashira, Y., Yoshida, M., Tosu, M., and Sekiguchi, T. (1982). *Exp. Cell Res.* **138,** 261.
Heby, O. (1981). *Differentiation* **19,** 1.
Heilporn, V., and Limbosch, S. (1971a). *Biochim. Biophys. Acta* **240,** 94.
Heilporn, V., and Limbosch, S. (1971b). *Eur. J. Biochem.* **22,** 573.
Hightower, M. J., Bruno, J., and Lucas, J. J. (1983). *Proc. Natl. Acad. Sci. U.S.A.* **80,** 5310.
Hogeboom, G. H., and Schneider, W. C. (1952). *J. Biol. Chem.* **197,** 611.
Holter, H. (1955). "Fine Structure of Cell," p. 71. Wiley (Interscience), New York.
Horst, J., Kluge, F., Beyreuther, K., and Gerok, W. (1975). *Proc. Natl. Acad. Sci. U.S.A.* **72,** 3531.
Hume, S. P., Lamb, J. F., and Weingart, R. (1975). *Nature (London)* **255,** 73.
Ivarie, R. D., Fan, W. J. W., and Tomkins, G. M. (1977). *J. Cell. Physiol.* **85,** 357.
Jacob, F., and Monod, J. (1961). *J. Mol. Biol.* **3,** 318.
Jeffreys, A., and Craig, I. (1974). *Biochem. J.* **144,** 161.
Jelinek, W., and Goldstein, L. (1973). *J. Cell. Physiol.* **81,** 181.
Jeon, K. W. (1975). *J. Protozool.* **22,** 402.
Jeon, K. W., Lorch, I. J., and Danielli, J. F. (1970). *Science* **167,** 1626.
Johnson, R. B., and Ackermann, W. W. (1953). *J. Biol. Chem.* **200,** 263.
Jones, C. W., Mastrangelo, I. A., Smith, H. H., Liu, H. Z., and Meck, R. A. (1976). *Science* **193,** 401.
Kao, F. T. (1983). *Intern. Rev. Cytol.* **85,** 109.
Karakashian, M. V., and Schweiger, H. G. (1976). *Proc. Natl. Acad. Sci. U.S.A.* **73,** 3216.
Kaufman, R. J., and Sharp. P. A. (1982). *J. Mol. Biol.* **159,** 601.
Kean, E. L., and Bruner, W. E. (1971). *Exp. Cell Res.* **69,** 384.
Keller, H. U., and Bessis, M. (1975). *Nature (London)* **258,** 723.
Klebs, G. (1889). *Biol. Zentralbl.* **7.**
Kloppstech, K., and Schweiger, H. G. (1973). *Exp. Cell Res.* **80,** 69.
Kloppstech, K., and Schweiger, H. G. (1975a). Protoplasma **83,** 27.
Kloppstech, K., and Schweiger, H. G. (1975b). *Differentiation* **4,** 115.

Koop, H. V. (1979). *Differentiation* **14**, 135.

Koop, H. V., Schmid, R., Heunert, H. H., and Spring, H. (1979). *Differentiation* **14**, 135.

Koritz, S. B., and Chantrenne, H. (1954). *Biochim. Biophys. Acta* **13**, 209.

Kurtz, D. T. (1981). *Nature (London)* **291**, 629.

Ladda, R. L., and Estensen, R. D. (1970). *Proc. Natl. Acad. Sci. U.S.A.* **67**, 1528.

Lafond, R. E., Woodcock, H., Woodcock, C. L. F., Kundahl, E. R., and Lucas, J. J. (1983). *J. Cell Biol.* **96**, 1815.

Laurila, P., and Stenman, S. (1982). *Exp. Cell Res.* **142**, 15.

Laurila, P., Virtanen, I., and Stenman, S. (1981). *Exp. Cell Res.* **131**, 41.

Laurila, P., Virtanen, I., Lehto, V. P., Vartio, T., and Stenman, S. (1982). *J. Cell Biol.* **94**, 308.

Ledoux, L., Huart, R., and Jacobs, M. (1971). *Eur. J. Biochem.* **23**, 96.

Lewis, W. H., and Srinivasan, P. R. (1983). *Mol. Cell. Biol.* **3**, 1053.

Linder, S., Zuckerman, S. H., and Ringertz, N. R. (1981). *Proc. Natl. Acad. Sci. U.S.A.* **78**, 6286.

Linder, S., Zuckerman, S. H., and Ringertz, N. R. (1982). *Exp. Cell Res.* **140**, 464.

Lipszyc, J. S., Phillips, S. G., and Miller, O. J. (1981). *Exp. Cell Res.* **133**, 373.

Loeb, J. (1899). *Wilhelm Roux' Arch. Entwicklungsmech. Org.* **8**, 689.

Lorch, I. J., and Danielli, J. F. (1950). *Nature (London)* **166**, 329.

Loyter, A., Scangos, G., Juricek, D., Keene, D., and Ruddle, F. W. (1982a). *Exp. Cell Res.* **139**, 223.

Loyter, A., Scangos, G. A., and Ruddle, F. H. (1982b). *Proc. Natl. Acad. Sci. U.S.A.* **79**, 422.

Lucas, J. J. (1977). *Methods Cell Biol.* **15**, 368.

Lüttke, A. (1981). *Exp. Cell Res.* **131**, 483.

Lüttke, A., and Bonotto, S. (1980). *Eur. Meet. Mol Genet. Biol. Unicell. Algae,* Book of Abstracts, p. 23.

Lüttke, A., and Bonotto, S. (1981). *Protoplasma* **105**, 358.

Lüttke, A., and Bonotto, S. (1982). *Int. Rev. Cytol.* **77**, 205.

McBride, W. O., and Ozer, H. L. (1973). *Proc. Natl. Acad. Sci. U.S.A.* **70**, 1258.

McCormick, F. (1977). *J. Cell. Physiol.* **93**, 285.

McNeill, C. H., and Brown, R. L. (1980). *Proc. Natl. Acad. Sci. U.S.A.* **77**, 5394.

Maschlanka, H. (1946). *Biol. Zentralbl.* **65**, 157.

Mazia, D. (1952). *In* "Modern Trends in Physiology and Biochemistry" (E. S. G. Barrón, ed.), p. 77. Academic Press, New York.

Mazia, D., and Hirshfield, H. (1950). *Science* **112**, 297.

Mergenhagen, D., and Schweiger, H. G. (1975). *Exp. Cell Res.* **94**, 321.

Miesfeld, R., Sollner-Webb, B., Croce, C., and Arnheim, N. (1984). *Mol. Cell. Biol.* **4**, 1306.

Miller, R. A., and Ruddle, F. H. (1974). *J. Cell Biol.* **63**, 295.

Mills, K. I., and Bell, L. G. E. (1981). *Exp. Cell Res.* **136**, 469.

Mukherjee, A. B., Orloff, S., Butler, J. D., Triche, T., Lalley, P., and Schulman, J. D. (1978). *Proc. Natl. Acad. Sci. U.S.A.* **75**, 1361.

Mulligan, R. C., and Berg, P. (1981). *Proc. Natl. Acad. Sci. U.S.A.* **78**, 2072.

Naumova, L. P., Pressmann, E. K., and Sandakhiev, L. S. (1976). *Mol. Biol.* **10**, 512.

Neuhaus, G., Neuhaus-Url, G., Gruss, P., and Schweiger, H. G. (1984). *EMBO J.* **3**, 2169.

Nizet, A., and Lambert, S. (1953). *Bull. Soc. Chim. Biol.* **35**, 771.

Nyman, U., Lanfranchi, G., Bergman, M., and Ringertz, N. R. (1984). *J. Cell Physiol.* **120**, 257.

Oliver, N. A., and Wallace, D. C. (1982). *Mol. Cell. Biol.* **2**, 30.

Onishi, T., Berglund, C., and Reeder, R. H. (1984). *Proc. Natl. Acad. Sci. U.S.A.* **81**, 484.

Ord, M. J. (1971). *J. Cell Sci.* **9**, 1.

Otteskog, P., Ege, T., and Sundqvist, K. G. (1981). *Exp. Cell Res.* **136**, 203.

Porter, K. R., and Bonneville, M. A. (1968). "An Introduction to the Fine Structure of Cells and Tissues." Lea & Febigor, Philadelphia, Pennsylvania.

Poste, G. (1972). *Exp. Cell Res.* **73**, 273.

Poste, G., and Reeve, P. (1971). *Nature (London), New Biol.* **229**, 123.
Potter, V. R., Lyle, G. C., and Schneider, W. C. (1951). *J. Biol. Chem.* **190**, 293.
Prescott, D. M., and Goldstein, L. (1969). *Ann. Embryol. Morphol., Suppl.* **181**.
Prescott, D. M., and Kirkpatrick, J. R. (1978). *Methods Cell Biol.* **7**, 198.
Prescott, D. M., and Mazia, D. (1954). *Exp. Cell Res.* **6**, 117.
Puiseux-Dao, S. (1970). *In* "Biology of Acetabularia." Academic Press, New York.
Puiseux-Dao, S., Dazy, A. C., Borghi, H., and Hoursiangou-Neubrun, D. (1980). *Eur. J. Cell Biol.* **22**, 356.
Rao, M. V., and Chatterjee, S. (1974). *Exp. Cell Res.* **88**, 371.
Rassoulzadegan, M., Binetruy, B., and Cuzin, F. (1982). *Nature (London)* **295**, 257.
Rechsteiner, M., and Catanzarite, V. (1974). *J. Cell. Physiol.* **84**, 409.
Rechsteiner, M., Hillyard, D., and Olivera, B. M. (1976). *Nature (London)* **259**, 695.
Reeves, R., Gorman, C. M., and Howard, B. (1983). *Nucleic Acids Res.* **11**, 2681.
Rifkind, R. A. (1972). *In* "Concepts of Development" (J. Lash and J. R. Whittaker, eds.), p. 156. Sinauer Assoc., Sunderland, Massachusetts.
Ringertz, N. R. (1983). *J. Cell Biol.* **97**, 138a.
Ringertz, N. R., and Bolund, L. (1969). *Exp. Cell Res.* **55**, 205.
Ringertz, N. R., and Savage, R. E. (1976). "Cell Hybrids." Academic Press, New York.
Ringertz, N. R., Carlsson, S. A., Bolund, L., and Ege, T. (1971). *Exp. Cell Res.* **67**, 256.
Robbins, D. M., Ripley, S., Henderson, A., and Axel, R. (1981). *Cell (Cambridge, Mass.)* **23**, 29.
Roos, D., Voetman, A. A., and Meerhof, L. J. (1983). *J. Cell Biol.* **97**, 368.
Rubin, J. S., Joyner, A. L., Bernstein, A., and Whitmore, G. F. (1983). *Nature (London)* **306**, 206.
Ruddle, F. H. (1981). *Nature (London)* **294**, 115.
Sandakhiev, L. S., Puchkova, L. I., and Pikalov, A. V. (1972). *In* "Biology and Radiobiology of Anucleate Systems" (S. Bonotto, R. Goutier, R. Korchman, and J. R. Maisin, eds.), Vol. 2, p. 297. Academic Press, New York.
Santulli, A., Casale, A., and Mazza, A. (1983). *J. Submicrosc. Cytol.* **15**, 843.
Scangos, G., and Ruddle, F. H. (1981). *Gene* **14**, 1.
Schaefer-Ridder, M., Wang, Y., and Hofschneider, P. H. (1982). *Science* **215**, 166.
Schaffner, W. (1980). *Proc. Natl. Acad. Sci. U.S.A.* **77**, 2163.
Scheer, U., Lanfranchi, G., Rose, K. M., Franke, W. W., and Ringertz, N. R. (1983). *J. Cell Biol.* **97**, 1641.
Schnaebli, W. H. P., and Burger, M. M. (1972). *Proc. Natl. Acad. Sci. U.S.A.* **69**, 3825.
Schröder, H., and Hsie, A. W. (1973). *Nature (London)* **246**, 58.
Schwartz, M. S., and Brumbaugh, J. A. (1982). *Exp. Cell Res.* **142**, 155.
Schweiger, E., Wallraft, H. G., and Schweiger, H. R. (1964). *Science* **146**, 658.
Schweiger, H. G. (1967). *Nature (London)* **216**, 555.
Schweiger, H. G. (1977). "Nucleic Acid and Protein Synthesis in Plants" (L. Bogorad and J. H. Weil, eds.), p. 65. Plenum, New York.
Schweiger, H. G., and Berger, S. (1979). *Int. Rev. Cytol., Suppl.* **9**, 11.
Schweiger, H. G., Master, R. W. ., and Werz, G. (1967). *Nature (London)* **216**, 554.
Scoggin, C. H., Gabrielson, E., Davidson, J. N., Jones, C., Patterson, D., and Puck, T. T. (1981). *Somatic Cell Genet.* **7**, 389.
Sells, B. H., Six, N., and Brachet, J. (1961). *Exp. Cell Res.* **22**, 246.
Shay, J., Porter, K. R., and Prescott, D. M. (1974). *Proc. Natl. Acad. Sci. U.S.A.* **71**, 3059.
Shay, J. W., Gershenbaum, M. R., and Porter, K. R. (1975). *Exp. Cell Res.* **94**, 47.
Shen, Y. M., Kirschhorn, R. R., Mercer, W. E., Surmacz, E., Tsutsui, Y., Soprano, K. J., and Baserga, R. (1982). *Mol. Cell. Biol.* **2**, 1145.
Shephard, D. C. (1965). *Exp. Cell Res.* **37**, 93.
Shephard, D. C., and Bidwell, R. G. S. (1973). *Protoplasma* **76**, 289.
Shoeman, R. L., Neuhaus, G., and Schweiger, H. G. (1983). *J. Cell Sci.* **60**, 1.

Slilaty, S. N., and Aposhian, H. V. (1983). *Science* **220**, 725.

Sperling, L., and Weiss, M. C. (1980). *Proc. Natl. Acad. Sci. U.S.A.* **77**, 3412.

Sperling, L., Tardieu, A., and Weiss, M. C. (1980). *Proc. Natl. Acad. Sci. U.S.A.* **77**, 2716.

Spring, H. (1978). *Exp. Cell Res.* **114**, 207.

Spring, H., Trendelenburg, M. F., Scheer, U., and Franke, W. W. (1974). *Cytobiologie* **60**, 1.

Spring, H., Scheer, U., Franke, W. W., and Trendelenburg, M. F. (1975). *Chromosoma* **50**, 25.

Spring, H., Krohne, G., Franke, W. W., Scheer, U., and Trendelenburg, M. F. (1976). *J. Microsc. Biol. Cell.* **25**, 107.

Spring, H., Grierson, D., Hemleben, V., Stöhr, M., Krohne, G., Stadler, J., and Franke, W. W. (1978). *Exp. Cell Res.* **114**, 203.

Steinert, G., and Van Gansen, P. (1971). *Exp. Cell Res.* **64**, 355.

Stich, H. (1955). *Z. Natuforsch., B: Anorg. Chem., Org. Chem., Biochem., Biophys., Biol.* **10B**, 281.

Sunkara, P. S., Al-Bader, A. A., and Rao, P. N. (1977). *Exp. Cell Res.* **107**, 444.

Tartar, V. (1953). *J. Exp. Zool.* **124**, 63.

Tartar, V. (1956). *In* "Cellular Mechanisms in Differentiation and Growth" (D. Rudnick, ed.), p. 73. Princeton University Press.

Townsend, C. O. (1897). *Jahrb. Wiss. Bot.* **30**.

Trendelenburg, M. F., Spring, H., Scheer, U., and Franke, W. W. (1974). *Proc. Natl. Acad. Sci. U.S.A.* **71**, 3626.

Tunnacliffe, A., Benham, F., and Goodfellow, P. (1984). *Trends Biochem. Sci.* **9**, 5.

Urbani, E. (1952). *Biochim. Biophys. Acta* **9**, 108.

Vanden Driessche, T. (1966). *Biochim. Biophys. Acta* **126**, 456.

Vanden Driessche, T. (1967). *Sci. Prog. (Oxford)* **55**, 293.

Vanden Driessche, T. (1973). *Subcell. Biochem.* **2**, 33.

Van der Veer, E., Barneveld, R. A., and Reuser, A. J. J. (1982). *Exp. Cell Res.* **142**, 235.

Van Wysselingh, C. (1908). *Beitr. Bot. Centralbl.* **24**, 133.

Varmus, H. E., Guntaka, R. V., Fan, W. J. W., Heasley, S., and Bishop, J. A. (1974). *Proc. Natl. Acad. Sci. U.S.A.* **71**, 3874.

Veomett, G., Prescott, D. M., Shay, J., and Porter, K. R. (1974). *Proc. Natl. Acad. Sci. U.S.A.* **71**, 1999.

Verworn, M. (1892). *Arch. Gesamte Physiol. Menschen Tiere* **51**, 1.

Vishniac, W., and Ochoa, S. (1952). *J. Biol. Chem.* **198**, 501.

Wallace, D. C. (1981). *Mol. Cell. Biol.* **1**, 697.

Wallace, D. C., Bunn, C. L., and Eisenstadt, J. M. (1975). *J. Cell Biol.* **67**, 174.

Warburg, O. (1909). *Hoppe-Seyler's Physiol. Chem.* **59**, 112.

Wejksnora, P. J., and Warner, J. R. (1981). *J. Biol. Chem.* **256**, 9406.

Werz, G. (1974). *Int. Rev. Cytol.* **38**, 319.

White, J. D., Bruno, J., and Lucas, J. J. (1983). *Mol. Cell. Biol.* **3**, 1866.

Wiberg, F. C., Sunnerhagen, P., Kaltoft, K., Zeuthen J., and Bjursell, G. (1983). *Nucleic Acids Res.* **11**, 7287.

Wigler, M., and Weinstein, S. (1974). *J. Cell Biol.* **63**, 371a.

Wigler, M. H., and Weinstein, B. (1975). *Biochem. Biophys. Res. Commun.* **63**, 669.

Wigler, M., Pellicer, A., Silverstein, S., Axel, R., Urlaub, G., and Chasin, L. (1979). *Proc. Natl. Acad. Sci. U.S.A.* **76**, 1373.

Willecke, K., and Ruddle, F. H. (1975). *Proc. Natl. Acad. Sci. U.S.A.* **72**, 1792.

Wilson, E. B. (1925). "The Cell in Development and Heredity." Macmillan, New York.

Wise, G. E. (1974). *J. Cell Biol.* **63**, 375a.

Wise, G. E. (1976). *J. Cell Sci.* **22**, 623.

Wise, G. E., and Larsen, R. (1976). *Exp. Cell Res.* **97**, 141.

Wise, G. E., and Prescott, D. M. (1973). *Exp. Cell Res.* **81**, 63.

Woodcock, C. L. F., and Bogorad, L. (1970). *J. Cell Biol.* **44,** 361.
Wright, W. E. (1984a). *J. Cell Biol.* **98,** 427.
Wright, W. E. (1984b). *Exp. Cell Res.* **151,** 55.
Wright, W. E., and Hayflick, L. (1975). *Proc. Natl. Acad. Sci. U.S.A.* **72,** 1812.
Yoder, J. I., and Ganesan, A. T. (1981). *Mol. Gen. Genet.* **181,** 525.
Yoshida, Y. (1956). *J. Fac. Sci. Niigata Univ., Ser. 2* **2,** 73.
Zorn, G. A., Lucas, J. J., and Kates, J. R. (1979). *Cell (Cambridge, Mass.)* **18,** 659.
Zuckerman, S. H., Linder, S., and Ringertz, N. R. (1982). *J. Cell. Physiol.* **113,** 99.

CHAPTER 2

NUCLEOCYTOPLASMIC INTERACTIONS IN OOCYTES AND EGGS

I. GENERAL BACKGROUND

In the preceding chapters, we have often mentioned oocytes and eggs as ideal systems for the study of fundamental cellular processes [endocytosis, DNA replication, microtubule (MT) assembly, etc.]. This chapter is devoted entirely to oocytes and eggs. Although some repetition of what we have already said is unavoidable, the same basic problems will be discussed, but from the viewpoint of the embryologist rather than that of the cytologist. However, since this is not an embryology textbook, we shall try to present the huge subject of nucleocytoplasmic interactions in oocytes and eggs using a general perspective. Details can be found in a previous book by the author (1974), in an excellent book by Davidson (1976), and in reviews by Gurdon (1977) and Woodland (1982).

Many experimental embryologists have been reluctant to accept intervention of the genes in primary morphogenesis (cleavage, gastrulation, organogenesis); according to them, genes would control cell differentiation only at later stages of development. For instance, while the color of the eyes is obviously under genetic control, the formation of the eye cup and the induction of the lens might not be directed by specific "eye genes." This idea was made clear by A. Brachet (1910, 1930) when he drew a sharp distinction between a general heredity controlling early morphogenesis and involving both the nucleus and the cytoplasm, and a special heredity of the Mendelian type.

The reason for this skepticism was that, in many instances, destruction or displacement of purely cytoplasmic areas (plasms) in fertilized eggs results in profound modifications of morphogenesis. In many species, germinal localizations are of fundamental importance for future development. They result from cytoplasmic heterogeneity with, as its simplest form, a polarity gradient where the concentration of the main constituents of the egg changes progressively from the animal (light) to the vegetal (heavy) pole of the egg (Fig. 1).

Figures 2 and 3 show how simple experimental interventions at early stages in the egg cytoplasm, without effecting the nucleus, can significantly modify development. The frog embryos shown in Fig. 2 result from low-speed centrifugation of fertilized eggs. Centrifugation disturbs the polarity gradient, and abnormal larvae, characterized by marked microcephaly, are obtained (Pasteels, 1940). Figure 3 shows that removal of a purely cytoplasmic region, the polar lobe at the

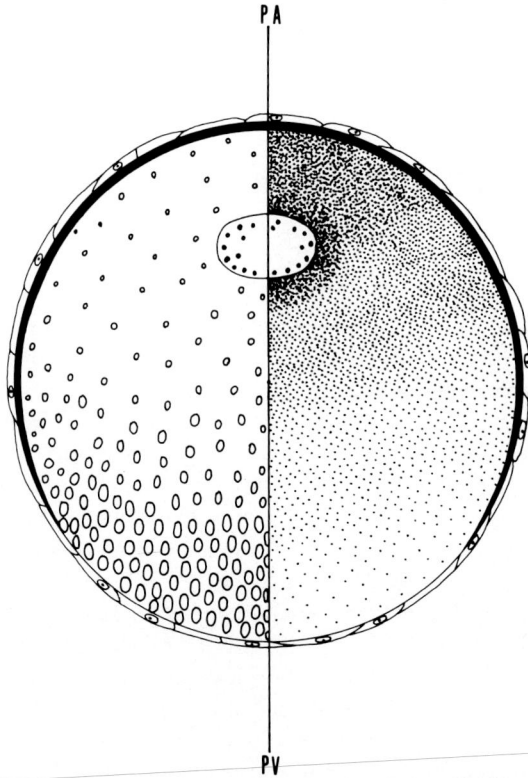

FIG. 1.   Schematic representation of polarity gradients in an amphibian oocyte surrounded by follicle cells. PA, animal pole; PV, vegetal pole. Left half: distribution of the yolk platelets. Right half: distribution of the ribosomes; note their accumulation around the nucleus (GV). In the cortex, gradient distribution of the pigment granules (melanosomes). [Drawn by P. Van Gansen.]

so-called trefoil stage of the mollusk *Ilyanassa,* results in the formation of very poorly differentiated embryos (Clement, 1952). Some material present in the polar lobe is thus necessary for normal embryogenesis. Other classic examples of germinal localization are the gray crescent of amphibian eggs (reviewed by Brachet, 1977), where the dorsoventral axis can be determined at will by gravity (Ancel and Vintemberger, 1948; Gerhart *et al.,* 1981; Neff *et al.,* 1984), and the ascidian egg, which is a mosaic of territories (plasms) giving rise to the various organs of the tadpole. The little we know about the chemical makeup of the germinal localizations will be presented when we deal with cleavage. In contrast to these mosaic eggs are those that, like sea urchin eggs, display embryonic regulation. If one separates the first two cells (blastomeres) from each other, each one is capable of giving rise to a small but complete larva (Fig. 4); Driesch, 1891). Regulation, which is the regeneration of a missing presumptive territory,

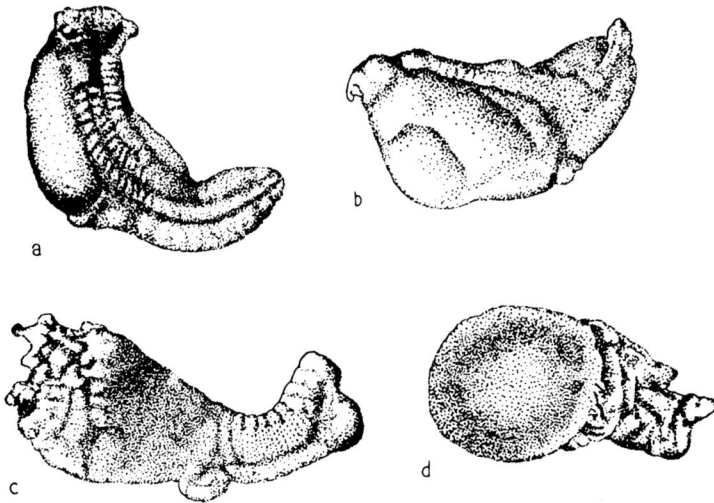

Fig. 2. Embryos obtained after moderate centrifugation of fertilized frog eggs. Microcephaly in (a) and (b), anencephaly in (c) and (d). The tails are better developed than the heads in embryos (a) to (c). [Pasteels, 1940.]

is a general characteristic of eggs (including mosaic eggs, where it is very discrete); its molecular bases are not yet understood.

Since the time of the experiments by Spemann (1928) and Seidel (1932), it has also been known that, in contrast to the heterogeneous cytoplasm, the nuclei are equipotential during the early cleavage of amphibian and insect eggs; exchange between nuclei located in the dorsal and ventral blastomeres has no effect on morphogenesis. These experiments disposed of the old Weissman–Roux theory, which held that the chromosomes of dorsal and ventral blastomeres possess different genetic determinants.

However, many experiments have shown that the integrity of the nucleus is required to obtain full and normal morphogenesis. Haploidy, aneuploidy, often hybridization, and mutations sooner or later prove lethal for the embryo. In 1934, T. H. Morgan, a leading experimental embryologist as well as the father of modern genetics, proposed an explanation for these apparently conflicting results. In his hypothesis, equipotential nuclei are distributed, during cleavage, in a heterogeneous cytoplasm; as a result, genetic activity (now generally called "gene expression") increases in certain parts of the egg and decreases in others. These changes in genetic activity would modify the chemical composition of the surrounding cytoplasm, with new changes in genetic activity as a consequence. Finally, both the nuclei and the cytoplasm would become more and more differentiated in the various parts of the embryos as a result of continuous dynamic interactions between the two. The final outcome would be organogenesis, followed by cell differentiation.

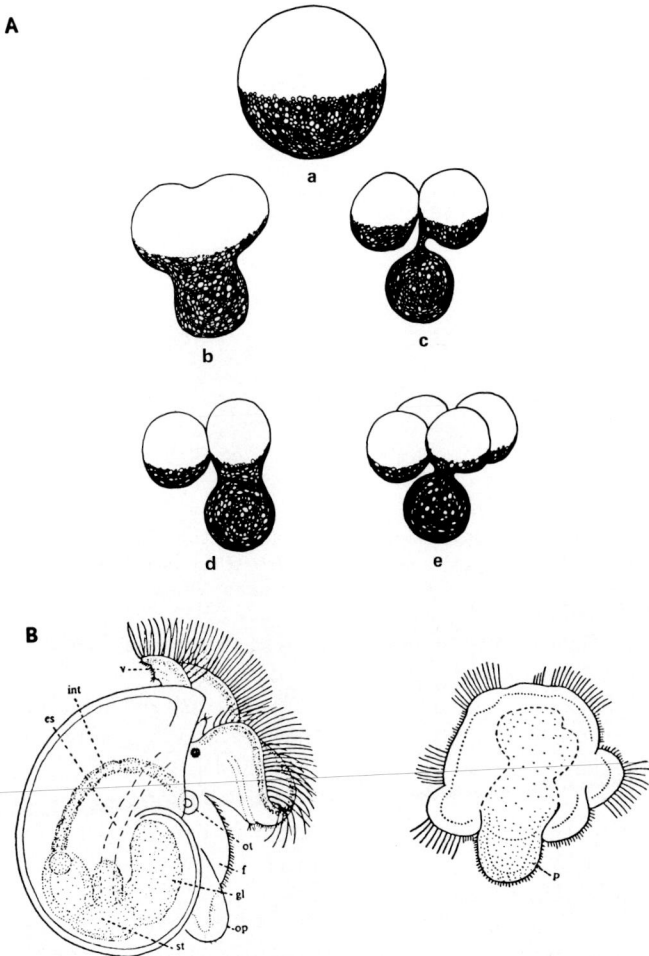

Fig. 3. (A) Cleavage stages of *Ilyanassa* (mollusk) eggs. (a) Fertilized egg; (b) formation of the first polar lobe; (c) trefoil stage: the material of the polar lobe is incorporated into one of the two blastomeres (called CD); (d) formation of the second polar lobe at the four-cell stage. (B) Left, a normal veliger larva after 9 days of development. Right, an abnormal "lobeless" embryo. The first polar lobe had been removed, and the egg had then been cultured for 9 days. [Clement, 1952.]

When does differentiation take place in the nuclei? Briggs and King (1952, 1953) performed what Spemann had called a "fantastic" experiment in order to answer this question: the transplantation of a nucleus removed from a blastula or a later developmental stage into an enucleated, unfertilized frog egg (Fig. 5). The question of nuclear transplantations cannot be dealt with here in great detail; we shall limit the discussion to the main results [see the reviews by Gurdon and Woodland (1970), Di Berardino (1979, 1980), Briggs (1979) and Di Berardino *et*

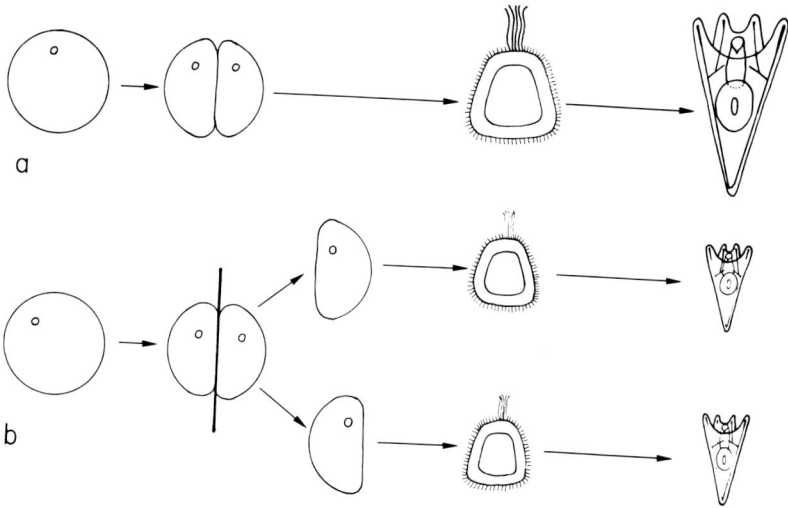

FIG. 4. Schematic representation of Driesch's experiment demonstrating regulation in sea urchin eggs at the two-cell stage. (a) Normal development leading to a ciliated blastula and a pluteus larva; (b) after separation of the first two blastomeres, two normal plutei of reduced size are formed. [Drawn by P. Van Gansen.]

al. (1984) for detailed discussions of these results]. It is agreed that nuclei at the blastula stage are still totipotent; after injection of such nuclei into enucleated, unfertilized eggs, these eggs can give rise to adults (Fig. 5). This is true of *Rana pipiens* (Briggs and King, 1953), *Xenopus laevis* (Gurdon *et al.*, 1958), *Drosophila* (Zalokar, 1971), but not mouse eggs (McGrath and Solter, 1984). For nuclei removed at later stages of development, there are discrepancies in the results. King and Briggs (1954) found a restriction in the potentialities of nuclei removed from advanced gastrulae in *Rana*. In *Xenopus*, development up to the tadpole (but not the adult) stage after injection of nuclei isolated from differentiated adult cells has been obtained (Gurdon and Laskey, 1970; Laskey and Gurdon, 1970; Gurdon *et al.*, 1975; Fig. 6). The discrepancy between the results obtained with different amphibian species is probably due in part to the fact that chromosomal abnormalities are frequent. In *R. pipiens,* when nuclei from "old" cells are transplanted into enucleated, unfertilized eggs (Di Berardino and Hoffner, 1970), "old" nuclei apparently can no longer follow the rapid DNA replication pace characteristic of cleaving eggs. Gurdon's experiments on *Xenopus* strongly suggest that nuclei from adult cells are still totipotent and that they still possess all of the genetic information required for organogenesis and differentiation of larval tissues. However, in the experiments of Gurdon *et al.* (1975), injection of nuclei removed from completely differentiated keratocytes into enucleated, unfertilized *Xenopus* eggs gave rise only to partial blastulas with

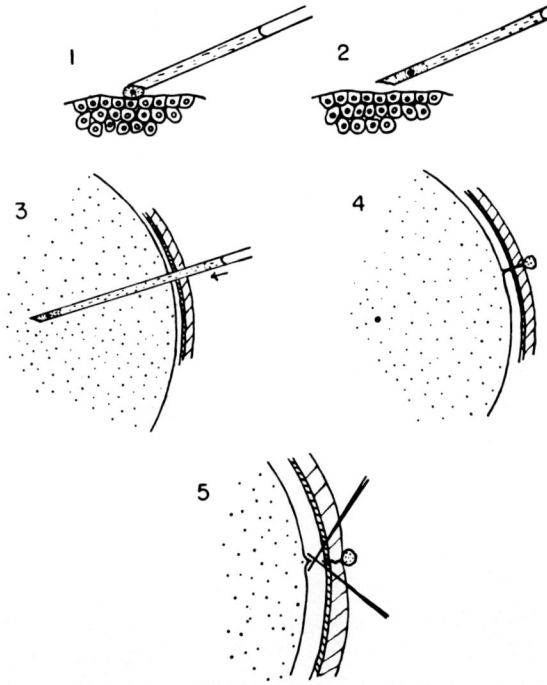

FIG. 5.   Method of Briggs and King for transplanting blastula cell nuclei into enucleated frog eggs. 1. Removal of a blastula cell. 2. The isolated cell breaks in the micropipette. 3. Injection of the nucleus into a previously activated and enucleated unfertilized egg. 4, 5. Removal of the exovate with glass needles. [Briggs and King, 1953.]

cells of unequal size; in order to obtain tadpoles, Gurdon resorted to "serial" nuclear transfers (Fig. 7). He removed the nucleus from small, healthy-looking cells of the partial blastula and injected it into another recipient enucleated, unfertilized egg. One cannot exclude the possibility that repeated replication of the adult keratinocyte nucleus before its final transfer produced a number of changes in its chromatin (for instance, DNA demethylation or methylation), since we have seen that an old *Acetabularia* nucleus can be rejuvenated by being transferred into young cytoplasm.

While Morgan's (1934) theory remains basically correct, more recently work-ers have tried to adapt it to some of the newer theories in molecular biology. In particular, many have tried to extend the famous Jacob–Monod (1963) model of negative gene regulation in bacteria to embryonic development. It is highly probable that, in eukaryotic cells (including eggs and embryos), the activity of the structural genes, as in bacteria, is controlled by regulatory genes producing repressors or activators. The most elaborate model explaining the genetic con-trol of embryonic differentiation has been proposed by Davidson and Britten

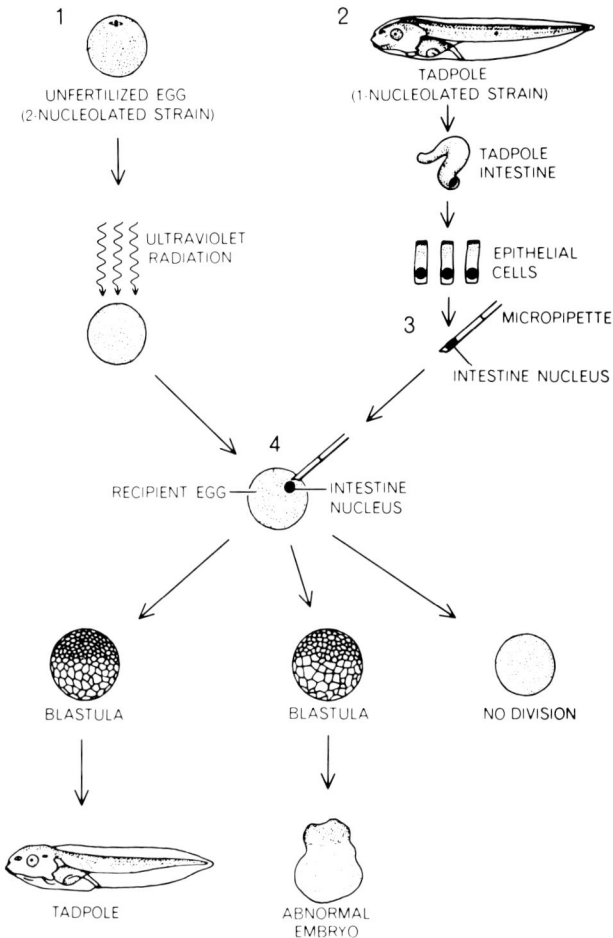

FIG. 6.   Transplantation of a tadpole intestinal nucleus into an enucleated unfertilized *Xenopus* egg. The nucleus is taken from a uninucleolated mutant tadpole. The presence, after nuclear transplantation, of a single nucleolus in the nuclei demonstrates that it originated from the injected nucleus. 1. An unfertilized egg is uv irradiated to destroy its chromosomes. 2. Isolation of an intestinal epithelial cell from a donor tadpole. 3, 4. Injection of its nucleus into a recipient uv-irradiated, unfertilized egg. The outcome of such experiments is shown at the bottom of the figure. Some of the eggs do not cleave, while others form abnormal embryos, but normal tadpoles are also obtained. [From *J. Embryol. Exp. Morphol.* **20**, 401, by J. B. Gurdon. W. H. Freeman and Company. Copyright © 1968.]

1979). This model takes into account the heterogeneity of the egg cytoplasm and proposes an original molecular mechanism for gene regulation in eukaryotes where heterogeneous nuclear RNA (hnRNA) and RNAs copied on middle re-

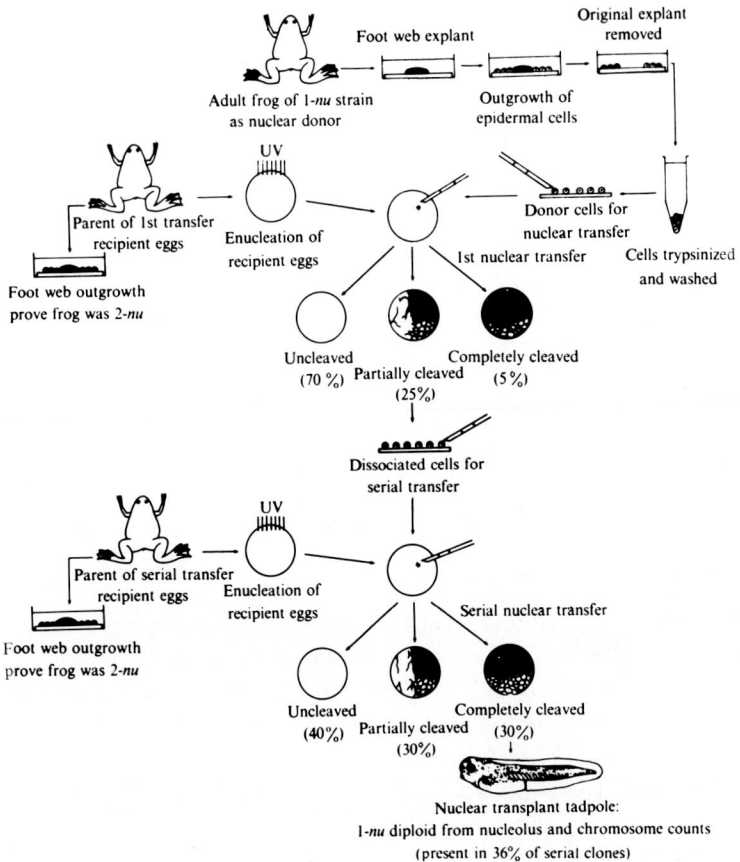

FIG. 7. Plan of serial transfer experiments in *Xenopus,* using nuclei from adult skin cells. The uninucleolate nuclear donor is an epidermal cell from an explanted and cultured foot web; the uv-irradiated egg is from a normal (binucleolate) female. Nuclear transplantation gives only 5% of completely cleaved blastulae. One of them is dissociated, and the nucleus of one of its cells is injected into a second uv-irradiated recipient egg. Thirty percent of these serial transfer recipient eggs develop into tadpoles; 36% of these tadpoles have uninucleolate nuclei. [Gurdon, 1975.]

petitive DNA sequences are supposed to play a major role. We shall not present this model here, despite its interest, because of its complexity and the difficulty of testing it experimentally.

In the past several years, more people have proposed the idea (first suggested by Scarano, 1969) that embryonic development might be controlled by changes in the DNA molecules themselves. A major difficulty seemed to be that, in Gurdon's experiments, some adult nuclei were still totipotent; however, we have just seen that one cannot rule out the possibility that biochemical changes take place in the transplanted adult nuclei when they replicate rapidly in a very young cytoplasm. Conceivably, these changes might affect the DNA of the adult nuclei that would return to the organization it had in recently fertilized eggs. However, it is easier to visualize changes in proteins or RNAs associated with DNA in chromatin than in DNA itself.

One of the possible changes in DNA molecules during embryogenesis is gene amplification. The work of Levine *et al.* (1981) has shown that, while the genes coding the chorion proteins are amplified in the follicle cells during oogenesis in *Drosphila,* no selective gene amplification can be detected during the development of this insect. It is, therefore, unlikely that gene amplification plays a major role in embryonic development.

Another possibility is the existence of gene rearrangements, as found for immunoglobin genes by Tonegawa (1979–1980) during the development of lymphocytes, during morphogenesis. In fact, according to Dickinson and Baker (1979), many translocations of inverted repeat sequences (hairpins) take place during sea urchin development. In discussing chromatin in Chapter 4, Volume 1, we saw that there is now overwhelming evidence for the existence in the eukaryotic genome of transposable elements (for instance, *copia* in *Drosophila*), but there is no compelling evidence that insertion of transposable elements into DNA plays an important role in early embryonic development. All we know is that the *copia* sequences become much more abundant after 10 hr of development (Scherer *et al.,* 1981). Obviously, the work of Dickinson and Baker (1978) should be repeated and extended to eggs of other species, and we should obtain more information about insertion of *copia*-like elements during early embryonic development before the idea that embryogenesis results from gene rearrangements can be accepted. That insertion of *copia* in the complex *white* genetic locus of *Drosophila* induces the apricot eye color phenotype (Goldberg *et al.,* 1982) does not alter this conclusion, since we are dealing here with what Brachet (1910, 1930) called "general" and not with "special" heredity. However, later in this chapter, we shall see that insertion of mobile genetic elements in the genome may have important effects on late morphogenesis, at least in *Drosophila.* Comparison of germ line (sperm) and differentiated cell DNA has so far failed to disclose gross structural rearrangements during development (Haigh *et al.,* 1982; Okada *et al.,* 1985). One cannot exclude the possibility that a finer analysis will demonstrate subtle DNA rearrangements during embryogenesis, but, if they exist, they are still to be discovered.

More convincing is the case for the idea, first proposed by Scarano (1969), that DNA methylation and demethylation might play an important role in embryonic differentiation (reviewed by Razin and Riggs, 1980; Burdon and Adams, 1980). Scarano and his colleagues presented some experimental evidence for the view that certain cytidine residues present in the DNA molecules are selectively methylated by specific DNA methylases localized in the nuclei during sea urchin development. Baur *et al.* (1978), Pollock *et al.* (1978), Bird *et al.* (1979), and

Bird and Southern (1980) failed to find measurable changes in overall DNA methylation during sea urchin egg development, but this does not mean that a finer analysis, using restriction endonucleases as a tool, might not allow the detection of changes in DNA methylation in limited regions of the genome. Adams *et al.* (1981) have studied DNA methylase activity during oogenesis in *Xenopus*. In young oocytes, activity is low and is located predominantly in the nucleus; in large oocytes, enzymatic activity is higher, but is found mainly in the cycoplasm. After fertilization, DNA methylase activity increases 40 times in the nuclei, suggesting that the enzyme migrates from the cytoplasm into the nuclei during cleavage and that nuclear DNA methylation might occur during this early period of development.

However, as mentioned in Chapter 4, Volume 1, and as pointed out by Razin and Riggs (1980) and by Burdon and Adams (1980), gene activity seems to be connected to DNA undermethylation (resulting from successive DNA replication cycles) rather than DNA methylation. The main argument for a connection between DNA methylation and differentiation is the existence of a tissue-specific pattern of DNA methylation, which was mentioned in Chapter 4, Volume 1. In tissues where a gene is strongly expressed (e.g., the ovalbumin gene in the oviduct, the hemoglobin gene in hematopoietic cells), DNA is, as a rule, under-methylated. However, as shown by Brown and Dawid (1968) and Dawid *et al.* (1970), while the amplified ribosomal DNA of *Xenopus* oocytes is not methy-lated, whereas *Xenopus* chromosomal rDNA is almost completely methylated, both DNAs are equally well transcribed. Bird *et al.* (1981) have pointed out that both sperm and egg DNA are fully methylated in *Xenopus* and that methyl groups are progressively lost during cleavage. Indirect evidence for the importance of DNA undermethylation for cell differentiation comes from the effects of 5-azacytidine, a cytidine analog that, if incorporated into DNA, inhibits the DNA methylases during its replication. As a result, the newly replicated DNA is undermethylated (Jones and Taylor, 1981). Azacytidine has interesting biolog-ical effects, but it is not yet proved that they result entirely from DNA under-methylation. The drug induces the differentiation of myotubes in cultures of fibroblasts (Constantinides *et al.*, 1977), and it is said to reactivate the inactive human chromosome X (Mohandas *et al.*, 1981; Jones *et al.*, 1981), but this effect has not been found by others (Wolf and Migeon, 1982). Finally, it should be mentioned that sperm DNA, which is genetically inactive, is highly methly-ated and that partial demethylation must necessarily occur during the repeated cell divisions that characterize embryonic development. For instance, the two genes coding for δ-crystalline in the chick embryo lens contain three CCGG sites that are undermethylated compared to sperm and red blood cells (Jones *et al.*, 1981). In terminally differentiated trophoblast cells of the mouse embryo, DNA is 35% undermethylated. In this case, DNA methylation seems to be linked to a loss of the differentiation potential (Manes and Menzel, 1981). As one can see, evidence is increasing for the view that DNA undermethylation is connected to gene expression and, as a consequence, to cell differentiation, but whether it plays a role in the early steps of morphogenesis and organogenesis, in particular, remains an open question.

Indeed, recent work shows that, as is so often the case in biology, situations are not as straightforward as they once seemed. It is now clear that the presumed correlation between undermethylation and gene expression has many exceptions. That such a correlation is not simple has been concluded by Kunnath and Locker (1983), who compared the methylation pattern of the albumin and α-fetoprotein genes in fetal livers (where only α-fetoprotein is expressed) and adult livers (where only the albumin gene is expressed). The vitellogenin genes of *Xenopus* are expressed in the liver of estrogen-treated animals, despite the fact that they are fully methylated (Gerber-Huber *et al.,* 1983). There are no differences in the methylation pattern of these genes in red blood cells, nonstimulated liver, and estrogen-stimulated liver (Folger *et al.,* 1983). Bower *et al.* (1983) have found that there are traces of γ-crystallin mRNA in the lung, liver, heart, and kidney of chick embryos. This mRNA is more abundant in embryonal neural retina, which is able to form lenses, but the γ-crystallin gene is more methylated in this tissue than in kidney. Bower *et al.* (1983) concluded that there is no correlation between undermethylation and gene expression; however, there might be a correlation with the state of differentiation of the tissue. Work by Grainger *et al.* (1983) on lens development in the chicken has shown that the situation is still more complex than was once believed. Many CCGG sites undergo undermethylation in the developing lens, but this occurs only 2 days after δ-crystallin synthesis has begun; however, a single site is undermethylated at about the time the synthesis of this lens protein begins. MacLeod and Bird (1983) have injected rDNA from *Xenopus* sperm into *Xenopus* oocyte nuclei; although the injected DNA was highly methylated, it was efficiently transcribed. Since it did not undergo demethylation in the oocyte nucleus, it must be concluded that demethylation is not necessary for effective transcription. Comparable results have been obtained by Pennock and Reeder (1984): *in vitro* methylation of *Xenopus* rDNA does not affect its transcription in the oocyte nucleus.

During early development, DNA methylation takes place and not, as one might have thought, demethylation (Razin *et al.,* 1984). In a recent review, Jaenisch and Jähner (1984) made the interesting suggestion that DNA methylation might be a means of turning off genes that were active during oogenesis; this might lead to the activation of a new set of genes.

The only conclusion one can draw from all of this work is that local undermethylation is probably one of the factors, but certainly not the only one, required for chromatin to assume an active configuration. As we have seen in Chapter 4, Volume 1, the main characteristic of transcriptionally active chromatin is hypersensitivity to DNase digestion. There is at least one case in which nuclease sensitivity correlates better with gene activity than hypomethylation. When the vitellogenin genes are activated by estrogen treatment, nuclease-hypersensitive sites become detectable in the chicken liver (Burch and Weintraub, 1983); the same occurs in *Xenopus* liver, where the hypersensitive sites disappear after hormone withdrawal (Folger *et al.*, 1983). As we have just seen, the methylation pattern of the vitellogenin genes remains unchanged in such experiments. How nuclease-hypersensitive regions form in chromatin is not known for sure, as we have seen in Chapter 4, Volume 1. Weintraub (1983) concludes that the major factor in the formation of DNase-hypersensitive structures in chromatin is the secondary structure of DNA, which should be affected by cytosine methylation and by B-to-Z DNA transitions. A more likely possibility is that DNase sensitivity or resistance results from the binding of specific protein factors to certain DNA sequences (Kaye *et al.*, 1984; Wu, 1984).

In "Biochemical Cytology," we devoted an entire section to plasmagenes, which are hypothetical cytoplasmic particles endowed with genetic continuity and supposed to determine hereditary characters (Brachet, 1959). Today only a few lines are sufficient because all the cases of cytoplasmic heredity that puzzled the geneticists and embryologists around 1950 have been explained by the intervention of viruses or mitochondria. The latter were believed to play a role in the differentiation of tunicate eggs because they accumulated in the plasm giving rise to the muscles. However, it is likely that, if mitochondria play any role at all in morphogenesis, it is by virtue of energy production rather than genetic continuity.

We shall see, especially when we study oocyte maturation, that translational controls at the level of the cytoplasmic protein-synthesizing machinery may be more important for early development than gene transcription. However, sooner or later, often by the end of cleavage, gene transcription becomes dominant, and the essential control mechanisms (transcriptional and posttranscriptional) then lie in the nucleus.

All we wish to show here is that embryonic development is too complicated a process to be explained by a single control mechanism. This point was made perfectly clear in Brown's excellent review (1981) of the large variety of biochemical and molecular mechanisms controlling gene expression in eukaryotes.

A last point should be made. In the eighteenth century, preformationists opposed epigeneticists. The former, who were the conservatives of that period, held that a miniature chicken is already present in the egg. The latter, who were progressives (suspected of atheism), said that the chicken develops progressively and that ontogenesis is a fact. They were, of course, right. Now, what might preformation and epigenesis mean in molecular terms? Preformation would be the genetic information that is stored in the unfertilized egg, while epigenesis would be the result of transcriptional events after fertilization that bring in new information. If one accepts this view, the amount and variety of genetic information stored in the egg and thus preformed is surprisingly large. In a *Xenopus* unfertilized egg, the complexity of the mRNA population is very high, amounting to about 20,000 different mRNAs, which are theoretically capable of directing the synthesis of as many different proteins (Davidson and Hough, 1971; Perlman and Rosbash, 1978). Similar values have been found for the much smaller sea urchin eggs. However, as we shall see, despite this large store of information (which results, of course, from intensive nuclear gene activity during oogenesis), the capacity for the development of parthenogenically activated anucleate fragments of eggs is exceedingly limited.

In the following section, the discussion will be limited to two favorite embryological materials: *X. laevis* for the study of oogenesis and maturation, and sea urchin eggs for that of fertilization and cleavage. Little will be said about later stages of development in these species and about work done with other species.

II. *XENOPUS* OOGENESIS

Oogenesis is the preparation for embryonic development. It is characterized by the progressive accumulation of reserve materials (glycogen, lipids, and proteins), ribosomes, and, as just pointed out, a variety of mRNAs. The oocytes are thus giant, highly differentiated cells. Maturation transforms the full-grown oocyte into an egg that, after fertilization or parthenogenic activation, can give rise to an embryo and finally to an adult of the same species. The egg is thus a totipotent cell. In addition, due mainly to the outstanding work of J. B. Gurdon and his colleagues, *Xenopus* oocytes have become an ideal system for the study of transcription, translation, assembly of nucleosomes, control of gene expression by cytoplasmic factors, and nuclear membrane permeability. The large size of these oocytes makes microinjections into their cytoplasm or nucleus an easy task.

Growth of the oocyte is due both to endogenous synthesis of a great variety of molecules and to the uptake of the yolk proteins by endocytosis. As shown in Fig. 8, three main stages in *Xenopus* oogenesis can be distinguished. Previtellogenic oocytes (0.1–0.3 mm in diameter) are transparent; vitellogenic oocytes (0.3–1.2 mm in diameter) are opaque due to the accumulation of yolk and pigment (melanin); and full-grown oocytes, which are ready for maturation (about 1.2 mm in diameter), display very distinct polarity. Mature oocytes have a pigmented animal half and a white vegetal half. We shall first examine the cytoplasm and then the nucleus of the occyte; finally we will discuss their use as "test tubes" for molecular biologists.

A. STRUCTURE AND COMPOSITION OF THE CYTOPLASM

Figure 8, Chapter 3, Volume 1 shows the outer layer of a full-grown oocyte. The cell surface is greatly increased by numerous large microvilli, and their axis, as in kidney or intestinal cells, is composed of actin filaments and villin (Franke

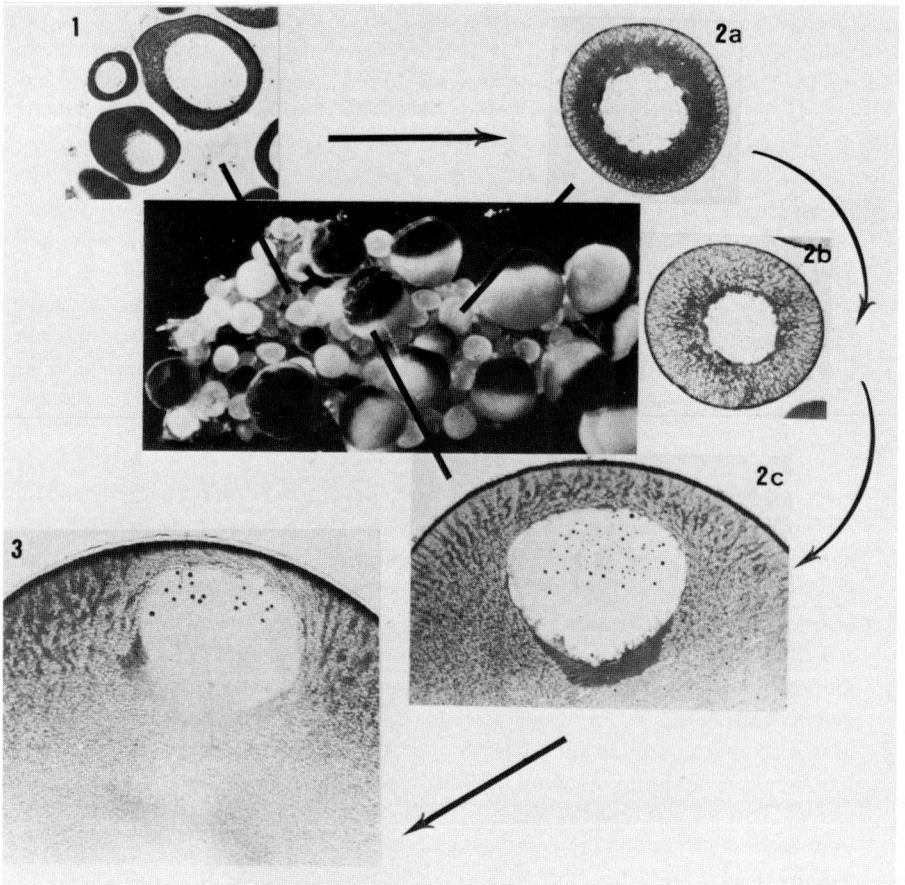

FIG. 8.   Oogenesis of *X. laevis*. Center: fragment of an ovary showing oocytes of varous sizes; the full-grown oocytes have a pigmented animal pole and a white vegetal pole. (1) Section through small previtellogenic oocytes. (2a,b) Two successive stages of vitellogenesis. (2c) full-grown oocyte showing, under the GV, which contains many nucleoli, a cap of RNA-rich material. (3) beginning of maturation. The nuclear membrane has broken down at its basal end; the nucleus, with its nucleoli, has moved toward the animal pole. The bars correlate the sections and the living oocytes at the various stages of oogenesis. [Brachet, 1979.]

et al., 1976). The oocyte is surrounded by a layer of follicle cells; between them and the oocyte, a thick protective membrane, the chorion or vitelline membrane, is perforated by the microvilli, which establish contact with the follicle cells by gap junctions. Beneath the cell membrane, in the so-called cortex, many cortical granules, pigment granules (melanosomes), and mitochondria can be seen. The cortical granules, which are composed mainly of glycoproteins, do not move into the cortex before the end of oogenesis; they play an important role at fertilization.

The cortex is more rigid than the endoplasm, which flows easily out of the oocyte (forming a so-called exovate) if the membrane and cortex are punctured with a needle. This rigidity is due to the presence in the cortex of *Xenopus* oocytes of a contractile structure that excludes the yolk but forms a network in the yolk mass (Merriam *et al.*, 1983). Gall *et al.* (1983) have shown that the cortex of *Xenopus* oocytes is so rigid that it can be isolated by microdissection. It possesses a layer of cortical actin microfilaments (MFs) that are continuous with the axis of the microvilli. A subcortical layer contains intermediate (9–13 nm) filaments (IFs) composed of cytokeratin (Franz *et al.*, 1983). The existence of this subcortical cytokeratin intermediate filaments network has been confirmed by Godsave *et al.* (1984a). The same authors (Godsave *et al.*, 1984b) discovered, by immunocytochemistry, a second intermediate filament network. It is made of vimentin and its localization during *Xenopus* oogenesis is entirely different from that of the cytokeratin cytoskeleton.

Sections through a large *Xenopus* oocyte show a very characteristic polarity gradient, which is depicted schematically in Fig. 1. In the endoplasm, the size of the yolk platelets gradually decreases from the vegetal to the animal pole; in contrast, the glycogen content, lipids, ribosomes, and endoplasmic reticulum (ER) vesicles decrease from the animal to the vegetal pole. As we have seen in Fig. 2, this polarity gradient is extremely important for morphogenesis, since mild centrifugation leads to embryos missing both brain and eyes. This is due to the fact that the polarity gradient corresponds approximately to the future cephalocaudal axis of the embryo, with the animal pole corresponding to the presumptive head region. The factors that establish the polarity gradient are not known for certain. One of them, as shown by Robinson (1979), is a transcellular electric current due to a movement of ions through the oocyte. Ions would be pumped at one end, and there would be permeability channels at the other end. It is also known, from membrane fluidity measurements, that the oocyte plasma membrane has a polarity that increases 100 times at fertilization (Dictus *et al.*, 1984).

Our early cytochemical observations (Brachet, 1942, 1950) on the distribution of total RNA during amphibian development have shown that there is a distinct RNA polarity gradient, decreasing from the animal to the vegetal pole, in unfertilized, fertilized, and cleaving eggs. At gastrulation, a secondary RNA gradient, decreasing from dorsal to ventral, is superimposed upon the initial animal–vegetal gradient, leading, during gastrulation and neurulation, to the formation of well-defined anteroposterior and dorsoventral RNA gradients (Fig. 9). The presumed role of these gradients in protein synthesis and morphogenesis has been discussed several times by the author (Brachet, 1957), but it should be pointed out that the RNAs that could be detected with the methods used in our early work (1942, 1950) were essentially ribosomal. In other words, the RNA polarity gradient present in amphibian oocytes reflects the existence of an animal–vegetal gradient in the distribution of the ribosomes (which are much more

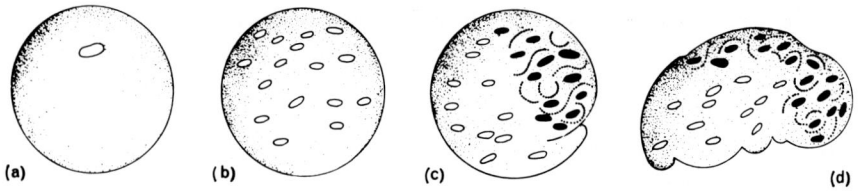

Fig. 9.   Schematic representation of protein synthesis and RNA distribution during amphibian development. (a) Distribution of the ribosomes (small dots) along an animal–vegetal gradient in a fertilized egg. (b) This gradient remains unchanged during cleavage. (c) During gastrulation, the nuclei become more active on the dorsal side; polysomes (dotted lines) are abundant on this side. (d) At the late neurula stage, the polysomes are distributed along dorsoventral and cephalocaudal gradients. [Redrawn from Brachet, 1967.]

abundant at the animal than at the vegetal pole). Interestingly, recent work with a more sophisticated method (*in situ* hybridization with tritiated polyuridylic acid) has shown that the poly(A)$^+$ RNAs have an entirely different distribution. In previtellogenic oocytes, poly(A)$^+$ RNA is uniformly distributed, while during vitellogenesis, it accumulates in the cortical and subcortical regions of the oocyte; the distribution becomes homogeneous again after germinal vesicle breakdown at maturation (Capco and Jeffery, 1982). However, in ripe *Xenopus* oocytes, where the animal–vegetal distribution of bulk RNA is particularly conspicuous, poly(A)$^+$ RNA is accumulated at the vegetal pole (Capco, 1982). Furthermore, if ovarian poly(A)$^+$ RNA is injected into *Xenopus* eggs, it accumulates along a concentration gradient with a maximum at the vegetal pole (Capco and Jeffery, 1981). Active migration of poly(A)$^+$ RNAs toward the vegetal pole seems to occur between fertilization and first cleavage, and it appears as if the RNAs microinjected into eggs recognize binding sites. Capco and Jäckle (1982) worked out a cytochemical method for the detection of protein synthesis in *Xenopus* oocytes. They found that it is homogeneous in previtellogenic oocytes; in full-grown oocytes, protein synthesis is most active in the cortex. This spatial pattern disappears during maturation. Therefore, the pattern of protein synthesis activities in the oocyte is exactly the same as that found by Capco and Jeffery (1982) for the distribution of poly(A)$^+$ RNAs. However, all of these results have been obtained by *in situ* hybridization; a different picture emerges when *Xenopus* eggs and early embryos are cut into pieces that can be analyzed with biochemical methods. According to Phillips (1982), both total and poly(A)$^+$ RNAs are distributed along the animal–vegetal polarity gradient; at a later stage (after fertilization), a dorsoventral gradient is superimposed on this initial polarity gradient. In a short note, Harsa-King (1982) reported that the animal pole of *Xenopus* eggs contains 16 times more poly(A)$^+$ RNAs than the vegetal one and that the patterns of protein synthesis are different at the two poles. More work is clearly needed before we can obtain a clear, unified picture of the patterns of poly(A)$^+$ RNA distribution and protein synthesis in large *Xenopus* oocytes; what is certain is that both vary along the animal–vegetal polarity gradient. It should be stressed that the existing data deal only with the total poly(A)$^+$ RNA population and are not necessarily true for individual mRNAs. Melton recently reported at a Congress that three mRNAs decrease in concentration from the animal to the vegetal pole while a fourth one displays the opposite distribution.

Quantitatively, the major constituent by far of the oocyte is the yolk phosphoprotein, a dimer of phosvitin (M_r 35,000), and two molecules of lipovitellin (M_r 200,000). As shown by R. Wallace and his co-workers, these yolk proteins are of exogenous origin. They are synthesized in the liver under estrogen stimulation as a large precursor called "vitellogenin" (M_r 470,000), which is secreted into the bloodstream. As we have seen in Chapter 3, Volume 1, vitellogenin is sequestered into the oocyte by receptor-mediated endocytosis (Wallace and Jared, 1976; Tucciarone and Lanclos, 1981). After binding to specific receptors, the vitellogenin–receptor complexes are internalized on the clathrin-coated regions of the cell surface; clustering of the complexes in coated pits is believed to be due to the action of transglutaminase. Some work has shown that there are several vitellogenins coded for by a small vitellogenin gene family. According to Wahli *et al.* (1981), there are two distantly related groups of vitellogenin genes called *A* and *B*, comprising genes A_1 and A_2 and B_1 and B_2. These four related genes produce four different mRNAs coding, in an *in vitro* protein synthesis system, for four polypeptides that differ in their primary structure. Of some interest is the fact that the vitellogenin genes contain as many as 33 introns that have widely diverged during evolution through mutation, deletion, insertion, and duplication. It should be added that, in addition to the major phosphoproteins (phosvitin and lipovitellin 1 and 2), yolk platelets contain smaller molecules (M_r 19,000 and 13,000), which have been called "phosvettes 1 and 2" by Wiley and Wallace (1981). They are probably cleavage products of vitellogenin during processing leading to the crystallization of phosvitin and lipovitellin in the yolk platelets (Fig. 10). According to Colombo *et al.* (1981), the yolk platelets are surrounded by an actin shell, which might result from cell surface constituents that have been dragged into the cytoplasm at the time of vitellogenin endocytosis. That vitellogenin intake is the major factor in the enormous increase in size that takes place during oogenesis has been clearly proved by Wallace and Misulovin (1978), who cultivated *in vitro* medium-sized (0.6 mm in diameter) vitellogenic oocytes of *Xenopus* in a medium containing vitellogenin and insulin. Such *in vitro*–grown oocytes finally become larger (1.43 mm in diameter) than the largest (1.2 mm in diameter) oocytes present in the ovary. Wallace *et al.* (1981) reported that even "full-grown" oocytes can take up vitellogenin added to the medium; their volume doubles within 2–3 weeks under suitable *in vitro* conditions. These giant *Xonopus* oocytes are appar-

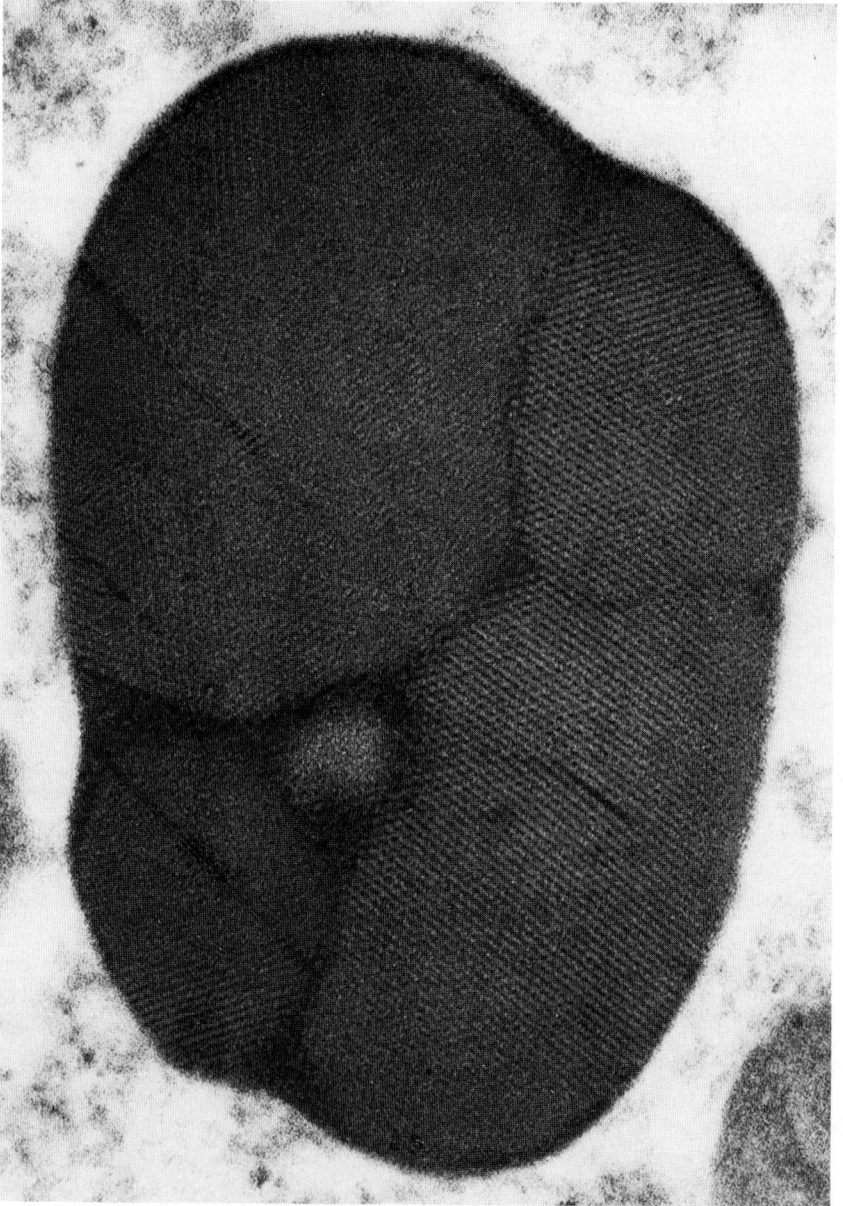

FIG. 10.   The yolk platelets of *Xenopus* oocytes have a crystalline structure; the crystals are composed of yolk phosphoproteins. [Original electron micrograph by P. Van Gansen.]

ently normal, since they undergo maturation after adequate stimulation. Unfortunately, their cytology has not yet been studied. It should be added that vitellogenin is quickly degraded when it is injected into *Xenopus* oocytes (Wallace *et al.*, 1973).

Yolk platelets contain small amounts of DNA and RNA in addition to the phosphoproteins. Due to the abundance of the yolk platelets, yolk DNA represents about 65% of the egg DNA (which is thus greater than the 4C value expected for chromosomal DNA). Yolk DNA is composed of linear, double-stranded molecules somewhat smaller than those obtained by the same methods from *Xenopus* chromatin (Hanocq *et al.*, 1972). It is probable that, like vitellogenin, it has an exogenous origin (Opresko *et al.*, 1979) and that it arises from the numerous red blood cells and hepatocytes that undergo cytolysis when the stimulated liver of the female synthesizes large amounts of vitellogenin. It is unlikely that yolk DNA plays any genetic role, more probably it is a reserve of deoxynucleotides to be used when the yolk platelets break down during embryonic development. In addition to this DNA store, oocytes possess a histone store. Curiously, the stored histones are not in the form of monomers, but rather a soluble complex similar to the histone octamer core of the nucleosomes (Earnshaw *et al.*, 1982).

The number of mitochondria greatly increases during amphibian oogenesis; this involves considerable synthesis of mitochondrial DNA, which finally amounts to about 10% of the total DNA (Dawid, 1966). However, biogenesis of the mitochondria stops almost completely when vitellogenesis begins; out of a total of 16–17 rounds of mitochondrial DNA replication, 12 take place before the onset of vitellogenesis (Callen and Mounolou, 1978; Webb and Camp, 1979; Callen *et al.*, 1980). In previtellogenic oocytes, most of the mitochondria are collected, together with many vesicles, in a cytoplasmic body called the "Balbiani yolk nucleus," "mitochondrial mass," or "mitochondrial cloud" (Fig. 11) (Billett and Adam, 1976). In addition to numerous mitochondria, the mitochondrial cloud of the previtellogenic *Xenopus* oocytes contains a granulofibrillar material which, according to Heasman *et al.* (1984), might be the precursor of the germ plasm granules. These granules accumulate at the vegetal pole of the unfertilized and fertilized eggs and probably play a role in the determination of the germ cells. Tubulin is also a major constituent of the yolk nucleus (Palacék *et al.*, 1985). The dispersal of the mitochondria during vitellogenesis has been described by Tourte *et al.* (1984) and by Heasman *et al.* (1984). Some of the mitochondria form a vegetal cortical layer, while others remain around the nucleus. This transient mitochondrial perinuclear crown results from active mitochondriogenesis. The vimentin IF cytoskeleton discovered by Godsave *et al.* (1984b) has the same distribution as the mitochondria. It has been estimated that a previtellogenic oocyte 30 μm in diameter possesses about 500,000 mitochondria (Marinos and Billett, 1981). This abundance of mitochon-

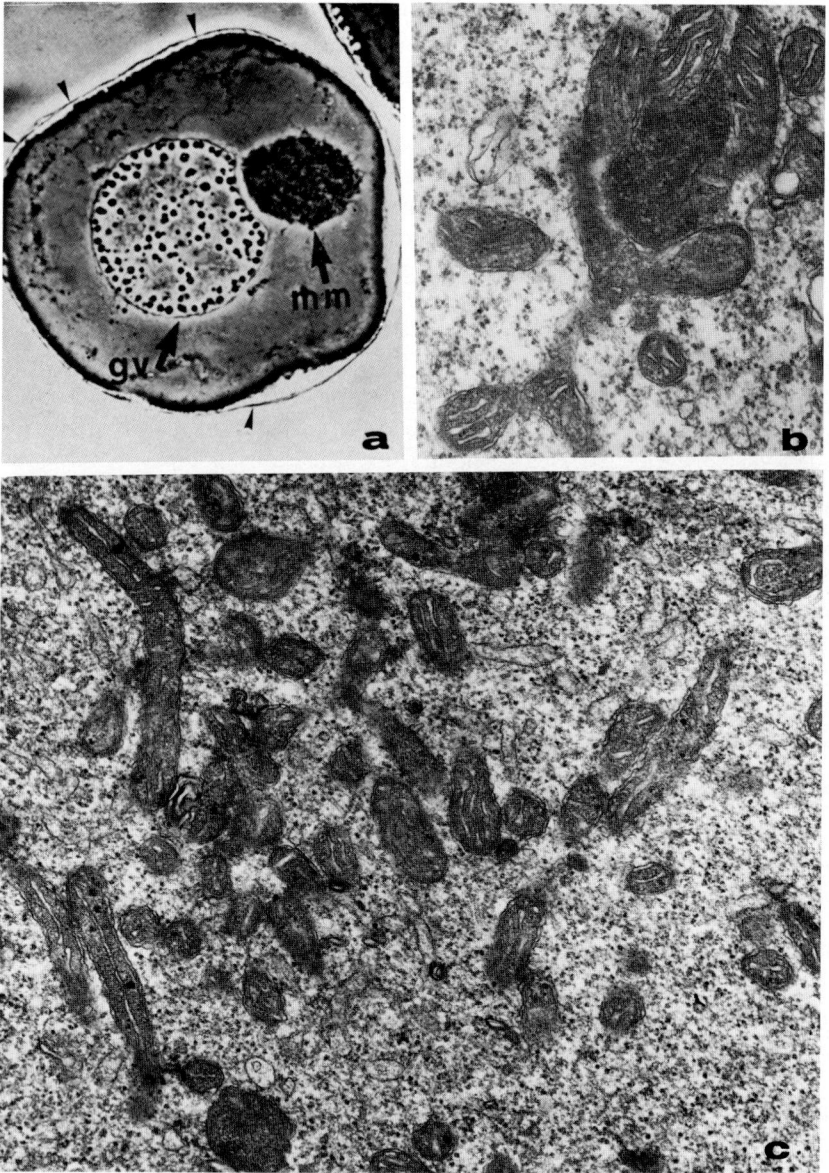

FIG. 11.   (a) Section through a small (150 μm in diameter) *Xenopus* oocyte. Arrowheads: follicular envelopes; gv, germinal vesicle with numerous nucleoli; mm, "mitochondrial mass"; this aggregate of mitochondria is also called a "mitochondrial cloud," "yolk nucleus," or "Balbiani body." (Callen *et al.*, 1980). (b,c) Ultrathin sections through the mitochondrial mass of young *Xenopus* oocytes. (b) A large, ramified mitochondrion surrounds some dense material. (c) Overview of the mitochondrial mass. [(a) Callen *et al.*, 1980; (b,c) original micrographs by P. Van Gansen.]

dria in *Xenopus* oocytes has allowed the characterization, for the first time, of a mitochondrial DNA by I. B. Dawid and his colleagues. Mitochondrial DNA molecules in *Xenopus* are circular, are about 5 μm long, and have an M_r of 11.7 $\times$ 10^6 [about 15.000 base pairs (bp)]. Their main transcription products are the 21 and 13 S mitochondrial rRNAs and about 25 tRNAs different from those coded by the nuclear genome. These RNAs associate with proteins to form mitochondrial miniribosomes (Dawid, 1966; Swanson and Dawid, 1970), which have a sedimentation constant as low as 58S. Almost 80% of the mitochondrial DNA in *Xenopus* oocytes is believed to be composed of untranscribed spacers that have diverged widely during evolution. However, *Xenopus* mitochondria also contain mRNAs that accumulate during the entire process of oogenesis (Golden *et al.*, 1980), in contrast to the cytoplasmic mRNAs, which will soon be discussed.

Synthesis of rRNAs by the oocyte nucleus [also called the ''germinal vesicle'' (GV) since its discovery by Purkinje in 1825 will be discussed later, when we deal with the GV. We have already described the distribution of the ribosomes in the cytoplasm of the full-grown oocytes. Previtellogenic oocytes contain very few ribosomes, as shown by electron microscopy and biochemical analysis (Thomas, 1970, 1974; Denis and Mairy, 1972). One finds instead fibrillar structures and small (7 and 42 S) particles where 5 S RNA is associated with proteins (Denis and Wegnez, 1977; Picard and Wegnez, 1979). The 42 S particles are composed of three molecules of tRNA, one molecule of 5 S RNA, two molecules of protein a, and one molecule of protein b; the 7 S particles are composed of 5 S RNA associated with protein b (M_r 35,000) in equal amounts. As we shall see later, this 5 S RNA binding protein b (now called TF IIIa) is unusually interesting because it plays an essential role in the regulation of 5 S gene activity. The 7 S particles were purified by Hanas *et al.* (1983), who confirmed that they contain 5 S RNA and the regulatory transcription protein factor TF IIIa. Both the 7 and 42 S particles of *Xenopus* oocytes are purely cytoplasmic, since they do not penetrate the nucleus after injection into the cytoplasm. This exclusively cytoplasmic localization is found in small (20–100 μm in diameter) previtellogenic oocytes (Mattaj *et al.*, 1983). According to Barrett *et al.* (1984), the most abundant proteins in young *Xenopus* oocytes have molecular weights of 48, 43 and 40 kDa (the last one is the 5 S DNA transcription factor TGF IIIa, which will be discussed later). All of them bind to 5 S RNA and display species specificity. Antibodies directed against *X. laevis* TF IIIa do not react with the *X. borealis* transcription factor. A likely function for these proteins is the transport of 5 S RNA into the various cell compartments. *Xenopus* oocytes display another peculiarity in that they possess two different kinds of 5 S genes producing a major and a minor type of 5 S RNA. In addition, these two types of ''ovarian'' 5 S RNAs are different from the 5 S RNA synthesized in somatic cells by another set of genes. In *Xenopus*, the ovarian 5 S genes are particularly active in very young oocytes; 60% of the 5 S RNA is already accumulated in oocytes less than 125 μm in diameter. At this very early stage, RNA polymerase III transcribes the 5 S RNA genes at a rate faster than 30 nucleotides per second (Arad and Beebee, 1981).

Typical 80 S ribosomes cannot be seen under the electron microscope before the onset of vitellogenesis. At that time, synthesis of 28 and 18 S rRNAs begins, while the 5 S component is provided by the 42 S particles (Mairy and Denis, 1971; Fig. 12). During vitellogenesis, there is a huge accumulation of ribosomes in the oocyte. More than 90% of these ribosomes are free; only a few are attached to vesicles that form a poorly organized ER. According to Woodland (1974), only 1–2% of the ribosomes in the oocyte are in the form of polyribosomes; in contrast, in hatching tadpoles, the majority of the ribosomes are associated with polyribosomes.

That *Xenopus* oocytes accumulate ribosomes far in excess of their own needs is shown by the studies of Brown and Gurdon (1964) on the *anucleolate* (*o-nu*) mutants. Deletion of both nucleolar organizers in a fertilized egg completely

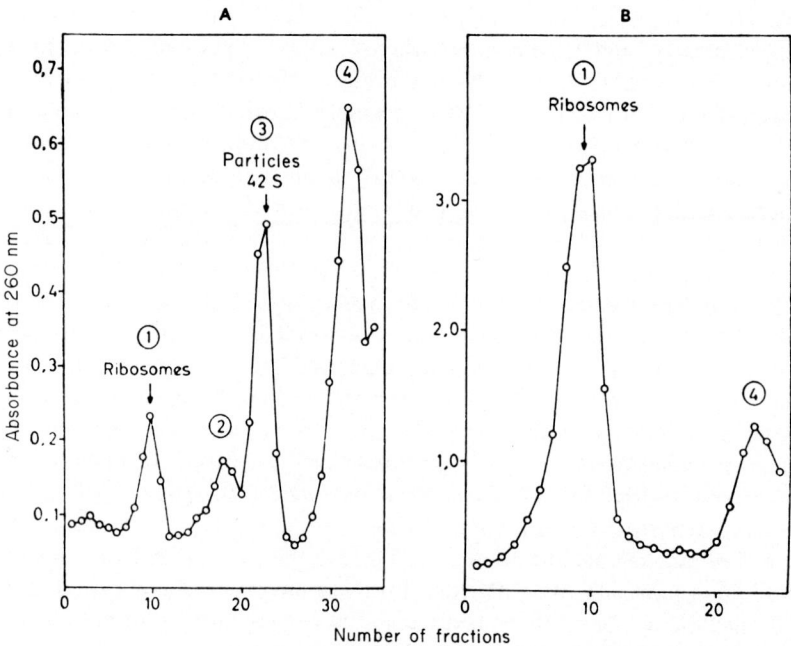

FIG. 12.   Fractionation of RNPs by gradient centrifugation in previtellogenic (A) and vitellogenic (B) *Xenopus* oocytes. ①, ribosomes; ②, RNP particles of unknown nature; ③, 42 S RNP particles; ④, soluble small RNPs. Note the increase in ribosomes and the disappearance of the 42 S particles during oogenesis. [Denis and Mairy, 1972.]

prevents 28 and 18 S rRNA synthesis, and thus the formation of new ribosomes. However, development remains possible for 1 week, and tadpoles with brain, eyes, muscles, etc. can be obtained. The store of ribosomes accumulated during oogenesis (about 10^{12} ribosomes per oocyte) is thus sufficient to ensure organogenesis and even cell differentiation. Interestingly, synthesis of 5 S RNA continues normally in *o-nu* mutants showing that it is not coordinated with that of the high molecular weight rRNAs and confirming that the 5 S genes are not localized in the nucleolus.

As already mentioned, *Xenopus* oocytes possess a store of mRNAs that form a highly heterogeneous population of about 20,000 different mRNA species. According to Rosbash and Ford (1974), the poly(A)$^+$ RNAs [thus, mRNAs provided with a poly(A) "tail"] represent 1% of the total RNA, and their amount remains constant during oogenesis. The maternal mRNAs stored in the oocyte correspond to 1–2% of the genome, according to Anderson *et al.* (1976). A remarkable feature of these maternal mRNAs is their exceptional stability. Poly(A)$^+$ RNAs synthesized before vitellogenesis begins can "survive" for as long as 18 months in the growing oocytes (Ford *et al.*, 1977). More recently, Golden *et al.* (1980) have shown that, in contrast to the mitochondrial mRNAs that accumulate during the entire process of oogenesis, the cytoplasmic mRNAs are synthesized and accumulate only during the previtellogenic period. During vitellogenesis the poly(A)$^+$ RNA store remains constant, and there are no marked changes in the composition of the mRNA population during oogenesis.

Another interesting feature of *Xenopus* oocytes concerns the histone mRNAs. In contrast to the general rule, i.e., that histone messengers are devoid of a poly(A) terminal sequence, *Xenopus* oocytes possess both poly(A)$^+$ and poly(A)$^-$ histone mRNAs (Levenson and Marcu, 1976; Ruderman and Pardue, 1977). In agreement with what has just been said about mRNA stability in oocytes, the nucleosome core (H2 to H4) histone messengers have already accumulated completely in previtellogenic oocytes (Van Dongen *et al.*, 1981). The poly(A) tract might play a role in ensuring histone mRNA stability. In keeping with this view is the fact that no poly(A)$^-$ mRNAs, except a part of the histone mRNA population, can be detected in *Xenopus* oocytes. The existence of a histone mRNA store in these oocytes raises a question about its function. There is apparently no need for histone production for nucleosome assembly since there is no chromosomal DNA synthesis during oogenesis. All one can guess is that the histone mRNAs and histones that have accumulated in the oocyte are used when, during the fast cleavages that follow fertilization, DNA replication becomes particularly rapid and intense. A similar situation has been found in *Drosophila* oocytes (Ruddell and Jacobs-Lorena, 1985).

Synthesis of all kinds of RNA is intense during oogenesis, and is, of course, a function of the nucleus. It is worth mentioning here that the cytoplasm of *Xenopus* oocytes contains four different protein factors that stimulate the activity of the RNA polymerases present in the nucleus; the concentration of these factors greatly decreases when the GV breaks down and RNA synthesis comes to a halt at maturation (Crampton and Woodland, 1979).

The stability of the mRNAs during *Xenopus* oogenesis is believed to be due to binding of the messengers to proteins. Such ribonucleoprotein complexes should be inactive in protein synthesis, since they are unable to bind to ribosomes. This explains why only 1–2% of the ribosomes are in the form of polyribosomes in *Xenopus* oocytes. However, these few polyribosomes are active, and during oogenesis the oocyte continuously synthesizes a large variety of proteins. As one can guess, previtellogenic oocytes, which contain very few ribosomes, have low protein synthesis activity. This activity increases considerably during vitellogenesis, together with a parallel increase in the size of the amino acid pool. There are no major qualitative changes in protein synthesis throughout the period of oogenesis, in agreement with the stability of the mRNA population.

Recent work has shown that 70% of the poly(A)$^+$ RNAs of *Xenopus* oocytes are not translated in *in vitro* systems of protein synthesis. These RNAs, which are interspersed with noncoding sequences such as the hnRNAs, have disappeared at the tadpole stage. The full-grown oocyte contains 90 ng of poly(A)$^+$ RNAs, but only 20 ng are translatable; they are sufficient to allow development up to the midblastula stage (Richter *et al.*, 1984). The nontranslating poly(A)$^+$ RNAs are associated with oocyte-specific proteins; their binding to the oocyte mRNAs prevents translation in *in vitro* systems. Oocyte-specific proteins presumably play an important role in the regulation of the stored maternal mRNAs (Richter and Smith, 1984).

Since small nuclear RNAs associated with proteins (snRNPs) play a role in the maturation (splicing) of the mRNA precursors, they deserve a few words here. Recent work by Fritz *et al.* (1984) has shown that accumulations of snRNAs and of the associated proteins are not coordinated during *Xenopus* oogenesis. U$_2$RNA reaches a plateau value at the onset of vitellogenesis; it does not accumulate before the blastula stage. It is very actively synthesized after this stage, since its content increases 10 times during the late blastula–early gastrula transition. In contrast, the proteins which are associated to U$_2$RNA increase during the whole vitellogenesis period and are in large excess over U$_2$RNA in full-grown oocytes. These proteins accumulate in the cytoplasm whereas U$_2$RNA is localized in the oocyte nucleus.

As one can see, the onset of vitellogenesis marks a sharp turning point in the biochemical activities of the oocyte. Multiplication of the mitochondria greatly slows down, 80 S ribosomes are synthesized and accumulated, and the rate of protein synthesis markedly increases. On the other hand, mRNA synthesis and accumulation are already completed when yolk platelets begin to form. Whether only the intake of vitellogenin by endocytosis is responsible for this biochemical revolution remains unknown. It seems more likely that many other still to be discovered control mechanisms operate when vitellogenesis begins.

B. STRUCTURE AND COMPOSITION OF THE NUCLEUS [GERMINAL VESICLE (GV)]

1. General Background

If the cytoplasm of a growing oocyte presents many features that are not found in common somatic cells, this is also true for the GV, where the large size and the presence of lampbrush chromosomes with many nucleoli and an abundant nuclear sap are striking and unusual features.

Oocytes derive from oogonia. After a last oogonial S phase, which brings the cell to the 4C level, meiosis begins at the time of tadpole metamorphosis. The early stages of meiosis (leptotene, zyotene, and pachytene) can be observed in froglets shortly after metamorphosis. In the adult, all oocytes (previtellogenic, vitellogenic, or full grown) are at the diplotene stage of meiosis; further progression in meiosis (meiotic divisions) requires maturation, which results from hormonal stimulation. In the following section, we shall deal mainly with oocytes from adult *Xenopus* females, as we have done for the cytoplasm.

Figure 13 shows a GV isolated from a full-grown oocyte. It is surrounded by a thick nuclear membrane and contains about 1000 spherical, refringent nucleoli. The very long, decondensed lampbrush chromosomes are poorly visible in *Xenopus* unless one breaks down the nuclear membrane and gently squashes the content of the GV. The large volume of the GV (up to 0.4 mm in diameter) is due to the abundance of the nuclear sap. It is very easy to dissect the GV out of an oocyte; methods for mass isolation by centrifugation of homogenates have also been described. Manually enucleated oocytes retain normal rates of respiration and protein synthesis, at least for a few hours.

In the following section, we shall deal successively with the nuclear membrane, lampbrush chromosomes, nucleoli, and nuclear sap.

2. The Nuclear Membrane

We can only add to our discussion of the nuclear membrane that our knowledge of the structure of the pore–lamina complexes (Fig. 5, Chapter 3, Volume 1) is due largely to the work done with *Xenopus* oocyte nuclei. The diameter of the pores remains constant (60 nm) throughout oogenesis, but their total number increases from 10×10^6 to 38×10^6 when the oocyte and its nucleus grow in size during vitellogenesis (Scheer, 1978). The RNAs synthesized on the lampbrush chromosomes and on the nucleolar organizers quickly move out of the nucleus into the cytoplasm through the nuclear pore complexes. As we have already seen, it has been calculated that three molecules of 28 and 18 S rRNAs

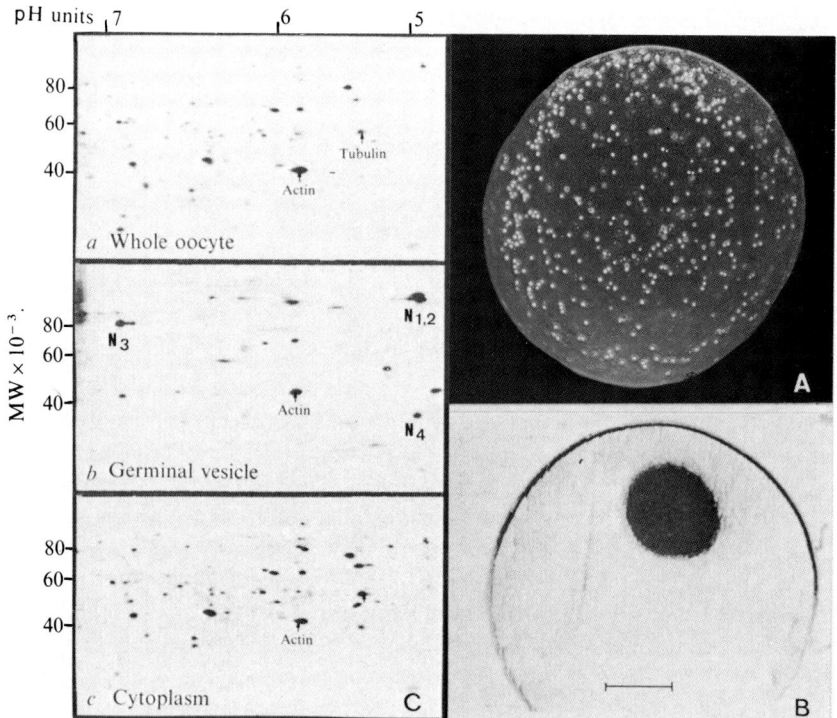

FIG. 13.   Injection of labeled proteins into *Xenopus* oocytes. (A) An isolated GV showing numerous nucleoli and lampbrush chromosomes. (B) Autoradiography of a *Xenopus* oocyte injected 24 hr earlier with a preparation of [^{35}S]methionine-labeled frog nucleoplasmic proteins: labeled proteins have accumulated in the nucleus. (C) Selective location of the radioactive proteins synthesized in an oocyte labeled for 24 hr in a C-amino acid mixture. Two-dimensional gel electrophoresis. (a) Preparation of ^{14}C protein synthesized by intact *Xenopus* oocytes; it contains actin, tubulin, and numerous other proteins. (b) The nucleus dissected out of the oocyte contains karyophilic nuclear proteins (Nr 2, 3, 4). (c) In the cytoplasm, actin is the predominant protein, and the nuclear proteins are undetectable. [(A) P. Van Gansen; (B) Reprinted by permission from *Nature* **295**, 572. Copyright © 1978 Macmillan Journals Limited.]

exit through each pore every minute. This means that about 300,000 molecules of rRNAs move out of the entire GV every second (Scheer, 1978). Krohne *et al.* (1981) isolated the pore–lamina complex from various cells of *Xenopus,* including oocytes. In the latter, they found a single major protein (M_r 68,000) that is continuously synthesized and phosphorylated. Lamina (the interporous material) isolated from *Xenopus* erythrocytes (which have very few pore complexes) are composed of two major proteins; in *Xenopus* liver nuclei, three major proteins are present in the lamina. This shows that the chemical composition of the pore–lamina complexes may differ from one cell type to another in the same animal species. However, the physiological meaning of these differences remains unknown.

Germinal vesicles, like somatic nuclei, have a nuclear matrix. After extraction of the nucleic acids, the pore complex lamina, a loose network of protein fibers and "ghosts" of the nucleoli, remain intact (Krohne *et al.* 1978, 1981). According to Krohne *et al.* (1982), *Xenopus* GVs contain two major "karyoskeletal" proteins. A 68,000-dalton protein is localized in the pore complex lamina structures; a larger protein (145,000 daltons) is found in the nucleoli, especially in their cortex. These two proteins are not specific for oocyte nuclei, since they can be detected by immunocytochemistry in many other cells. It is believed that the 145,000-dalton nucleolar protein is involved in the storage and transport of the preribosomal particles (Benavente *et al.*, 1984).

A major advantage of *Xenopus* oocytes over smaller cells is the possibility of studying nuclear membrane permeability in an intact cell (reviews by De Robertis, 1983 and Dingwall, 1985). Experiments on isolated nuclei suffer from the drawback that their permeability might be altered as soon as they have been isolated. A better approach, as mentioned in Chapter 4, Volume 1, is to inject into the oocyte cytoplasm labeled substances, particularly proteins, and to follow their penetration into the GV by autoradiography (Gurdon, 1979; Bonner, 1975a,b). The first experiments led to the conclusion that proteins of M_r 45,000 or smaller readily enter the nucleus after injection into the cytoplasm of the oocyte; bovine serum albumin (M_r 62,500) penetrates very slowly, and larger proteins do not move into the GV. These early conclusions had to be modified, however, after the experiments of De Robertis *et al.* (1978), who showed that proteins normally located in the nucleus accumulate in the GV after injection into the cytoplasm (Fig. 13). At the end of the experiment, it was demonstrated by autoradiography and chemical analysis of the labeled proteins present in isolated nuclei and in the remaining cytoplasm that all of the nuclear proteins quickly accumulated in the GV of an injected oocyte. This is true not only of small proteins, such as the histones, but also of much larger proteins, in particular, a nuclear protein named N1 by De Robertis *et al.* (1978). This protein, with an M_r as high as 120,000, very quickly reenters the nucleus after injection into the cytoplasm; it is soon concentrated 100 times in the GV. Actin, which is present in both the cytoplasm and the nucleus of *Xenopus* oocytes, is found in both cell compartments after injection into the cytoplasm. These experiments have led to the conclusion that there is a selective permeability of the nuclear membrane for nuclear proteins. Nuclear membrane permeability is unrelated to the size and electrical charge of the cytoplasmic proteins. Experiments by Feldherr and Ogburn (1980) cast serious doubts on the very existence of nuclear permeability. They found that the selective uptake of proteins by the GV remains unaltered when a hole is bored in the nuclear membrane by pricking with a fine needle. The logical conclusion is that the uptake of cytoplasmic proteins into the GV is conditioned by a specific binding of "karyophilic" proteins to nuclear sap constituents, and not to selection by the nuclear membrane. If this is so, it should be possible to study the specific binding of a number of proteins to isolated GVs and to identify the cell sap constituents responsible for their selective binding. It is likely that nuclear proteins bear a "karyophilic signal" that directs them toward the oocyte nucleus (De Robertis, 1983). A kinetic analysis of the entry of the large karyophilic N1 protein (M_r = 148,000 in R. pipiens) into frog GVs has confirmed that its entry is due to some sort of mediated transport, and not to simple diffusion (Feldherr et al., 1983). However, a recent EM study by Feldherr et al. (1985) has shown that the transfer of a karyophilic protein (nucleoplasmin) coated on gold particles takes place through the nuclear pores.

The cytoplasm of Xenopus oocytes contains a large variety of karyophobic proteins, which are not found in the germinal vesicle. They amount to 20% of all soluble proteins, and their molecular weights range from 15,000 to 230,000. Similar proteins have been found in the cytoplasm of HeLa cells (Dabauvalle and Franke, 1984).

As we have seen in our discussion of nucleocytoplasmic interactions in Acetabularia and A. proteus (Chapter 1), there are good reasons to believe that small RNAs can move from the cytoplasm into the nucleus. A study by De Robertis et al. (1982) has shown that, if labeled snRNAs (the U_1, U_2, U_4, U_5, and U_6 snRNAs) are injected into the cytoplasm of Xenopus oocytes, they quickly migrate into the nucleus, where they accumulate. Their final concentration is 30–60 times higher in the GV than in the cytoplasm. In contrast, injected tRNAs and a 7 S RNA remain in the cytoplasm, while 5 S RNA accumulates in the nucleoli. It is likely that the RNAs that move into the nucleus are associated with still unidentified specific RNA-binding proteins.

3. Lampbrush Chromosomes

The general structure of the lampbrush chromosomes (reviewed by Callan, 1963, 1972, 1982) has already been presented (Fig. 14) in Chapter 5. Figure 15 showed the classic model presented by Gall in 1956. The lampbrush chromosomes are at the diplotene stage of meiosis and are thus present as bivalents that are held together by synaptinemal complexes. Each homolog consists of two chromatids, each containing a single DNA molecule of considerable length. These continuous DNA fibers are folded to form Feulgen-positive chromomeres (Brachet, 1940), from which about 10,000 loops extend in the nuclear sap. The total length of all of the loops is about 50 cm, but their DNA represents only 5% of the lampbrush chromosome DNA. Lampbrush chromosomes are much larger and easier to study in urodeles than in anurans, due to the fact that the DNA content per nucleus (C value) is 7–10 times higher in the former than in the latter. As has been shown by Rosbash et al. (1974), the number of unique potentially coding sequences is the same in Triturus and in Xenopus, indicating that urodeles contain a much higher proportion of noncoding DNA than anurans. In lampbrush chromosomes, the "C paradox," i.e., a 10-fold difference in the C value among amphian species, is explained by the fact that only the loops are genetically active; the chromomeres are composed of repetitive DNA sequences that are not transcribed. In other words, the loops are composed of extended euchromatin fibers, and the chromomeres, which play a structural role, are a special form of condensed heterochromatin.

A paper by Scheer and Sommerville (1982) has presented interesting information about a correlation between the C value and the length of the loops and the attached transcription units. The dimensions of the loops are 5–10 μm in *Xenopus,* 30–50 μm in *Triturus,* and more than 100 μm in *Necturus;* the C values (picograms of DNA per haploid genome) are 3.1, 23, and 78, respectively, for the three species. However, there are no large quantitative differences as such when one looks at the hnRNA molecules isolated from the three species. While more DNA sequences are transcribed if the C values are high, processing of the RNA transcripts takes place while they are still attached to the chromosome loops as nascent ribonucleoprotein fibrils.

Pioneer work by H. G. Callan (who first isolated lampbrush chromosomes from dissected nuclei) and by J. G. Gall (who treated such isolated chromosomes with various hydrolytic enzymes) demonstrated both the continuity of the DNA fiber along the full length of the chromatids and the presence of a ribonucleoprotein matrix on the loops. That the RNA present on this matrix results from transcription of the DNA fiber, which forms the axis of each of the 10,000 loops, was conclusively demonstrated in 1962 by the autoradiography studies of Gall and Callan.

It was first believed that each of the 10,000 chromomeres corresponds to a single gene; later, it was thought that each loop represents a single gene. However, the situation turned out to be much more complicated. In 1975, Vlad and MacGregor pointed out that the chromomeres were not genetic units and that the loops contained both unique and repeated sequences. For Sommerville and Malcolm (1976), each loop was a transcription unit that synthesized many molecules of a primary transcription product. Thus, while 4% of the DNA in *Triturus* is transcribed during oogenesis, only 0.1–0.2% is expressed in coding sequences. This would correspond to the 10,000–20,000 genes that are believed to be necessary for embryonic development, further implying that most, if not all, of the genes are simultaneously active during oogenesis. What is perhaps the most striking and unexpected consequence of this calculation is that 99.8% of the lampbrush chromosome DNA does not code for mRNAs. The role of this "selfish" or "junk" DNA (Orgel and Crick, 1980) remains a complete mystery. Calculations also showed that only 30% of the unique sequences were actually transcribed in the lampbrush chromosomes.

The primary transcription products of the loops are very large (10–30 μm long) RNA molecules with an M_r as high as 5×10^6. During transcription, they associate with nuclear proteins to form ribonucleoprotein particles that detach from the loops and move into the nuclear sap. It is in the nuclear sap that these large precursors are believed to be processed to yield the mRNAs that will be exported to the cytoplasm (Sommerville *et al.*, 1978). We have no information about the mechanisms of this processing, which presumably implies the removal of introns by endonucleases present in the nuclear sap. Are the primary transcripts similar to the hnRNAs of somatic cells? As we shall now see, this is not certain. *In situ* hybridization experiments by Old *et al.* (1977) and Scheer and Sommerville (1982) have disclosed an unusual feature of RNA processing in lampbrush chromosomes: it starts before transcription is complete, and thus while the RNA transcript is still attached to the loop. In somatic cells, there is no evidence thus far for processing of hnRNA while it is still bound to its DNA template.

The already mentioned development by Oscar Miller (reviewed by Hamkalo and Miller, 1973) of the powerful chromosome spreading technique allows us to visualize under the electron microscope the growth of the ribonucleoprotein nascent fibrils on the loops. As shown in Fig. 2, Chapter 2, Volume 1, the fibrils originate from the DNA fiber and progressively grow; many nascent nucleoprotein fibrils are coiled or folded. This technique has been used very successfully by U. Scheer and his colleagues for a refined analysis of transcription in the loops. In 1976, Scheer *et al.* (1976a) showed that several transcription units may be present on the same loop, and that they may be different or identical and have the same or opposite orientation, as shown in Fig. 14. The important conclusion drawn from these experiments was that a loop is not a gene, but rather a coordinated transcription unit of one or a set of genes. The biggest surprise for those who believed in the "one loop-one gene" hypothesis came when it became clear that highly repetitive DNA was transcribed on many loops of the lampbrush chromosomes (Varley *et al.*, 1980). According to Scheer (1981), tandemly repeated genes are actively transcribed on the lampbrush chromosomes of *Pleuro-*

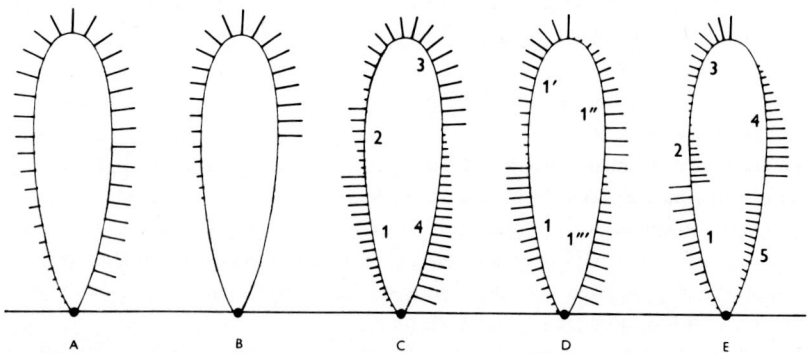

FIG. 14.   Various alternatives for arrangements of transcription units within individual loops of lampbrush chromosomes. The numbers 1–1''' denote units of equal; 1–5 denote units of different lengths. This schematic representation is based on the analysis, under the electron microscope, of spread *X. laevis* lampbrush chromosomes. [Scheer *et al.*, 1976a.]

deles, forming clusters of more than 100 copies separated by spacers. More recently, Sommerville and Scheer (1982) have presented new information on this question. The repeated sequences are transcribed in the GV from families of DNA sequences that are each represented thousands of times in the genome. These repeat sequences are spread throughout the genome and differ in various amphibian species. There is no correlation between their size and the C values, but their complexity increases with these values. Transcription of repeated sequences in *Xenopus* lampbrush chromosomes has also been observed by Jamrich *et al.* (1983). Possibly connected with these observations is the finding that, in *Triturus,* 10–15 loop pairs synthesize rRNA, a function that was believed to be the exclusive role of the nucleolus and that is carried on by highly repetitive DNA sequences. Thus, during oogenesis, there would be two distinct systems (nucleolar and chromosomal) for rRNA synthesis (Morgan *et al.,* 1980). This situation is reminiscent of the distribution of the histone genes in sea urchins. As we have seen, they are clustered, but there are also ''orphons'' dispersed elsewhere in the genome.

The transcription products of the repetitive genes are found in the cytoplasm. Curiously, 68% of the poly(A)$^+$ RNA sequences renature easily after heating, showing that there are complementary interspersed sequences on different RNA molecules; this percentage decreases to 15% in the hatching tadpole (Anderson *et al.,* 1982). These maternal poly(A) RNA molecules are located in the cytoplasm, not in the mitochondria. Anderson *et al.* (1982) concluded that the lampbrush chromosomes supply a continuous flow of a complex set of maternal poly(A)$^+$ RNAs to the cytoplasm. The question of repetitive sequence transcription during development has been reviewed by Davidson and Posakony (1982). Repetitive sequences are frequently found, in addition to the maternal poly(A)$^+$ RNAs stored in eggs, in hnRNA and a few mRNAs. Their role remains completely unknown.

If we recall that, according to Golden *et al.* (1980), the complete mRNA population, comprising about 20,000 different species, is synthesized and accumulated in previtellogenic oocytes, the meaning of the high transcriptional activity of the lampbrush chromosomes loops during vitellogenesis becomes a complete mystery. If the results of Golden *et al.* (1980) are correct—they have already been substantiated for histone mRNAs by Van Dongen *et al.* (1981)— vitellogenic oocytes do not synthesize mRNAs in appreciable amounts. What then is the meaning of the transcription units seen in spread chromosomes? All we know is that no transcription units can be seen at the pachytene stage of meiosis and that they appear at diplotene when the nucleus reaches a diameter of 25–40 μm. The pattern found in vitellogenic oocytes is already present in GVs measuring 50 μm (Hill and MacGregor, 1980). Golden *et al.* (1980) have proposed that the transcripts seen on the lampbrush chromosome loops might play a role in controlling translation in the cytoplasm. There is no doubt that oocytes

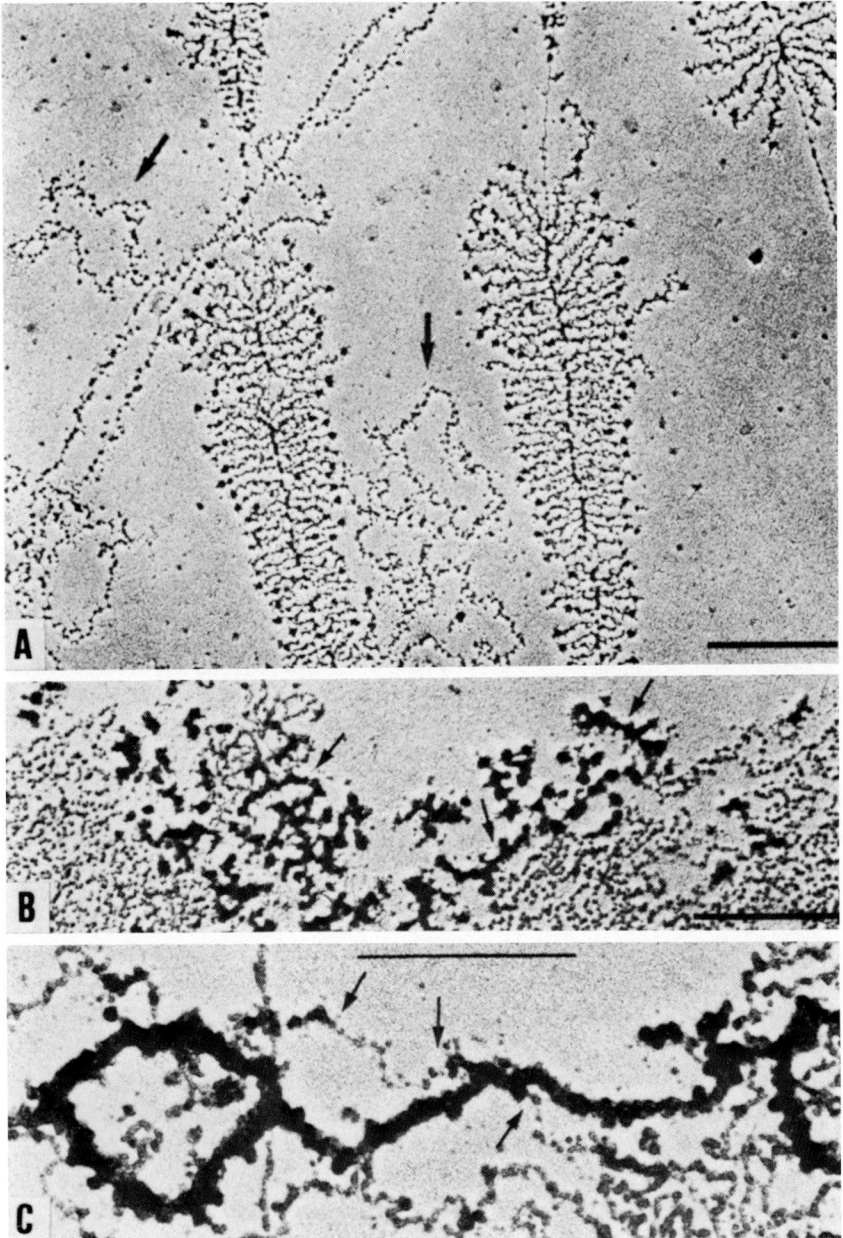

Fig. 15. Appearance of transcriptionally active chromatin in the nucleus of a *Pleurodeles* (urodele) oocyte. (A) A yeast rDNA plasmid had been injected into this nucleus. The arrows show two plasmid rings with a beaded nucleosomal appearance. Fully active rRNA genes in an extended, largely nonnucleosomal configuration occur side by side with inactive, nucleosomally arranged chromatin strands of the injected plasmid DNA. (B, C) In the transcriptionally inactive chromomeres

possess very accurate mechanisms for the selection and activation of stored mRNAs (see Section III and IV) and that we know very little about the control of selective translation in cells that possess 10,000–20,000 different mRNAs in their cytoplasm. It must be admitted that, for the time being, the nature and role of the numerous transcription units seen on the loops during vitellogenesis are not understood. This should stimulate further research on the fascinating lampbrush chromosomes, which were discovered a century ago by Flemming (1882).

Little is known about the mechanisms that preside over the maturation (splicing) of the pre-mRNAs synthesized on the lampbrush chromosomes. It is likely that, as elsewhere, small nuclear ribonucleoproteins (snRNPs) are involved in this process. A recent paper by Forbes *et al.* (1984) suggests that this process might be particularly complex in oocytes. *Xenopus* oocytes and embryos have seven different U RNA species. All have the same length (165 nucleotides), but their sequences differ. Two U_1 RNA genes are transcribed in late blastulas and early gastrulas, but not in oocytes. Other U_1 RNA species are already expressed during late oogenesis: there is obviously no coordination in the control of U_1 genes expression. We have just seen that the same conclusion has been drawn regarding the synthesis of U_2 RNA and its associated proteins.

That RNA synthesis is an essential function of the loops has been conclusively established by the fact that inhibition of RNA synthesis by actinomycin D causes their collapse. This was first observed by phase-contrast microscopy and has been reinvestigated, on spread chromosomes, by Scheer *et al.* (1979). Interestingly, they found that arrest of transcription in lampbrush chromosomes by actinomycin D treatment or by induction of maturation leads to extensive changes in chromatin structure. Although nucleosomes are absent in transcribed loops, they become apparent when transcription is suppressed (Fig. 15). Similar results were obtained by Scheer *et al.* (1979) when they injected into the GV an antibody directed against histone H2B. Retraction of the loops quickly followed the arrest of transcription, which was probably due to blockage of the movement of RNA polymerase II molecules along the axis of the loops. At the ultrastructural level, the contracted lampbrush chromosomes acquired a "superbead" structure. Scheer *et al.*'s (1979) suggestion that the collapse of the lampbrush structure in these experiments was due to interference with RNA polymerase II movement along the loops by the antibody was substantiated by the observations of Bona *et al.* (1981). After injection into the GV of a *Xenopus* oocyte of an antiserum against RNA polymerase II of *Drosophila,* they observed the disap-

of the lampbrush chromosomes, two different forms of chromatin packing can be seen: nucleosomes and large supranucleosomal globules approximately 30 nm in diameter. These globules form chains of ca. 30-nm beads (arrows in B) or thicker strands (40–60 nm) (c). Arrows in (C) show transitions between higher-order globules and nucleosomes. Bars: 0.5 μm. [Scheer *et al.,* 1980.]

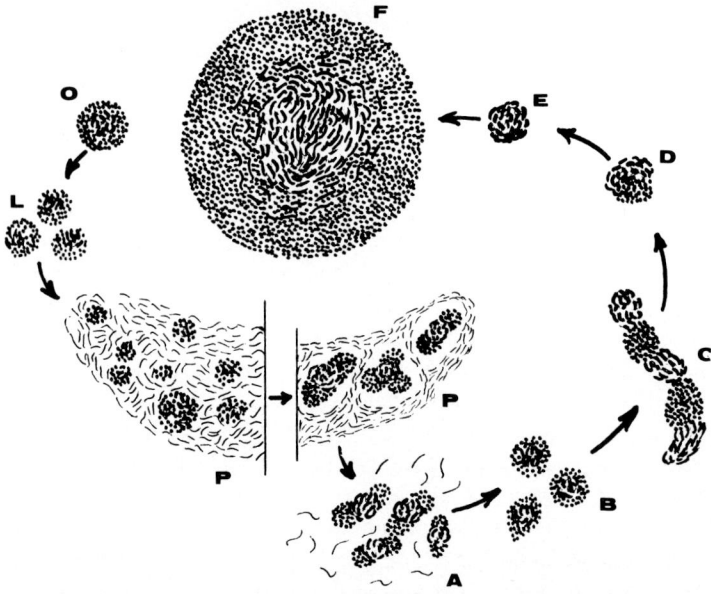

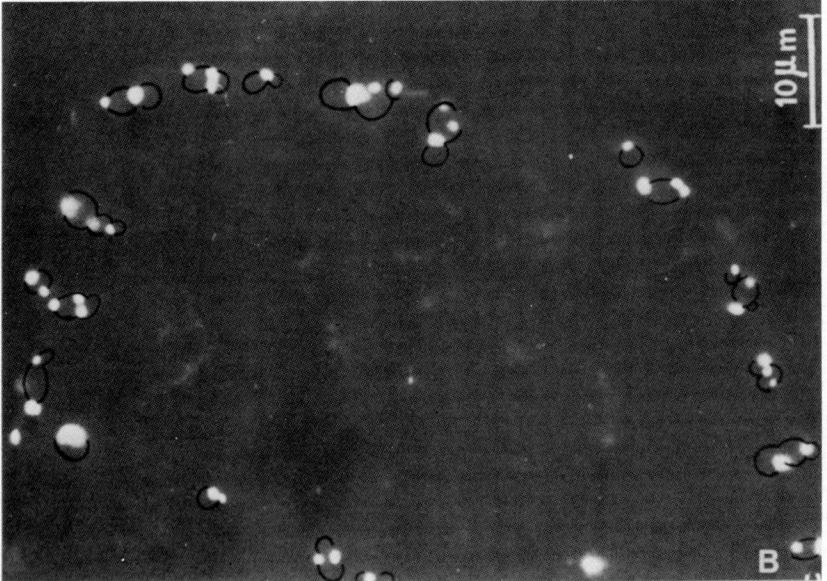

Fɪɢ. 16. Nucleoli and nucleolar organizers in *Xenopus* oocytes. (A) Schematic representation of the changes in structure and size of the nucleoli during oogenesis. Dots correspond to the granular regions and small lines to the fibrillar regions. O, oogonia; L, leptotene; P, pachytene (left, maximal size of the rDNA cap; right, cap in regression). (A–E) Previtellogenic diplotene oocytes. (F) Vitellogenic oocytes (300–1200 μm in diameter). [Original drawing by P. Van Gansen.] (B) Detection of the nucleolar organizers in a previtellogenic (170 μm in diameter) *Xenopus* oocyte. Pho-

pearance of the nascent transcripts, followed by the collapse of the loops. Injection of an antibody against a high mobility group protein called HMG-A (see Section II,B,5) also produces loop retraction and inhibition of nonnucleolar gene transcription (Kleinschmidt *et al.*, 1983).

Another way to induce the contraction of the lampbrush chromosomes is by addition of ATP (Karsenti and Gounon, 1979). The reaction is Ca^{2+} independent and Mg^{2+} dependent. It was therefore suggested that the lampbrush chromosomes were associated with an actomyosin matrix endowed with Mg^{2+}-ATPase activity. Previous immunocytochemical work by Karsenti *et al.* (1978) showed that lampbrush chromosomes contain tubulin in both the chromomeres and the loops; actin was found only in the loops where it might contribute to chromatin condensation. Indeed, injection into the GV of an antiactin serum inhibits the contraction of the lampbrush chromosomes that takes place when maturation is induced by hormonal stimulation; injection of the same antiserum into the cytoplasm has no effect. In addition, it was found that an antimyosin serum has no effect on chromosome condensation in the same system after injection into the GV (Rungger *et al.*, 1979). It thus seems that actin and probably actin-associated proteins play an important role when the giant lampbrush chromosomes undergo enormous condensation in order to become metaphase I chromosomes during maturation. Scheer *et al.* (1984) demonstrated recently that nuclear actin plays an important role in lampbrush chromosome transcription: they injected antibodies against actin and one of the actin-binding proteins (fragmin) into germinal vesicles of *Pleurodeles* oocytes and observed a complete inhibition of transcription in the lampbrush chromosomes, but not in the nucleolar genes. After injection of ribonuclease into the germinal vesicle, an extensive meshwork of actin filament bundles remained associated to the lampbrush chromosomes.

4. Nucleoli

The nucleoli of *Xenopus* oocytes should be as famous, in the history of cell biology, as the lampbrush chromosomes of *Triturus* oocytes. Their development during oogenesis has been carefully described by Van Gansen and Schram (1972) (Fig. 16). As we already know, vitellogenic and full-grown *Xenopus* oocytes contain about 1200 nucleoli. This number is not constant, since fusion or fragmentation of nucleoli are frequent events. However, the number of nucleolar organizers tends to be constant and, as a consequence, large nucleoli resulting from fusion of normal-sized nucleoli may contain several nucleolar organizers (Fig. 16). They originate, as shown in Fig. 17, from a Feulgen-positive cap that

tographs taken with a fluorescence microscope after Feulgen staining. The nucleoli have been delimited according to their localization by phase contrast in the same section. [Thomas and Schram, 1977.]

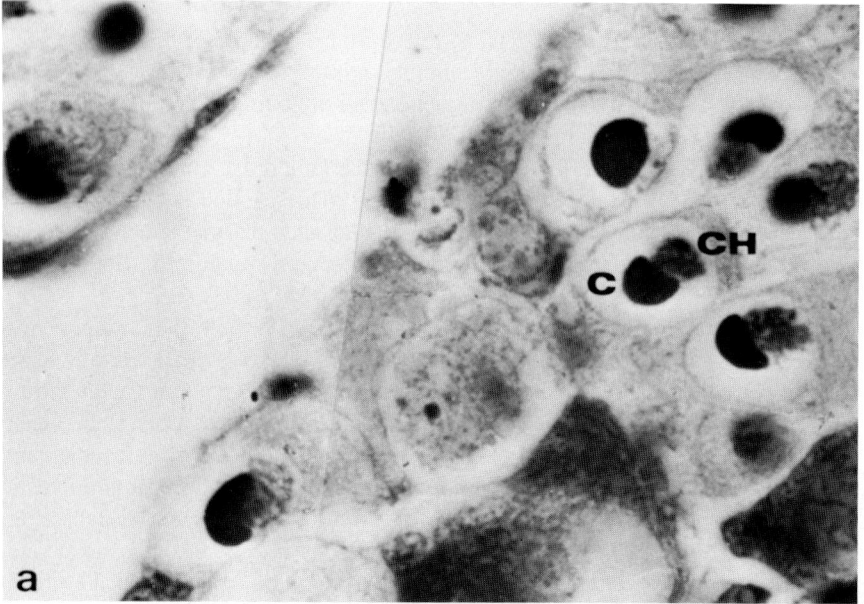

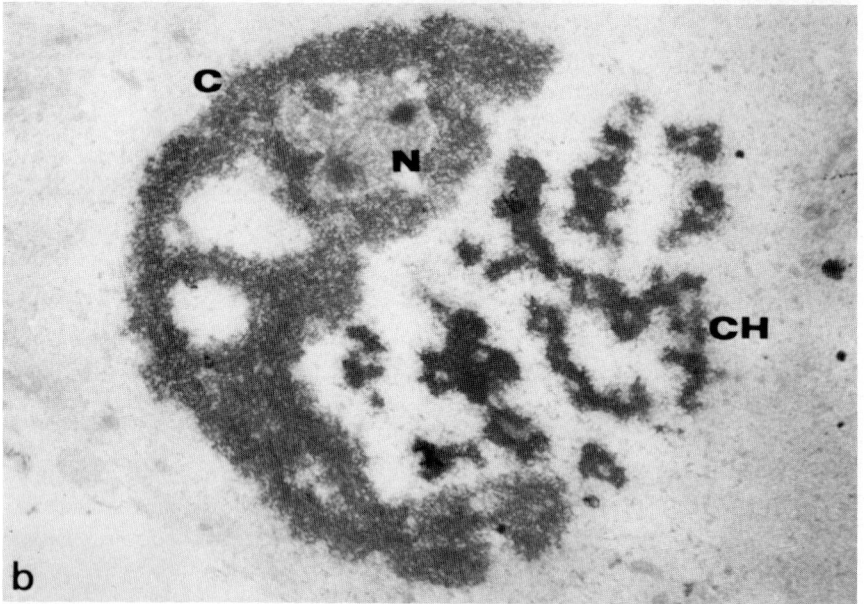

FIG. 17. Section through very young *Xenopus* oocytes at the pachytene stage of meiosis. (a) Light micrograph. C, cap containing the ribosomal genes (rDNA) in the process of amplification; CH, pachytene chromosomes. (b) Electron micrograph. C and CH as in (a); N, a nucleolus appears in the cap. [(a) A. Ficq; (b) P. Van Gansen.]

is conspicuous in very young *Xenopus* oocytes (at the pachytene stage of meiosis). The DNA present in this cap replicates 1 week after the replication of chromosomal DNA is concluded. In addition, it stains red (instead of green) with methyl green pyronine, which is due, as shown by Ficq (1970), to the abundance of single-stranded DNA molecules in the cap. *In situ* hybridization has shown that the DNA present in the cap hybridizes specifically with radioactive 28 and 18 S RNAs (Gall and Pardue, 1969; Birnstiel *et al.,* 1969). These cytochemical studies clearly showed that the DNA localized in the cap is rDNA (composed of ribosomal genes coding for the 28 and 18 S rRNAs) and that this DNA is replicated independently from chromosomal DNA. This intensive replication leads to a 1000-fold amplification of the ribosomal genes, a process to which we shall soon return. We have already mentioned that, in contrast to the 28 and 18 S genes, the 5 S genes are not localized in the cap, but instead at the distal end of most chromosomes (Pardue *et al.,* 1973). Since these genes are not amplified during oogenesis, oocytes, and somatic cells contain the same number of copies of the 5 S genes (about 24,000).

In nucleoli, the number of nucleolar organizers increases from 30 to 1000 in the cap at pachytene; amplification of the ribosomal genes is thus complete in previtellogenic oocytes. Oocytes at that stage synthesize low molecular weight RNAs (5 S RNA, tRNA), but no macromolecular rRNAs (Thomas, 1970, 1974). In connection with this failure of the nucleoli to synthesize 28 and 18 S rRNAs during previtellogenesis, the nucleoli possess only the central fibrillar region at that stage (Fig. 18), lacking the peripheral granular part. Purely fibrillar nucleoli seem to be a characteristic of cells that are unable to synthesize the high molecular weight RNAs, since they are found in mammalian oocytes and in cleaving sea urchin eggs, where ribosomal RNA synthesis is maintained at a very low level. During vitellogenesis, the nucleoli acquire their classic structure (a fibrillar core surrounded by a granular cortex, Fig. 31, Chapter 4, Volume 1). This change in nucleolar ultrastructure coincides with the onset of 28 and 18 S rRNA synthesis. Why synthesis of these large ribosomal RNAs is repressed in previtellogenic oocytes that are exceedingly active in mRNA and 5 S RNA synthesis remains unexplained. A study by Williams *et al.* (1982), who applied the nucleolar organizer silver staining method to *Xenopus* oocytes, failed to disclose differences between previtellogenic and vitellogenic oocytes. The fibrillar nucleolar core stains strongly during the whole oogenesis. In addition, silver staining is due to a single nucleolar protein of M_r 195,000; the authors suggested that this protein might be the large subunit of RNA polymerase I. Similar results have been obtained by Boloukhère (1984), who followed under the electron microscope the silver staining of the nucleoli during all stages of *Xenopus* oogenesis. Very strong staining was observed in the nucleoli of previtellogenic oocytes, although these oocytes are inactive in rRNA synthesis. This shows that one

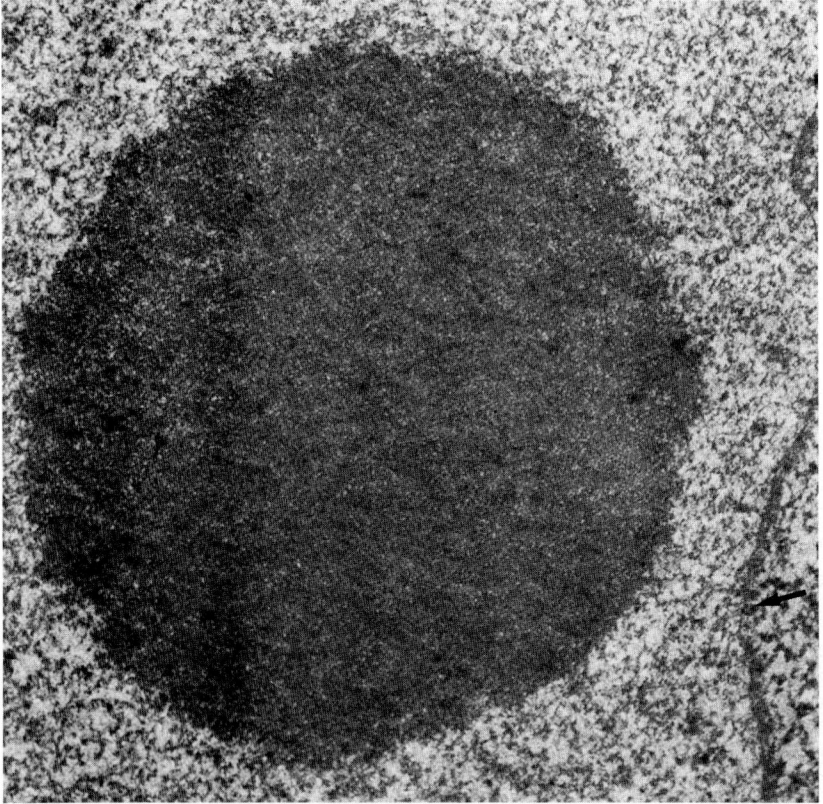

Fig. 18.   At the end of the previtellogenic period, rRNA synthesis is suppressed and the nucleoli have a purely fibrillar appearance. Arrow, section through the nuclear membrane. [Original photograph by P. Van Gansen.]

should be careful in generalizing the concept that silver staining of the nucleolar organizers indicates that these organizers are active in transcription.

We have seen that amplification of rDNA in pachytene *Xenopus* oocytes was discovered by Brown and Dawid in 1968, based on the fact that rDNA contains more G + C sequences than chromosomal DNA. Because of this property, rDNA can be separated from bulk DNA by ultracentrifugation in a CsCl gradient as a heavy satellite (Fig. 19). This satellite hybridizes specifically with 28 and 18 S radioactive rRNAs (Gall, 1968) and it represents as much as 25 pg of DNA. It can be calculated that the tetraploid *Xenopus* oocyte contains as many as 2×10^6 copies of the ribosomal genes; since a diploid *Xenopus* somatic cell possesses only 900 copies of these genes, the amplification is at least 2000 times. In

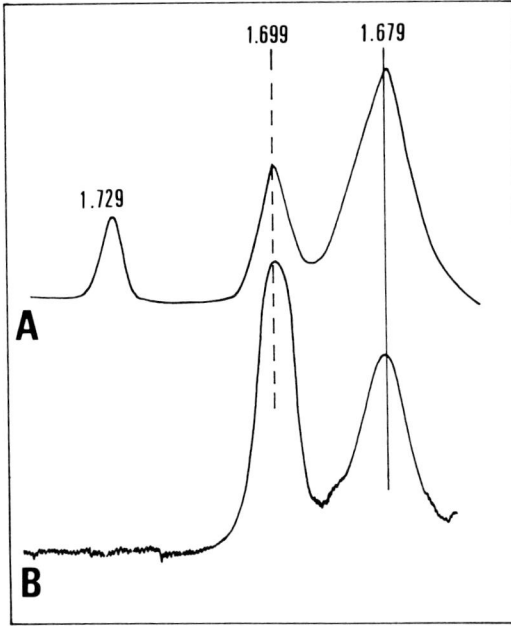

Fig. 19.   (A) DNA of *Xenopus* GVs is separated into two peaks of different density, 1.729 and 1.699, by centrifugation at equilibrium in the analytical centrifuge. The heavy peak (1.729) corresponds to the GC-rich amplified ribosomal genes, the light peak (1.699) to bulk DNA. The 1.679 peak corresponds to a dAT_n marker. (B) In somatic cells, where there is no ribosomal gene amplification, the heavy peak (1.729) is not detectable. The density of the main band is the same as in GVs. [Brown and Dawid, 1968. Copyright 1968 by the AAAS.]

keeping with this high selective gene amplification is the increase in the number of nucleoli during oogenesis (from 2 to about 1200).

Electron microscopy has been extremely useful in uncovering the mechanisms of gene amplification at the molecular level and the organization of the ribosomal genes at later stages of oogenesis. In the cap, a very heterogeneous population of rDNA molecules can be observed during amplification. The majority of the molecules are linear; however, replication forms that are unusual for eukaryotic DNA (Cairns' circles, rolling circles), but are well known in bacteria and phages, are observed (Hourcade *et al.*, 1973; Bird, 1977). Rolling circles (lariates) longer than 65 μm can be mistaken for linear DNA molecules, since these circular DNA molecules of 4–60 μm have a long "tail" (Fig. 20). Circles and rolling circles are probably intermediates in rDNA replication during amplification; the replication mechanisms seem to be very complex and quite different from those of chromosomal DNA replication during the S phase. However, in both cases, replication is inhibited by aphidicolin, indicating that DNA poly-

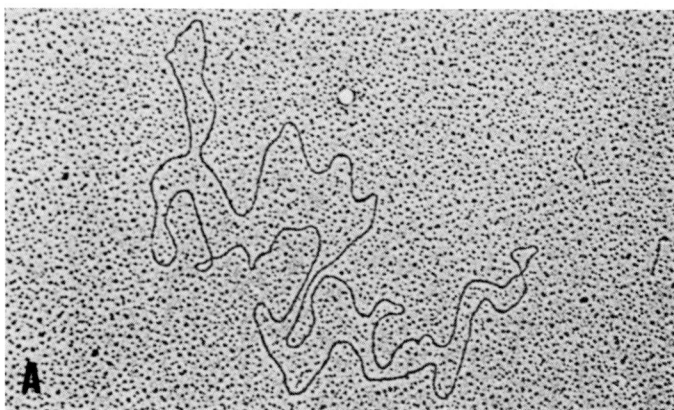

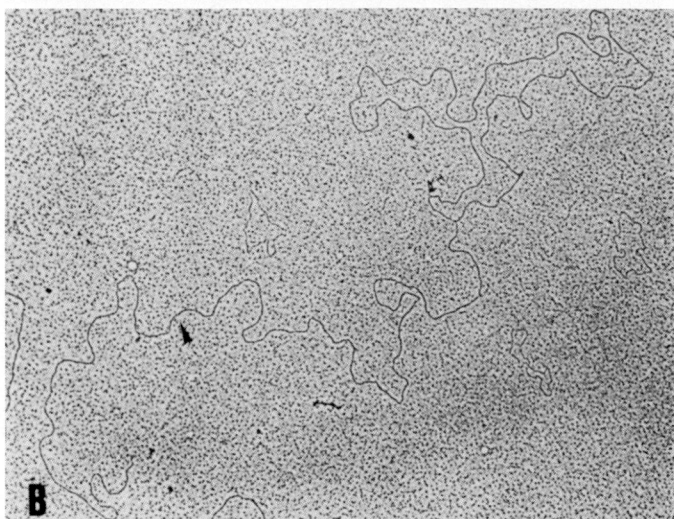

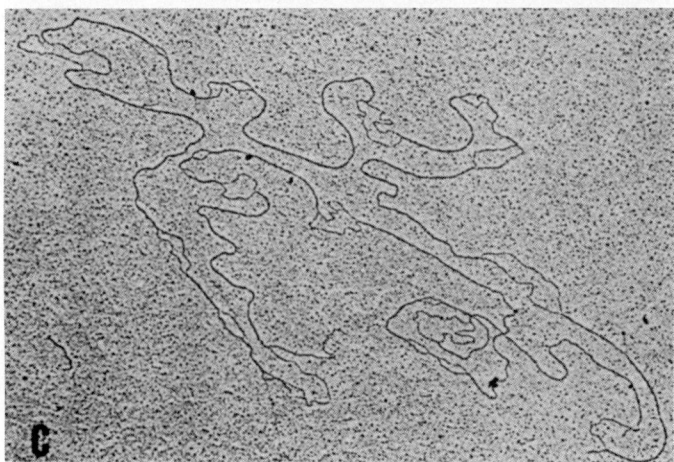

merase-α is implicated in the two processes (Zimmerman and Weissbach, 1981).

In 1969, for the first time, Miller and Beatty spread the nucleolar organizers of nucleoli isolated from large *Xenopus* oocyte GVs. As shown in Fig. 23, Chapter 1 (a,b), the ribosomal genes are tandemly repeated and separated by non-transcribed spacers. Transcription starts at a promoter site and the RNA precursor chains, which are associated with proteins, grow continuously, giving rise to the now classic "Christmas tree" configuration. There are about 450 ribosomal genes in each nucleolar organizer. The end product of transcription in *Xenopus* is a 40 S rRNA precursor that is processed after it has been released from its rDNA template. The small granules that can be seen at the origin of the growing ribonucleoprotein fibrils are believed to be molecules of RNA polymerase I. If this is the case, there are about 100 molecules of this enzyme per transcription unit, and thus per ribosomal gene (Scheer *et al.*, 1976a,b). Scheer *et al.* were able to identify precisely the origin and termination of rDNA transcription. It could be seen that part of the spacer is transcribed, in agreement with biochemical data that were obtained in D. Brown's laboratory and that will be briefly discussed later. Even the untranscribed spacers in some females of *Xenopus* may display small transcription units (Trendelenburg, 1981). Before electron microscopy was used for biochemical studies, there were some discrepancies regarding the fine structure of the transcription units and the spacers that separated them. Scheer *et al.* (1979) and Scheer (1981) concluded that there were no nucleosomes in the regions where transcription was active, and that the spacers were not compacted into nucleosomes but were covered with proteins that did not shorten the DNA. Labhart and Koller (1982) concluded that very few proteins were associated with rDNA during transcription and that the apparent lack of histones precluded the existence of nucleosomes. In contrast, according to Pruitt and Grainger (1981), the nontranscribed spacers were in the form of supranucleosomal structures where the DNA fiber was contracted at least 20 times. The ribosomal genes have a much more extended structure, even in previtellogenic oocytes, where, as we have seen, they are not transcribed. Therefore, the contraction of DNA in the gene is only 1.4 times, which would be due to a mononucleosomal structure. These divergent opinions are probably due to technical differences in the handling of the material during the delicate spreading technique. Furthermore, according to Scheer *et al.* (1976a,b), the Christmas trees do not always display the perfect regularity shown in Fig. 23, Chapter 1: electron microscopy often shows images of incomplete or abnormal transcrip-

---

FIG. 20.   Amplified rDNA forms seen under the electron microscope. (A) rDNA circle with a circumference corresponding to four rRNA genes. (B) An rDNA rolling circle; the circumference of the template ring corresponds to three rRNA genes and the attached tail (arrowhead) to 3.4 rRNA genes. (C) A partially denatured rDNA circle containing six rRNA genes. [Hourcade *et al.*, 1973.]

tion. It would indeed be a miracle if all of the 2 million ribosomal genes worked perfectly together all the time.

The characteristics of *Xenopus* ribosomal genes, i.e., amplification, high G + C content, and lack of 5-methylcytosine, made it possible to isolate them in a pure form long before the advent of recombinant DNA technologies allowed gene isolation. This was done by D. Brown and his colleagues from 1970 to 1980, and led to very precise analysis of the structure of the ribosomal cistrons and processing of the rRNA precursor. This discussion will be limited to a very brief description of these findings, since their main interest lies in the field of molecular biology rather than molecular cytology or embryology and since this subject has already been touched upon in this book.

Figure 35, Chapter 4, Volume 1, summarizes the ribosomal gene structure. The 28 and 18 S genes are separated by nontranscribed spacers; they are flanked at their 5′-side by a transcribed spacer (600,000 daltons). The entire transcription unit is a 5×10^6-dalton DNA segment of constant length and composition. In contrast, the length and composition of the nontranscribed spacers are so variable that they may differ from one *Xenopus* individual to another (Reeder *et al.*, 1976; Wellauer *et al.*, 1976). This variability is largely due to the number of repeating subunits of fewer than 50 nucleotides located at the two ends of the spacer; its length can vary from as much as 1.8 to 5.5×10^6 daltons. It has been suggested (Wellauer *et al.*, 1976) that spacer heterogeneity results from unequal crossing over, and this is certainly an attractive possibility.

It has been known, since the work of Brown and Blackler (1972), that the rDNA spacers of two closely related *Xenopus* species, *X. laevis* and *X. borealis,* are so different in base composition that the two rDNAs can be separated by analytical ultracentrifugation; in hybrids between the two species, *X. laevis* rDNA is dominant over *X. borealis* rDNA. La Volpe *et al.* (1983) have reported that, in *Xenopus,* spacer rDNA contains a DNase-hypersensitive site; in the hybrids, this site is absent in *X. borealis* spacer rDNA. It was concluded, in agreement with other work, that well-conserved regions of the spacer are important for the transcription of the ribosomal genes. Reeder and Roan (1984) have shown that the *X. laevis* spacer possesses a larger number of copies of an enhancer sequence (see Chapter 4, Volume 1) than the *X. borealis* spacer. In addition, the *X. borealis* promoter reacts poorly with the transcription machinery. All this provides a satisfactory explanation for the dominance of the *laevis* ribosomal genes in hybrids between the two species.

The proteins present in the nucleoli of *Xenopus* oocytes are not yet well known. However, we know that they are different from those of the lampbrush chromosomes and the nuclear sap (Maundrell, 1975). Of major importance for transcription is the presence of RNA polymerase I in the nucleoli. It is probable that ribosomal proteins are also present in the *Xenopus* oocyte nucleoli. It is interesting to mention, in this respect, that the genes that code for these proteins are not amplified in *Xenopus* oocytes. There are two copies per haploid genome for three of the ribosomal proteins and four or five for three of the others (Bozzoni *et al.*, 1981).

As we have seen on several occasions, there is no coordination, in *Xenopus* at least, between 5 S RNA synthesis and synthesis of the 28 and 18 S high molecular weight species. In previtellogenic oocytes, 5 S RNA synthesis proceeds actively, but there is no production of 28 and 18 S rRNA. The same situation is found in anucleolate mutants of *Xenopus*. Finally, we already know that the 5 S genes are localized in the telomeres of the chromosomes, not in the nucleolar organizers. Nevertheless, since the three kinds of rRNA are ultimately incorporated into the ribosomes, the interesting problems raised by 5 S rRNA synthesis during *Xenopus* oogenesis will be discussed here. Again, it is mainly in D. Brown's laboratory that they have found solutions.

A first interesting point about the 5 S RNA genes in *Xenopus* is that this organism possesses three different major kinds of 5 S RNA. In the ovary, a major and a minor 5 S RNA are encoded by 24,000 and 2000 genes respectively (Brown *et al.*, 1977). They are slightly different from each other and can be easily distinguished from the somatic 5 S RNA, which is also coded by 24,000 gene copies per haploid genome. This somatic 5 S RNA differs from the ovarian 5 S RNA nucleotide sequence by only a few bases. The three kinds of 5 S RNA have 120 bases (M_r 40,000), which have been sequenced (Korn and Brown, 1978). As shown in Fig. 21, the 5 S genes are organized like the ribosomal genes in tandem repeats of a 500,000-dalton unit; 83% of each unit is a nontranscribed spacer. The larger of the two spacers is composed of A-T–rich repeated subunits of 15 bp; the length of the spacer depends upon the number of these subunits (Carroll and Brown, 1976). According to Brown and Brown (1976), the precur-

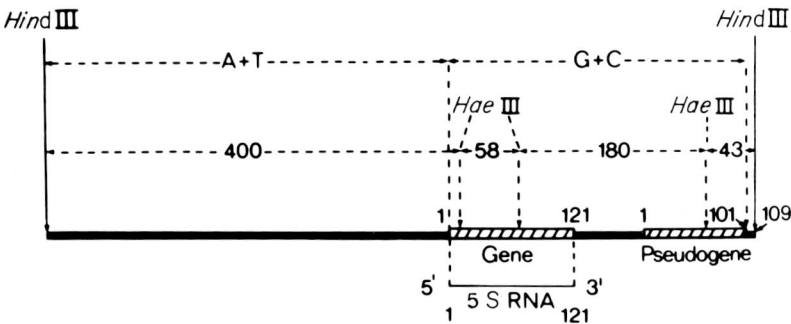

FIG. 21. Location of a 5 S RNA gene and a pseudogene of *Xenopus*. The gene has 121 residues, the pseudogene only 101. The location of the restriction sites after digestion with the *Hae*III and *Hind*III endonucleases are shown, as well as the (A + T) and (G + C)-rich regions. The gene region 1–121 codes for mature oocyte 5 S RNA; the pseudogene is not transcribed. [Jacq *et al.*, 1977. Copyright 1977 by M.I.T.]

sor of the 5 S RNA has 135 nucleotides. One of the 15-bp subunits of the repetitive spacer is transcribed together with the gene. In addition to the 5 S gene, the repeating unit contains a pseudogene that differs from the true gene by six bases and by the sequence of the 20 terminal nucleotides (Jacq et al., 1977). This pseudogene is not transcribed in vivo, but Miller and Melton (1981) made the unexpected finding that it is correctly transcribed in vitro by RNA polymerase III isolated from Xenopus GVs. The reasons for this difference remain to be elucidated. Miller and Melton (1981) suggest that the termination of transcription is inefficient in vivo. It thus seems that different regulatory mechanisms are at work in the GV and in the in vitro system.

Picard and Wegnez (1979), we mentioned, showed that in cytoplasmic 7 S particles present in previtellogenic oocytes, 5 S RNA is associated with an M_r 35,000 protein. This protein (which constitutes as much as 15% of the soluble proteins at that stage) has turned out to be of unusual interest, because it binds to an internal region of the 5 S gene. This binding results in gene activation (Pelham and Brown, 1980; Engelke et al., 1980; Honda and Roeder, 1980). Thus, binding of the M_r 35,000 protein to a control region located in the interior of the 5 S RNA gene stimulates its correct transcription by RNA polymerase III. It has therefore been called "transcription factor (TF) IIIA" by Engelke et al. (1980). This protein factor serves a dual function. It binds to a 50-bp intragenic region necessary for accurate transcription and, by binding to 5 S RNA, allows its storage in the cytoplasm. Interestingly, this protein disappears or is inactivated at maturation and is absent from somatic tissues. Loss of the protein at maturation stops ovarian 5 S rRNA synthesis and allows its replacement by somatic 5 S RNA. Its appearance in previtellogenic oocytes activates 5 S RNA synthesis and allows its accumulation. The stored 5 S RNA is incorporated in the ribosomes when synthesis of 28 and 18 S rRNAs begins at the onset of vitellogenesis. Work by Bogenhagen et al. (1982) has provided interesting information on the interactions between the 5 S RNA genes and their specific transcription factor. They found that when 5 S rRNA genes are added to an extract of Xenopus GVs, they assemble within a few minutes into stable, active transcription complexes. Stable, inactive transcription complexes are formed if the nuclear extract is deprived of the 5 S DNA-specific transcription factor and added with histones. Such stable, inactive transcription complexes are present in somatic cells, and their formation explains why the ovarian 5 S rRNA genes are repressed after maturation. Such stable transcription complexes might play an important role in maintaining the differentiated state in eukaryotic cells. As one can see, Xenopus oocytes have provided us, for the first time, with pure genes and with a system of gene regulation in which the control region is in the middle of the gene and not, as with the genes transcribed by RNA polymerase II, in the 5' flanking sequences. Finally, the unusual properties of the M_r 35,000 protein explain how previtellogenic oocytes accumulate 5 S RNA and why, at maturation and fertilization, the ovarian 5 S genes become silent and the somatic 5 S genes are activated.

D. Brown (1984) has recently discussed the biochemical and embryological role of stable complexes which repress or activate eukaryotic genes. Transcription is controled by a competition between positive transcription factors and general repressors like the histones. The various RNA polymerases bind to transcription complexes, which are probably replicated at the same time as genomic DNA. This multiplication of the complexes could be either symmetric or asymmetric; the latter would lead to the formation of two different cells and thus cause the diversity which is characteristic of embryonic differentiation. Determination (commitment) would result from the establishment of a stable active transcription complex, differentiation from the influence of diffusible effector molecules on a previously determined gene.

5. The Nuclear Sap (Nucleoplasm)

The main constituents of the nuclear sap are proteins of cytoplasmic origin. As we have seen from the experiments of De Robertis *et al.* (1978), many newly synthesized proteins selectively accumulate in the nucleus after injection of [^{35}S] methionine into the cytoplasm of *Xenopus* oocytes. Besides these "nuclear-specific" (karyophilic) proteins, many others are common to both the nucleus and the cytoplasm. A few years ago, about 70 distinct proteins were separated from isolated GVs by two-dimensional gel electrophoresis; this number is likely to increase in the future with improved methods of protein separation.

Among the proteins that are shared by the GV and the cytoplasm, actin is the most common. The central core of the GV is much more viscous than the rest of the nuclear sap; this is due, largely or entirely, to the presence of actin in this central "nuclear gel." According to Clark and Merriam (1977), actin represents 6% of the total proteins present in the nuclear sap; its concentration reaches 25% in the nuclear gel, where it is the major protein. Cytochalasin B treatment does not destroy the gel. A paper by Clark and Rosenbaum (1979) has shown that 63% of GV actin is in the unpolymerized G form; the remaining 37% is in filaments of F-actin and is located in the nuclear gel. Finally, Gounon and Karsenti (1981) reported that, in *Pleurodeles,* 90% of the actin present in the GV is G-actin. Chelation of calcium ions induces its polymerization, so that 50% of GV actin is transformed in the F form; addition of phalloidin induces the formation, in the nuclear sap, of a network of actin cables that are composed completely of F-actin. These cables are composed of actin MFs associated with RNP-like particles, which are believed to contain myosin. Finally, according to Vandekerckhove *et al.* (1981), *Xenopus* oocytes contain three distinct molecular forms of actin; they are found in the same proportions in the GV and the oocyte cortex.

The major protein of the GV (where it was discovered) is nucleoplasmin, which represents almost 10% of nuclear sap proteins (Krohne and Franke, 1980a,b; Mills *et al.*, 1980). This very interesting protein has already been mentioned in Chapter 4, Volume 1, because it is a nucleosome assembly factor. It is an acidic protein with M_r 35,000 existing in various forms that differ in their degree of phosphorylation. According to Mills *et al.* (1980), nucleoplasmin behaves like a typical nuclear-specific or karyophilic protein. After injection into the cytoplasm, it moves quickly into the GV, where it accumulates. Dingwall *et al.* (1982) have reported that the nucleoplasmin molecule is composed of two different domains: a core and a tail; the former is necessary for the entry of nucleoplasmin, but not for its retention in the GV.

Histones are needed for nucleosome assembly. Kleinschmidt and Franke (1982) have found that the core histones exist far in excess of what is required for this purpose in the GVs of *Xenopus* oocytes. These abundant histones are in solution in the nuclear sap and are not bound to chromatin or nucleoplasmin. Histones H3 and H4 are associated with a couple of very acidic proteins (M_r = 110,000). Amphibian oocytes also possess a large pool of high mobility group (HMG) proteins; they are not bound to chromatin, and are equally distributed between the nucleus and the cytoplasm. A high mobility group protein called HMG-A (M_r = 25,000) accumulates in the GV during oogenesis. We have seen that injection of antibodies against this protein induces loop retraction and arrest of transcription in the lampbrush chromosomes (Kleinschmidt *et al.*, 1983).

The GV contains another nucleosome assembly factor, topoisomerase I, which is absent from enucleated *Xenopus* oocytes (Attardi *et al.*, 1981). The enzyme that plays the major role in DNA replication, DNA polymerase-α, is also accumulated in the GV, despite the fact that there is no DNA synthesis in the oocyte nucleus after rDNA amplification is completed; 80–90% of the enzymatic activity has been found in the GVs dissected out of *Xenopus* oocytes (Martini *et al.*, 1976).

The three RNA polymerases (I, II, and III) are particularly abundant in the GV of *Xenopus* oocytes, where there are as many as 4×10^9 molecules of RNA polymerase III, according to Roeder (1983); only part of the enzymatic activity is bound to the chromosomes and the nucleoli. The excess is in solution in the nuclear sap. Isolated, crushed GVs are, therefore, becoming a favorite system for biochemists interested in the *in vitro* transcription of any kind of DNA. The *Xenopus* oocyte easily provides them with very active preparations without the necessity of lengthy enzyme isolation and purification. One of the functions of RNA polymerase III in all nuclei is the synthesis of the tRNAs. tRNAs require the addition of a terminal -CCA sequence to function. The enzyme that adds -CCA to the tRNA (tRNA-nucleotidyltransferase) is more abundant in the cytoplasm of *Xenopus* oocytes than in their nuclei (Solari and Deutscher, 1982).

In addition, the GV contains small amounts of very heterogeneous RNAs, with M_r values ranging from 5×10^6 to 30,000. They are a complex mixture of primary transcription products, fragments arising from the processing of macromolecular precursors, and finished molecules of mRNAs, rRNAs, tRNAs, which are on their way for export in the cytoplasm, and snRNAs. Many years ago, we detected the presence of RNase activity in isolated frog GVs (Brachet, 1939). It seems likely that in the future, many different species of RNases will be found in isolated GVs where they probably have a role in RNA processing: Solari and Deutscher (1983) reported that *Xenopus* oocytes contain multiple RNases of different specificities.

C. *Xenopus* Oocytes as Test Tubes

Because of the intelligence, skill, and ingenuity of John B. Gurdon, *Xenopus* oocytes have become excellent test tubes for molecular biologists (Gurdon, 1974). The methodology is simple. The substance of interest is microinjected into either the cytoplasm or the GV of a full-grown oocyte. This first step does not require exceptional skill or costly micromanipulation equipment because of the oocyte's large size (1.2 mm in diameter). Following this, the results can be determined using standard biochemical or cytochemical techniques.

We have already seen that microinjection of labeled proteins into *Xenopus* oocytes yielded very important results in the study of nuclear membrane permeability. We have also seen that more than 90% of the ribosomes are "unprogrammed" in these oocytes and that there are about 10^{12} ribosomes in an oocyte, meaning that if a purified mRNA is injected into their cytoplasm, it should easily bind to free ribosomes and be translated. Finally, we have mentioned that the GV is a rich source of RNA polymerases and nucleosome assembly factors. Therefore, injection of DNA molecules into the GV should be followed by transcription and assembly into chromatin. We shall examine briefly the main results obtained with these approaches and conclude with experiments in which adult nuclei were injected into *Xenopus* oocytes.

1. Translation of mRNAs after Injection into Cytoplasm

About 10 years ago, Gurdon *et al.* (1971) and Lane *et al.* (1971) reported that injection of rabbit hemoglobin mRNA into *Xenopus* oocytes is followed by intensive synthesis of rabbit hemoglobin (Fig. 22); each mRNA molecule can give rise to as many as 100,000 hemoglobin molecules (Gurdon *et al.*, 1973). Synthesis continues for weeks even if the oocyte was enucleated before mRNA injection. These experiments show that the oocyte possesses the machinery needed for the translation of a "foreign" mRNA, as well as mechanisms that protect the injected mRNA against degradation by endogenous RNases.

This experiment was a major breakthrough at the time. Since 1971, *Xenopus* oocytes have become classic models for the analysis of all kinds of mRNA preparations. mRNAs for crystallins, immunoglobulins, ovalbumin, interferon, plasminogen activator, acetylcholine receptor, thyroglobulin, and a host of other

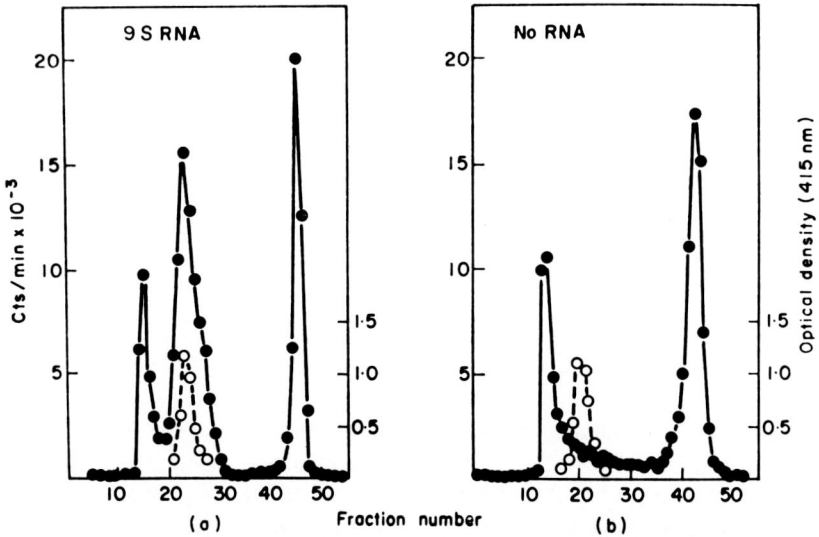

FIG. 22. *Xenopus* oocytes injected into the cytoplasm with rabbit globin 9 S mRNA and hemin synthesize rabbit hemoglobin. (a) Oocytes injected with globin 9 S RNA and cultured for 6 hr in [³H]histidine; (b) controls. Open circles correspond to the optical density of rabbit hemoglobin added as a marker; closed circles refer to counts per minute. Separation of the supernatant from homogenized and centrifuged oocytes on G100 Sephadex. [Lane *et al.*, 1971.]

proteins have been tested in this way. As already mentioned, the *Xenopus* oocyte system has also proved useful in the analysis of the role played by the terminal poly(A) sequence found in almost all mRNAs (reviewed by Littauer and Soreq, 1982). Experiments by Huez *et al.* (1974) and Marbaix *et al.* (1975) have shown that hemoglobin synthesis stops very quickly in injected oocytes when the poly(A) tail of globin mRNA is enzymatically removed. Addition of a terminal poly(A) sequence restores the stability of the injected globin mRNA. One of the functions of the terminal poly(A) sequence, therefore, is to ensure greater stability to the mRNAs, which have moved from the nucleus into the cytoplasm. However, it should be pointed out that the results obtained with hemoglobin mRNA are not valid for all messengers. According to Soreq *et al.* (1981), mRNA stability if poly(A) is removed before injection into *Xenopus* oocytes varies from one mRNA to another. It has also been shown by McCrae and Woodland (1981) that the half-life of reovirus RNA, which has no poly(A) tract, is more than 3 days in the *Xenopus* oocyte cytoplasm. They suggest that the 5'-terminal cap might play a role in mRNA stability, but there may be important differences between a viral RNA and a eukaryotic mRNA.

Another interesting approach is the injection of suppressor tRNAs into *Xenopus* oocytes (Bienz *et al.*, 1980, 1981; Martin *et al.*, 1981). Suppressor tRNAs allow the reading of the termination codons *amber, ochre,* and *opal* in bacteria and lower eukaryotes. Simultaneous injection of globin mRNA and yeast suppressor tRNAs allows the reading of the UGA termination codon with the accumulation of a large read-through product (i.e., globin with an additional polypeptide sequence). We have seen that, in the mitochondrial DNA code, the triplet UGA is read "tryptophan" instead of "stop." Thus, injection in the oocyte of yeast mitochondrial tRNA^{trp} together with an mRNA suppresses the termination of the growing polypeptide chain, provided that an acylation enzyme from *Escherichia coli* is simultaneously injected into the oocytes. Thus, the *Xenopus* oocyte provides interesting possibilities for a refined analysis of translational control in a living system.

There has been an increasing interest in the fate of the proteins synthesized by *Xenopus* oocytes following injection of heterologous mRNAs (reviewed by Lane, 1981 and Colman, 1982) because it has been discovered that, in contrast to hemoglobin, certain proteins are excreted in the medium by the oocytes. Protein excretion is inhibited by a mixture of colchicine and cytochalasin B, and under such conditions the newly synthesized proteins are degraded (Colman, 1981). Lane (1981) and Colman *et al.* (1981a,b) have shown that secretory proteins synthesized by the oocytes after injection of a heterologous mRNA are modified (glycosylated, for instance), taken up by ER vesicles, and excreted. As in secretory cells, a signal sequence is required for entry of the protein into a vesicle and, ultimately, for secretion by exocytosis. Many proteins, either endogenous or synthesized after injection of an mRNA into *Xenopus* oocytes, are excreted by the oocytes, according to Mohun *et al.* (1981). The proteins that will be secreted and those that will not are found in different compartments in mRNA-injected oocytes. For instance, while rabbit globin is found in the cytosol, carp proinsulin is recovered from membranes after injection of their respective mRNAs (Rapoport, 1981). The processing of the protein synthesized by the oocytes in response to mRNA injection may be very precise and complex. For instance, if *Xenopus* oocytes are injected with a mixture of the two forms of albumin mRNAs, they secrete two different albumins with M_r 68,000 and 74,000 (Westley *et al.*, 1981). More complex (as one would expect) are the consequences of immunoglobin (Ig) mRNA injection into oocytes. It is followed by synthesis and excretion of tetrameric Ig, but there is no excretion of free light chains. Injection of Ig heavy chain mRNA is followed by accumulation of heavy chain dimers without secretion unless one simultaneously injects light chain mRNA (Valle *et al.*, 1981). More recently, Colman *et al.* (1982) injected the mRNAs coding for the heavy (H) and light (L) chains of immunoglobins at the opposite poles of the oocyte; after 24 hr, they observed the synthesis of tetrameric (H_2L_2) Ig. Immunoglobulin production was marooned if only one of the two mRNAs was injected, but was rescued if the mRNA coding for the complementary chain was injected after 24 hr. The *Xenopus* oocyte system is becoming more and more popular among immunologists and neurobiologists. Without going into details, I should mention the work of Severinsson and Peterson (1984) on class I transplantation antigens; Liu and Orida (1984) on the IgE receptor; Pure *et al.* (1984) on Fc receptors (they bind the Fc domain of the immunoglobulins to the cell membrane); Sumikawa *et al.* (1984) on the induction of chloride channels and acetylcholine receptors after injection of RNA fractions from *Torpedo* electric organs; and Houamed *et al.* (1984) on the expression of functional receptors for GABA, glycine and glutamate. The proteins synthesized by the injected oocyte move to its plasma membrane where the receptors are incorporated and function efficiently. Very curious also are the observations of Hurkman *et al.* (1981). They found that after injection of the mRNA coding for zein, a reserve protein of maize, zein is synthesized and deposited in protein bodies that are strikingly similar to those present in maize endosperm. Injection of poly(A)$^+$ RNAs from bean cotyledons is followed by the synthesis of two storage proteins (legumin and vicilin). However, in this case, the translation products, the propolypeptides, are secreted by the oocyte without proteolytic cleavage. Secretion is not inhibited by tunicamycin, an inhibitor of protein glycosylation (Bassüner *et al.*, 1983).

Also worth mentioning are the results of Sumikawa *et al.* (1981) because they were obtained with an mRNA coding for a protein of great physiological interest, the acetylcholine receptor. After translation in *Xenopus* oocytes, the protein is glycosylated and sequestered inside membranes. After cleavage of the propeptide, the receptor is incorporated into the plasma membrane in a physiologically active form that is capable of binding α-bungarotoxin (Barnard *et al.*, 1982; Miledi *et al.*, 1982; Gundersen *et al.*, 1984). As one can see, it is not easy to fool a *Xenopus* oocyte; it responds in a clever and faithful way to all the tricks we imagine could perturb its protein synthesis and processing abilities. At first sight, it displays little originality in its responses. The oocyte seems to be incapable of selecting a specific mRNA. Does it use, in an indiscriminate way, all of the messages it has received, whether they were of exogenous or endogenous origin? In order to answer this question, two distinct mRNAs were injected together into a *Xenopus* oocyte. Cutler *et al.* (1981) simultaneously injected ovalbumin and lysozyme mRNAs and found that the secretion rate of lysozyme was 12-fold higher than that of ovalbumin. A similar approach was used by Richter and Smith (1981), who injected into *Xenopus* oocytes mRNAs coding for hemoglobin and zein and found that the synthesis of globin is at least five times greater than that of zein. This is due to the fact that injected globin mRNA is found in the polysome fraction, while zein mRNA is located in a membrane fraction. The limiting factor in mRNA translation is the amount of rough ER present in the oocyte, and not message availability (Richter *et al.*, 1982a,b). This is shown by the fact that, if dog pancreas ER membranes are added to zein mRNA prior to injection, zein synthesis is greatly stimulated (Richter and Smith, 1981). These experiments show that mRNAs injected into oocytes are distributed between two different compartments before the initiation of translation: membrane bound or free. Another more intriguing possibility is that the oocyte possesses message-specific factors. It is likely that continuation of such experiments will lead to a better understanding of the still mysterious molecular mechanisms of translational control.

Xenopus oocytes are also capable of directing the newly synthesized protein toward one cellular compartment or another. After injection of RNAs extracted from either *Xenopus* liver or locust fat body, respectively, *Xenopus* and locust vitellogenins are synthesized. After synthesis, both are sequestered in vesicles and excreted, but only *Xenopus* vitellogenin is accumulated in the yolk platelets after conversion into lipovitellin and phosvitin (Lane *et al.*, 1983a). According to Lane *et al.* (1983b), the cytoplasm of *Xenopus* oocytes can correct errors in the compartmentation of proteins; this is often facilitated by the presence in the newly synthesized protein of a detachable signal sequence (see Chapter 3, Volume 1). Miscompartmentized secretory proteins are degraded by proteases.

In conclusion, *Xenopus* oocytes are an ideal system for the study of mRNA translation and the fate of the newly synthesized proteins (which, of course, cannot be followed in *in vitro* systems of protein synthesis). Its advantage over oocytes from other species is that the *Xenopus* oocyte's size is ideal. It is large enough to allow injection, and it is not overloaded with inert yolk platelets, like the larger oocytes of other amphibians, fishes, and birds. Mammalian eggs could also be used, since Ebert and Brinster (1983) have obtained the translation of α-globin mRNA after its injection into mouse eggs. However, they are so small that they will never become a favorite material for biochemists who need an abundant material. The main—and very real—significance of Ebert and Brinster's experiments is to show that mouse eggs, like *Xenopus* oocytes, possess a store of unprogrammed ribosomes.

2. Transcription and Chromatin Assembly after DNA Injection into the GV[1]

In 1977, it was discovered that injection of cloned eukaryotic genes into the GV of *Xenopus* oocytes is followed by intense and prolonged (lasting for several days) transcription (Mertz and Gurdon, 1977; De Robertis and Mertz, 1977; Brown and Gurdon, 1977). For instance, in the pioneer work of Brown and Gurdon (1977), it was shown that transcription of 5 S DNA after injection into the GV is perfectly faithful, while injection into the cytoplasm is useless. The injected DNA is not transcribed and is progressively degraded. These results could have been expected, since, as we have seen, the three major forms of RNA polymerase are accumulated in the nucleus. In some cases, the mRNA transcribed on the injected gene moves into the cytoplasm, where it is correctly translated into the corresponding protein. This occurs, for instance, after injection of SV40 virus DNA (De Robertis and Mertz, 1977). It should be mentioned

[1]Reviewed by Etkin (1982).

that injection of the cloned 5 S gene into the GV or female pronucleus of mouse oocytes or fertilized eggs is also followed by correct transcription (Brinster *et al.*, 1981).

The consequences of injection into the GV of cloned ribosomal (28 and 18 S) genes have been analyzed by Trendelenburg and Gurdon (1978) using Miller's spreading technique (Fig. 23). One can see that part of the circular DNA molecule has the typical Christmas tree configuration and obviously corresponds to the ribosomal cistrons. The rest is made up of the untranscribed spacer and plasmid DNA; plasmid DNA is organized in nucleosomes, while spacer DNA is not. Single-stranded and circular DNA, including mitochondrial DNA, are not transcribed after injection into the GV, which is in contrast to eukaryotic genes (Zentgraf *et al.*, 1979). Particularly interesting among the latter are the ovalbumin (Wickens *et al.*, 1980) and histone (Grosschedl and Birnstiel, 1980) genes because they could be subjected to various manipulations (*in vitro* mutations) prior to injection into the GV. The ovalbumin gene is completely transcribed in the GV. The oocyte nucleus then splices the primary transcript, accurately removing the seven introns; processing continues by addition of the cap sequence at the 5'-end and of a poly(A) tract at the 3'-end. Ovalbumin mRNA

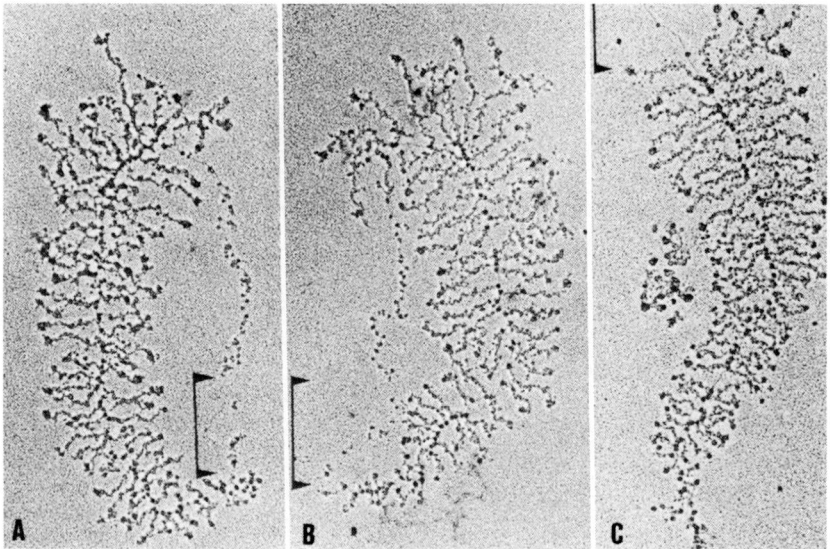

Fig. 23. Ribosomal DNA transcription patterns of injected (into the GV) recombinant plasmid circles (A, B) and an endogenous rDNA segment (C). Only the rDNA gene portion shows transcription units (gradient arrangement of lateral fibrils ending with dense knob-like structures). Plasmid DNA has the beaded appearance characteristic of nucleosome organization. The spacer segments (arrows) have a predominantly unbeaded structure. [Reprinted by permission from *Nature* **276,** 147. Copyright © 1978 Macmillan Journals Limited.]

then moves into the cytoplasm, where it is translated; ovalbumin is taken up in vesicles and finally excreted from the oocyte. If one deletes the 5′-flanking sequences, including the TATA box (see Chapter 4, Volume 1), and the cap sequence before injection, normal transcription takes place. These control sequences are thus not required for transcription of the ovalbumin gene in the GV, while they are needed for its *in vitro* transcription. Curiously, if, instead of injecting the ovalbumin gene, one injects ovalbumin cDNA (i.e., the DNA copy by reverse transcriptase of ovalbumin mRNA) into the GV, it is transcribed, but there is no ovalbumin synthesis. This cDNA differs from the gene by the absence of the seven introns. It might be that splicing of the introns, in the case of the ovalbumin gene at least, is necessary for the passage of the mRNAs from the nucleus to the cytoplasm (Wickens *et al.*, 1980). Using a comparable approach, Grosschedl and Birnstiel (1980), Grosschedl *et al.* (1981), and Birchmeier *et al.* (1982) analyzed in great detail the control of transcription in sea urchin histone genes by injecting partially deleted or chemically mutated genes into the *Xenopus* oocyte GV. These authors have facilitated the methodology by working on centrifuged oocytes in which the GV can be seen as a transparent spherule at the surface of the centripetal pole. In addition to identifying three control regions upstream from the sea urchin histone genes (see Chapter 4), Birnstiel and his colleagues have done interesting coinjection experiments. Transcription of the sea urchin histone H3 gene after injection into the *Xenopus* oocyte nucleus is incomplete. It stops before it reaches the 3′-end of the gene. Coinjection of the gene and a 12 S protein isolated from the chromatin of sea urchin morulae corrects the inefficient termination of the transcription (Stunnenberg and Birnstiel, 1982). More recently, Galli *et al.* (1983) achieved the same result by injecting a small (60-nucleotide) poly(A)$^-$ RNA together with the sea urchin histone H3 gene. This termination factor, which is probably a small nuclear ribonucleoprotein, allows complete transcription of the histone gene. Site-directed *in vitro* mutagenesis, followed by injection into the *Xenopus* oocyte GV, allowed McKnight and Kingsbury (1982) to demonstrate the existence of three different control regions in the 105 nucleotides upstream from the initiation codon of the thymidine kinase gene. This extends to this gene the findings of Birnstiel's group on sea urchin histone genes.

A refined analysis of tRNA genes transcription and processing, using similar techniques, has been carried on, mainly by E. De Robertis (De Robertis and Olson, 1979; Melton *et al.*, 1980; De Robertis *et al.*, 1981). Transcription of the tRNA genes takes place in the GV, where it is catalyzed by RNA polymerase III. Processing of the transcribed tRNA precursor also takes place in the GV, which involves removal of sequences at the 5′- and 3′-ends, removal of an intron, and addition of the terminal CCA sequence (Fig. 24). Even processing of the dimeric precursor for two different tRNAs can occur in the GV, which obviously contains the machinery needed for faithful tRNA production.

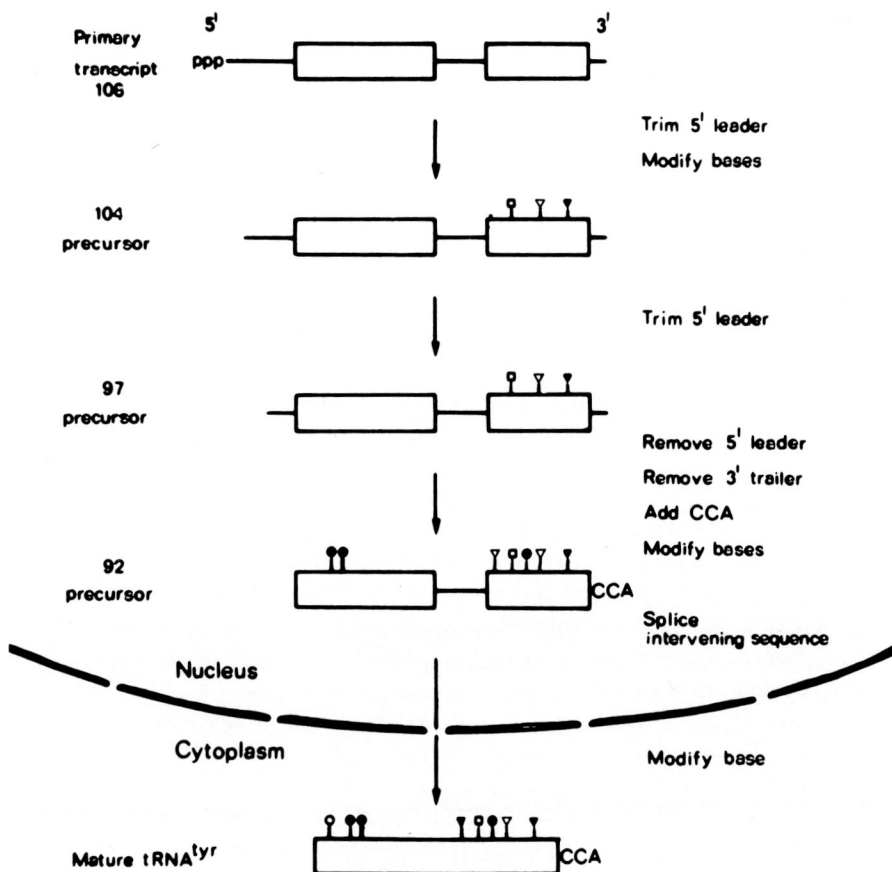

FIG. 24.   RNA processing of tRNA^tyr precursors into mature tRNA^tyr in injected *Xenopus* oocytes. Almost all of the steps occur in the oocyte nucleus. The primary transcript is a precursor 108 nucleotides in length with a 5′ leader, an intervening sequence (intron), and a 3′ trailer. The 5′ leader is removed in three stages, the last of which is accompanied by excision of the 3′ trailer and addition of the 3′ CCA. The base modifications that have been analyzed are indicated (□, 5 methylcytosine; ▽, pseudouridine; ▼, 1-methyladenosine; ●, dihydrouridine; ○, an uncharacterized G modification. [Reprinted by permission from *Nature* **284**, 143. Copyright © 1980 Macmillan Journals Limited.]

In vitro induction, by chemical manipulation, of specific mutations in pure genes prior to their injection into the GV of *Xenopus* oocytes promises to become a powerful tool for the analysis of the control of gene activity. For instance, using this methodology, it has been shown that the promoter of the tRNA^pro gene is formed by three discontinuous regions inside the coding sequence (Ciliberto *et al.,* 1982) and that this gene is poorly transcribed in the GV after *in vitro* mutations in the coding region (Ciampi *et al.*, 1982). It has also been possible to follow the effects of site-directed *in vitro* mutations on the movement from the nucleus to the cytoplasm of injected tRNAs. Zasloff *et al.* (1982) found that a point mutation (a G to T transversion induced *in vitro* prior to injection) prevents a tRNA^{met} from leaving the nucleus. More recently, Zasloff (1983) showed that the passage through the nuclear membrane of injected tRNAs is a carrier-mediated process and not a simple diffusion through the nuclear pores. After injection of optimal amounts of tRNAs into the GV, as many as 190×10^7 molecules leave the nucleus every minute; an A to G base substitution in loop IV of the tRNA reduced the rate of transport 20 times. The authors suggested that this loop might bind to the eight ribosomes that surround the octogonal pores. In any event, site-directed *in vitro* mutagenesis, combined with injection of the modified RNA into the *Xenopus* GV, is likely to provide important information on the transport of nucleic acids from the nucleus to the cytoplasm.

Bakken *et al.* (1982) injected into *Xenopus* oocyte nuclei fragments of rDNA that had been digested with specific endonucleases and analyzed the outcome using Miller spreading. This allowed them to identify the promoter and the termination signal for rDNA transcription. According to Moss (1983), the rDNA spacer is transcribed both *in vivo* and after injection into the *Xenopus* GV; the transcription products would modulate the transcription of the ribosomal genes located upstream. Another interesting result is that the rDNA of *X. borealis* is perfectly transcribed after injection into a *X. laevis* GV; this was shown by chromatin spreading, according to Oscar Miller (Morgan *et al.*, 1982). Since RNA polymerase I present in the GV does not recognize a difference between the two rDNAs, this cannot be the reason why only *X. laevis* rDNA is expressed in the hybrids between the two species. On the other hand, RNA polymerase II recognizes the U_1 and U_2 snRNA genes from their pseudogenes. Only the first ones are transcribed after injection into the GV (Nojima and Kornberg, 1983). Finally, Vardimon *et al.* (1982) used *in vitro* mutagenesis for the study of DNA methylation. They injected into the GV viral DNA (from an adenovirus) in both methylated and unmethylated forms. They found that the viral DNA remains the same, suggesting that the GV has little or no DNA methylase activity. Interestingly, only unmethylated viral DNA was transcribed in the oocyte nucleus.

Injection into the oocyte nucleus of an mRNA precursor can provide valuable information about its processing, as shown by the work of Green *et al.* (1983). They found that human β-globin mRNA is accurately spliced in the GV. Splicing requires the capping of the pre-mRNA, while polyadenylation at the 3'-end is not necessary. This demonstrates that transcription and splicing are not necessarily coupled. It was also shown that a point mutation in the splicing region prevents splicing of the pre-mRNA in the oocyte nucleus.

As one can see, injection of genes of pre-mRNA, combined with *in vitro* mutagenesis ("surrogate genetics"), has already provided a wealth of information about transcription control sequences, splicing of the primary transcript, passage of RNAs through the nuclear pores, etc. There is no doubt that continued progress in this line of research will lead to unexpected and exciting results.

The assembly of injected DNA molecules that cannot be transcribed inside the GV can be followed in nucleosomes. We have already seen this in regard to plasmid DNA in the experiments of Trendelenburg and Gurdon (1978). Further work on circular plasmid DNA by Scheer *et al.* (1980) has shown that, after injection into the GV, supranucleosomal globular structures made of dense, higher-order packing forms (Fig. 26, Chapter 4, Volume 1) can be seen under the electron microscope. After injection of mitochondrial DNA into the *Xenopus* oocyte nucleus, the injected molecules appear as circles of chromatin that are devoid of transcription units (Zentgraf *et al.*, 1979, 1982). Another result may be obtained after injection of circular DNAs into oocyte nuclei. After injecting SV40 DNA, Mertz and Miller (1983) observed the catenation of this circular viral DNA in a DNA network; decatenation occurred after some time. Catenation was also observed after injection into the oocyte cytoplasm; it probably results from the action of DNA topoisomerases. Finally, Carroll (1983) has shown that the oocyte nucleus possesses the machinery required for genetic recombination. He injected into the GV two phages that were genetically marked and observed recombination between them. Injection of cloned 5 S RNA and H4 genes into the oocyte nucleus is followed by the formation of active and stable minichromosomes; this requires the binding to the injected DNA of endogenous factors present in the germinal vesicle (Gargiulo *et al.*, 1984).

As already discussed, the main factors involved in nucleosomes assembly are DNA topoisomerase I and nucleoplasmin, which are both abundant in the GV. Self-assembly of nucleosomes can take place when DNA and histones are mixed together under adequate conditions, but such *in vitro* assembly is greatly facilitated by addition of DNA topoisomerase I and nucleoplasmin, which react with DNA and histones, respectively. According to Laskey and Earnshaw (1980), it is the pentamer of nucleoplasmin that is the active form. The same authors suggest that, when DNA molecules are injected into the GV, they are first relaxed by topoisomerase I; this would allow DNA to react with the large store of histones (sufficient for 20,000 *Xenopus* nuclei) present in the oocyte. The main role of nucleoplasmin, which remains active in the presence of excess histones, would be to make the nucleosomes insoluble in the nuclear sap environment.

Other factors required for chromatin assembly will probably be discovered. For instance, Gottesfeld and Bloomer (1982) reported that the transcription factor TF IIIA is required for the *in vitro* assembly in chromatin of 5 S genes mixed with histone in chromatin.

In conclusion, the GV of *Xenopus* oocytes is a very favorable, if not ideal, material for the study of DNA transcription and hnRNA processing. The results obtained so far, with the ovalbumin and histone genes, show that injection of chemically or enzymatically modified DNA molecules into the GV might greatly improve our understanding of the mechanisms that control gene transcription. Thus far, the *Xenopus* system is unique for the analysis of nucleosome assembly in the nucleus of a living cell.

3. Injection of Whole Nuclei from Adult Cells into Xenopus Oocytes[2]

The first experiments, which were designed to improve our understanding of the control mechanisms by the cytoplasm on nuclear activities, were performed by J. Gurdon in 1968. He injected *Xenopus* adult brain nuclei (which synthesize RNA *in situ,* but not DNA) into the cytoplasm of *Xenopus* oocytes and eggs and obtained the following results (Fig. 25). If the brain nuclei are injected into full-grown oocytes, they swell and their nucleoli become more conspicuous; they incorporate uridine, but not thymidine. If the nuclei are injected into oocytes undergoing maturation (where there is neither DNA nor RNA synthesis and where the lampbrush chromosomes are undergoing active condensation), they incorporate neither uridine nor thymidine, but their chromatin condenses and some of the nuclei may form chromosomes and mitotic spindles. Finally, if the nuclei are injected into unfertilized eggs (where there is almost no RNA synthesis, but active DNA synthesis if the egg is activated by pricking), RNA synthesis ceases, the nucleoli become inconspicuous, and thymidine is incorporated into the chromatin of the injected nuclei. In other words, the injected nuclei behave exactly like the oocyte's nucleus; clearly, cytoplasmic factors decide whether the foreign nucleus will synthesize RNA, DNA, or no nucleic acid at all. These important results are in complete agreement with what has been said in the preceding chapter about the reactivation of chick erythrocyte nuclei after fusion with cells that synthesize either RNA alone or both kinds of nucleic acids. They agree with the conclusions drawn by A. Brachet (1922) some 60 years ago based on studies of polyspermic sea urchin eggs, and can be extended to the *Xenopus* system. Sperm nuclei in these eggs always assume the same morphology as the egg nucleus (Fig. 26). They remain condensed if the egg nucleus is dividing and swell if the egg nucleus is in a swollen stage. It had already been concluded that the morphology of the nuclei is controlled by cytoplasmic events and the phenomenon was called "mise à l'unisson" by Brachet, but Gurdon's (1968) work showed marked progress over the early pioneer studies. Due to the advent of autoradiography, the analysis had shifted from morphology to molecular biology.

Study of the control exerted by the nucleus on nuclear activities made remarkable progress with the experiments of Gurdon *et al.* (1976a). They injected HeLa cell nuclei into *Xenopus* oocytes and found that they could survive and synthesize

[2]Reviewed by Etkin (1982).

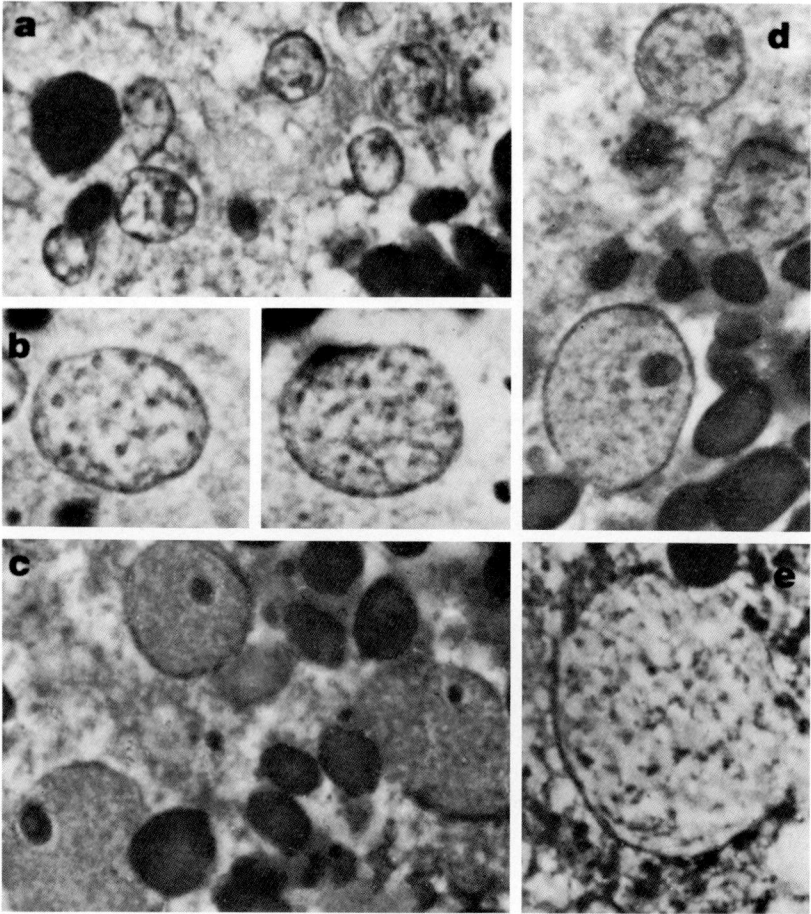

FIG. 25.   Nuclei from adult frog brain injected into *Xenopus* oocytes. (a) Unenlarged nuclei 3 hr after injection. (b) Nuclei 36 hr after injection. The nuclei are somewhat enlarged and the chromatin partly dispersed; the nucleoli are not yet apparent. (c, d) Nuclei 48 hr after injection. Enlarged nuclei with dispersed chromatin and very large nucleoli. (e) Much enlarged nucleus from the brain of a recently metamorphosed frog; oocyte fixed 72 hr after injection. [Gurdon, 1968.]

for several weeks all types of RNA; proteins present in the oocyte's cytoplasm were transported into the nuclei, resulting in swelling. Continuous production of mRNAs by the injected HeLa nuclei allowed analysis of the proteins synthesized by these nuclei in the oocyte cytoplasm. The result was that only 3 HeLa proteins out of 25 were expressed in the injected oocytes, whether or not they had been previously enucleated (De Robertis *et al.*, 1977). Thus, the synthesis of many proteins (but not all) that the HeLa genes were directing in their own cytoplasm was suppressed after nuclear transplantation into a *Xenopus* oocyte, indicating

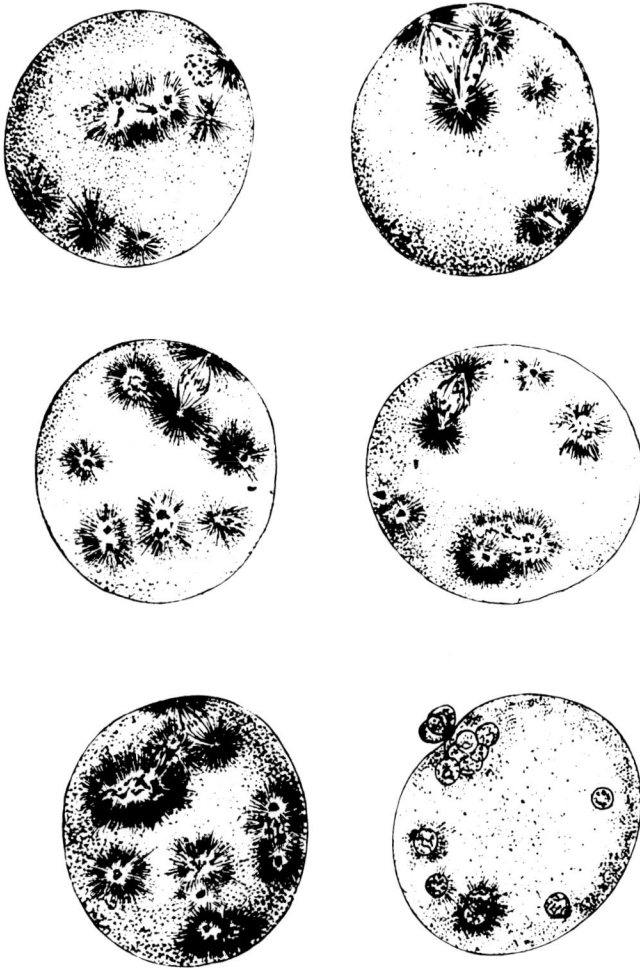

Fɪɢ. 26. Polyspermic sea urchin eggs. When the egg nucleus is in metaphase, all of the supernumerary sperm nuclei are surrounded by asters and have formed chromosomes. When (bottom, right) the egg nucleus is in interphase, all of the supernumerary sperm nuclei are swollen and are surrounded by an intact nuclear membrane. The morphology of the nuclei thus depends on cytoplasmic factors. [A. Brachet, 1922.]

that the oocyte cytoplasm has the remarkable property of reprogramming the expression of individual HeLa genes. Reprogramming gene expression in injected nuclei is better demonstrated in the experiments of De Robertis and Gurdon (1977), in which adult kidney nuclei of *Xenopus* were injected into oocytes of *Pleurodeles*. Analysis of the proteins synthesized by the injected oocytes showed that the *Xenopus* adult nuclei synthesized some of the *Xenopus* oocyte's typical proteins. These proteins had not been synthesized by the kidney cells for a number of months. Simultaneously, synthesis of the specific kidney proteins came to a halt. The cytoplasm of the oocyte thus contains factors that are capable of reactivating certain genes (in particular, genes that were active during oogenesis) and inactivating others (that are active in the adult). Whether these factors are proteins or RNAs remains to be established, but their identification would greatly increase our understanding of the control of gene expression in eukaryotic cells.

Weisbrod *et al.* (1982) and Wakefield *et al.* (1983) have studied the biochemical consequences of the injection of adult nuclei into the GV of *Xenopus* oocytes. They observed a selective activation of the few adult genes that synthesize low-molecular-weight RNAs due to RNA polymerase III. In particular, a gene family called OAX is activated; its transcripts are about 180 nucleotides long. This stimulation of RNA synthesis is not due to displacement of the histones and HMGs from the chromatin of the injected nuclei.

Finally, Korn and Gurdon (1981) have tried to shed some light on the control of 5 S RNA synthesis in oocytes and adult cells. As we have seen, ovarian and somatic 5 S RNAs differ from each other and are coded by different sets of genes. Korn and Gurdon first worked out a method for the accurate separation of the two 5 S RNA species. They then injected 100–200 adult *Xenopus* nuclei into the GVs of recipient oocytes, where the 5 S genes were accurately transcribed. The results were not as clear-cut as was hoped. The ovarian 5 S genes were reactivated in the GVs of oocytes from certain females, but not in those of others. If the nuclei were treated with 0.35 M NaCl (in order to remove some of the non-histone proteins) prior to injection into the GV, reactivation of the ovarian 5 S genes took place in oocytes from all females. These experiments clearly demonstrate that inactivation of the ovarian 5 S genes in the adult is labile. They further suggest that non-histone proteins are involved in their inactivation after fertilization. As pointed out by Korn and Gurdon (1981), the fact that reactivation occurs in oocytes from only some females is not necessarily a drawback. Comparison of the two kinds of females might help us to identify the factors that control the activity of the 5 S genes. This possibility has been explored by Korn *et al.* (1982). They injected *Xenopus* erythrocyte nuclei together with cytoplasmic extracts from oocytes that reactivate or fail to reactivate the oocyte-type 5 S genes injected into *Xenopus* oocytes. They found that only the extracts from reactivating oocytes induce the transcription of these 5 S genes. Since the extracts of nonreactivating oocytes had no inhibitory effects, it can be concluded that the cytoplasm of the reactivating oocytes exerts a positive regulatory effect on the transcription of oocyte-type 5 S genes in adult nuclei. The active factor is probably a protein. It is not, as one might have guessed, the transcription factor TF IIIA. In a more recent study, Wakefield and Gurdon (1983) found that the genes coding for oocyte-type and somatic 5 S RNAs are both expressed in advanced blastulas of *Xenopus*. However, the level of oocyte-type 5 S RNA synthesis is 100 times lower in the blastula than in the oocyte; there is a further

20-fold decrease in transcription during gastrulation. These findings led to the following experiment. Wakefield and Gurdon (1983) transplanted neurula nuclei (in which the oocyte-type 5 S genes are inactive) into enucleated oocytes and followed 5 S RNA synthesis during development. They found that the oocyte-type 5 S genes were reactivated in blastulas and inactivated during later development. This experiment shows that the cytoplasm contains factors that control the expression of the oocyte-type 5 S genes and that these cytoplasmic regulatory mechanisms disappear during the repeated mitoses that take place during development.

III. MATURATION OF *XENOPUS* OOCYTES[3]

Maturation is the progression of meiosis from the oocyte to the fertilizable egg stage, a stage that varies greatly among different animal species. In amphibians, only the first polar body is expelled in the maternal organism. Elimination of the second polar body, and thus completion of meiosis, does not take place before fertilization or parthenogenetic activation (which can easily be obtained by pricking the egg with a glass needle). One might say that maturation is a rejuvenative process. The oocyte is an old cell, since it has the same age as the mother (eventually more than 40 years in the human species). The unfertilized egg is a very young cell; it can be considered as zero time for embryogenesis.

A. MORPHOLOGY

Morphologically, the most spectacular process during maturation is the breakdown of the huge GV (GVBD) after a lag period of several hours. In amphibians, it starts at the basal pole of the GV (Fig. 27a,b) where a temporary network of MT forms (Huchon *et al.,* 1985). The disintegration of the nuclear membrane spreads toward the apical end of the GV, where it is completed about 6–7 hr after hormonal stimulation. Simultaneously, the lampbrush chromosomes strongly condense and attach to a spindle. The nucleoli disintegrate, with the exception of the nucleolar organizers that fuse together to form large, Feulgen-positive bodies scattered throughout the cytoplasm of the animal pole (Brachet, 1965a,b). (Fig. 27e). Steinert *et al.* (1976) have demonstrated by *in situ* hybridization with rRNA that these bodies actually contain rDNA. They progressively disappear, but molecular hybridization shows that the rDNA that has been cast off in the cytoplasm during maturation remains detectable during early cleavage (Thomas *et al.,* 1977; Busby and Roeder, 1981, 1982). The spindle that carries the strongly condensed meiotic chromosomes moves toward the animal pole and anchors itself in the cortex (Fig. 27c). The first meiotic division follows, and the first polar body is expelled (Fig. 27d), reducing by one-half the number of chromosomes. Maturation stops at the metaphase II stage of meiosis in amphibians.

---

[3]See the detailed review by Masui and Clarke (1979).

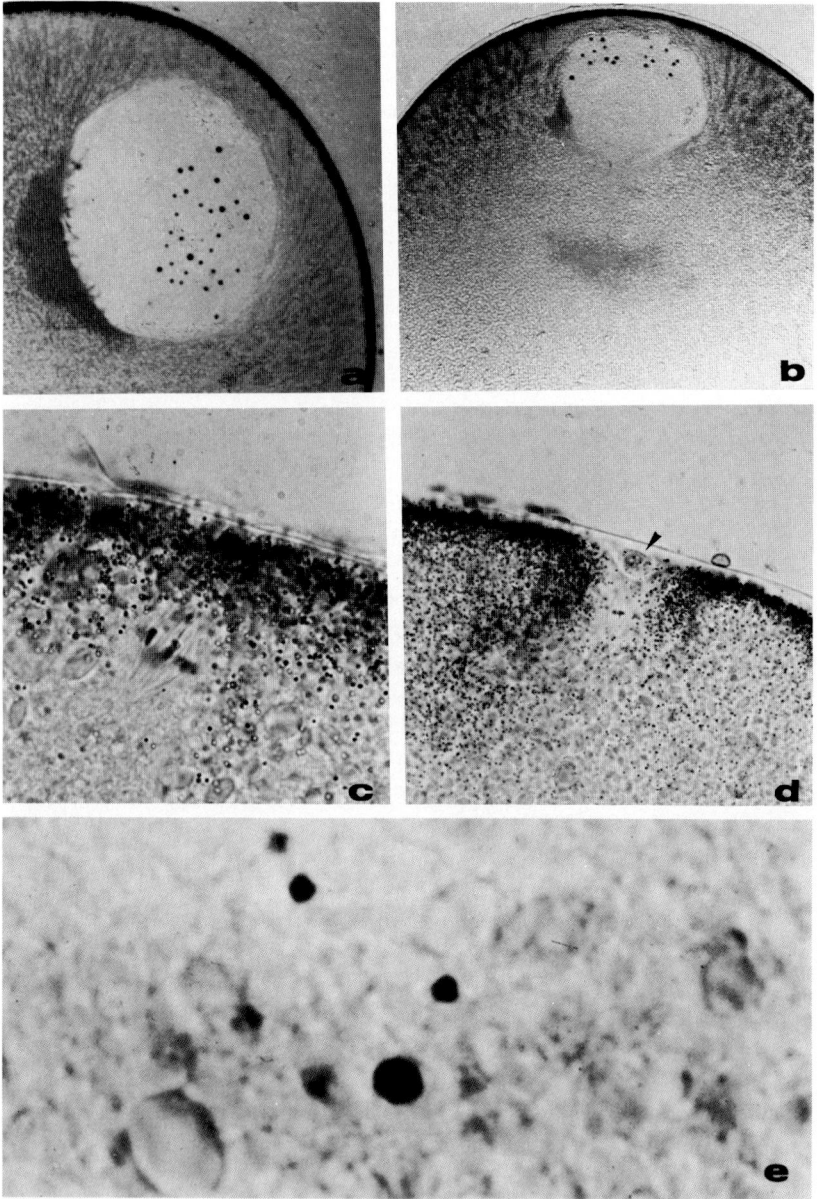

FIG. 27. Morphological changes during *Xenopus* oocyte maturation. (a) Full-grown oocyte with intact GV; (b) breakdown of the nuclear membrane at its basal pole; (c) first maturation spindle; (d) second maturation spindle under the first polar body (arrowhead); (e) at much higher magnification, Feulgen-stained coalescent nucleolar organizers. The Feulgen-positive bodies are composed of DNA, as shown by *in situ* hybridization at the ultrastructural level (Fig. 3, Chapater 2, Volume 1). [Photographs by P. Van Gansen, F. Hanocq, and J. Brachet.]

B. Induction of Maturation

Maturation is induced in many animal species by hormonal stimulation. In amphibians, the natural inducer is progesterone, which is synthesized by the follicle cells that surround the oocyte under pituitary hormone stimulation. Maturation takes place when follicles containing full-grown oocytes are dissected out of the ovary and exposed to either a frog pituitary extract or progesterone. In the first case, maturation does not take place in the presence of actinomycin D; in contrast, progesterone induction of maturation is actinomycin D insensitive.

In other target cells (in the uterus, for instance) the response to steroid hormones is inhibited by actinomycin D. This implies that RNA synthesis is involved in the response to the hormonal stimulus. Figure 28 describes schematically the response of a "classic" target cell to a steroid hormone. Briefly, the hormone binds to a cytosol receptor. The hormone–receptor complex is internalized, moves through the cytoplasm, and penetrates the nucleus, where the receptor undergoes changes in composition and conformation. The nuclear hormone–receptor complex binds to chromatin and, ultimately, to one or a few genes that are activated. But recent work by King and Greene (1984) and Welshons *et al.* (1984) indicates that this classic mechanism might be completely wrong. Using two different approaches (immunofluorescence and separation of the cells into karyoplasts and cytoplast), they came to the unexpected conclusion

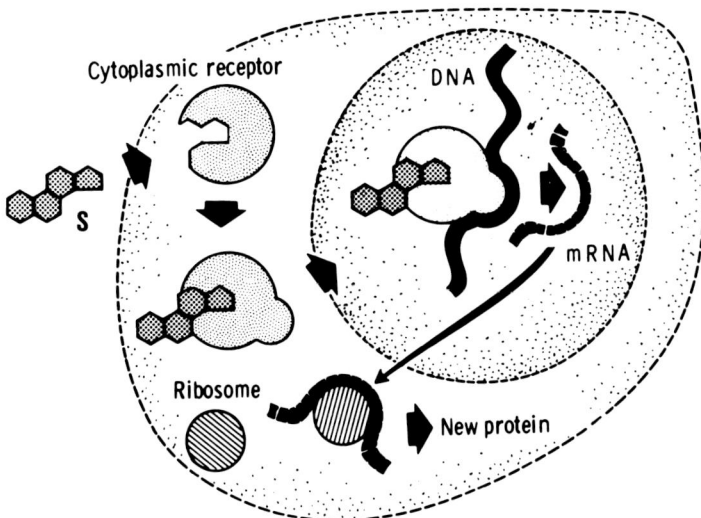

Fig. 28.   Classic model of steroid hormone action on target cells. A steroid hormone S binds to a cytoplasmic receptor; the hormone–receptor complex moves into the nucleus and binds to specific DNA sequences. The corresponding gene is activated and transcribed; the hormone-induced mRNA moves into the cytoplasm, binds to ribosomes, and is translated into a specific protein. [Imperato-McGinley and Peterson, 1976.]

that the estrogen receptors are completely localized in the nucleus of the target cells. This does not alter the essential fact that formation of the nuclear receptor–hormone complex in chromatin induces the transcription of large amounts of the corresponding mRNA; as a result, one or a few proteins are synthesized in very large amounts in the cytoplasm. This is what happens, for instance, in the liver of *Xenopus* when vitellogenin synthesis is induced by estrogens, or in the hen oviduct, where estrogens induce the synthesis of large amounts of ovalbumin. This does not occur when *Xenopus* oocytes are treated with progesterone, since the induction of maturation is actinomycin D insensitive and thus does not require RNA synthesis. On the other hand, progesterone production by the follicle cells after pituitary hormone stimulation probably occurs by the classic mechanism shown in Fig. 28.

There are other differences between the oocyte and other target cells in their response to progesterone. The hormone must act directly on the cell membrane, since it is inactive if injected into the oocyte, unless the hormone is dissolved in an oil droplet (Tso *et al.*, 1982). In fact, maturation can be induced by synthetic derivatives of steroids that are unable to cross the cell membrane (Godeau *et al.*, 1978). The presence of a surface receptor for progestin has been reported (Sadler and Maller, 1982). Specific progesterone receptors have been found on the surface of frog and *Xenopus* eggs (Kostellow *et al.*, 1982; Blondeau and Baulieu, 1984); they are absent from yolk platelets, melanosomes, cytosol, and the nucleus. In addition, progesterone lacks the specificity found in other systems. For instance, corticosteroids and testosterone also induce GVBD in isolated *Xenopus* oocytes.

These findings were puzzling and became even more mysterious when we found that organomercurials, which block free —SH groups without penetrating the cells, induce maturation as efficiently as progesterone in *Xenopus* oocytes (Brachet *et al.*, 1975a,b). This led to a search for other agents that might induce maturation, although chemically unrelated to steroids: Lanthanum chloride, propranolol, and several other pharmacological agents, insulin, etc. (Schordeeret-Slatkine *et al.*, 1976, 1977). As correctly pointed out by Baulieu *et al.* (1978), all of these agents have one thing in common: they set free membrane-bound calcium and thus increase the free Ca^{2+} concentration in the cytoplasm. In the particular case of the organomercurials, it was found that they are inactive in a Ca^{2+}-free medium and increase the Ca^{2+} influx into oocytes cultured in normal medium.

C. EARLY BIOCHEMICAL CHANGES

There is now overwhelming evidence for the view that a localized increase in the free Ca^{2+} concentration is indeed a prerequisite for the induction of maturation. For instance, the slow introduction of Ca^{2+} by iontophoresis into the oocyte cortex is followed by maturation; the latter is inhibited if the oocytes have been previously treated with a Ca^{2+}-chelating agent. Finally, Ca^{2+} is inactive if it is injected into the endoplasm instead of the cortex (Moreau *et al.*, 1976). The divalent cation ionophore A23187 also induces maturation, provided that the outer medium contains excess Ca^{2+} or Mg^{2+} (Wasserman and Masui, 1976). We have even obtained maturation by placing the oocytes in Ringer's solution containing excess $CaCl_2$ or $MgCl_2$. However, in these experiments, Mg^{2+} was somewhat more efficient than Ca^{2+} (Baltus *et al.*, 1977). Conversely, agents that impede the intake of Ca^{2+} (such as papaverine) or inhibit the liberation of Ca^{2+} bound to the membrane phospholipids (gammexane) prevent GVBD in progesterone-treated oocytes. More recently, it has been shown that *Xenopus* oocytes contain calmodulin and that this calcium-binding protein increases by 70% at the time of GVBD. Injection of calmodulin induces maturation, which is inhibited by the calmodulin inhibitor trifluoperazine (Mulner *et al.*, 1980; Wasserman and Smith, 1981). However, it has been reported that injection of an antibody against calmodulin speeds up progesterone-induced maturation (Cartaud *et al.*, 1981). Finally, Wasserman *et al.* (1980) demonstrated the reality of a Ca^{2+} burst at maturation by injecting aequorin into the oocytes and measuring the flash of luminescence produced by the liberation of free Ca^{2+}. These experiments demonstrate that, as one might expect, Ca^{2+} release is one of the very early events induced by progesterone addition and that Ca^{2+} plays the role of "second messenger" in the induction of maturation. However, injection of inositoltrisphosphate (IP_3), which should set free Ca^{2+} from its intracellular stores (see Chapter 3, Volume 1), fails to induce maturation in *Xenopus* oocytes (Picard *et al.*, 1985): another mechanism should operate to indicate Ca^{2+} release.

It would be unwise, however, to think that Ca^{2+} is the only ion involved in maturation. We have already seen that Mg^{2+} can be more efficient than Ca^{2+} in inducing maturation in *Xenopus* oocytes. There is some evidence that K^+ probably also plays a role. Maturation is accelerated when *Xenopus* oocytes are placed in a K^+-free medium or treated with ouabain, the classic inhibitor of the sodium pump (Vitto and Wallace, 1976; Kofoid *et al.*, 1979). We found that the K^+ ionophore valinomycin induces maturation when the oocytes are maintained in a K^+-free medium in order to induce a K^+ efflux. However, in this case, maturation is abortive in the sense that GVBD is not followed by the assembly of a meiotic spindle. This failure is probably due to the fact that valinomycin is an inhibitor of energy production in the mitochondria and that it might decrease the internal pH of the oocyte's cytoplasm (Baltus *et al.*, 1977). On the other hand, another monovalent cation ionophore, monensin, inhibits progesterone-induced maturation (Marot *et al.*, 1984). It has no effect on the biochemical changes that take place after progesterone addition, but it prevents the breakdown of the nuclear envelope.

That the maintenance of or increase in intracellular pH (pH_i) is important in maturation is shown by the fact that CO_2 inhibits progesterone-induced maturation (Bellé *et al.*, 1982). Work by Cicirelli *et al.* (1983), Wasserman and Houle

(1984), and Morrill *et al.* (1984) has shown that the pH_i increases by 0.3–0.4 during maturation. The mechanisms that increase the pH_i of progesterone-treated oocytes (Na^+/H^+ exchange, correlated with the increase in free Ca^{2+}) and the possible role of this increase in the stimulation of protein synthesis and respiration remain controversial. The pH_i, which rises as high as 7.6–7.8 soon after progesterone stimulation, slowly decreases in the course of maturation and drops to about 7.1–7.2 in unfertilized eggs. In *Ambystoma mexicanum* (axolotl) oocytes, maturation raises the internal pH from 7.15 to 7.40 within 3 hr; the internal concentration in sodium ions (Na^+_i) increases from 6 to 17 mM in 10 hr. An increase in pH_i thus occurs during maturation of both anuran and urodele oocytes (Baud and Barish, 1984). The absence of Na in the medium does not prevent progesterone-induced maturation, but amiloride and benzamil, which inhibit Na^{2+} influx and H^+ efflux, make it impossible (Baltus *et al.*, 1977; Cameron *et al.*, 1982). Finally, Koide *et al.* (1979) have reported that $MnCl_2$ is also capable of inducing maturation, but our own unpublished cytological observations indicate that $MnCl_2$ is definitely toxic for oocytes. Although there are contradictory reports in the literature, there appear to be no major reproducible electrophysiological changes, despite these ionic movements during the first steps of maturation in *Xenopus*. Modifications of the electric properties cannot be detected until the time when the long microvilli characteristic of the oocytes undergo retraction, a phenomenon to which we shall soon return. This takes place halfway between progesterone addition and GVBD (i.e., about 3 hr after hormone addition) (Kado *et al.*, 1981).

Before we discuss the morphological and biochemical ''late'' changes, which occur by the time of GVBD (thus, several hours after progesterone addition to isolated oocytes), it is important to mention a change that, like Ca^{2+} release, occurs very soon after hormonal stimulation, namely, a 50% decrease in membrane adenylate cyclase activity (Sadler and Maller, 1981; Finidori-Lepicard *et al.*, 1981; Baltus *et al.*, 1981). This inhibition can be detected in both intact oocytes and membrane fractions isolated from homogenates. The molecular mechanisms of adenylate cyclase inhibition by progesterone are not known. However, Goodhardt *et al.* (1984) recently showed that inhibition is not mediated by the guanine nucleotide-binding regulatory protein N_i (which inhibits the enzyme). According to Sadler and Maller (1981), when intact oocytes are dissected out of the ovary, their adenylate cyclase is activated after a 30-min lag; this activation is inhibited by progesterone. This hormone also inhibits the stimulation of the enzymatic activity that can be induced by sodium fluoride or the nonhydrolyzable analog of GTP, (GppNHp). Since the activity of cAMP diesterase remains unchanged after progesterone addition (Baltus *et al.*, 1981), it follows that the cAMP level should drop at the onset of maturation. Although the data concerning changes in the cAMP content of progesterone-treated amphibian eggs were contradictory until recently, the most recent results of Maller *et al.* (1979) and Schorderet-Slatkine *et al.* (1982) appear to show that they are not.

These authors reported that there is a strong decrease in the cAMP content when *Xenopus* oocytes are treated with progesterone. However, this is a very rapid and transient phenomenon. There is a sharp drop (40–60%) in the cAMP content 15–60 sec after progesterone treatment, with a return to the initial level within 1 hr. This drop does not exceed 20% according to a recent report by Cicirelli and Smith (1985). In agreement with these findings, it has also been shown that progesterone decreases the rate of *in vivo* cAMP synthesis (Mulner *et al.*, 1979). Finally, agents that increase the cAMP content by either inhibiting phosphodiesterase (theophyllin, isobutyl methylxanthine) or stimulating its synthesis by cAMP cyclase (choleratoxin) prevent the induction of maturation by progesterone (O'Connor and Smith, 1976; Bravo *et al.*, 1978; Schorderet-Slatkine *et al.*, 1978; Mulner *et al.*, 1979; Maller *et al.*, 1979). Although side effects of these inhibitors on other parameters, such as ionic fluxes or protein synthesis, have not been ruled out, one conclusion seems to overwhelmingly emerge: one of the earliest effects of progesterone is to decrease almost immediately both cAMP synthesis and the cAMP level (Schorderet-Slatkine *et al.*, 1982); this decrease is absolutely required for a successful response to progesterone. Thus, the initial steps after progesterone addition to *Xenopus* oocytes are not merely binding to a receptor and endocytosis of a progesterone–receptor complex, but also a Ca^{2+} surge and a decrease in the cAMP level resulting from partial inhibition of cAMP cyclase. However, the significance of the rapid inhibition by progesterone of membrane-bound adenylate cyclase activity for maturation itself is not clear. Jordana *et al.* (1982) have reported that the same inhibition occurs even in small vitellogenic oocytes (0.7 mm in diameter) that do not respond to progesterone by GVBD. This suggests that the progesterone-induced temporary inhibition of adenylate cyclase activity is an initial step in maturation, but is not sufficient for the induction of the complete process. As we shall now see, other biochemical changes, triggered by the Ca^{2+} burst and the cAMP decrease, are necessary for the completion of maturation, in particular for GVBD.

D. LATE BIOCHEMICAL CHANGES

We now come to the morphological and biochemical changes that take place a few hours after progesterone stimulation. The cytology of *Xenopus* oocytes undergoing maturation has been described by Brachet *et al.* (1970). The major steps, at the light and EM levels, are shown in Fig. 27. As already mentioned, the large microvilli so conspicuous on the cell surface of *Xenopus* oocytes (Fig. 8, Chapter 3, Volume 1) progressively retract, leaving a smooth surface over the cortical granules (which will not break down before fertilization or activation). Retraction of the microvilli, which is correlated with the arrest of vitellogenesis, is presumably due to depolymerization of the actin filaments that are abundant in their core, as was shown by Franke *et al.* (1976). The basal end of the nuclear membrane becomes indented before its breakdown: its folds contain many ribosomes and mitochondria at the time this breakdown begins. When the nuclear sap and the cytoplasm mix together, a thickened "basal plate," from which hairy structures spread into the nuclear sap, progressively forms; its main ultra-structural characteristic is an accumulation of vesicles (Huchon *et al.*, 1981c) and microtubules (Huchon *et al.*, 1985). It is in this region that, after Feulgen staining, the partially condensed chromosomes and nucleolar organizers are found. Finally, the chromosomes become attached to a spindle that moves toward the animal pole, leaving behind a trail of fused large nucleolar organizers composed of rDNA and surrounded by mitochondria (Fig. 3, Chapter 1) (Steinert *et al.*, 1976). According to a paper by Williams *et al.* (1981), who used Miller's spreading technique, rDNA transcription in the nucleoli remains intense for as long as 200 min after progesterone treatment; it does not decrease until the onset of GVBD. Finally, we should mention the abundance of the annulate lamellae in maturing oocytes (Fig. 34, Chapter 3, Volume 1). In the frog *R. pipiens* and in *Xenopus*, they disappear from the cytoplasm shortly after GVBD, according to Kessel and Subtelny (1981), who think that these prominent annulate lamellae originate from the nuclear membrane and disintegrate together with the nuclear envelope. In full-grown *Xenopus* oocytes, there are about 260 stacks of annulate lamellae. After progesterone stimulation, their number increases and they tend to accumulate in the subcortical layer and at the vegetal pole. Their number reaches a maximum (320 stacks of lamellae) 6 hr after hormone addition. All of the annulate lamellae suddenly disappear when the nuclear membrane breaks down (Imoh *et al.*, 1983).

Progesterone treatment of *Xenopus* oocytes is followed by a number of bio-chemical changes that are slower and more continuous than the aforementioned transient increase in free Ca^{2+}, and decrease in adenyl cyclase activity and cAMP content. These should thus be considered late biochemical events, although they are already detectable before GVBD. Among these changes are a progressive increase in oxygen consumption (Brachet *et al.*, 1975b) and possibly in some RNA synthesis (Morrill *et al.*, 1975) or stability (Gelfand and Smith, 1983). In contrast, 5 S rRNA synthesis stops (Wormington and Brown, 1983). This increase in RNA synthesis is not required for maturation, since this process is actinomycin D insensitive. The RNAs synthesized during maturation might perhaps play some role in the processes that follow fertilization. Much more important is the strong increase in protein synthesis, which was discovered by L. D. Smith and R. E. Ecker in their pioneer studies on the induction of maturation in *R. pipiens* oocytes by progesterone addition (reviewed by Smith and Ecker, 1970). In *Xenopus*, as shown in Fig. 29, the rate of protein synthesis steadily increases after progesterone addition, reaching a peak after 6 to 8 hr (thus, around the time of GVBD and meiotic spindle formation). This level then decreases to one that, after 24 hr, is lower than that of unstimulated oocytes. There has been some argument about the qualitative aspects of protein synthesis during maturation. Is the synthesis of all proteins stimulated to the same extent? Do new species of proteins appear after progesterone stimulation? The most recent evi-

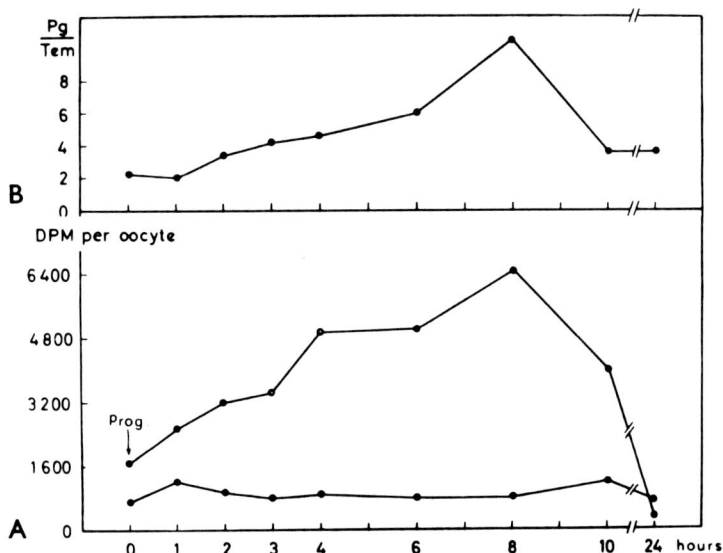

FIG. 29. Increase in protein synthesis in *Xenopus* oocytes after progesterone treatment. (A) [^{14}C]phenylalanine incorporation into progesterone-treated and control oocytes. The arrow shows the time of progesterone addition. (B) Ratio between progesterone-treated and control oocytes. [Brachet *et al.*, 1974.]

dence available (Younglai *et al.*, 1981, 1982) indicates that the synthesis of only 5 out of 600 polypeptides selectively increases at the time of GVBD; one of these proteins is probably phosphorylated. Two were not detectable in unstimulated oocytes. There is a slight excretion of the neosynthesized proteins into the medium. Wasserman *et al.* (1982) and Richter *et al.* (1982a,b) indicated that progesterone or MPF treatments double the rate of protein synthesis without the occurrence of major qualitative changes; this increase would be due to the recruitment of stored mRNA molecules. The changes observed by Younglai *et al.* (1982) are amplified in the presence of actinomycin D, which agrees with earlier results showing that this inhibitor of RNA synthesis speeds up progesterone-induced maturation (Baltus *et al.*, 1973; Brachet *et al.*, 1974).

The work of Adamson and Woodland (1977) on histone synthesis in progesterone-treated *Xenopus* oocytes leaves no doubt about selectivity in the stimulation of protein synthesis during maturation. There is a 20- to 50-fold increase in "core" histone synthesis (thus, histones H2A, H2B, H3, and H4) without a concomitant increase in histone H1 synthesis (Fig. 30). This was confirmed more recently by Flynn and Woodland (1980), who showed that the synthesis of histone H1 catches up during cleavage. Its rate becomes faster than that of the nucleosome core histones, which accumulate during maturation. Thus, the synthesis of histone H1 and core histones is not coordinated during maturation and cleavage. For Flynn and Woodland (1980), the condensation of chromatin that

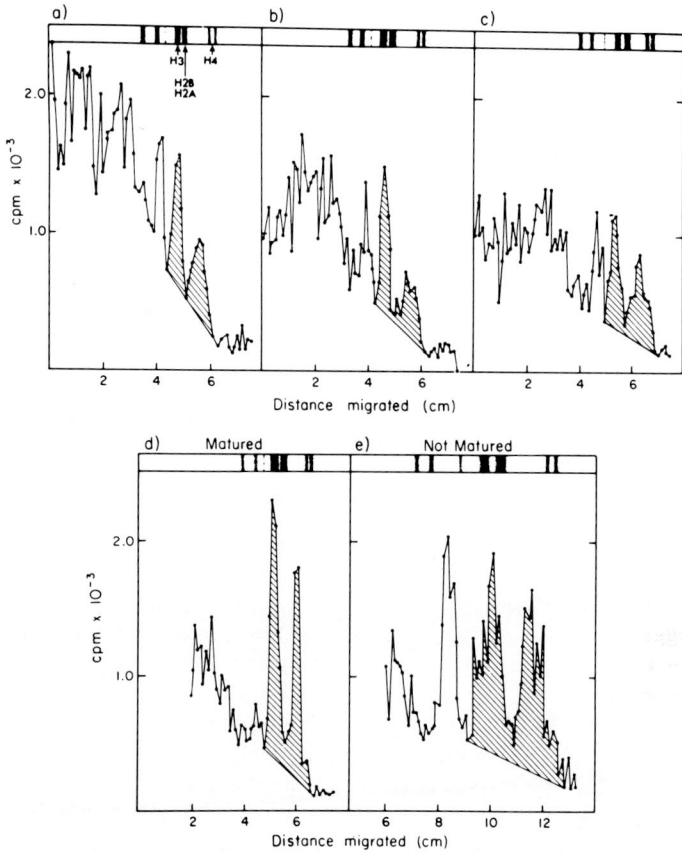

FIG. 30.   Histone synthesis during maturation in *Xenopus*. Radioactive profiles after [³H]lysine incorporation and separation on acid-urea gels. Radioactivity in the region of the histones is shaded. (a) Control oocytes without progesterone; (b) oocytes treated with progesterone and labeled for 1–5 hr; (c) same as (b), but labeled between 5 hr and 6 hr 40 min; (d) oocytes that had matured at 7 hr and were then labeled for 3 hr; (e) oocytes that had not matured at 7 hr and were then labeled for 5 hr (intermediary profile between (a) and (d). There is an increase in histone synthesis 7 hr after progesterone addition, whether or not the GV had broken down. [Adamson and Woodland, 1977.]

takes place between the early and late blastula stages might be due to this increased rate in H1 histone synthesis during cleavage. The requirement for core histone synthesis during cleavage is much less stringent, since a large store of these histones (800–3200 pg per oocyte) has been accumulated during oogenesis and maturation and is thus immediately available to the rapidly dividing nuclei of the cleaving egg.

A very important point, since we are discussing nucleocytoplasmic interactions in this chapter, is that both the overall increase in protein synthesis discovered by Smith and Ecker (1970) and the selective synthesis of the core histones found by Adamson and Woodland (1977) still take place when *Xenopus* oocytes are enucleated prior to progesterone addition. Protein synthesis during maturation is thus controlled by selective posttranscriptional mechanisms about which we still know very little. One of these mechanisms might be adenylation or deadenylation of given mRNAs, with ensuring changes in the stability of the preformed mRNAs. Ruderman *et al.* (1979) found that histone H1 mRNA is deadenylated during maturation, and thus that some of its stability is lost. Other controls presumably exist at the level of the translational machinery, perhaps by phosphorylation of some of its components.

The independence of cytoplasmic protein synthesis from nuclear control in full-grown *Xenopus* oocytes is further shown by the experiments of Bienz and Gurdon (1982) on the production of heat shock proteins (hsp). Heating the oocytes to 31°C induces the synthesis of the same major hsp in both nucleate and anucleate oocytes. Injection of α-amanitin (the classic inhibitor of RNA polymerase II) into full-grown oocytes does not prevent the synthesis of the hsp at 31°C. We are thus left with the conclusion that the oocytes must contain preformed hsp mRNA molecules and that heating leads to their selective translation. This is a very unusual situation. In all of the systems studied so far, including sea urchin eggs, the synthesis of hsp is controlled at the level of gene transcription.

Recent work by Bienz (1984) has confirmed that the molecular responses of *Xenopus* oocytes and eggs to a heat shock are unusual. *Xenopus* adult cells in culture respond to heat shocks by synthesizing two proteins (hsp 70 and 30); unheated cells do not possess the corresponding mRNAs that are synthesized as a response to a heat shock. In contrast, the oocytes possess hsp 70 mRNA; this RNA does not increase after a heat shock, and disappears from the egg at the time of fertilization. Induction by a heat shock of hsp 70 mRNA cannot be detected before the end of cleavage. On the other hand, hsp 30 mRNA is undetectable in oocytes and embryos before the tadpole stage. These experiments show clearly that the hsp genes are not coordinately regulated in *Xenopus*.

The heat shock response (synthesis of hsp 70) is not affected by progesterone-induced maturation, or activation of unfertilized *Xenopus* eggs by the Ca^{2+} ionophore A23187 (Baltus and Hanocq-Quertier, 1985); the work of Bienz (1984) has shown that morulae do not respond to a heat shock by the synthesis of hsp 70; it is not yet known at which early stage after fertilization the egg becomes unable to respond to a heat shock by the production of heat shock proteins.

Unexpectedly we observed, in still unpublished work, that heat shocks induce in *Xenopus* matured oocytes, unfertilized and recently fertilized eggs the appearance of numerous cytasters in the cytoplasm; this cytological response to heat shocks is no longer seen when the eggs have undergone their first cleavage.

Studies with inhibitors of protein synthesis (Dettlaff, 1966; Brachet *et al.*, 1975a) and energy production (Brachet *et al.*, 1975b) have disclosed an interesting fact. These inhibitors impede GVBD if they are allowed to act during the first 3 hr after progesterone treatment, but they have no effect after that time. There is thus a 3-hr initial period during which energy production and protein synthesis are essential. GVBD takes place about 6 hr after the beginning of progesterone treatment, and thus at a time when energy production and protein synthesis are no longer required.

There is no chromosomal or nucleolar DNA synthesis during maturation. All one can detect at that time is a low level of mitochondrial DNA synthesis (Hanocq *et al.*, 1974). However, maturation is characterized by the accumulation of the machinery which is required for the synthesis of DNA. This involves a selective increase in the production of a number of proteins. This DNA-synthesizing machinery, together with the store of histones accumulated during maturation, is required for rapid, intensive chromosome replication and assembly after fertilization or parthenogenetic activation. As we have seen, the experiments in which Gurdon (1974) injected brain nuclei into *Xenopus* oocytes and eggs clearly demonstrated that a switch from RNA to DNA synthesis takes place during maturation. The arrest of RNA synthesis, in addition to the aforementioned inactivation of the nucleolar organizers, is partly due to the disappearance, during maturation, of the protein factors that stimulated RNA polymerases activity in the oocyte (Crampton and Woodland, 1979). On the other hand, progesterone treatment induces the synthesis or activation of several enzymes and factors involved in DNA replication. A new form of DNA polymerase (Grippo and Lo Scavo, 1972; Benbow *et al.*, 1975), as well as of ribonucleotide reductase (Tondeur-Six *et al.*, 1975) and thymidine kinase (Woodland and Pestell, 1972), which are required for the accumulation of a sufficient deoxyribonucleotide pool, become detectable. The overall activity of DNA polymerase (which is associated with a DNA primase activity) increases four times during maturation. Interestingly, this increase does not take place if the oocyte was enucleated before the addition of progesterone (Grippo *et al.*, 1977). In addition, a protein factor that stimulates the initiation of DNA synthesis appears during maturation (Benbow and Ford, 1975). Possibly related to the accumulation of the DNA-synthesizing machinery is a strong increase in ornithine decarboxylase activity, which is followed by an accumulation of putrescine (Sunkara *et al.*, 1981).

One well-established fact is that there is a strong wave of protein phosphorylation about 20 min before GVBD (Morrill and Murphy, 1972; Maller *et al.*, 1977; Bellé *et al.*, 1978) whereby many proteins, both soluble and membrane-bound, are simultaneously phosphorylated. One of them is the S_6 ribosomal protein (Hanocq-Quertier and Baltus, 1981; Nielsen *et al.*, 1982; Kalthoff *et al.*, 1982), which is phosphorylated whenever cells grow actively. There might be a correlation between this burst in protein phosphorylation and the breakdown of the nuclear membrane, although this takes place in previously enucleated oocytes. It has not yet been proved that phosphorylation of the pore–lamina proteins, which form the GV membrane, leads to their solubilization and is instrumental to GVBD, but this is a very likely possibility.

The changes in cAMP content and protein phosphorylation that take place during maturation in *Xenopus* oocytes led Maller and Krebs (1977) to develop the model shown in Fig. 31. Briefly, this model assumes that, as the result of Ca^{2+} liberation and a decrease in the cAMP content, a key phosphorylated protein of the oocyte would be dephosphorylated. This protein (Mp-P) would prevent maturation as long as it remained in its phosphorylated form. Its dephosphorylation would result from an increase in phosphodiesterase activity due

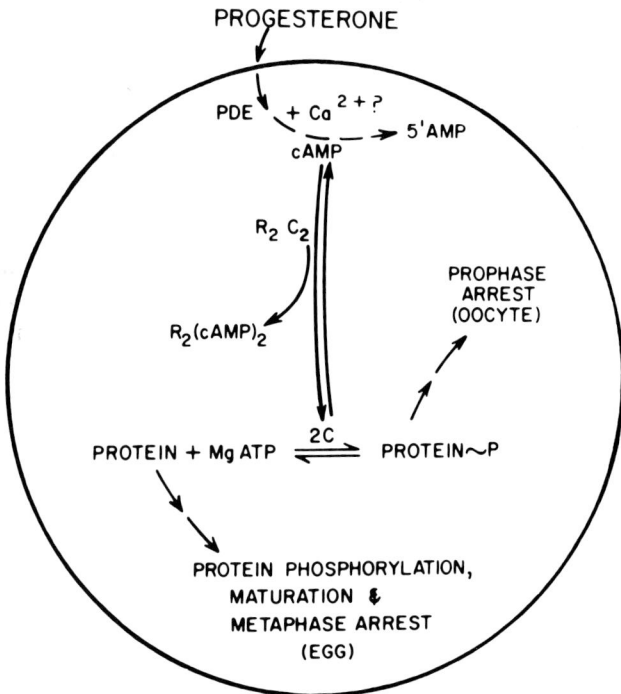

FIG. 31. Proposed model for *Xenopus* oocyte maturation. PDE, phosphodiesterase. Dotted arrows refer to steps for which no direct evidence was available at the time. Solid arrows refer to steps presented in the paper. The prophase block of the oocyte would be due to phosphorylation of a protein by the catalytic subunit C of the cAMP-dependent protein kinase. After progesterone addition, there is an increase in free Ca^{2+} and a decrease in cAMP content. This results in a decrease in the cAMP-dependent protein kinase activity by promoting the formation of inactive R_2C_2 (R is the regulatory subunit of the enzyme). The concentration of the particular phosphoprotein maintaining the prophase block would be reduced, permitting maturation to proceed. It is now known that there is indeed a rapid Ca^{2+} burst and that progesterone inhibits adenylate cyclase activity. [Maller and Krebs, 1977.]

to Ca^{2+} release and a decrease in cAMP-dependent protein kinase. However, Huchon *et al.* (1981a) have shown that injection of a specific inhibitor of protein phosphatase inhibits progesterone-induced maturation, demonstrating that this enzyme is involved in the dephosphorylation of Mp-P. Recent work in Ozon's laboratory has brought fresh evidence that the phosphorylation status of a phosphoprotein (Mp-P) which prevents the oocyte from undergoing maturation is an important step in the control of maturation. Injection of various inhibitors of alkaline phosphatase speeds up progesterone-induced maturation, whereas injection of alkaline phosphatase inhibits it (Hermann *et al.*, 1984). Bellé *et al.* (1984) reported that injection of ATP-γ-S inhibits maturation in progesterone-treated eggs; this would be due to a stabilization of the maturation protein Mp-P in its phosphorylated, maturation-inhibitory form. This protein probably has a M_r of 32,000. Progesterone inhibits the phosphorylation of a 48K protein localized in the oocyte plasma membrane (Blondeau and Baulieu, 1985). This might be the still-hypothetical Mp-p protein that supposedly inhibits maturation when in its phosphorylated form. When the cAMP content returns to normal levels, cAMP-dependent protein kinases would be activated and produce a wave of protein phosphorylation that somehow triggers GVBD and the subsequent events of maturation. Injection into the oocytes of the cAMP-dependent protein kinase thermostable inhibitor of muscle induces maturation but does not prevent the burst in protein phosphorylation (Bellé *et al.*, 1978). In addition, no changes in cAMP-dependent protein kinase activity could be detected during maturation. It therefore seems likely that the burst of protein phosphorylation that precedes GVBD results from the activity of a Ca^{2+}-dependent protein kinase (possibly protein kinase C) rather than that of a cAMP-dependent kinase. This conclusion does not imply that the cAMP level does not play a key regulatory role in the control of maturation. On the contrary, the elegant experiments of Maller and Krebs (1977) have shown that injection of the regulatory subunit of protein kinase (which binds to cAMP) induces GVBD; injection of the catalytic subunit of the enzyme (which should be followed by an increase in the cAMP level) inhibits progesterone-induced maturation. These results can be explained by the following classic equation in which R is the regulatory and C the catalytic subunit of the enzyme:

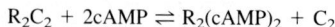

$$R_2C_2 + 2cAMP \rightleftharpoons R_2(cAMP)_2 + C_2$$

Comparable results have been reported by Schorderet-Slatkine *et al.* (1981). They found that 5'-deoxy-5'-S-isobutyl-thioadenosine (SIBA), an inhibitor of nucleic acid and protein methylation, inhibits GVBD, but that this negative effect on maturation is due to the increased activity of a membrane-bound cAMP cyclase. Thus, SIBA increases the cAMP content whether or not the oocytes have been treated with progesterone. The Maller–Krebs (1977) model remains a very useful guideline for further investigations but requires modifications, since, as we have seen, it is doubtful that the burst in protein phosphorylation that precedes GVBD is due to a cAMP-dependent protein kinase. Furthermore, Hanocq-Quertier and Baltus (1981) failed to find changes in phosphodiesterase activity shortly after progesterone addition.

An injection of a purified preparation of a histone cAMP-dependent kinase into *Xenopus* or *Ambystoma* oocytes induces the condensation of the lampbrush chromosomes (Wiblet *et al.*, 1975), as shown in Fig. 32. It is probable that histone H1 phosphorylation, in conjunction with the already discussed involvement of actin (Rungger *et al.*, 1979), nucleoplasmin, and DNA topoisomerase I, plays a role in the very extensive condensation of the meiotic chromosomes that is one of the main features of maturation.

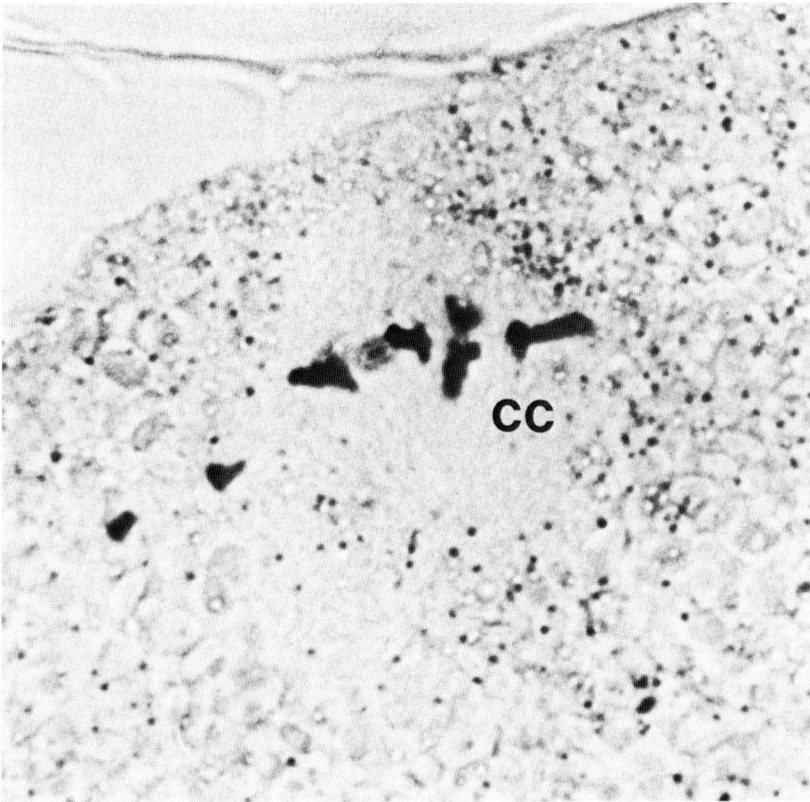

FIG. 32.   Injection of a histone kinase preparation from *Xenopus* oocytes that have undergone maturation into a recipient axolotl oocyte may induce its maturation. CC, condensed metaphase chromosomes. [Wiblet *et al.*, 1975.]

Apart from these biochemically well-characterized changes, progesterone-treated oocytes synthesize a number of biologically active "factors" of still unknown chemical nature. The most important of these is the maturation-promoting factor (MPF), discovered by Smith and Ecker (1970) and by Masui and Markert (1971). As shown in Fig. 33, if cytoplasm from a progesterone-treated oocyte is injected into a normal recipient oocyte, GVBD takes place within 1–2 hr; this is due to the presence of MPF in the cytoplasm. The burst in protein phosphorylation and an increase in protein synthesis occur almost immediately after injection of an MPF-containing cytoplasm. It is believed that MPF does not appear unless the oocyte Mp-P phosphoprotein has been dephosphorylated (Huchon *et al.*, 1981a,b; Maller *et al.*, 1977). The cytoplasm of the MPF-injected oocyte synthesizes additional MPF, which can be transferred from one oocyte to another by serial transfers of ctyoplasm (Masui and Markert, 1971). In such experiments, MPF undergoes a kind of autocatalytic amplification. MPF synthesis takes place, even in enucleated oocytes, during the first 3–4 hr after progesterone addition, which is the cycloheximide-sensitive period of maturation. As expected, MPF production requires protein synthesis. After MPF injection, GVBD is no longer inhibited by cycloheximide, at least in *Xenopus* oocytes. When maturation is induced by MPF injection, its initial phase, which requires energy and protein synthesis, is bypassed. Within 5 min after injection,

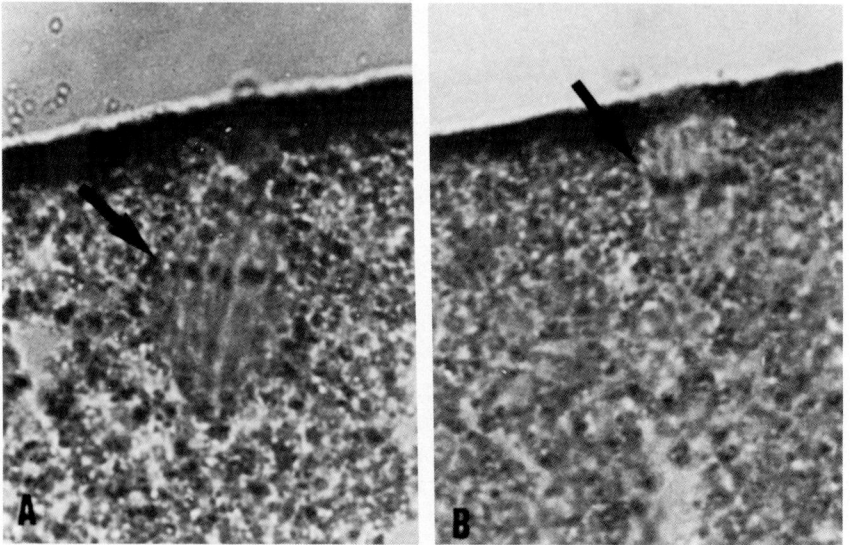

FIG. 33.   Both progesterone addition (A) and injection of MPF-containing endoplasm (B) induce maturation in amphibian oocytes. Arrows point to meiotic metaphases. [Masui and Markert, 1971.]

MPF also induces the breakdown of the nuclear membrane in cleaving *Xenopus* eggs arrested in G_2 by treatment with cycloheximide. In this system, there is no amplification of MPF (Miake-Lye *et al.*, 1983).

MPF has no species specificity. We found (unpublished) that *Xenopus* MPF induces maturation in axolotl oocytes. It is now clear, as discussed in Chapter 5, Volume 1, that MPF-like factors exist not only in cleaving eggs (Wasserman and Smith, 1978) but also in HeLa cells (Sunkara *et al.*, 1979). The role of these probably ubiquitous factors is to induce, directly or indirectly, nuclear membrane breakdown and chromosome condensation at prophase. Unfortunately, we know nothing about the mode of action of MPF and very little about its chemical nature. Wasserman and Masui (1976) showed that it is a protein (or a mixture of proteins) that is inactivated by Ca^{2+} and activated by Mg^{2+}. More recently, Wu and Gerhart (1980) obtained a 20- to 30-fold purification of MPF and found that it is a 100,000-dalton, Ca^{2+}-sensitive protein that copurifies with a protein kinase activity. This fraction is capable of phosphorylating four endogenous proteins of M_r 0.7–2.5 × 10^5 daltons in the presence of ATP, which doubles its biological activity. This activity is prolonged by the addition of NaF, which can have various complex effects on cell enzymatic machinary. There are thus good indications, but no final proof, that MPF is involved in protein phosphorylation.

The purification of MPF has been hampered for several years by the fact that homogenates of progesterone-treated *Xenopus* oocytes contain a "pseudo-maturation"-inducing factor (PIF). Pseudomaturation (Fig. 34) is an abortive maturation whereby the nuclear membrane breaks down at several places simultaneously; there is no condensation of the lampbrush chromosomes, and the nucleoli remain intact. Pseudomaturation always ends with cytolysis. Toxicity is due to the fact that PIF, which is located in the egg cortex 1 hr after progesterone addition, is a very powerful inhibitor of protein synthesis after injection into full-grown *Xenopus* oocytes (Brachet *et al.*, 1974). It is not known whether PIF plays a physiological role or is an artifact produced by homogenization of pro-gesterone-treated oocytes.

More important is the cytostatic factor (CSF) discovered by Masui (1974), which appears much later than MPF, after the expulsion of the first polar body. If cytoplasm from an oocyte arrested at metaphase II (thus, of a fertilizable egg) is injected into blastomeres of a cleaving egg, mitotic activity at metaphase is arrested in the injected balstomere (Fig. 35). As shown by Meyerhof and Masui (1977, 1979), CSF, like MPF, is inactivated by Ca^{2+} and activated by Mg^{2+}. This explains why CSF activity ceases to be detectable when the eggs are fertilized (Masui, 1974), since there is a free Ca^{2+} surge at fertilization. The activity of both MPF and CSF in the oocyte is thus apparently controlled by the free Ca^{2+}/Mg^{2+} ratio. According to Meyerhof and Masui (1979), CSF is not responsible for chromosome condensation at metaphase, but rather plays a role in MT assembly, stabilizing the spindle MTs and preventing their depolymerization

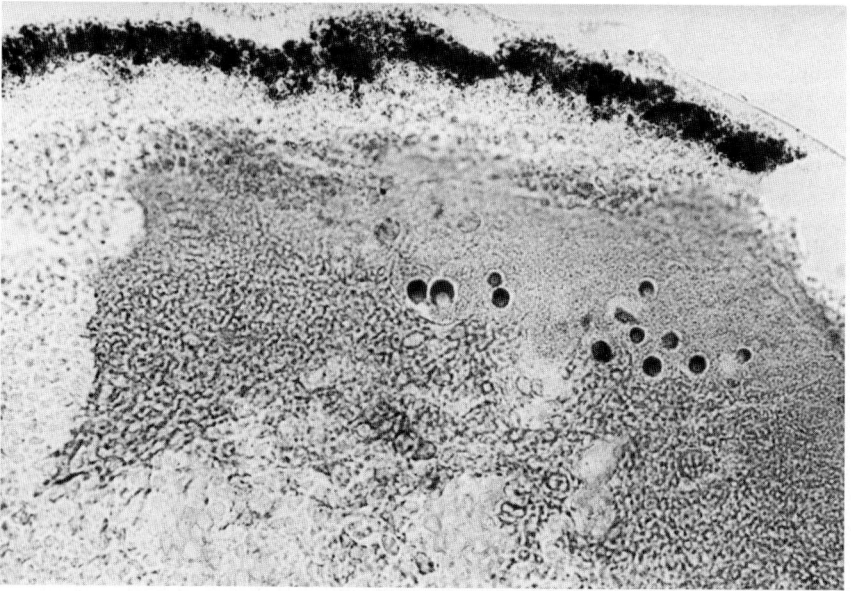

FIG. 34.   Pseudomaturation in a *Xenopus* oocyte injected with the supernatant of a homogenate from *in vitro* matured oocytes. The nuclear membrane has broken down, and ribosomes have invaded the nuclear sap. However, the nucleoli are still present, and the lampbrush chromosomes, which have not condensed, are invisible. [Brachet, 1976.]

at anaphase, acting as an MT assembly promoter. In the unfertilized egg, the physiological roel of CSF is probably to arrest meiosis in metaphase II.

The interactions between MPF and CSF have been recently clarified by Gerhart *et al.* (1984). Due to the partial purification of MPF, they have worked out a biological test for the titration of this factor. They found that MPF appears in the oocyte before GVBD, disappears at the end of the first meiotic cycle, and reappears before metaphase II. Injection of cycloheximide does not prevent MPF amplification, nor does it suppress its oscillations during the meiotic cycle. MPF cycling is accelerated by treatment with the MT inhibitors colchicine and nocodazale. Injection of CSF suppresses MPF cycling. MPF disappears 8 min after fertilization or activation by an electric shock. Activated or fertilized eggs contain an inactivating agent that destroys injected MPF; this inhibitor is activated when MPF reappears at prophase. The reciprocal cycling of MPF and its inactivating agent is not affected by colchicine and is thus independent of aster formation in fertilized and activated eggs. Gerhart *et al.* (1984) have confirmed previous findings that cycloheximide inhibits the appearance of MPF before the initiation of meiosis; they found, in addition, that CSF injection blocks the disappearance of MPF at fertilization. They made the interesting suggestion that

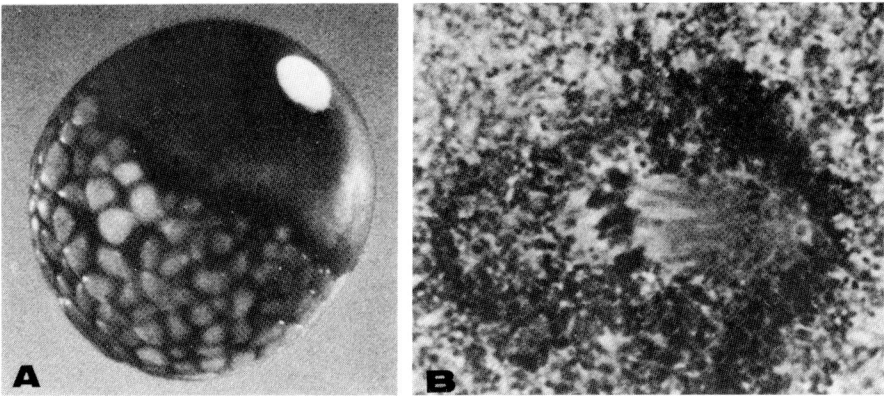

FIG. 35. Injection of the CSF present in unfertilized eggs into one of the two blastomeres of a cleaving frog arrests its cleavage. (A) Arrested injected blastomere on top, normal cleavage of the uninjected blastomere below. (B) Arrested mitosis (in metaphase) in the CSF-injected blastomere. [Masui, 1974.]

MPF activity is linked to the "cell cycle oscillator" observed in cleaving sea urchin and *Xenopus* eggs (see Section V). This endogenous cytoplasmic oscillator is believed to be responsible for the rhythmic contraction waves that take place in activated and fertilized eggs. Gerhart *et al.* (1984) suggested that this cytoplasmic oscillator is blocked in nonmaturing oocytes by an arrest system that is eliminated by progesterone treatment. Finally, since MPF-like molecules are found in all dividing cells, they propose to call MPF "M-phase–promoting factor" rather than "maturation-promoting factor."

Morphologically, the most impressive events during maturation are the agony and death of the huge GV. It appears as if the oocyte nucleus, exhausted by its continuous activity during months of oogenesis, breaks down and plays only a passive role as soon as the oocyte has been challenged with progesterone. As already shown, if enucleated oocytes are treated with the hormone, the usual increase in protein synthesis (including preferential synthesis of the core histones) and protein phosphorylation occurs. After progesterone stimulation, these enucleate oocytes also synthesize the all-important MPF (Masui and Markert, 1971). They undergo the same changes in ion permeability as progesterone-stimulated normal oocytes (Ziegler and Morrill, 1977). Oocytes enucleated before progesterone treatment display the same cortical changes (retraction of the microvilli, opening up of the cortical granules after parthenogenetic activation by pricking or treatment with the divalent ions ionophore A23187) as normal progesterone-treated oocytes. There is no doubt that the control of protein synthesis in amphibian oocytes takes place at the level of translation of preexisting mRNAs and not, as in the estrogen-stimulated amphibian liver (or hen oviduct), at the level of new transcriptional events.

Does this mean that the mixing of the nuclear sap with the cytoplasm from which it has been separated for months by the nuclear membrane has no biological importance? The answer is no, as we shall now see. First of all, the proteins that were present in the GV are not degraded when they mix with the surrounding cytoplasm. They are taken up by the nuclei during embryonic development, and they can still be detected in the nuclei of the tadpole cells (Dreyer and Hausen, 1983). This is in contrast to the disappearance of the nucleolar RNAs at the beginning of GVBD and suggests that the GV proteins play an important, but still unknown, role in development.

The nuclear sap of the GV is absolutely required for the swelling of injected spermatozoa (Katagiri and Moriya, 1976; Moriya and Katagiri, 1976; Skoblina, 1976) and for the condensation of chromatin in injected brain nuclei (Ziegler and Masui, 1973). These classic reactions of nuclei injected into the oocyte cytoplasm no longer occur if the oocytes have been enucleated first. Particularly important is the mixing of nuclear sap and its surrounding cytoplasm for MT assembly, which can easily be induced in *Xenopus* unfertilized eggs by treatment with D_2O or by injection of basal bodies. Figure 36, Chapter 5, Volume 1, shows the formation of numerous cytasters in D_2O-treated *Xenopus* unfertilized eggs (Van Assel and Brachet, 1968). As mentioned in Chapter 5, Volume 1, the simultaneous formation of hundreds of cytasters (which never divide) does not require protein synthesis. Oocytes with an intact GV never form cytasters when they are treated with heavy water (Heidemann and Kirschner, 1975). In the same paper, Heidemann and Kirschner showed that injection of basal bodies [microtubule-organizing centers (MTOC)] isolated from protozoa (*Tetrahymena* and *Chlamydomonas*) into the cytoplasm of unfertilized *Xenopus* eggs, but not of oocytes, induces the assembly of asters; this may be followed by the formation of anarchic furrows. Similar results have been obtained after injection into *Xenopus* unfertilized eggs of sea urchin sperm heads containing the proximal centriole (Maller *et al.*, 1976) or mitotic apparatuses isolated from cleaving sea urchin eggs (Forer *et al.*, 1977). Finer analysis by Heidemann *et al.* (1977) showed that induction of asters by injection of basal bodies is especially successful if the MTOC have been injected into the hyaloplasm of centrifuged *Xenopus* unfertilized eggs. Pretreatment of these MTOCs with protease or RNAse inactivates them. Finally, Heidemann and Kirschner (1978) showed that basal bodies do not induce asters in enucleated, progesterone-treated *Xenopus* oocytes. Injection of nuclear sap from a full-grown oocyte into such enucleate oocytes fails to restore the capacity of the injected MTOC to induce MT assembly. The obvious conclusion is that the nuclear sap changes when it mixes with the cytoplasm after the oocyte has been treated with progesterone. There is thus a cytoplasmic maturation in addition to the morphologically visible nuclear maturation. Promotion of aster formation by injected basal bodies and other MTOC clearly requires the mixing of the nuclear sap and the surrounding cytoplasm. There is no contradiction between this conclusion and the results obtained by Masui *et al.* (1978), who injected mitotic apparatuses of cleaving sea urchin eggs into enucleated unfertilized frog eggs. They obtained 40–60% partial blastulas, even if the eggs had been previously enucleated by pricking. However, in this case, enucleation means removal of the maturation spindle, not of the whole GV. The nuclear sap was mixed together with the cytoplasm several hours before the experiment was performed. The most interesting point in these experiments is that the sea urchin chromosomes apparently replicated and participated in cleavage in the frog egg.

Before leaving the subject of *in vivo* MT assembly, we should mention an interesting paper by Heidemann and Gallas (1980), who injected taxol (this substance, like D_2O and probably CSF, promotes the initiation and assembly of MTs) into oocytes and eggs of *Xenopus*. Taxol has no visible effect if it is injected into full-grown oocytes. However, if it is injected into unfertilized eggs, aster-like structures become visible in the cytoplasm and abnormal cleavage furrows are formed. Finally, if taxol is injected into fertilized eggs, cleavage is arrested; the injected blastomeres possess asters, but no spindles. Heidemann and Gallas (1980) conclude that the sperm centriole provides the missing ingredient required for the formation of true asters instead of fuzzy cytasters. We have obtained similar results (G. Steinert and J. Brachet, unpublished) by injecting taxol into progesterone-treated *Xenopus* oocytes. As shown in Fig. 36, taxol induces the formation of cytasters, provided that GVBD has taken place; maturation is abortive in the sense that the maturation spindle fails to assemble. These effects of taxol are very similar to those of heavy water (Hanocq-Quertier *et al.*, 1978). D_2O induces GVBD in full-grown oocytes, but the chromosomes do not

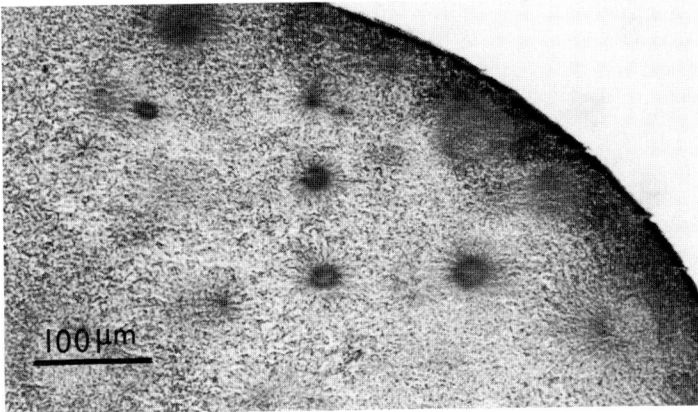

FIG. 36. Taxol induces the appearance of cytasters in oocytes undergoing maturation. In this case, taxol was injected in a full-grown *Xenopus* oocyte, in which maturation was then induced by injection of 50 ml cytoplasm taken out of a ripe oocyte. Fixation was performed 90 min after the injection. [Courtesy of Dr. D. Huchon.]

undergo complete condensation and the meiotic spindle does not assemble. It appears as if a critical degree of chromosome condensation is required for the kinetochores to become "competent" for spindle MT assembly. However, as already mentioned, hundreds of cytasters appear in the cytoplasm of D_2O-treated eggs as soon as the nuclear membrane of the GV has broken down.

Although maturation *sensu stricto* ends in *Xenopus* when the first polar body has been expelled and the egg is fertilizable, a few experiments on unfertilized, activated (by pricking with a fine needle or by an electric shock), and recently fertilized *Xenopus* eggs deserve mention. One of the most surprising experiments is that of Forbes *et al.* (1983a). They injected phage DNA into unfertilized *Xenopus* eggs and observed the appearance of structures similar to eukaryotic nuclei in the vicinity of the injected material. These structures are surrounded by a typical double membrane provided with the classic pore complex-lamina structure; they break down if mitosis is induced in the injected egg. Like cleavage nuclei, these "nuclei" swell after treatment with cycloheximide and break down after injection of CSF-containing cytoplasm. These experiments clearly demonstrate that nuclear assembly and breakdown are independent of specific DNA sequence information. Forbes *et al.* (1983a) also mentioned that formation of the nucleus-like structures can even be obtained *in vitro* with extracts of *Xenopus* eggs. No less interesting results have been obtained by Lohka and Masui (1983a,b). They treated demembranated *Xenopus* spermatozoa with extracts from activated eggs of *R. pipiens* and observed the formation of a nuclear envelope and the decondensation of chromatin; this was followed by DNA synthesis and condensation of chromosomes. To obtain this complete series of events, both soluble and particular components of the extract are needed. The 150,000-g supernatant produces only the initial dispersion of chromatin; cytoplasmic vesicles are needed to form the nuclear membrane. These experiments, as well as those of Forbes *et al.* (1983a), should make it possible for biochemists to identify factors responsible for nuclear membrane formation and breakdown, initiation of DNA synthesis, chromosome condensation, etc. Their impact on our understanding of the molecular mechanisms of mitosis might be decisive.

More recently, Lohka and Masui (1984) found that extracts from activated frog eggs transform demembranated *Xenopus* spermatozoa into pronuclei that resemble those present in fertilized eggs. The sperm chromatin first disperses, and a typical nuclear membrane (with pores and lamina) later surrounds it. Again, it is the particulate fraction of the homogenate, which probably contains fragments of the ER, that provides the nuclear membrane. These artificial pronuclei synthesize DNA; after 3 hr, their chromatin condenses in chromosome-like structures. Similar results have been obtained by Iwa and Katagiri (1984), who worked with the cytosol of toad oocytes that had undergone maturation. They found that the cytosol of unfertilized eggs is inactive unless the eggs have been activated; addition of Ca^{2+} to the extract from nonactivated eggs makes their cytosol very active. There is no chromatin-decondensing activity in the cytosol of full-grown oocytes and in that of enucleated, progesterone-treated oocytes. Development of this activity thus requires progesterone-induced GVBD. Iwa and Katagiri (1984) also reported that addition of EGTA (a Ca^{2+} chelator) or of serine protease inhibitors suppresses the ability of cytosol to decondense the sperm nuclei; this suggests that a serine protease might be involved in the decondensation of sperm chromatin. These experiments are important because they make it possible to manipulate chromatin condensation *in vitro* and pave the way for biochemical studies.

Another important paper (Karsenti *et al.*, 1984) deals with the role of the centrosomes and the nucleus in aster formation in *Xenopus* oocytes and eggs. Purified centrosomes were injected into oocytes or eggs in metaphase or interphase. The result was that no asters formed around the centrosomes in unfertilized eggs arrested in metaphase; activation of these eggs was followed by the appearance of asters around the injected centrosomes. In other experiments, Karsenti *et al.* (1984) injected cytoskeletons containing a nucleus and a centrosome into oocytes or eggs in metaphase; this time, mitotic spindles provided with centrosomes were formed. If karyoplasts devoid of centrosomes were injected, only anastral MT arrays formed around the condensing chromatin. Coinjection of a nucleus and a centrosome was followed by the formation of spindle-like structures with well-defined poles. Injection of centrosomes into activated, unfertilized eggs was followed by the parthenogenetic development of blastulas and even haploid tadpoles. Injected centrioles do not induce asters in oocytes, nonactivated unfertilized eggs, or fertilized eggs arrested by CSF injection. Finally, it was shown that the concentration of D_2O required to induce the formation of cytasters is higher in nonactivated (40%) than in activated (30%) eggs. This suggests that the critical tubulin concentration required for tubulin polymerization is higher in nonactivated than in activated eggs. The main conclusion to be drawn from all of these experiments is that association with the nucleus activates the centrosome; this important fact had been overlooked in the past.

Still speaking of freshly fertilized eggs, a few other findings deserve mention. Busby and Reeder (1982) found that both amplified (unmethylated) and chromosomal (methylated) rDNAs are equally well replicated after injection into *Xenopus* eggs. Regarding the initiation of DNA synthesis, Ford *et al.* (1983) showed that cycloheximide does not inhibit it, but does prevent its reinitiation; protein synthesis during the 30 min following fertilization is necessary for cleavage. Méchali *et al.* (1983) found that one of the phorbol esters (that promote carcinogenesis, TPA, see Chapter 3) induces DNA replication in *Xenopus* eggs when it is added. It has no effect if injected, and it must therefore act on the plasma membrane; it has no visible effect on oocytes. Analysis of these experiments suggest that DNA replication is controlled by a cytoplasmic clock in

Xenopus eggs. Sakai and Shinagawa (1983) have injected demembranated spermatozoa into anucleate fragments of fertilized *Xenopus* eggs; these fragments undergo a series of contraction and relaxation cycles. It was found that the injected sperm nuclei swell when the anucleate fragment contracts and remain condensed when it relaxes. Such experiments support the already mentioned cytoplasmic clock hypothesis.

Our studies on the induction of maturation in young, still vitellogenic oocytes have provided some information on *in vivo* MT assembly (Hanocq-Quertier *et al.*, 1976, 1978; Brachet *et al.*, 1976; Baltus *et al.*, 1977). *Xenopus* oocytes do not respond to progesterone stimulation unless they are full grown (1.1–1.3 mm in diameter). Smaller oocytes (0.9–1 mm in diameter) often respond by complete maturation (first polar body elimination) to progesterone or to the A23187 ionophore, provided that the medium contains excess Mg^{2+} or Ca^{2+}. Still smaller oocytes (0.6–0.8 mm in diameter) no longer respond to these stimuli, but undergo GVBD after injection of MPF-containing cytoplasm removed from full-grown, progesterone-treated oocytes. However, as shown in Fig. 37, maturation is abortive in such small, MPF-injected oocytes. GVBD takes place, the chromo-

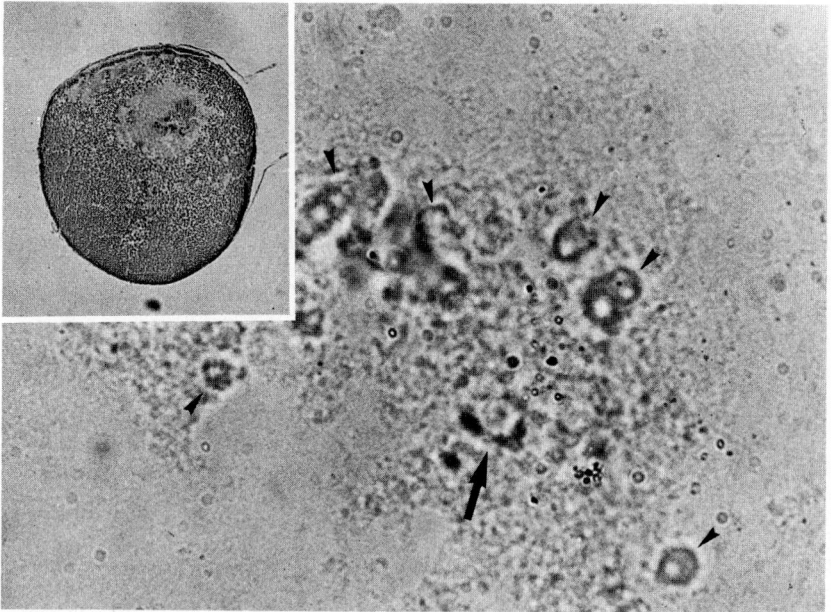

Fig. 37.   Partial maturation in a small (0.45 mm in diameter) *Xenopus* oocyte 5 hr after injection with MPF-containing cytoplasm from a full-grown, progesterone-treated oocyte. The nuclear membrane has broken down, but the nuclear sap has remained in its central position (see insert) and has not mixed with the surrounding cytoplasm. Arrows show condensed chromosomes, and arrowheads show degenerating nucleoli. [Brachet *et al.*, 1976.]

somes condense, and the nucleoli slowly disappear, but chromosomes remain scattered in the cytoplasm, the cortical microvilli do not retract, and a typical maturation spindle never develops. If cytoplasm from such a young oocyte, in which GVBD has been induced by MPF injection, is injected into a full-grown, progesterone-stimulated oocyte, a secondary spindle, carrying many chromosomes, is often found at the injection site (Fig. 38). Thus, polymerization of tubulin, which is already present in excess in vitellogenic oocytes (Pestell, 1975), does not take place in MPF-injected young oocytes. It does occur when their cytoplasm and chromosomes are transferred into full-grown, hormone-stimulated oocytes; the lack of spindle formation in the young oocytes must be due to some factor required for tubulin polymerization. In fact, in still unpublished experiments, we have obtained the formation of asters or spindles in some MPF-injected young oocytes by simultaneous injection of substances that are known to favor *in vitro* tubulin polymerization: protamines, MT-associated proteins (MAP), guanosine triphosphate (GTP), and nonhydrolyzable analogs of GTP (Fig. 39). Incidentally, GTP and its analogs inhibit maturation in progesterone-treated, full-grown oocytes, but not in MPF-injected large oocytes. This is probably due to the fact that GTP and its analogs, like cholera toxin, stimulate adenylate cyclase activity and increase the cAMP of the oocytes.

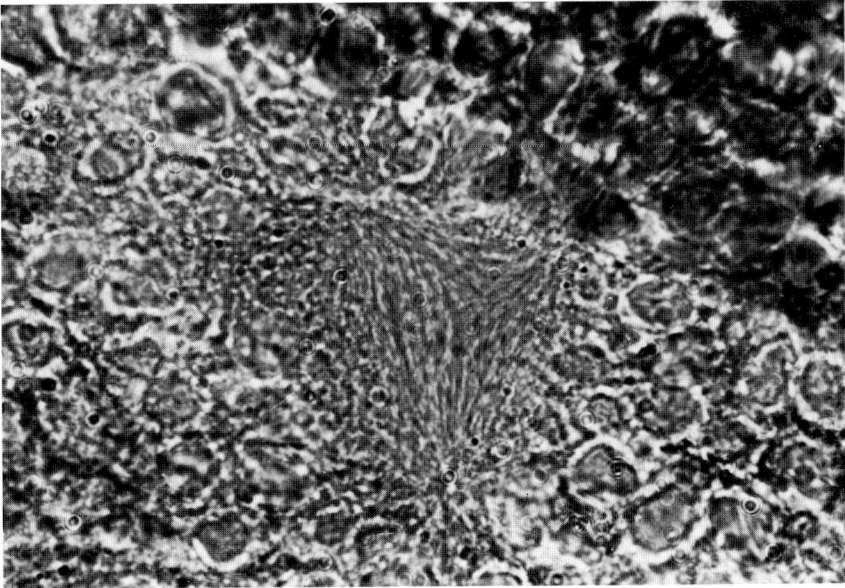

FIG. 38.   After transfer of cytoplasm from an oocyte, as in Fig. 37, into a full-grown, progesterone-treated oocyte, a secondary spindle carrying chromosomes may be found at the site of injection. [Brachet *et al.*, 1976.]

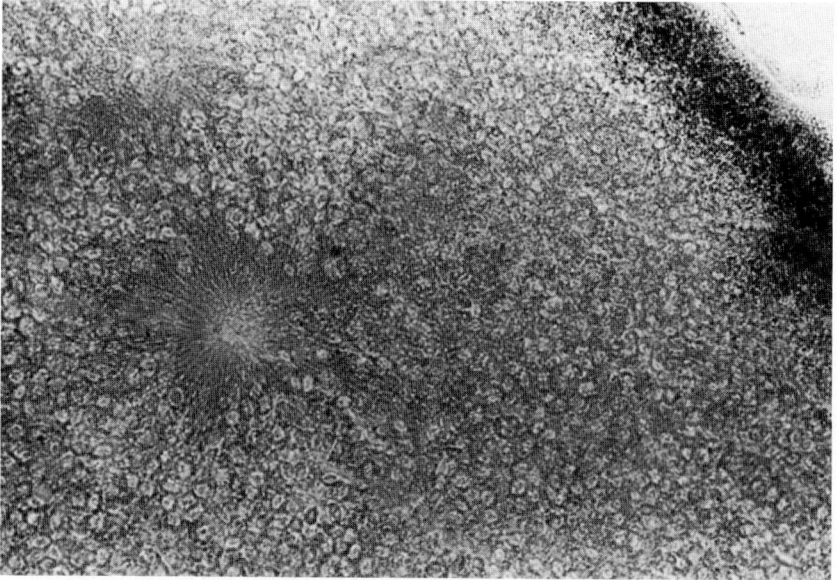

FIG. 39.   Small oocytes do not form spindles or asters after MPF injection. However, asters may appear in their cytoplasm if a substance that favors tubulin polymerization (in this case, protamine sulfate) is injected together with MPF. [Hanocq-Quertier *et al.*, 1978.]

As shown in Fig. 40, these experiments on cytoplasm transfer from young to full-grown oocytes show that the chromosomes of the young oocytes replicate in the cytoplasm of progesterone-treated large oocytes; in contrast, there is no replication of the recipient oocyte chromosomes [and of the nucleolar organizers DNA (Thomas *et al.*, 1977; Busby and Reeder, 1982)] that display the typical morphology of metaphase I. Similar results have been obtained in another kind of still unpublished experiments (E. Hubert and J. Brachet). Germinal vesicles were dissected out of very young previtellogenic oocytes and transferred into large oocytes; progesterone was then added. These experiments showed that the nuclear membrane of the "young" nuclei breaks down when the recipient progesterone-treated oocytes produce MPF. The nuclear membrane is thus "competent" to MPF in very young as well as in full-grown oocytes. While the large recipient oocytes form typical metaphase I spindles, the chromosomes of injected young GVs replicate and become attached to anarchic spindles and asters. Why chromosomal DNA in normal oocytes does not replicate and escapes the action of the DNA-synthesizing machinery developed during maturation is not known. It is probable (but remains to be proved) that the chromosomes of the full-grown oocytes are more condensed than those of the young oocytes that have been

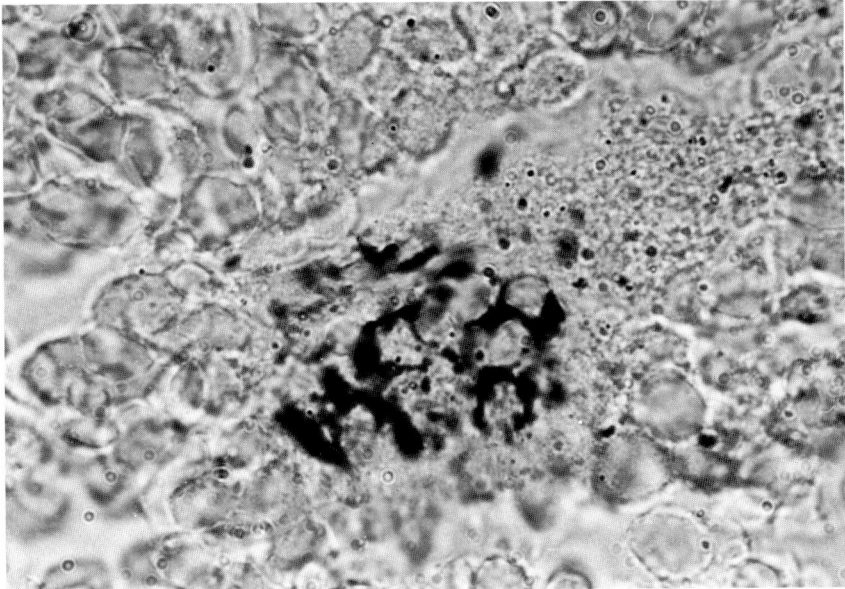

FIG. 40.   If the scattered chromosomes of MPF-injected young oocytes are transferred into large, progesterone-treated oocytes, they replicate and undergo pycnotic degeneration. [Brachet *et al.*, 1976.]

introduced into their cytoplasm. The molecular organization of metaphase I chromosomes remains to be studied.

Unfortunately, we know little about the biochemistry of small, still vitellogenic oocytes. A paper by Sadler and Maller (1983) provides some useful information. They found that vitellogenic oocytes, like full-grown oocytes, possess progesterone receptors, but there are five times fewer receptors on the surface of small oocytes than on that of large ones. Addition of progesterone induces the same inhibition of adenylate kinase activity at both stages; however, injection of the heat-stable inhibitor of a cAMP-dependent protein kinase induces GVBD only in the large oocytes. The conclusion was that the nonresponsiveness of the small oocytes to progesterone is due to the absence of factors located between the cAMP fluctuations and MPF action. More recently, Wasserman *et al.* (1984) have shown that vitellogenic oocytes, in contrast to full-grown oocytes, do not respond to progesterone by alkalinization of their cytoplasm, increase of S6 ribosomal protein phosphorylation and stimulation of protein synthesis. If their cytoplasm is made more alkaline by treatment with trimethylamine, procain or methylamine, S6 phosphorylation and protein synthesis are stimulated, but GVBD does not occur. Increases in these biochemical parameters are thus insufficient to induce maturation. Injection of MPF, as we had already shown, induces GVBD in small oocytes; there is, as in large oocytes, an amplification of the injected MPF; small oocytes must therefore possess a store of inactive MPF.

We now turn from *Xenopus* oocytes and eggs, from which we have learned so much, to sea urchin eggs. These eggs are not as bulky as those of the amphibians. Although this is an obvious disadvantage for injection experiments, it is a great advantage for more classic biochemical work in which yolk platelets and pigment granules are often a source of difficulty.

However, before we leave the subject of maturation, it should be added that *Xenopus* oocytes are by no means an exception. What has been said about them in this chapter remains valid, *mutatis mutandis,* for other animal species. For instance, injection of flagella and centrioles from sea urchin spermatozoa into unfertilized eggs of the fish *Oryzyas* induces the formation of small and large asters and of irregular cleavage (Iwamatsu and Ohta, 1974; Iwamatsu *et al.*, 1976; Ohta and Iwamatsu, 1980).

Extensive work (reviewed by Masui and Clarke, 1979, and by Meijer and Guerrier, 1984) has been done on maturation in starfish oocytes since Kanatani *et al.* (1969) discovered that it can be induced by the addition of small amounts of 1-methyladenine (1-MA). This purine, like progesterone in nonmammalian vertebrates, is produced by the follicle cells that surround starfish oocytes. 1-MA, again like progesterone, induces the formation of an MPF. It has been shown that this MPF is formed even when anucleate starfish oocytes are treated with the hormone. However, it is produced in smaller amounts than in normal oocytes, and it is not amplified in the absence of the nucleus (Kishimoto *et al.*, 1981). As expected, starfish, *Xenopus,* and mammalian cell MPFs display no species specificity (Kishimoto *et al.*, 1982). An early response to 1-MA, as in amphibians, is an increase in free Ca^{2+} (Moreau *et al.*, 1978). The importance of calcium ions for starfish maturation, which was discovered by A. Dalcq in 1928, is shown by the fact that GVBD, after treatment with 1-MA, is delayed in Ca^{2+}-free seawater. However, 1-MA–induced maturation does not occur in the absence of Mg^{2+}. In addition, there is an increase in free Mg^{2+} as well as free Ca^{2+} after hormone addition (Rosenberg and Lee, 1981). As in amphibian eggs treated with progesterone, 1-MA binds to the cell surface of starfish oocytes. After this binding, maturation is no longer hormone dependent. Meijer and Guerrier (1981) have observed that starfish maturation is inhibited by trifluoperazine, a calmodulin inhibitor, and by vinblastine during the early hormone-dependent phase (Dorée *et al.*, 1982). There are indications that calmodulin is active at the cell membrane level. It seems that cAMP and cAMP-dependent protein kinases play a less important role in starfish than in amphibian maturation. More important is a Ca^{2+}-calmodulin protein kinase present in hormone-treated starfish oocytes (Mazzei *et al.*, 1981). According to Dorée *et al.* (1981), Ca^{2+} release, which is an essential step, according to Picard and Dorée (1983), continues until the transduction (intramembrane transfer) of the hormone is complete; this transduction is modulated by cAMP and by protein membrane phosphorylation. Indeed 1-MA induces the selective phosphorylation of an M_r 60,000 protein. Finally, 1-MA increases the rate of protein synthesis in both intact and anucleate oocytes; however, in starfish oocytes, inhibition of protein synthesis does not prevent the production (and amplification) of MPF and its effects on GVBD (Dorée et al., 1982). The pattern of protein synthesis changes during maturation in starfishes; as in amphibians, these changes are the same in anucleated halves and whole oocytes after 1-MA treatment (Martindale and Brandhorst, 1984). Clearly, there are great similarities and minor differences when one compares maturation in amphibian and starfish oocytes. One of the differences is that, as we have seen, maturation can be induced in *Xenopus* by treatment with organomercurials, which block the cell surface —SH groups (Brachet et al., 1975a). However, the reverse occurs with starfish oocytes, in which GVBD can be induced by treatment with dithiols, dithiothreitol in particular (Kishimoto and Kanatani, 1973). Such thiols are toxic for *Xenopus* oocytes, in which they merely induce a swelling of the GV (Pays et al., 1977). It thus seems that the sulfhydryl–disulfide equilibrium is important for the production of MPF and the ensuing GVBD, but the respective roles of —SH and —SS are opposite in starfishes and amphibians.

There is another difference between amphibian and starfish oocytes. Pricking the GV of a full-grown *Xenopus* or *Rana* oocyte with a needle has only detrimentral effects; cytolysis ensues sooner or later. In contrast, mechanical disruption of the GV induces "cytoplasmic maturation" in starfish oocytes. Emission of the polar bodies takes place. If such oocytes are fertilized, an aster forms around the sperm head. The sperm nucleus swells and becomes a pronucleus. This happens without treatment with 1-MA (Guerrier et al., 1983). Finally, Meijer et al. (1982) have reported that maturation in the echiuroid worm *Urechis* is accompanied, as in *Xenopus,* by an increase in protein phosphorylation; in *Urechis,* there is a concomitant increase in protein kinase activity.

The little we know about fish eggs (induction of maturation by steroids, production of MPF) indicates great similarities to the better-studied *Xenopus* oocytes. However, a different situation occurs in mammals (reviewed by Eppig and Downs, 1984) because their oocytes undergo spontaneous maturation as soon as they are removed from their follicles. It appears as if the follicle fluid contains a maturation inhibitor. As in amphibians, maturation can be prevented by the addition of cAMP or agents that increase the cAMP level (Cho et al., 1974; Schultz et al., 1983; Sato and Koide, 1984). Delicate experiments of fusion between young or full-grown oocytes with oocytes at metaphase I or blastomeres at the four-cell stage have shown that, as in amphibians, there is no MPF in small oocytes and that it is present only in large ones (Bałakier, 1978; Bałakier and Crołowska, 1977; Tarkowski and Bałakier, 1980). It is, therefore, likely that maturation, which is a very important and complex stage of development, follows the same general course in the entire animal kingdom. There are, of course, species differences in both morphology and biochemistry during maturation, but there is unity in the achievement of the final result: the production of a fertilizable egg.

IV. FERTILIZATION OF SEA URCHIN EGGS: ACTIVATION OF ANUCLEATE FRAGMENTS

A. INTRODUCTION

One of the main purposes of this section is to compare, from the morphological and biochemical viewpoints, nucleate and anucleate fragments of sea urchin eggs. However, the biological importance of fertilization is such that a brief description of the major changes that take place after egg insemination is necessary. Fertilization restores diploidy to the egg. This is a necessity in most animal species, since haploids obtained by parthenogenetic activation of unfertilized eggs are generally lethal. The spermatozoon brings to the egg the paternal heredity characteristics, which is of crucial importance to the life of the zygote. In many species, sex determination is genetically controlled and the sex of the offspring is already established when the two gamete nuclei fuse together at amphimixy. Last, but not least, fertilization marks the onset of embryonic development.

Unfertilized sea urchin eggs can easily be cut into two halves (nucleate and anucleate) by high-speed centrifugation in a density gradient (Harvey, 1933 (Fig. 41). Large amounts of heavy, pigmented anucleate halves can be obtained with this simple method. This works particularly well with *Arbacia* eggs because their pigment-containing granules have a very high density. Harvey's method has been used to cut in half unfertilized eggs of the sea urchins *Sphaerechinus* (Aimi, 1974) and *Paracentrotus* (Rinaldi *et al.*, 1979a,b), as well as those of the starfish

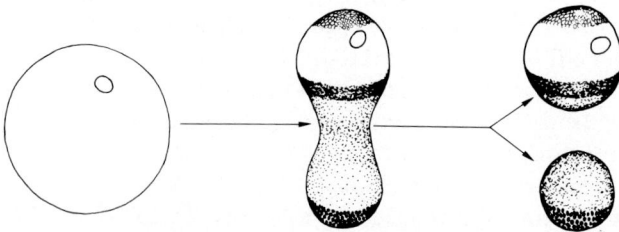

FIG. 41.   High-speed centrifugation of unfertilized *Arbacia* eggs in a density gradient leads to sedimentation of the various organelles in the intact egg; the centrifuged egg develops a dumbbell shape and finally breaks into two fragments. The light fragment contains the nucleus, lipids, mitochondria, a little yolk, and the majority of the hyaloplasm. The heavy fragment is anucleate; it contains mitochondria, most of the yolk, and, at the centrifugal end, red pigment granules. [Drawn by P. Van Gansen after Harvey, 1936.]

Astropecten (Dale *et al.,* 1979). Both nucleate and anucleate fragments can easily be fertilized or treated with parthenogenetic agents.

In this section, we shall deal almost exclusively with sea urchin eggs, because they have long been and still remain a favorite material for embryologists. We have obtained a wealth of information about sea urchin egg fertilization, which can be generalized to a certain extent, to eggs of other species. However, it should be pointed out that the morphological and biochemical changes that take place at fertilization are particularly intense in sea urchin eggs, which do not represent a universal model.

B. DESCRIPTION[4]

When they are shed, unfertilized sea urchin eggs have completed their two meiotic divisions and are thus at the ootide stage. They possess a single haploid nucleus: the female pronucleus. First, when the spermatozoa come in contact with the mucous jelly that surrounds the unfertilized eggs, they undergo the acrosome reaction (Fig. 42a,b). The acrosome, as we have seen, is a specialized lysosome located at the tip of the sperm head. It opens up when the swimming spermatozoa penetrate the egg jelly coat. The acrosome contains an acrosomal granule composed of unpolymerized actin (G-actin). When its membrane has broken down, G-actin undergoes polymerization and forms a thin, flexible filament, the acrosome filament. Simultaneously, the acid hydrolases present in the acrosome are released and contribute to the digestion of the envelopes that surround and protect the egg (jelly coat and vitelline membrane). Binding of the spermatozoon to the egg surface, which is immediately followed by fusion between the acrosome membrane and the egg plasma membrane, results from an interaction (of the lectin receptor type) between specific receptors present on the vitelline membrane and a specific glycoprotein (bindin) located on the acrosome membrane. [See the extensive review by Vacquier (1981), who discovered bindin.] The egg surface receptor is also a glycoprotein. Kinsey and Lennarz (1981) have discovered a glycopeptide (M_r 6000) that they believe to be an active fragment of the egg receptor. At very low concentrations, it inhibits sperm–egg binding and fertilization. Thus, the inhibitory effect is directed against the spermatozoa, not the eggs. Since both the sperm bindin and the egg surface receptor display species specificity, it is understandable that cross-hybridization between sea urchin species does not occur in nature unless the species are very closely related.

Addition of sperm to unfertilized sea urchin eggs is followed, within a few seconds, by a spectacular reaction of the egg (which is easily visible under the light microscope). It is the cortical or activation reaction that ends with the uplifting of a fertilization membrane around the egg (Fig. 43). Between the egg

[4]See the review by Monroy and Rosati (1983) of sperm–egg interactions.

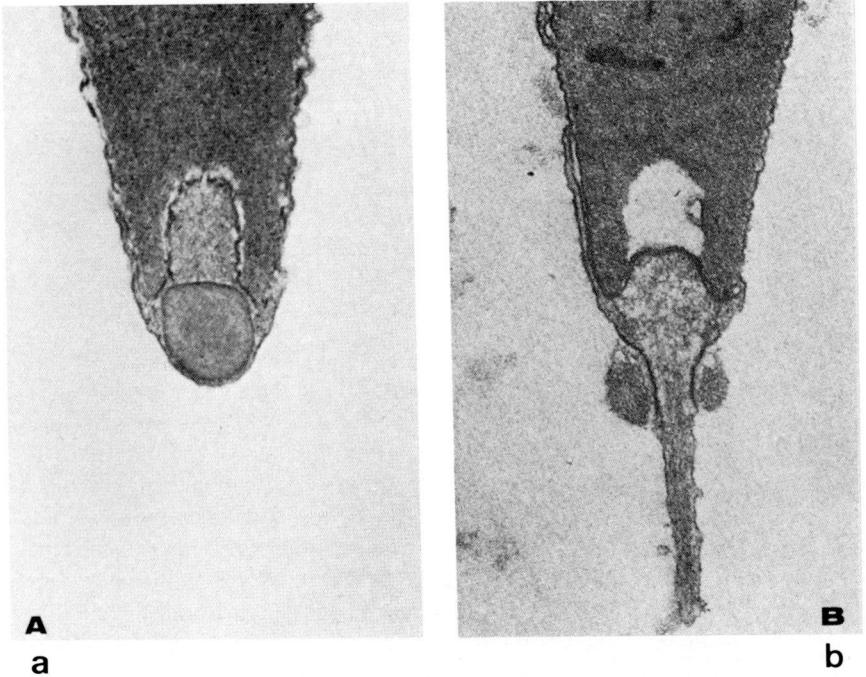

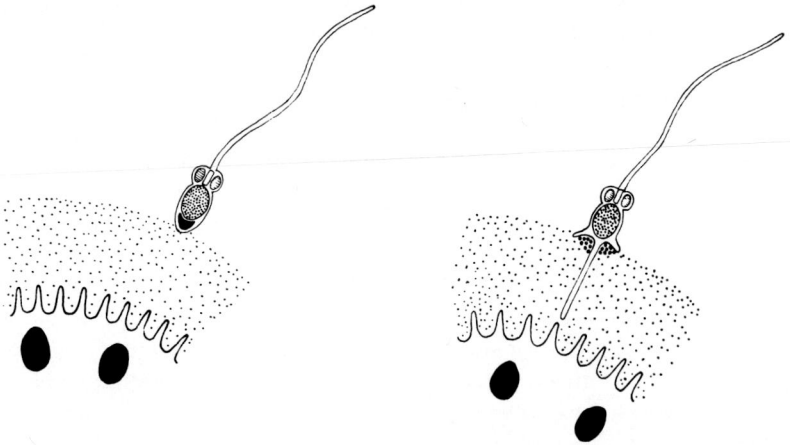

Fig. 42. The acrosome reaction (A, B). Electron micrographs of sperm heads. (A) Unreacted acrosome. The acrosome granule contains a protease (acrosin) and unpolymerized actin. (B) Reacted acrosome. G-actin has polymerized, and the acrosomal filament is composed mainly of F-actin. [Epel, 1977.] (a, b) Schematic representation of the acrosome reaction. (a) A spermatozoon comes in contact with the jelly that surrounds the egg. The egg surface is covered with microvilli. Cortical granules (in black) lie under the egg plasma membrane. (b) Contact with the jelly coat elicits the acrosome reaction. [Drawn by P. Van Gansen.]

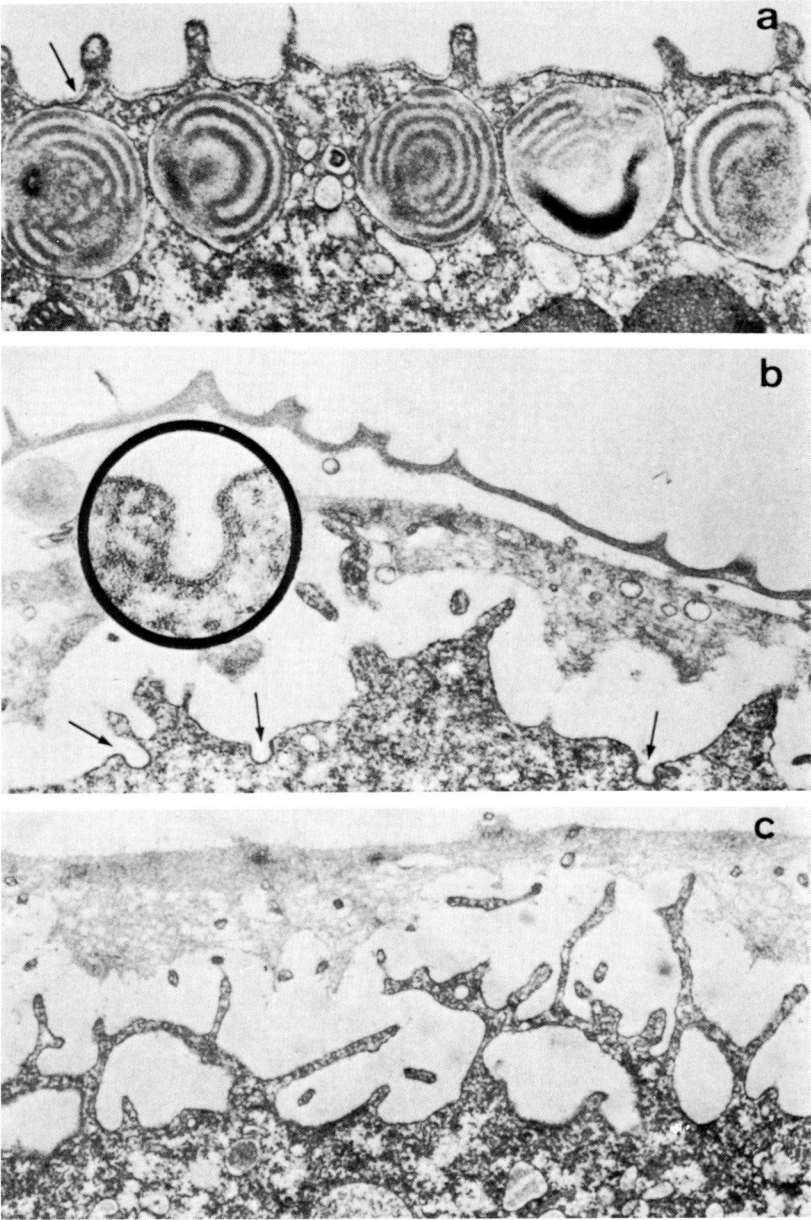

FIG. 43. The cortical reaction. (a) Electron micrograph of the cortex of an unfertilized sea urchin egg. Cortical granules (also called "cortical vesicles") lie in a single layer just beneath the plasma membrane and have typically striated contents. The vitelline layer is the thin, fuzzy coat just outside the plasma membrane (arrow). (b) Two minutes after insemination, numerous invaginations (arrows) are observed between highly arborized microvilli; they resemble the coated pits typical of endocytosis (inset). (c) Five minutes after insemination, the microvilli have long, thin branches that may join to form tunnels or cavities. [Reproduced from *J. Cell Biol.* **83,** 94, by copyright permission of the Rockefeller University Press.]

surface and the fertilization membrane the perivitelline space is formed, which is filled with perivitelline fluid. The activation reaction (which can also be induced by classic parthenogenetic treatments) results from complex, dynamic changes in the egg cortex (which have been reviewed in detail by Vacquier, 1981). The most important of these are a general contraction of the egg, a change in the number and size of the microvilli (Fig. 44) [due to local increases in free Ca^{2+}

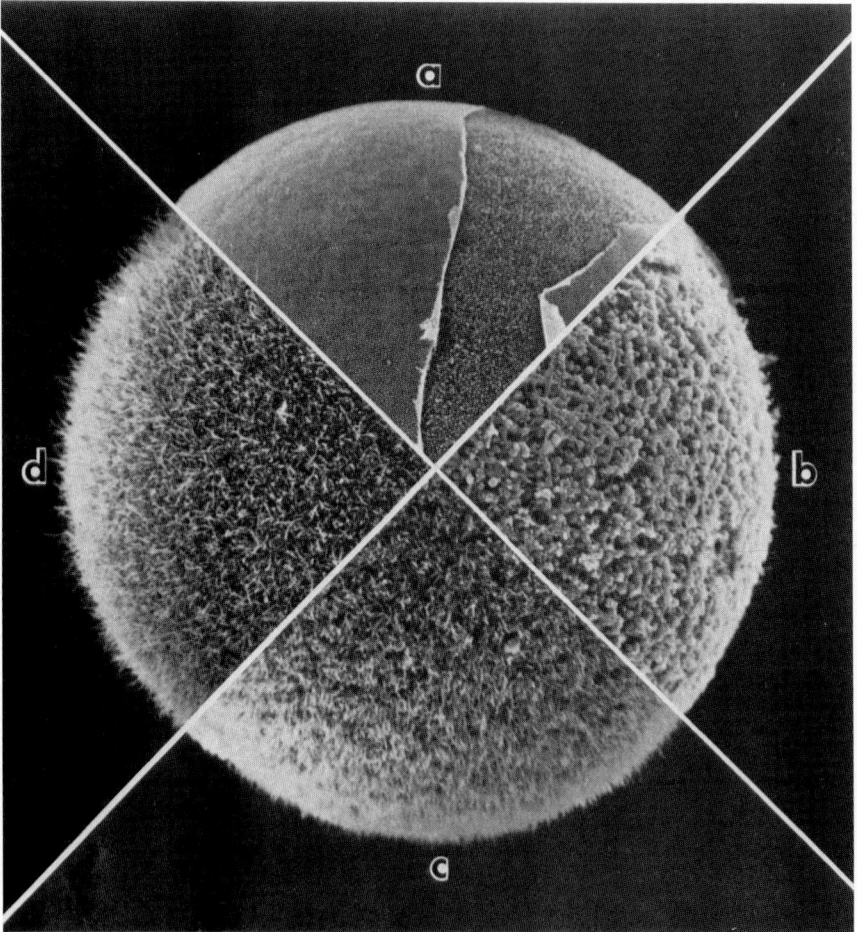

FIG. 44. A composite of scanning electron microscope (SEM) images of plasma membrane surfaces of sea urchin eggs at selected stages. (a) Before fertilization; (b) 1 min after fertilization, when the surface is obscured by globules from the cortical vesicles; (c) 5 min after fertilization; (d) 13 min after fertilization, when the presence of a few long microvilli give a distinctive fuzzy appearance. [Schroeder, 1979.]

and pH, according to Begg *et al.*, 1982] (Fig. 45) and the exocytosis of the contents of the cortical granules that lie immediately underneath the egg plasm membrane (Fig. 43). All of these changes begin at the point of sperm entry and spread to the opposite pole within a few seconds in sea urchin eggs (the cortical reaction takes 1 hr or more in the much larger amphibian eggs). The exocytosis of the cortical granules is followed, within 3–5 min, by membrane retrieval; this occurs through clathrin-coated pits and is probably a Ca^{2+}-dependent event (Fisher and Rebhun, 1983).

The main constituents of the cortical granules are glycoproteins. When they are released by exocytosis in the surrounding medium, some of them become part of a thin, transparent layer that closely adheres to the egg surface (the so-called hyaline layer); others remain in the perivitelline fluid. A majority of the cortical granule glycoproteins contribute to the formation of the fertilization membrane. Once outside the egg, they undergo precipitation, polymerization by the formation of disulfide linkages, and, finally, crystallization on the former vitelline membrane (Fig. 43). Treatments with reducing agents such as di-

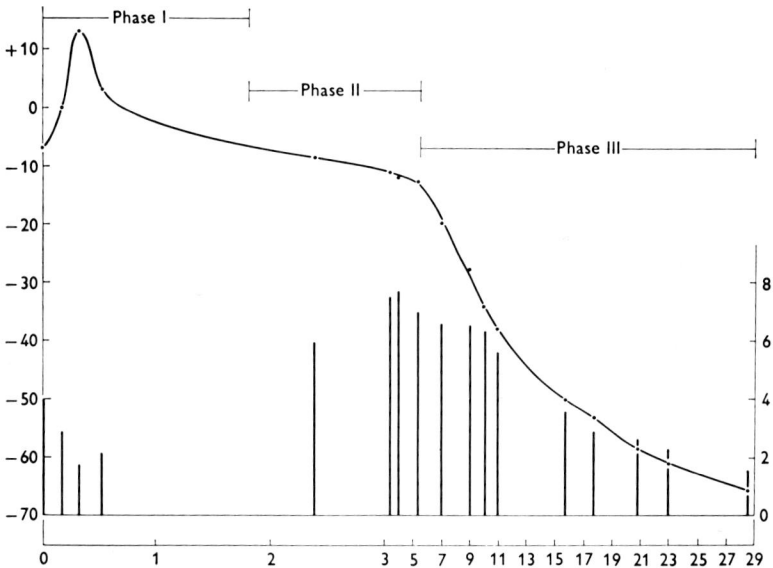

FIG. 45. Membrane potential and resistance during the activation of a sea urchin egg. Abscissa: time (minutes). Ordinate (left): Vm (mv); (right): Rm (KΩ cm²). Vm refers to the line graph of membrane potential in millivolts; Rm refers to the bar graphs of membrane resistance in kilo-ohms per square centimeter. The bar graph at time zero indicates the resistance of the unfertilized egg. After a rapid depolarization (phase I), the membrane potential remains almost constant during the cortical reaction (phase II); this is followed by a marked hyperpolarization due to the development of a K^+ conductance (phase III). [Steinhardt *et al.*, 1972.]

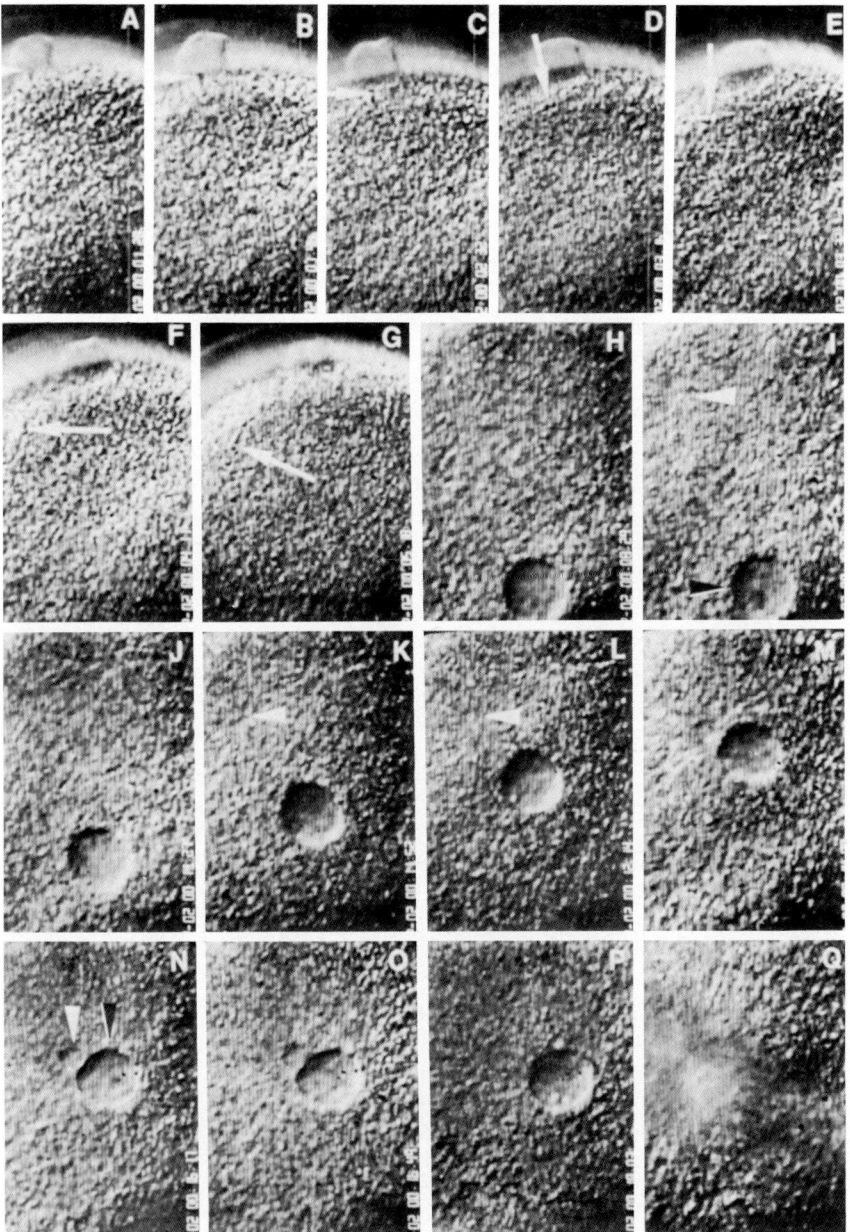

FIG. 46. Amphimixy in sea urchin eggs shown by microcinematography. Movements after fertilization. Sperm (white triangle) incorporation involves first the extension of the fertilization cone around the sperm head and midpiece (A, B) and then the rotation (B–D) and lateral displacement of the sperm along the subsurface region of the egg (D–G). The formation of the spermaster (white triangle) moves the rotated male pronucleus centripetally (I). The female pronucleus (black triangle)

thiodiglycol or with trypsin prevent the formation of the fertilization membrane; denuded fertilized eggs are thus obtained. The experiments on isolated sea urchin egg cortices have demonstrated that exocytosis of the cortical granules occurs if there is an increase in the free Ca^{2+} concentration of the medium. One of the functions of the cortical reaction is to prevent polyspermy, the entrance of more than one spermatozoon into the egg. In particular, if one removes the hyaline layer by treatment with a Ca^{2+}-chelating agent or with alkaline seawater, an already fertilized egg can be refertilized (Sugiyama, 1951). However, this block to polyspermy is a relatively late event; it follows an early block to polyspermy due to very rapid changes in the electrical properties of the egg surface.

Soon after the spermatozoon has penetrated the egg cortex a powerful monaster forms around its proximal centriole. This spermaster grows, occupying the major part of the egg, and regresses (Fig. 46) prior to fusion of the male and female pronuclei (amphimixy). The formation and regression of the spermaster have been accurately described by Harris et al. (1980), who detected polymerized tubulin by immunofluorescence in whole sea urchin eggs. According to Bestor and Schatten (1981), there are no MTs in unfertilized eggs. They appear around the sperm head and are required for the migration of the pronuclei, which is inhibited by colchicine, griseofulvin, and taxol (Schatten et al., 1981). Entry of the spermatozoon is inhibited by cytochalasin B but not by colchicine. It is thus mediated by the cortical MFs, while the motility of the pronuclei is MT mediated (Schatten and Schatten, 1981). The deep changes in the organization of the cytoskeleton which take place in sea urchin eggs at fertilization have been reviewed in great detail by Schatten (1984). In particular, the role of the microtubules and of the microfilaments in the movements which bring the pronuclei together at amphimixy are discussed in detail.

Amphimixy marks the end of fertilization, since the zygote diploid nucleus immediately enters the prophase of first cleavage. Its far-reaching consequences will last until senility and death.

C. PHYSICAL AND CHEMICAL CHANGES AT FERTILIZATION

1. Early Changes

After this brief outline of the morphological changes that take place at fertilization in sea urchins, we shall examine the major physical and chemical events that occur at that time. More information can be found in the excellent review by Epel (1978).

---

migrates to the center of the spermaster (J–L). The adjacent pronuclei are then moved to the center of the egg by the continued growth of the sperm's astral rays (N). Syngamy occurs when the male pronucleus (white triangle) coalesces with the female pronucleus (black triangle, N–P). First division occurs thereafter (Q). [Schatten and Schatten, 1981.]

The normal inducer of the acrosome reaction is the mucous jelly that surrounds the unfertilized eggs. However, the acrosome reaction can be obtained by a variety of agents, such as treatment with alkaline seawater, with the ionophore nigericin (which exchanges potassium ions for protons), or by the mere contact of the spermatozoa with glass. Schackmann et al. (1981) have shown that the pH_i of the acrosome is acid, as it is for lysosomes. Induction of the acrosome reaction by addition of egg jelly raises the acrosome's pH_i by 0.1–0.2 and produces the collapse of the membrane potential. Another effect of the egg jelly is to dephosphorylate almost immediately a 160,000-dalton phosphoprotein present in the membrane of the sperm flagellum (Ward and Vacquier, 1983).

In the egg, the first recorded change is a brief depolarization of the membrane potential. This action potential [which is preceded by a rapid but weak electrical transient, [according to Hülser and Schatten (1982)] lasts for 10 sec and is followed by a recovery of the initial resting potential within 100 sec (Fig. 45). It is believed to be due to an influx of Na^+ ions. During the cortical reaction (discharge of the content of the cortical granules and uplifting of the fertilization membrane), the membrane potential remains constant. This plateau is followed by a strong hyperpolarization due to an increase in permeability to potassium ions (development of a K^+-conductance) (Steinhardt et al., 1972). The initial Na^+-dependent depolarization of the plasma membrane potential is usually believed to be responsible for the rapid block to polyspermy. Grey et al. (1982) showed that, in Xenopus as well as in sea urchin eggs, monospermy is due to a rapid electrical block (membrane depolarization). Treatment with salts (NaI, in particular) in Xenopus eggs produces a membrane hyperpolarization leading to polyspermy, which cannot be prevented by the slow block to polyspermy correlated with the elevation of the fertilization membrane. However, electrical blocks to polyspermy are not universal. No changes in membrane potential can be recorded in mouse eggs, although monospermy is the rule in mammals (Jaffe et al., 1983).

The sequence of early events at sea urchin egg fertilization has been recently reinvestigated by Eisen et al. (1984): membrane depolarization and surface contraction are the first recorded events; they are followed by the increase in free Ca^{2+} which will now be discussed; it takes place 23 to 40 sec. after insemination and is followed by an increase in the NADPH content (51 sec. after sperm addition). All these early changes precede the elevation of a fertilization membrane.

Among the ionic changes occurring at fertilization (reviewed by Whitaker and Steinhardt, 1982) is the mobilization of sequestered calcium ions. This increase in free Ca^{2+} (much larger than that which follows progesterone addition to Xenopus oocytes) has been monitored in eggs that had been injected with aequorin and then fertilized; Fig. 47 shows the large flash of luminescence that can be registered under such conditions. It is now generally accepted that the breakdown of the cortical granules and the exocytosis of their contents are due to a

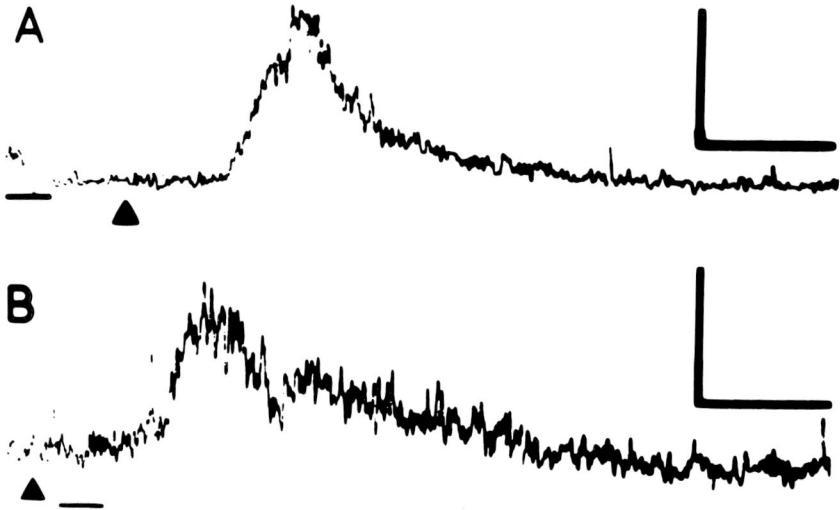

FIG. 47. Calcium release at fertilization. Between 16 and 27 unfertilized eggs were injected with aequorin and then fertilized; the fluorescence was recorded continuously. Sperm addition at arrows. Time base is a horizontal bar equal to 1 min. (A) Best aequorin emission. Vertical bar, 2 nA. (B) Typical aequorin emission. Vertical bar, 1 nA. In both figures, the lower left horizontal bar indicates zero light level. [Steinhardt et al., 1977.]

local increase in free calcium ions (see Vacquier, 1981, for details). Steinhardt and Alderton (1982) have further shown that calmodulin localized in the plasma membrane is responsible for cortical granule exocytosis. The importance of this Ca^{2+} surge for fertilization was first demonstrated by Steinhardt and Epel (1974) when they discovered that the Ca^{2+} ionophore A23187 is an excellent activating agent not only for sea urchin eggs but also for those of many animal species. It induces the cortical reaction and the formation of monasters as a result of the liberation of free Ca^{2+}. However, the presence of Ca^{2+} in the medium is not required if the eggs are fertilized with acrosome-reacted sperm; this shows that sperm releases Ca^{2+} from intracellular storage (Schmidt et al., 1982).

The biochemical mechanisms that lead to Ca^{2+} release and the localization of the calcium intracellular stores are not yet known in detail. By analogy with other systems, an early response of the egg to the sperm or activating agents may be the hydrolysis of membrane phospholipids (phosphoinositides) in diacyl glycerol and inositol trisphosphate (reviewed by Berridge, 1984). Inositol trisphosphate mobilizes intracellular Ca^{2+}; diacetyl glycerol activates the phospholipid, Ca^{2+}-dependent protein kinase C, which phosphorylates many proteins and plays an important role in cell proliferation. The calcium store is more probably localized in ER vesicles then in the mitochondria. This hypothesis has been verified by Whitaker and Irvine (1985). Injection of inositol trisphosphate into unfertilized sea urchin eggs induces a cortical reaction. The source of calcium is the Er. Free Ca^{2+} then accumulates in the mitochondria (Eysen and Reynolds, 1985). In *Xenopus* eggs also, the free Ca^{2+} concentration increases three times soon after insemination. A free Ca^{2+} wave moves from the animal to the vegetal pole at a rate of 10 μm/sec, coincident with cortical granules breakdown (Busa and Nuccitelli, 1985).

Interestingly, activation of sea urchin eggs can also be obtained, but with very different consequences, by treatment with ammonia (Mazia and Ruby, 1974; Wilt and Mazia, 1974). In this case, there is no cortical reaction and the cortical granules remain intact; nevertheless, chromosome condensation and DNA replication leading to polyploidy take place in the absence of cytokinesis. Ammonia treatment thus bypasses the calcium increase step, which is necessary for fertilization membrane formation and uplift. Ammonia easily penetrates unfertilized sea urchin eggs and raises their pH_i. This increase has far-reaching consequences for the synthesis of macromolecules. An important point is that, as found by Johnson *et al.* (1976), normal fertilization also induces, within 1–4 min, an increase in the pH_i of the egg as great as 0.3–0.5. This increase is due to an exchange between extracellular Na^+ and intracellular protons, which are ejected into the outer medium. This exchange reaction explains why fertilization is no longer possible when the NaCl content of the seawater drops below 2.5 mM as a result of substitution of NaCl by choline. Furthermore, amiloride (which inhibits passive Na^+ uptake) blocks both Na^+/H^+ exchange and activation of sea urchin eggs. In an analysis of these processes using improved techniques, Johnson and Epel (1981a,b) concluded that the pH_i of sea urchin eggs rises from 6.8 to 7.23 after fertilization and does not change during the first mitotic cycles. This pH is controlled by an Na^+-independent mechanism in unfertilized eggs and by an Na^+-dependent mechanism after sperm addition. The authors mention that sea urchin eggs possess lysosomes that are broken down by treatment with ammonia or nigericin. Whether an increase in pH_i at fertilization is a general phenomenon is not yet known. That this might well be the case has been shown by Webb and Nuccitelli (1981), who reported that the pH_i of unfertilized *Xenopus* eggs is 7.38, reaching 7.67 1 hr after fertilization. As in sea urchins, the pH_i does not change during the first mitotic cycles. However, in contrast to sea urchin eggs, the increase at fertilization of the pH_i is not affected by amiloride and is thus probably not Na^+ dependent in *Xenopus* eggs. It should be added that during maturation and fertilization of starfish eggs, the increase in the pH_i is modest. Sea urchin eggs are an extreme case, because the pH_i of the unfertilized eggs is particularly low (Johnson and Epel, 1983). We shall see that increases in oxygen consumption and protein synthesis at fertilization are exceptionally great in sea urchin eggs; it is likely that the low pH_i of the unfertilized sea urchin eggs is responsible for their very low metabolism.

Thus, in addition to the classic method of J. Loeb for the induction of parthenogenesis in sea urchin eggs (successive treatments with hypertonic seawater and butyric acid), we now possess two very valuable methods for activating eggs: (1) treatment with the ionophore A23187, which allows the analysis of early events linked to the Ca^{2+} surge (cortical reaction); (2) treatment with ammonia, which bypasses the Ca^{2+} increase step, permitting the study of later events (induction of protein and DNA synthesis). Treatment of sea urchin eggs with procaine (Vacquier and Brandriff, 1975; Mazia et al., 1975) has the same effect as ammonia (DNA replication without cortical granule breakdown).

Of interest for cell biologists is the finding that, although eggs treated with all of these methods develop asters, only the asters formed after treatment with Loeb's two-step method possess centrioles (Moy et al., 1977; Miki-Noumura, 1977). With all of the other techniques of activation that, in contrast to Loeb's method, do not lead to further parthenogenetic development, the aster MTs originate from osmiophilic foci and the condensed chromosomes. Closer ultrastructural study of sea urchin eggs treated with a variety of methods inducing either activation or true parthenogenetic development is already yielding interesting information about the mechanisms of centriole self-assembly in eggs. For instance, Kallenbach (1982) and Kallenbach and Mazia (1982) have followed the progressive appearance of centrioles in unfertilized eggs continuously treated with hypertonic seawater. At first, no monasters are formed, but asters can be detected after a 3-hr treatment; they derive from osmiophilic centriolar precursor bodies in the vicinity of the nuclear membrane. Centrioles mature in the center of cytasters and finally form diplosomes soon before prophase. Kuriyama and Borisy (1983) have studied the effects of a variety of parthenogenetic agents on centriole and aster formation in sea urchin eggs. Agents that act in a single step (acid or alkaline seawater with the addition of procaine or thymol) induce the formation of monasters devoid of centrioles; a second step (D_2O, ethanol, hypertonic seawater) induces the appearance of cytasters centered on one to eight centrioles. Obviously, centrioles can be dispensed with in MTOC. Similar results have been reported by Kallenbach (1983). Two-step parthenogenesis induces the formation of cytasters and centrioles; all of the cytasters have one or more centrioles. Kallenbach (1983) has suggested that parthenogenetic agents, which tend to decrease protein synthesis, inhibit the synthesis of certain proteins that repress centriole formulation in sea urchin eggs; however, experimental proof for that hypothesis is still lacking.

Although not strictly related to centriole formation in eggs, two further observations deserve mention. Hirano (1982) isolated a sea urchin sperm centriolar fraction and found that it loses its MTOC activity after treatment with trypsin; it is not inactivated by RNase or DNase digestion. Injection of RNase into fertilized sea urchin eggs inhibits the breakdown of the nuclear membrane at the onset of the first cleavage division; this is in agreement with the results we obtained many years ago on amphibian eggs treated with this enzyme (Brachet and Ledoux, 1955; see Chapter 5, Volume 1, and Fig. 31). The second observation was made by Wadsworth and Sloboda (1983). They injected a fluorescent derivative of tubulin into fertilized sea urchin eggs and found that it was readily incorporated into the astral fibers (except when the asters reached their maximal size at anaphase). Treatments that inhibited tubulin polymerization (cold, colchicine) decreased the fluorescence of the astral fibers. These experiments clearly show that the asters grow at the expense of monomeric tubulin molecules and that there must be a mechanism for stopping further polymerization when the asters have reached a given size. Incorporation of fluorescent tubulin is much faster in the mitotic apparatus than in the MT cytoskeleton (Salmon *et al.*, 1984).

2. Later Biochemical Changes

From the biochemical point of view, two processes have been the subject of extensive study: strong increases in oxygen consumption (known since the pioneer work of Warburg, 1908) and in protein synthesis (Monroy, 1960). These take place a few minutes after sea urchin eggs have been activated or fertilized. It should be pointed out that such immediate large increases are not the rule for the eggs of invertebrates. In most cases, respiration and protein synthesis do not change much after fertilization. In the polychaete *Chaetopterus,* there is even a transient, but marked, decrease in both respiration (Whitaker, 1933) and protein synthesis (Zampetti-Bosseler *et al.*, 1973) when the eggs are either activated or fertilized (Fig. 48). Whitaker (1933) correctly concluded from his extensive studies that the oxygen uptake of unfertilized eggs may be abnormally high or low according to the species and that the effect of fertilization is to normalize the respiratory rate, not to stimulate it.

In unfertilized sea urchin eggs, the rate of cellular oxidation is abnormally low for reasons that are not yet clear. What seems to be well established is that the oxygen consumption of sea urchin oocytes with an intact GV is as high as that of fertilized eggs (Fig. 49). As shown by Lindahl and Holter (1941), the respiratory rate decreases during maturation and returns to the high value characteristic of full-grown oocytes shortly after fertilization. The main enzymatic change that has been detected so far at fertilization is a marked and rapid increase in (NAD) kinase activity (Epel, 1964). Phosphorylation of NAD in (NADP) is one of the earliest biochemical changes so far recorded after sea urchin egg fertilization and is concomitant with the increase in free Ca^{2+}. Epel *et al.* (1981) have shown that the two processes are linked together at fertilization. NAD kinase is activated by calmodulin, but not by ammonia, which, as we have seen, bypasses the Ca^{2+} increase step.

The identity of the substrate that is oxidized by sea urchin eggs at fertilization has remained elusive for many years. Only recently has it been discovered that this "substrate" is water! The increase in oxygen consumption at fertilization is essentially due to the formation of H_2O_2, which is used by an egg ovoperoxidase for hardening the fertilization membrane by crosslinking tyrosine residues (Foerder *et al.*, 1978). More recently, Perry and Epel (1981) found that H_2O_2 production results from a cyanide-insensitive oxidation of the sea urchin egg

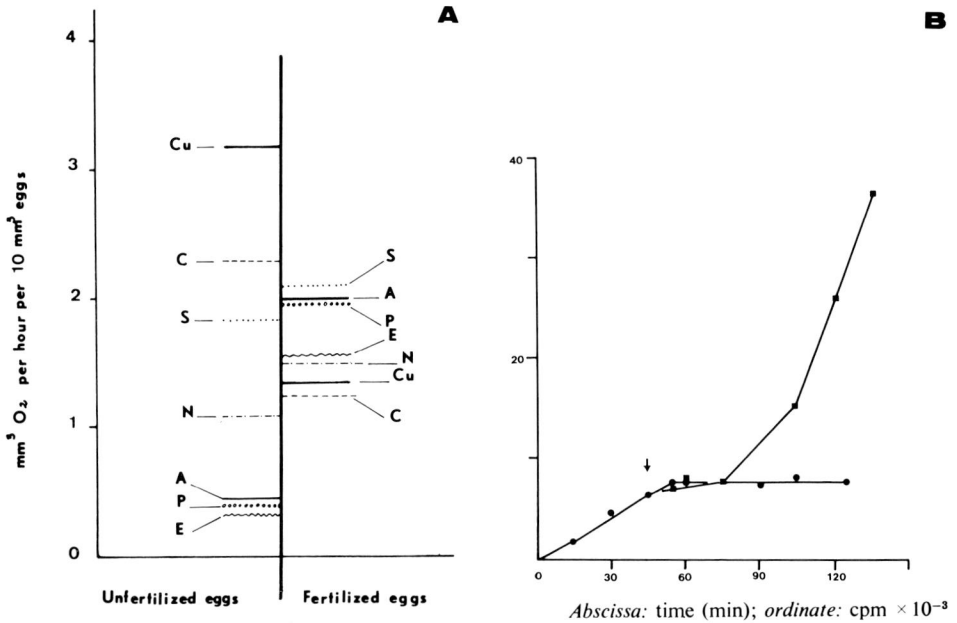

FIG. 48. (A) Effects of fertilization on the oxygen consumption of a few invertebrate eggs. A, P, and E are eggs from three different sea urchin species. N, *Nereis;* S, *Sabellaria;* C, *Chaetopterus;* Cu, *Cumingia.* Fertilization may increase (A, P, E) or decrease (Cu, C) egg respiration. (B) Incorporation of [³H]leucine during maturation and activation of *Chaetopterus* oocytes. The arrow indicates the time of excess KCl addition in order to induce activation. Circles: incorporation in normal seawater; squares: incorporation in the presence of KCl. There is a 30-min lag after KCl addition before protein synthesis begins to increase. [(A) Whitaker, 1933; (B) Zampetti-Bosseler *et al.*, 1973.]

pigment echinochrome. The echinochrome-containing granules of the sea urchin eggs might thus be related to the peroxisome family. The role of H_2O_2 is probably more complex than was first thought. Besides its function in hardening the fertilization membrane by the ovoperoxidase-catalyzed reaction, it inactivates excess spermatozoa and in this way contributes to the prevention of polyspermy (Boldt *et al.*, 1981). Indeed, addition of catalase at the time of fertilization results in polyspermy (Coburn *et al.*, 1981).

Runnström discovered in 1928 that there was a strong burst of CO_2 soon after insemination of sea urchin eggs resulting from the production of a "fertilization acid." For many years, Runnström and many others tried unsuccessfully to identify the chemical nature of this acid. The mystery was solved when Johnson *et al.* (1976) discovered that there was a proton efflux at fertilization. The mysterious acid of fertilization was not CO_2 that was stored in unfertilized eggs

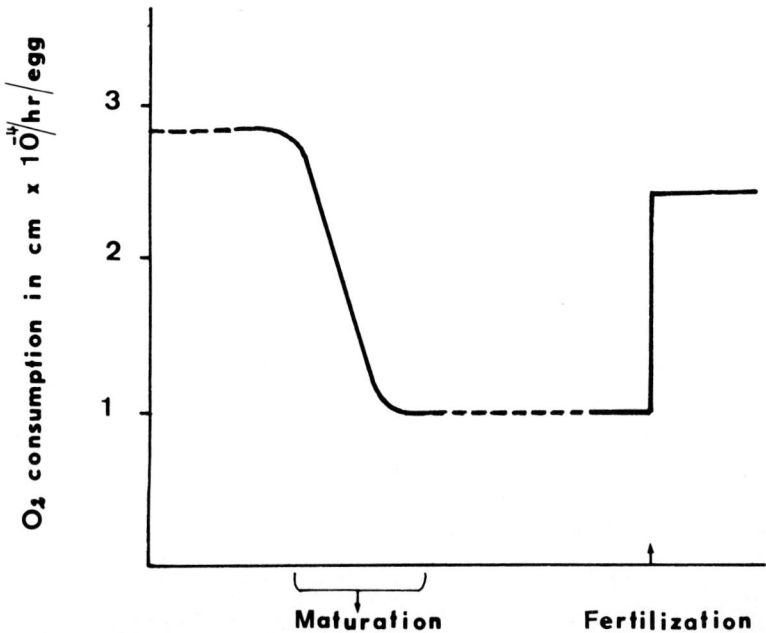

Fig. 49. Changes in oxygen consumption during the maturation and fertilization of sea urchin eggs. Fertilization returns the respiratory rate to the value it had in oocytes. [Lindahl and Holter, 1941.]

in the form of CO_3Ca (Gillies *et al.*, 1981), but simply protons (Holland and Gould-Somero, 1982).

We come now to the synthesis of macromolecules. DNA synthesis begins a few minutes after the penetration of the spermatozoon into the egg, when the sperm head loses its compactness and becomes the male pronucleus. Swelling of the sperm head is correlated with the loss of sperm-specific basic proteins. Carroll and Ozaki (1979) have shown that three histone-like proteins are lost and replaced by egg maternal histones in the zygote nucleus. DNA synthesis immediately follows chromatin decondensation, taking place simultaneously in both male and female pronuclei; it is completed within 10–20 min—thus, before amphimixy (at least in several sea urchin species). In the case of polyspermy, DNA synthesis occurs simultaneously in all sperm nuclei and in the female pronucleus; its initiation thus results from cytoplasmic changes. It is unlikely that DNA synthesis is triggered by the basic proteins changes that take place in the sperm nucleus, since the female pronucleus, which already contains maternal histones, does not replicate DNA unless parthenogenesis is induced. As we have seen, initiation of DNA synthesis is not related to the cortical reaction and Ca^{2+} release, since it takes place in ammonia-activated eggs. It is likely that DNA synthesis is triggered, after fertilization, by the rise in pH_i and, as suggested by

Whitaker and Steinhardt (1981), by the production of NADPH. They found that the NADPH content increases in ammonia-treated eggs and have proposed that NADPH, together with Ca^{2+} and pH_i, are causal agents of sea urchin development.

The reason why the G_1 phase, which precedes DNA synthesis and lasts for several hours in cultured cells, is very short (if it exists at all) in sea urchin eggs is understandable. We already know that during oogenesis and maturation, eggs develop the complex machinery required for DNA synthesis. However, preparation for DNA synthesis during oogenesis and maturation is insufficient to support completely the intensive DNA synthesis that characterizes cleavage. In both sea urchin (Noronha et al., 1972; De Petrocellis and Rossi, 1976) and Xenopus (Tondeur-Six et al., 1975) eggs, fertilization is followed by a sharp increase in ribonucleotide reductase activity. This increase is inhibited by puromycin but not by actinomycin D, indicating that the enzyme neosynthesis is directed by preexisting maternal RNAs. Since reduction of ribonucleotides to deoxyribonucleotides by ribonucleotide reductase requires NADPH as the hydrogen donor, it is conceivable that the increase in NADPH at fertilization discovered by Whitaker and Steinhardt (1981) contributes to the initiation of DNA synthesis in fertilized eggs. However, the key enzyme for DNA replication at fertilization is not ribonucleotide reductase but DNA polymerase-α. Inhibitors of ribonucleotide reductase, such as hydroxyurea or deoxyadenosine, (Brachet, 1967, 1968), do not arrest the cleavage of sea urchin eggs before the 8- to 16-cell stage. In contrast, aphidicolin, which inhibits DNA polymerase-α, stops their development at amphimixy and completely prevents DNA replication (Ikegami et al., 1978, 1979; Brachet et al., 1981). Another interesting enzyme involved in DNA synthesis is DNA ligase, which has been studied in axolotl eggs by Signoret et al. (1983, 1984). They found that this enzyme is present in a "light" form in oocytes and in a "heavy" form soon after fertilization. Nuclear transplantation experiments have shown that production of heavy DNA ligase is controlled at the level of its structural gene; simultaneously, the light DNA ligase gene is repressed. Species-specific exclusion between nonallelic genes thus takes place soon after fertilization in axolotl eggs.

Possibly related to the stimulation of DNA synthesis and cell division is an increase in the activity of a tyrosine protein kinase soon after fertilization of sea urchin eggs. As we shall see in the next chapter, tyrosine-specific protein kinases are involved in malignant transformation, which leads to continuous cell proliferation. In sea urchin eggs, tyrosine protein kinase activity is also stimulated by treatment with the Ca^{2+}-ionophore A23187. The enzymatic activity still increases during cleavage, a period of intense mitotic activity (Kinsey, 1984).

There is no burst in RNA synthesis at fertilization. Very little RNA synthesis is detectable in fertilized sea urchin eggs, and it is mainly of mitochondrial origin. Furthermore, the fact that fertilization and early cleavage are not affected by actinomycin D excludes the possibility that new RNA synthesis plays an important role in these early embryological events (Gross and Cousineau, 1964). This lack of RNA synthesis at fertilization is probably correlated with the presence of a large store of all kinds of RNAs in unfertilized sea urchin eggs. As already mentioned, they possess a large, complex population of poly(A)$^+$ RNAs that is theoretically capable of coding for as many as 15,000–30,000 different proteins of average size (Galau *et al.*, 1976; Anderson *et al.*, 1976, reviewed by Hough-Evans and Anderson, 1981; Davidson *et al.*, 1982). The sequence complexity of these maternal messages is 87×10^6 nucleotides, and there are about 1000 copies of each mRNA sequence (Hough-Evans *et al.*, 1979). It has been found that, in addition to "mature" mRNAs, unfertilized sea urchin eggs contain polyadenylated transcripts of repetitive and nonrepetitive DNA sequences covalently linked into long interspersed molecules; 70% of the poly(A)$^+$ RNA is present in this form, which might (at least in part) be unprocessed precursor-like molecules (Thomas *et al.*, 1981). A sea urchin maternal mRNA has been analyzed in detail by Posakony *et al.* (1983); it contains numerous members of a dispersed, repeat-sequence (3000–5000 bases) family. These repeated sequences persist during development and contain stop codons; they display extensive polymorphism when individual sea urchins are compared. On the whole, maternal mRNAs in sea urchins, as well as in *Xenopus*, are more similar to nuclear RNAs than to the mRNAs of later cells.

The maternal mRNAs are bound to proteins in the form of RNP particles smaller than the ribosomes (Gross *et al.*, 1973). These 60 S RNP particles have been studied by Kaumeyer *et al.* (1978) and Jenkins *et al.* (1978). The found that they contain translatable maternal mRNAs that cannot be translated in an *in vitro* system unless the protein moiety of the RNA particles has been removed. Clearly, the maternal mRNAs of unfertilized sea urchin eggs are in a masked, inactive form as long as they are bound to proteins. However, Moon *et al.* (1982) disagree with the generally accepted view that the mRNAs present in mRNP particles cannot be translated in an *in vitro* system; they obtained direct translation of the mRNPs in a wheat germ system. Even if mRNPs are perhaps not masked *in vitro,* they are masked *in vivo,* as shown by the fact that unfertilized sea urchin eggs contain few polyribosomes; their number increases during development (Humphreys, 1971). As we shall see when we study protein synthesis in anucleate fragments of sea urchin eggs, the maternal mRNAs are mobilized after fertilization or parthenogenetic activation; this allows them to bind to the preexisting ribosomes and to form fully active polyribosomes.

Griffith *et al.* (1981) have studied rRNA synthesis during sea urchin oogenesis. They found that its rate is very high (40 times greater than in sea urchin embryos). The GV synthesizes as many as 110,000 ribosomes per hour. Since sea urchin oogenesis lasts for 4–5 months, the oocyte accumulates up to 4×10^8 ribosomes. The authors found no evidence for ribosomal gene amplification and, therefore, concluded that all of these genes are fully active during oogenesis.

Although there is no stimulation of overall RNA synthesis at fertilization, this event is quickly followed by a doubling of the poly(A) sequences in the preexisting mRNAs (Wilt, 1973; Slater and Slater, 1974). The average length of the poly(A) chains increases from 100 to 200 nucleotides. This polyadenylation also occurs after activation of the unfertilized eggs with ammonia and is thus not a Ca^{2+}-dependent proces. Its role is presumably to stabilize the maternal mRNAs when they are released from the inactive mRNP particles and to protect them against degradation by cytoplasmic enzymes. However, the biological significance of polyadenylation after fertilization remains unknown, since cordycepin (an inhibitor of polyadenylation) does not affect fertilization and cleavage in sea urchins (Spieth and Whiteley, 1981). Another puzzling fact is that emetin, which very efficiently inhibits protein synthesis and arrests development before first cleavage, doubles the amount of polyadenylation (Slater and Slater, 1979). Concomitant with the doubling of the poly(A) sequences, there is a decrease at fertilization in oligo(U) sequences in large maternal mRNAs. This might be one of the factors involved in their activation at fertilization, since the oligo(U) tracts give a compact structure to the mRNAs (Duncan and Humphreys, 1983).

It is clear from this description that sea urchin eggs possess an extensive machinery for protein synthesis (ribosomes and mRNAs, in particular) that has been accumulated during oogenesis. This is probably a very general phenomenon, since it occurs in oogenesis and maturation in *Xenopus* and since, according to Bachvarova and De Leon (1977), 80% of the ribosomes are inactive in unfertilized mouse eggs.

An intriguing possibility is that, in sea urchin eggs, excess ribosomes and mRNPs are stored in the form of large basophilic granules, called "heavy bodies" (Afzelius, 1957; Pasteels *et al.,* 1958; Harris, 1969), because they are easily displaced by mild centrifugation. Figures 50a and 50b show their structure at the light and EM levels. They are composed of closely packed granules, sometimes surrounded by an annulate lamella; the granules have a high RNA content, as shown by cytochemical tests. Heavy bodies appear in the cytoplasm at maturation and disappear during first cleavage (Harris, 1969). If cleavage is inhibited by the treatment of fertilized eggs with emetin, aphidicolin, valinomycin, etc., the heavy bodies greatly enlarge, presumably as the result of fusion with ribosomes, since this process is not affected by actinomycin D (submitted for publication). It has been repeatedly proposed that the heavy bodies might be a store of maternal mRNAs, but the nature of their RNA component remains unknown. *In situ* hybridization experiments still in progress in our laboratory show that heavy bodies are not simple aggregates of ribosomes or glycogen particles and that they are devoid of basic proteins and acid mucopolysaccharides. Heavy bodies hybridize *in situ* with radioactive probes specific for rRNA, but only after treatments that remove part of the proteins. It thus appears that ribosomes were present in a masked form in the heavy bodies;

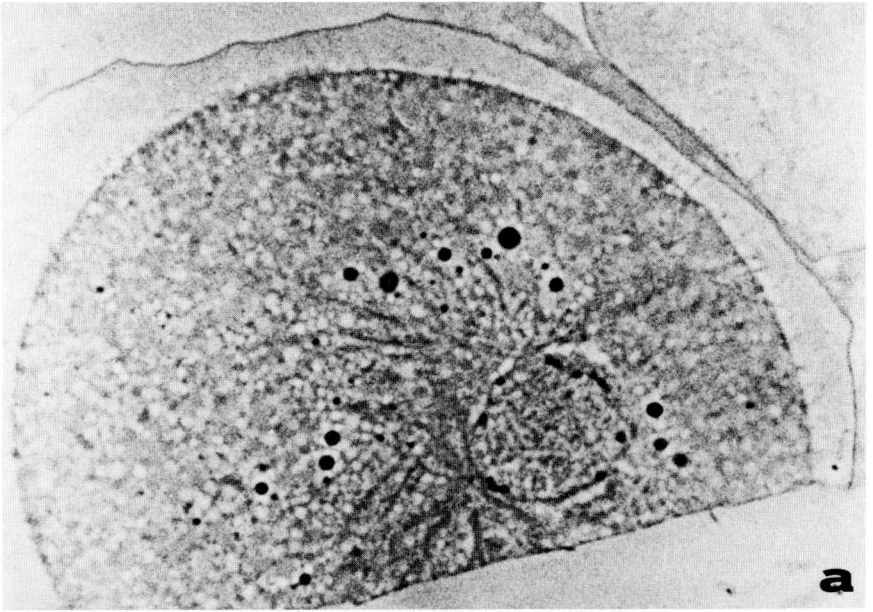

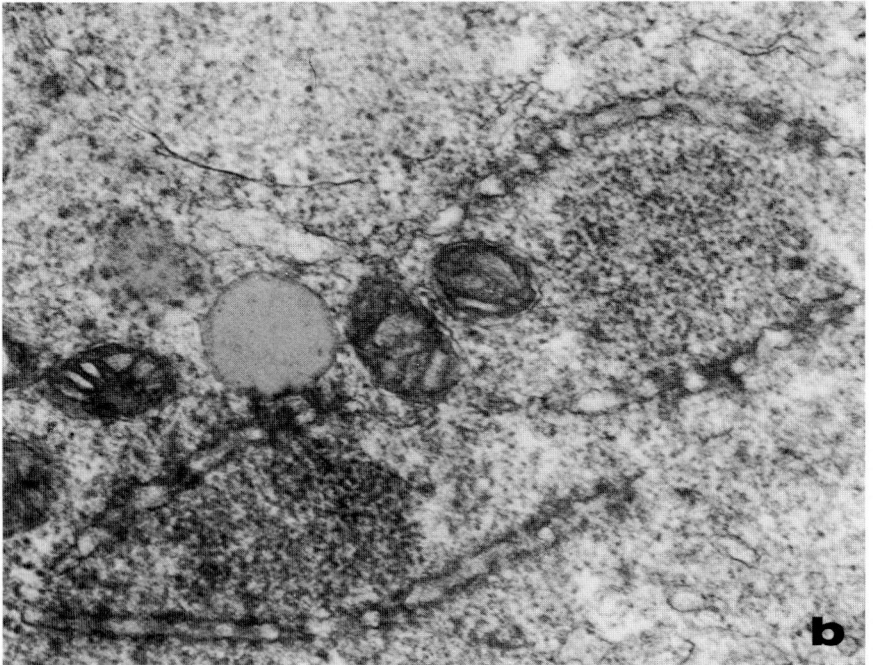

FIG. 50. Heavy bodies in fertilized sea urchin eggs. (a) Staining of a semithin section with the basic dye toluidine blue. (b) Two heavy bodies seen under the electron microscope; they are aggregates of small granules partly surrounded by an annulate lamella. [Courtesy of G. Steinert.]

whether they also contain masked mRNAs is not yet known (Steinert *et al.*, 1984).

A considerable amount of work has been done to elucidate the molecular mechanisms responsible for the large increase in protein synthesis that takes place a few minutes after insemination (Monroy, 1960) (Fig. 51). In unfertilized sea urchin eggs, both the uptake of amino acids and their incorporation into polypeptides take place at a very low, but measurable, rate. According to Rinaldi and Parente (1976), the rate of protein synthesis is already very low in full-grown oocytes; repression of protein synthesis is thus not a direct consequence of maturation. It has been reported that both unfertilized and fertilized sea urchin eggs synthesize at least 400 different proteins, which seem to be identical in the two kinds of eggs; it is thus the rate of protein synthesis that increases considerably at fertilization. There are no major qualitative changes in protein synthesis, but there is a general increase in the synthesis of a large number of polypeptides. However, work by Evans *et al.* (1983) leads to a different conclusion. They found that three or four new proteins are synthesized soon after sea urchin egg fertilization. One of them, which has been called "cyclin," disappears and reappears at each division cycle synchronously with MPF activity. There are no cyclic oscillations of cyclin in eggs activated by ammonia or the ionophore

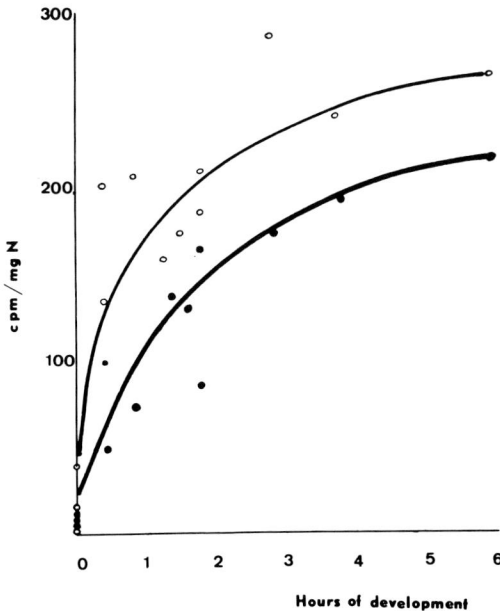

FIG. 51. Rapid labeling of the microsomal (white circles) and soluble (black circles) protein fractions following fertilization of sea urchin eggs. The unfertilized eggs were prelabeled by injection of [^{35}S]methionine into female sea urchins. [Monroy, 1960.]

A23187 where there is no cell division; cytochalasin, colchicine, and taxol slow down the disappearance of cyclin. In the clam *Spisula,* where protein synthesis increases slowly after fertilization (Rosenthal *et al.,* 1983), two new proteins appear at fertilization and are destroyed at certain points during the cell cycle. A similar situation seems to prevail in mouse eggs, where fertilization strongly stimulates the synthesis of a particular set of six "fertilization proteins"; their synthesis stops at the eight-cell stage. Experiments with actinomycin D indicate that these fertilization proteins are synthesized on maternal mRNA templates (Cascio and Wassarman, 1982).

Numerous attempts to identify the factors that strongly stimulate protein synthesis a few minutes after fertilization have been made. Since most of this work, which was based in general on the utilization of polyuridylic acid as an artificial messenger, has only historical interest, it will be summarized only briefly here. The first hypothesis was that fertilization induces the synthesis of mRNA molecules in the pronuclei. These newly synthesized mRNAs would bind to preexisting ribosomes and form active polyribosomes (Hultin, 1964). The control mechanisms would thus operate at the transcriptional level. The fact that there is no large increase in mRNA synthesis at fertilization, that fertilization and cleavage are actinomycin D insensitive, and (as we shall see) that protein synthesis increases in the anucleate halves of unfertilized eggs after parthenogenetic activation has ruled out this simple and apparently logical hypothesis.

Another possibility, which has been proposed by Monroy *et al.* (1965), is that the ribosomes are abnormal in unfertilized eggs, where they would be unable to bind properly to mRNAs; mild digestion with trypsin restores the ability of ribosomes of unfertilized egg to bind *in vitro* polyuridylic acid. In addition, the ribosomal "wash" (extract with salts of a ribosomal pellet that contains, among other things, initiation, elongation, and termination factors) from unfertilized eggs apparently contains factors that inhibit *in vitro* protein synthesis by ribosomes of fertilized eggs (Metafora *et al.,* 1971). One can also easily imagine that release of a protease at fertilization (for which there is some experimental evidence) could remove inhibitory proteins from ribosomes of unfertilized eggs and allow them to bind mRNAs with increased efficiency. However, recent data are contradictory. The most recent report at our disposal (Danilchik and Hille, 1981) states that there is a 15-min lag before ribosomes from unfertilized sea urchin eggs become active in *in vitro* protein synthesis, while there is no lag if the ribosomes have been isolated from fertilized eggs. Perhaps related to this finding is the fact that, 4 min after sea urchin egg fertilization, a selective phosphorylation of one of the ribosomal proteins (called rp3) takes place; protein rp3, which is present in the 40 S ribosomal subunit, is the equivalent of the already mentioned ribosomal S_6 protein of vertebrates. We have seen that this protein is phosphorylated when *Xenopus* oocytes undergo maturation and when cultured mammalian cells are induced to multiply at a rapid rate. The role played by S_6 phosphorylation is not yet clear, but Duncan and McConkey (1984) have shown that phosphorylated ribosomes are preferentially localized in polysomes (in HeLa cells), where they form initiation complexes more easily. This role of S_6 phosphorylation would fit very well with the increased binding of the maternal mRNAs to the ribosomes at fertilization. Electron microscopy of spread polysomes shows that they are shorter 1 hr after fertilization than they were in unfertilized eggs; between 3 min and 1 hr after fertilization, many polysomes have an unusual morphology: they display gaps or tails. The large size of the polysomes in unfertilized eggs might be due to their reduced translation efficiency (Martin and Miller, 1983). Coming back to S_6 phosphorylation, it is interesting to note that sea urchin eggs possess a specific phosphatase for the homologous rp3 protein and that this enzyme is inactivated at fertilization (Ballinger and Hunt, 1981). However, Ward et al. (1983) have pointed out that phosphorylation of S_6 (or rp3) protein is not universal in sea urchins. However, changes other than phosphorylation probably modify the ribosomes at fertilization. Takeshima and Nakano (1983) have reported that five ribosomal proteins are transformed in embryo-specific ribosomal proteins at sea urchin fertilization. Taken together, all of these observations strongly suggest that the ribosomes undergo changes when sea urchin eggs are fertilized, and it is very likely that these changes are important for the increase in protein synthesis. However, it seems unlikely that changes in ribosome conformation or composition (phosphorylation) are the only factors involved in the stimulation of protein synthesis at fertilization.

In fact, it is now generally believed that the stimulation of protein synthesis is closely linked to the previously mentioned increase in pH_i that takes place shortly after fertilization. This has been shown by Johnson et al. (1976), by Shen and Steinhardt (1978), and particularly by Grainger et al. (1979). A close correlation between the pH_i and the rate of protein synthesis has been clearly demonstrated by elegant experiments in which the pH_i value has been manipulated by the addition of weak bases (ammonia) or acids (acetate) to seawater. While ammonia stimulates both DNA and protein synthesis, acetate lowers the pH_i, inhibits protein synthesis, and blocks the development of fertilized eggs at amphimixy, like the classic inhibitors of protein synthesis, puromycin (Hultin, 1961) and emetin (Hogan and Gross, 1971). In contrast, ammonia increases the pH_i by 0.3–0.5. As a result, protein synthesis also increases, and the chromosomes condense and are replicated (Mazia and Ruby, 1974). Incidentally, intracellular pH also controls the development of the aforementioned new potassium conductance after the fertilization of sea urchin eggs (Shen and Steinhardt, 1980).

There is no doubt that the increase in pH_i that follows fertilization in sea urchin and Xenopus eggs is a very important event, triggering not only protein and DNA synthesis but even cleavage and the following stages of development, at least in sea urchins. However, a word of caution is needed. In a note published only in abstract form, Johnson and Epel (1981b) point out that all of the work showing a correlation between pH_i and rate of protein synthesis has been performed on a single species of sea urchins, *Lytechinus;* according to the authors of this abstract, no such correlation can be found for eggs of another species, *Strongylocentrotus.* To this we wish to add a more general remark. So many chemical reactions are pH dependent that the finding of a correlation between pH_i and an increased rate of protein synthesis or initiation of DNA synthesis does not reveal much about the molecular mechanisms that control these events. More work is needed in order to better understand the changes in the protein-synthesizing machinery at fertilization. A step in that direction has been made by Raff *et al.* (1981), who carefully compared protein synthesis in unfertilized and fertilized sea urchin eggs and concluded that two factors control the increase in protein synthesis that follows insemination: increased availability of translatable maternal mRNAs and a change in the polypeptide chain elongation rate. That these two factors are pH dependent would, of course, be no surprise.

Before we leave the subject of macromolecule synthesis at fertilization, we must return to the paper of Rosenthal *et al.* (1983) on the clam *Spisula.* As we have seen, protein synthesis increases after fertilization of *Spisula* eggs; this increase is slow, in contrast to the abrupt change that occurs in sea urchins. Rosenthal *et al.* (1983) found that four mRNAs that were inactive in oocytes are completely or partially recruited by the ribosomes in the fertilized eggs. As in sea urchins, their poly(A) tail increases in size after fertilization. In contrast, α-tubulin mRNA is translated in the oocytes and is lost from the polysomes after fertilization; simultaneously, this mRNA loses its poly(A) terminal sequence. Curiously, the poly(A) tracts of the two mitochondrial RNAs increase in length at fertilization. These findings are in good agreement with those of Evans *et al.* (1983), who reported, as we have seen, the cyclic appearance and destruction of two proteins in fertilized *Spisula* eggs. They raise the important problem of selective translation of maternal mRNAs, a question previously raised in the discussion of protein synthesis in nucleate and anucleate *Xenopus* oocytes. Rosenthal *et al.* (1983) suggest that the translation of specific mRNAs might be modulated by the proteins associated with the mRNAs, a possibility reviewed by Ehrenfeld (1982). The same explanation was proposed by Fruscoloni *et al.* (1983), who followed a specific mRNA during development in *Drosophila.* This mRNA is translated on polysomes in oocytes, excluded from the polysomes after 3–5 hr of development, and translated again in 20-hr embryos. It was suggested that this translational regulation might be due to changes in the proteins associated with this particular mRNA.

D. MORPHOLOGY AND BIOCHEMISTRY OF ANUCLEATE FRAGMENTS OF SEA URCHIN EGGS

We now return to the nucleate and anucleate halves of unfertilized sea urchin eggs that were shown in Fig. 41. Electron microscopy has shown that both halves contain, except for the pigment and pronucleus, all of the major egg constituents. However, due to centrifugation, their relative proportions between the two halves vary. The light nucleated halves contain more lipids, ribosome-coated vesicles, and mitochondria than the heavy anucleated halves, while the latter possess more yolk granules than the former. Both kinds of fragments can give rise to larvae (almost normal plutei) after fertilization. After parthenogenetic treatment by the method of Loeb (successive treatments with hypertonic seawater and butyric acid), only the nucleate halves can produce larvae; at best, after parthenogenetic activation, anucleate fragments form asters and undergo irregular cleavage (Fig. 52). The results look more like anarchic fragmentation than genuine cleavage of the egg (Harvey, 1936). Activated anucleate fragments never hatch, suggesting that they are unable to synthesize the hatching enzyme (a protease); they never form cilia, like normal blastulae.[5] Similar results were obtained by Lorch et al. (1953), who removed, with a micromanipulator, the nucleus of one of the blastomeres during early cleavage of sea urchin eggs. For a while, the enucleated blastomere showed irregular and abortive cleavages, but it never took part in morphogenesis, despite the fact that it remained in contact with normal cells. As one can see, development of sea urchin eggs deprived of their nucleus is extremely rudimentary; it is limited to anarchic furrowing, and true morphogenesis does not occur.

The same conclusions can be drawn from work done on amphibian eggs, where cleavage in the absence of the nucleus has been obtained many times (Dalcq and Simon, 1932; Fankhauser, 1934; Stauffer, 1945; Briggs et al., 1951). In general, only partial blastulae, where only the animal pole has cleaved, are obtained when both the egg and sperm nuclei have been destroyed by pricking or irradiation. The morphogenetic potencies of these "achrosomal" blastulae were tested by Briggs et al. (1951). Nonnucleated "cells" from the arrested blastulae were transplanted to inductive sites of normal embryos. The result was that the graft survived up to 4 days, but showed absolutely no sign of differentiation (Fig. 53). In conclusion, cleavage (usually partial and abnormal) is possible in the complete absence of the nucleus, but morphogenesis has never been obtained, even when the anucleate cells remain in close contact with normal nucleate cells. Whether intercellular communication is established between an anucleate cell and its neighbor is, unfortunately, not known.

If we recall what has been said about the great deal of information present in unfertilized sea urchin and amphibian eggs, it is somewhat surprising that anucleate fragments are unable to give rise to something better than a partial, anarchic morula or blastula. Unfortunately, we do not know how the stored

[5]The author had the privilege to watch, in 1938, Ethel Brown Harvey working on anucleate fragments of sea urchin eggs in Princeton and in Woods Hole. Although he had good eyesight in those days, he could never see hatching or ciliation in activated anucleate halves, to Harvey's great disappointment.

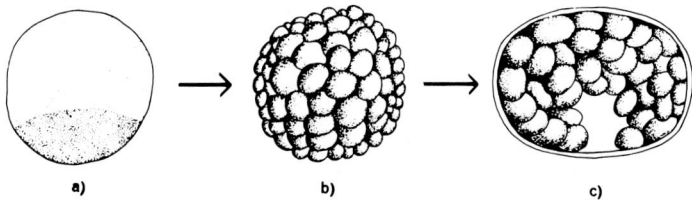

FIG. 52. Development of parthenogenetically activated anucleate fragments of sea urchin eggs. (a) Anucleate half at the time of activation; (b) anucleate morula after 13 hr; (c) abnormal anucleate blastula after 3 days. [Redrawn after Harvey, 1936.]

maternal mRNA sequences are distributed (evenly or unevenly) in nucleate and anucleate halves during centrifugation. This problem does not exist for anucleate amphibian eggs in which the nuclei have been destroyed by pricking or irradiation. Nor do we know whether the stability of the maternal mRNAs is modified by removal of the nucleus or whether the maternal mRNAs are translated with the same efficiency in the two halves. Experiments presently done in our laboratory (A. Pays and J. Brachet) indicate that the total amount of maternal mRNAs is higher in anucleate than in nucleate fragments and that these RNAs remain stable in the absence of the nucleus during at least 15 hr. We also find that anucleate halves still produce a large variety of proteins after this lapse of time,

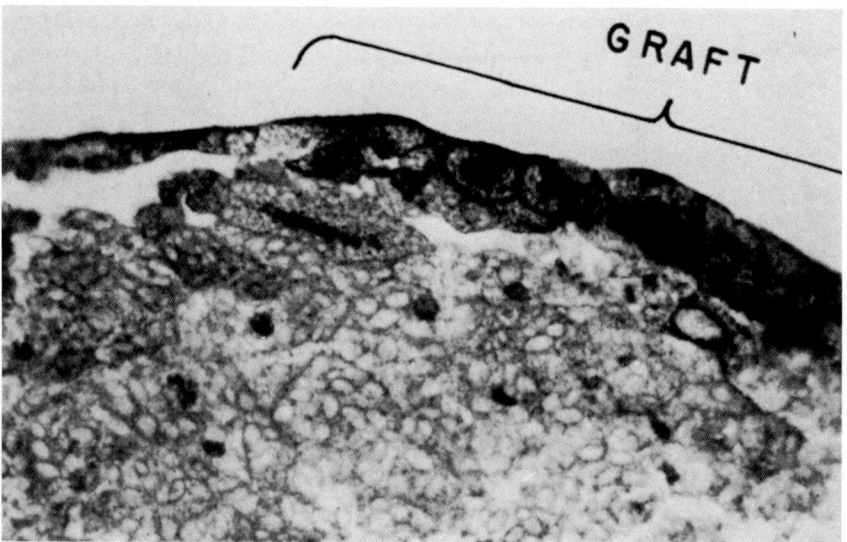

FIG. 53. Absence of differentiation when nonnucleated ectoderm (graft) is grafted to a normal embryo. [Briggs et al., 1951.]

but it is not yet possible to say whether a few key proteins cease to be synthesized in anucleate halves because their mRNAs have been degraded. These gaps in our knowledge are unfortunate, since maternal mRNAs and their translation products represent, in molecular terms, the "preformation" of the old embryologists. For the present, all we can say is that preformation amounts to very little, since it allows, at best, only the blastula stage to be attained. In order for the blastula to hatch, form cilia, and undergo gastrulation, fresh information must be added to that which was stored in the unfertilized egg. This new information necessarily results from the activation and expression of genes at or before the blastula stage. This conclusion agrees with the already mentioned fact that actinomycin D treatment does not stop development before this stage (Gross and Cousineau, 1964). Since it is known that, during sea urchin egg cleavage, mRNA synthesis takes place, it seems clear that the mRNAs synthesized during cleavage are not necessary for cleavage itself, but are required for further development (Gross and Cousineau, 1964). These conclusions apply equally well to amphibian eggs.

A few cytological observations of some interest should be mentioned. That anarchic cleavage after parthenogenetic stimulation of anucleate halves is due to aster formation was first shown by Harvey (1936); anucleate fragments are thus capable of MT assembly. It has indeed been shown that cytasters appear in anucleate halves treated with D_2O, and it has been reported that they are centered on a small centriole that has apparently formed *de novo* (Kato and Sugiyama, 1971). More recently, Hirai *et al.* (1981) have studied polyspermy in insemi-nated, immature starfish oocytes. No asters could be seen, but they developed if the fertilized oocytes were treated with 1-MA; an amphiaster was formed when meiosis (induced by 1-MA) was complete. Polyspermic immature starfish oocytes were cut into two halves and treated with MA. It was found that anucle-ate halves formed only monasters, while their nucleated counterparts developed amphiasters. This suggests that, in starfish eggs, material present in the GV is required for the edification of an amphiaster, but not for the assembly of monasters.

Other studies have dealt with the contractility of the cortical layer. Yoneda *et al.* (1978) have shown, by microcinematography, that after activation with butyric acid, anucleate fragments of sea urchin eggs undergo periodic tension and relaxation in their cortex. There is also a cyclic lifting of the hyaline layer that closely surrounds the plasma membranes after fertilization or activation. According to Yoneda *et al.* (1978), these changes are due to a rhythmic thicken-ing of the intrahyaloplasmic space. More recently, Yamamoto and Yoneda (1983) have extended these studies to starfish oocytes and eggs. They found that cyclic changes in tension do not appear in anucleate fragments unless the GV has broken down; this suggests that its nuclear sap might trigger the cyclic activity. Similar observations have been made on amphibian eggs. Sawai (1979) found that anucleate fragments of fertilized *Triturus* eggs undergo the same increase in stiffness in their cortex as normal eggs do at furrowing. Hara *et al.* (1980) observed a cyclic contraction of the animal half of *Xenopus* eggs even when cleavage was arrested by colchicine treatment. The same rhythmic contractions were found in anucleate fragments of fertilized *Xenopus* eggs, and the authors concluded that a cytoplasmic biological clock controls the cell cycle. However, according to Sakai and Kubota (1981), these surface contraction waves are slower in anucleate fragments than in nucleate halves, where a regular cell cycle is operating. More recently, Shinagawa (1983) confirmed that the rigidity cycle is 30% slower in anucleate fragments of fertilized *Xenopus* eggs than during the normal cleavage cycle. Injection into fertilized eggs of colchicine or vinblastin slows down their cycle, which becomes the same as in anucleate fragments; it was concluded that the rigidity cycle in *Xenopus* eggs is modulated by the assembly and disassembly of the mitotic apparatus but that the factor responsible for the cycling activity is located in the egg endoplasm (Shinagawa, 1985). A rhythmic cortical activity, independent of the presence of a nucleus, seems to be a general phenomenon. Waksmundzka *et al.* (1984) have reported that both nucleate and anucleate fragments of mouse eggs display such an activity. This activity is even higher in anucleate fragments, where it leads to furrowing and fragmentation; it might be that the female pronucleus or the second polar body moderate cortical activity in mouse eggs. That autonomous contractility of the cortical layer continues for a long time is not surprising. Small anucleate fragments of *Chaetopterus* eggs (Brachet and Donini-Denis, 1978), as well as cytoplasts from leukocytes (Keller and Bessis, 1975), display considerable ameboid activity for many hours. What is more surprising is that the contractile activity of the egg cortex is an oscillatory process. Whether it is due to a cytoplasmic clock that has the same period as the mitotic cycle in fertilized eggs remains to be seen.

Shapiro (1935) showed that oxygen consumption of anucleate halves of *Arbacia* eggs is much higher than that of their nucleate counterparts (Fig. 54). This demonstrated, for the first time, that Loeb's theory about the leading role of the cell nucleus in biological oxidations cannot be correct. The experiments of Shapiro (1935) were extended by Ballentine (1939), who showed that anucleate halves possess 70% more dehydrogenase activity than nucleate fragments of equal volume. This difference might be due to the fact that gradient centrifugation of intact eggs segregates, in the two halves, two different populations of mitochondria that can easily be distinguished under the electron microscope (Geuskens, 1965). However, it seems to us more likely, in view of the work on the formation of H_2O_2 at fertilization, that the higher oxygen consumption of the heavy anucleate halves is due to the accumulation at the centrifugal pole of echinochrome-containing granules. As we have seen, KCN-insensitive oxidation of echinochrome leads to H_2O_2 production (Perry and Epel, 1981).

The work on protein synthesis in anucleate fragments of *Arbacia* eggs still retains some historical interest. As in intact unfertilized eggs, the rate of protein

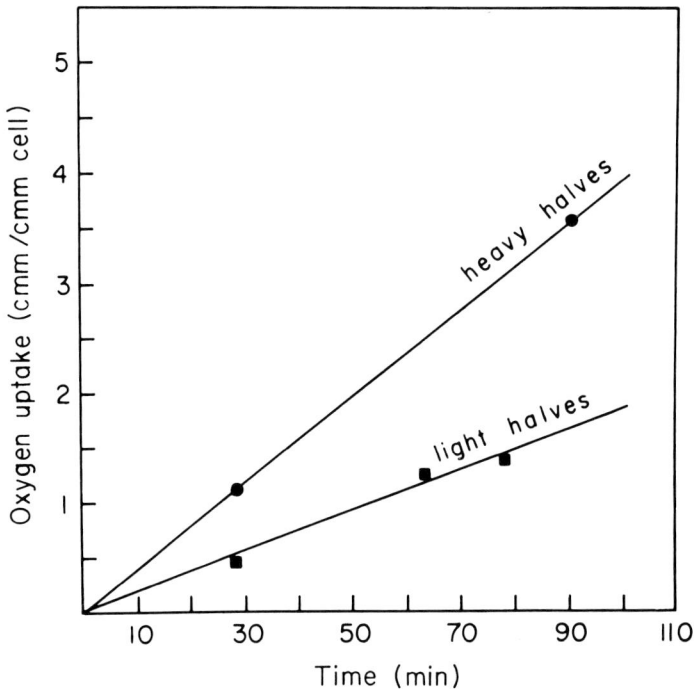

FIG. 54.   Oxygen consumption is higher in heavy (anucleate) than in light (nucleate) halves of unfertilized sea urchin eggs. [After Shapiro, 1935.]

synthesis is very low in the fragments obtained by centrifugation, but, if these fragments are activated by treatment with hypertonic seawater, strong stimulation of protein synthesis is observed in both halves (Brachet *et al.*, 1963; Denny and Tyler, 1964; Baltus *et al.*, 1965). Since this stimulation is even stronger in nucleate than in anucleate halves, the results ruled out the possibility that the increase in protein synthesis that takes place at fertilization could be due to the *de novo* synthesis of mRNA molecules by the pronuclei. In addition, it was shown that the number of polyribosomes greatly increases in anucleate fragments after activation with hypertonic seawater (Brachet, 1968). This proved that the mRNAs that bind to the ribosomes after activation preexisted in the egg cytoplasm. These experiments were the first to lead to the now universally accepted view that unfertilized sea urchin eggs contain a store of masked mRNAs that were synthesized during oogenesis.

It is known that many proteins are synthesized when anucleate fragments of sea urchin eggs are activated. However, the number of proteins is not known, since no one has yet compared the patterns of protein synthesis in nucleate and anucleate halves using modern analytical methods. This is unfortunate, since this kind of experiment could provide valuable data on the stability and translatability of the maternal mRNA population in the presence or absence of the female pronucleus.

However, two specific mRNAs have been identified among the maternal mRNAs stored in anucleate fragments of sea urchin eggs. They code for tubulin (Raff et al., 1971, 1972) and histones (Gross et al., 1973; Skoultchi and Gross, 1973), respectively. Tubulin mRNA stored in anucleate halves of unfertilized sea urchin eggs is not translated unless the anucleate fragments have been activated. It is not known whether translation of tubulin mRNA is required for the formation of asters in activated anucleate fragments. It is well known that unfertilized sea urchin eggs possess a large pool of tubulin dimers, and it seems likely that the formation of aster MTs after activation is linked to the appearance of MTOCs rather than to an increase in tubulin resulting from maternal tubulin mRNA translation. This is only a guess, since we have no facts. Why activated anucleate halves never form cilia is another enigma, since, according to Bibring and Baxandall (1981), only 2–4% of the tubulin present in the MTs of sea urchin blastulae cilia are synthesized during cleavage. Probably some minor, but necessary, proteins constitutive of cilia or basal bodies are missing in activated anucleate halves; again, this is only a guess.

Similar questions may be asked about the function—or, rather, the lack of function—of histone mRNA in anucleate halves. Azceci and Gross (1977) have conclusively shown that histones are synthesized on maternal mRNAs when anucleate fragments of sea urchin eggs are activated. In this case, there is, of course, no coordination between histone production and nuclear DNA synthesis. When an anucleate fragment of an egg synthesizes new histone molecules, it is a curious enigma.

We know little about the distribution of the various kinds of RNAs in nucleate and anucleate halves of sea urchin eggs. Showman et al. (1982) have made the first (and interesting) contribution to this question. They found that, in contrast to total RNA, total poly(A)$^+$ RNA, actin mRNA, and α-tubulin mRNA, histone mRNA is accumulated in the nucleate halves of Stranglylocentrotus eggs. This accumulation of histone mRNA is observed even when noncentrifuged, unfertilized eggs have been cut manually into two halves. According to the authors, this is not due to an accumulation of mRNA in the female pronucleus, since nuclei isolated from the eggs contain little histone mRNA. In addition, Showman et al. found that the stored histone mRNAs move into the polyribosomes after activation, even if breakdown of the nuclear membrane has been prevented by treatment with 6-dimethylaminopurine. However, De Leon et al. (1983), using in situ hybridization, reported that the female pronucleus in whole sea urchin eggs contains the majority of the histone mRNAs; these mRNAs are shed in the cytoplasm at first cleavage. The opposite results of Showman et al. (1982) might well be due to leakage of the histone mRNAs during isolation of the pronuclei.

We have seen that fertilization is followed by a doubling of poly(A) sequences in preexisting mRNAs; polyadenylation also occurs in anucleate halves after activation with hypertonic seawater (Wilt, 1973) or ammonia (Wilt and Mazia, 1974). More recently, it has been shown that the enzyme responsible for polyadenylation [poly(A) polymerase] is accumulated in the heavy anucleate half after centrifugation (Slater *et al.*, 1978); the same mRNAs are polyadenylated *in vitro* in whole unfertilized eggs and in anucleate fragments (Slater and Slater, 1979). In the latter, the poly(A) tail obviously cannot serve, as has sometimes been suggested, for the transfer of mRNAs from the nucleus to the cytoplasm. Its role might, of course, be, as in *Xenopus* oocytes, to stabilize the maternal mRNAs and protect them against degradation by cytoplasmic enzymes (Huez *et al.*, 1974). Finally, it should be mentioned that, according to Morris and Rutter (1976), sea urchin eggs contain enough DNA and RNA polymerase molecules to reach the 1000-cell stage, and that their anucleate halves display strong RNA polymerase activity. We are again faced with unanswered questions. Does this enzyme play a role in anucleate cytoplasm? Is it involved in mitochondrial RNA synthesis in the absence of the nucleus? This does not seem very likely, since, in general, mitochondria possess their own RNA polymerase, which differs from the nuclear enzymes. As we have seen before, both nucleate and anucleate fragments of *Arbacia* eggs possess a large population of mitochondria, which contain appreciable amounts of DNA. In fact, the DNA content of nucleate and anucleate halves is about the same, because the haploid female pronucleus represents only a small proportion of the total DNA (Baltus *et al.*, 1965). This raised the possibility that the stimulation of protein synthesis after activation of anucleate fragments is not due to a store of masked maternal mRNAs, but rather to the transcription of mitochondrial DNA by an RNA polymerase. In fact, anucleate fragments after activation synthesize small amounts of mitochondrial RNAs whose sedimentation constants are 11, 13, and 15 S (Chamberlain and Metz, 1972). The 11 and 13 S species are undoubtedly mitochondrial rRNAs, while the 15 S mitochondrial RNA might be either a precursor of mitochondrial rRNA or a mitochondrial mRNA. At the same time, Selvig *et al.* (1972) found that in anucleate fragments of *Arbacia*, mitochondria synthesize an RNA that can diffuse out into the cytoplasm. However, this RNA of mitochondrial origin is unable to bind to cytoplasmic ribosomes, ruling out the possibility that the strong increase in protein synthesis that follows activation of anucleate halves is due to transcription of mitochondrial DNA.

So far, we have paid very little attention to the nucleus of the nucleate fragments, since at first it appears to be negligible. In somatic cells, the main functions of the nucleus are DNA transcription and replication. However, the female pronucleus contains very little DNA compared to the cytoplasm; DNA and RNA polymerases in unfertilized sea urchin eggs are predominantly located in the cytoplasm. A few observations indicate that the female pronucleus in sea urchin eggs and in nucleate fragments should be treated with "more respect."

For instance, Venezky *et al.* (1981) have reported that the pronuclei of sea urchin eggs contain as much as 12% of the histone mRNAs and we have seen that a still greater accumulation has been reported by De Leon *et al.* (1983). Histone mRNAs are still accumulated in the pronuclei at amphimixy, but are lost from the nuclei at the mid-two-cell stage. These nuclear histone transcripts have a long life in the pronuclei, where they are selectively accumulated; in contrast, no accumulation of overall mRNAs and poly(A)$^+$ RNAs can be detected in the pronuclei. That the pronuclei contain little poly(A)$^+$ RNA compared to the cytoplasm has also been found by Angerer and Angerer (1981), who used *in situ* hybridization with tritiated poly(U) of sections from sea urchin eggs. How and why histone mRNAs selectively accumulate in the pronuclei of sea urchin eggs is not known.

No less curious are the results reported by Rinaldi *et al.* (1977, 1979a,b) concerning the control exerted by the pronuclei on mitochondrial RNA and DNA synthesis in *Paracentrotus*. In contrast to *Arbacia,* nucleate fragments of *Paracentrotus* eggs contain very few mitochondria; the bulk of the mitochondria is thus accumulated in the heavy anucleate halves. Rinaldi *et al.* (1977) first found that mitochondrial RNA synthesis is much more strongly stimulated in activated anucleate fragments than in fertilized or activated whole eggs. Since fertilization of the anucleate fragments prevents this increase in mitochondrial RNA synthesis, the authors concluded that in whole eggs the nucleus exerts a negative control on mitochondrial RNA synthesis. More recently, Rinaldi *et al.* (1979a) studied mitochondrial DNA synthesis in fragments of *Paracentrotus* eggs. It is generally believed that, in fertilized sea urchin eggs, there is no mitochondrial DNA synthesis until the pluteus stage (Matsumoto *et al.,* 1974; Bresch, 1978). In contrast, Rinaldi *et al.* (1979a,b) found that mitochondrial DNA synthesis takes place in activated anucleate fragments of *Paracentrotus* eggs. This was shown rather convincingly by analysis of thymidine incorporation and by an EM demonstration of replicating mitochondrial DNA molecules. In anucleate activated fragments, electron microscopy also showed figures that resemble dividing mitochondria (Fig. 55) (Rinaldi *et al.,* 1979b). In such fragments, mitochondrial protein synthesis is also increased (Rinaldi *et al.,* 1983). The authors conclude that the nucleus (in the present case, the female pronucleus) exerts a negative control on the activity of the mitochondrial genome. This conclusion has been extended to *Xenopus* unfertilized eggs. If they are enucleated and then activated, there is a 5- to 10-fold increase in mitochondrial DNA synthesis (Rinaldi *et al.,* 1981). Recall that in the discussion of regeneration of anucleate fragments of *Acetabularia,* we pointed out that the nucleus may exert negative as well as positive controls on the cytoplasm. As we have seen, regeneration and protein synthesis are speeded up in enucleated stalks. Nothing is known about the molecular mechanisms involved in negative nuclear control in eggs and in *Acetabularia;* this is another promising field for future research.

Two other instances in which the nucleus exerts unexpected control over

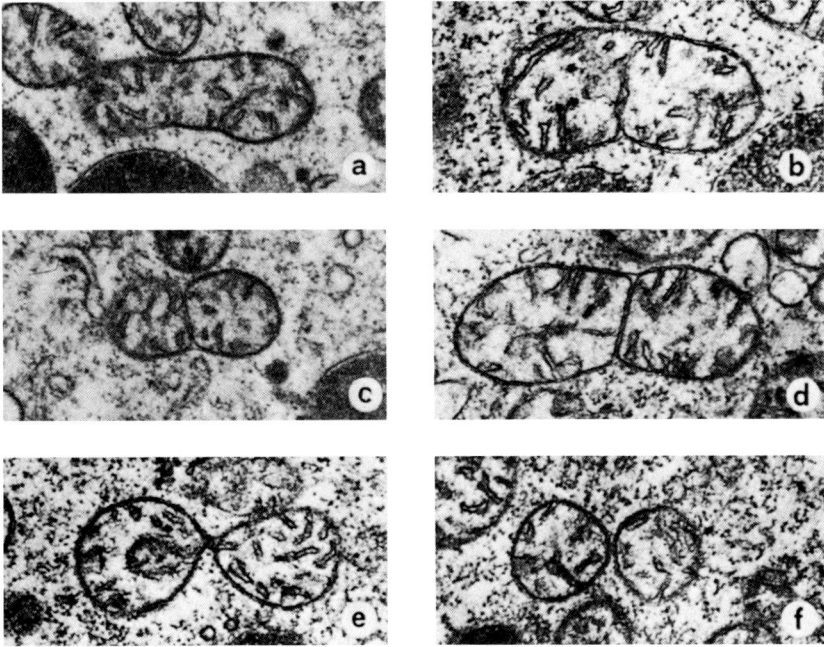

FIG. 55.   Possible sequence of mitochondrial division in an activated anucleate fragment of a sea urchin egg. (a) Elongated mitochondrion; (b) a central crista is growing across the mitochondria; (c, d) the transecting crista fuses with the opposite inner membrane, dividing the matrix into two compartments; (e) the constriction tightens uniformly around the mitochondrion; (f) the mitochondrion is completely divided. [Rinaldi *et al.*, 1979b.]

cytoplasmic activities in echinoderm eggs have been brought to light. Working with sea urchin eggs, Krystal and Poccia (1979) extended the classic experiments of A. Brachet (1922) on the fertilization of immature eggs. He found, as already mentioned, that in these polyspermic eggs the nuclei of all spermatozoa assume the same morphology as the egg nucleus and concluded that chromatin condensation or decondensation is controlled by cytoplasmic factors. Krystal and Poccia (1979) combined polyspermy with ammonia treatment and worked on nucleate and anucleate halves (obtained by either centrifugation or manual dissection), as well as on whole eggs. Their experiments showed that such treatments induce premature chromosome condensation (PCC) in the spermatozoa at the time chromosome condensation takes place in the female pronucleus. Unexpectedly, PCC is much less stable in anucleate than in nucleate halves; this shows that "a component of the maternal pronucleus may modulate the stability of the chromosome condensing environment." In this system PCC is inhibited by a phidicolin in nucleate, but not in anucleate halves. The conclusion was that the unreplicated female pronucleus exerts a negative control on chromosome condensation in maternal cytoplasm (Killian *et al.*, 1985). Other experiments by Krystal and

Poccia (1981), in which eggs were treated with emetin, have further shown that histone H1 phosphorylation and chromosome condensation in the male pronucleus can be dissociated, indicating that H1 phosphorylation is not sufficient for the induction of chromosome condensation. Thus, the long-neglected female pronucleus seems to exert a negative control over the activities of the mitochondrial genome and over PCC.

The other unexpected effect of the egg nucleus has been reported by Dale *et al.* (1979), who studied the changes in the resting potential that take place when GVBD is induced in starfish oocytes by the addition of 1-MA. They found that induction of maturation produces an important shift in the membrane potential at the time of (or soon after) GVBD; this shift is probably due to the closure of K^+ channels. The same changes are found in nucleate halves obtained either manually or by centrifugation, but they do not occur in anucleate fragments treated with 1-MA. The resting potentials of the anucleate halves are much more variable than those of their nucleate counterparts. This suggests that enucleation alters the electrical properties of the plasma membrane, preventing the response of the oocytes to 1-MA. It would be interesting to know whether the electrical response to activating agents is identical in nucleate and anucleate fragments of sea urchin eggs; this has not yet been studied.

In this section, we have tried to show that nucleate and anucleate halves of sea urchin eggs have contributed to the solution of important problems (role of the cell nucleus in biological oxidation, control of protein synthesis at the translational level at fertilization) and that they still raise many unanswered questions. It is our hope that asking these questions will promote research in this field.

V. CLEAVAGE (SEGMENTATION) OF FERTILIZED EGGS[6]

Cell division was described in detail in Chapter 5, Volume 1. Cleaving eggs afford unique opportunities for the study of DNA replication, entry into prophase (nuclear membrane breakdown, chromosome condensation), assembly of the mitotic apparatus, chromosome movement at anaphase, and cytokinesis.

Here we shall look at egg cleavage through the eyes of an embryologist, but without forgetting that this book deals mainly with ordinary cells. We shall thus briefly examine the various types of cleavage and the resultant segregation, in "mosaic" eggs, of the heterogeneous cytoplasmic regions called "germinal localizations" by the embryologists who discovered them (it is now more fashionable to call them "cytoplasmic determinants". We shall also point out some biochemical particularities that distinguish cleaving eggs from dividing somatic cells. We will conclude with the blastula–gastrula transition, since the following stages of development are mainly of embryological interest; in addition, a late blastula cell is not fundamentally different from an adult cell.

---

[6]For a review of the molecular biology of sea urchin embryos, see Davidson (1982).

A. THE VARIOUS TYPES OF EGG CLEAVAGE

Figure 56 describes schematically the main types of egg cleavage. In eggs of many species, including those of sea urchins, amphibians and mammals, cleavage is complete and equal. The first cleavage furrow divides the egg into two blastomeres of the same size. After the second cleavage, which divides the egg into four blastomeres of equal size, differences between species may already be observed. For instance, in amphibian eggs, the third cleavage is not equatorial, dividing the egg into micromeres at the animal pole and macromeres at the vegetal pole. This is due to the existence, in amphibian eggs, of the polarity gradient. The yolk, which has accumulated at the vegetal pole, is an obstacle to furrow progression. In sea urchin eggs, where there is much less yolk, the first eight blastomeres are the same size. At the 16-cell stage, the egg is divided into 8 mesomeres (at the animal pole), 4 macromeres, and 4 tiny micromeres; the last result from unequal cytoplasmic division at the vegetal pole.

In the eggs of worms and mollusks, unequal cleavage is the rule. The first

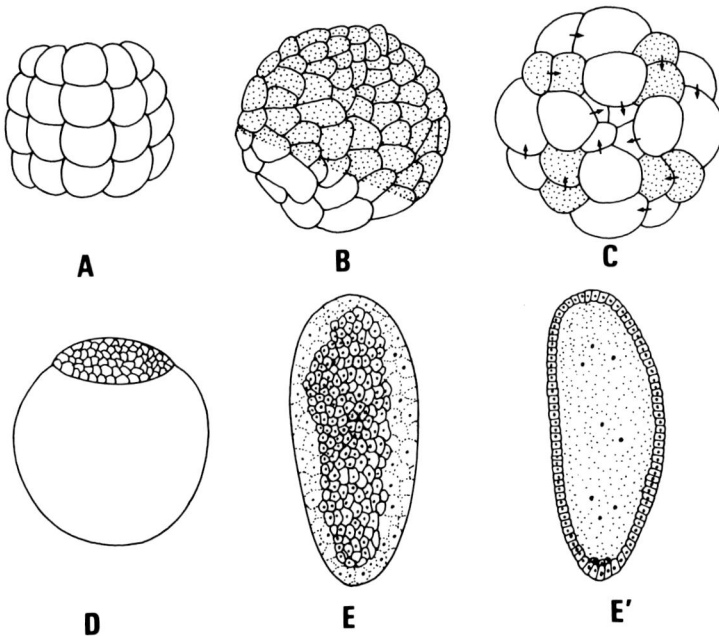

FIG. 56. Schematic representation of the main cleavage types. (A) Radial, equal and total cleavage (*Amphioxus*, prochordates); (b) radial, unequal and total cleavage (frog eggs); (C) spiral, unequal and total cleavage (worms, mollusks); (D) partial cleavage (fish, reptiles, birds); (E, E') superficial cleavage of insect eggs; ventral side (E) and longitudinal section (E'). [Drawn by P. Van Gansen.]

furrow separates a small AB blastomere from a larger CD blastomere. CD is larger than AB because it has incorporated a cytoplasmic polar lobe that is present, between two blastomeres of equal size, at the preceding "trefoil stage" (Figs. 30 and 57). Just prior to the second cleavage, a second polar lobe (again, purely cytoplasmic) can be seen; it fuses with one of the blastomeres (called D), which will become larger than the three others (A, B, and C). The following cleavage leads to the formation of micromeres and macromeres. The four micromeres are at right angles on the macromeres. The complex type of cleavage that follows is called "spiral cleavage." Its patterning is determined by the orientation of the mitotic spindles at each division cycle.

In the mollusk *Lymnaea*, dextrality or sinistrality is due to a maternal mode of inheritance involving a single locus with dextrality dominant. The product of the dextral gene during oogenesis influences the cleavage pattern. As shown by Freeman and Lundelius (1982), injection of dextral egg cytoplasm into uncleaved sinistral eggs changes their cleavage pattern into sinistral cleavage. Thus, while the dextral gene product is synthesized during oogenesis, it does not function before cleavage begins. This is an interesting example of gene–cytoplasm interactions at a very early stage of development.

The importance of the yolk mass in determining the pattern of egg cleavage is particularly apparent in large, yolk-laden eggs. In insects, the yolk is located in the center of the egg and is surrounded by a thin layer of cytoplasm. After fertilization, the nuclei repeatedly and rapidly divide, so that there is no furrowing; the yolk mass becomes surrounded by a syncytium. At the blastula stage, cell membranes appear to separate the nuclei present in the peripheric syncytium. Cleavage is thus superficial in insect eggs.

Finally, in very large eggs such as those of the fish and birds, cleavage is only partial. The furrows are unable to penetrate deeply into the compact yolk mass, and a small blastula forms at the animal pole. It is separated from the uncleaved yolk mass by a blastocele cavity filled with fluid.

After their first cleavages, eggs with equal or spiral cleavage (but not those of the insects) become morulae, in which a layer of blastomeres surrounds a blastocele cavity containing, in addition to water and salts, glycoproteins and sometimes glycogen. The blastocele results from secretion by the inner (basal) surface of the blastomeres that surround it (P. Van Gansen, personal communication). Our own unpublished studies agree with the idea that the organic substances (particularly glycoproteins) present in the blastocele fluid result from secretion by the surrounding cells. I found that fertilized sea urchin eggs treated with monensin, which prevents the secretion of glycoproteins (see Chapter 3, Volume 1), fail to form a blastocele cavity during cleavage. At each division, the volume of the blastomeres decreases. As a result, the ratio between the volumes of the nucleus and the cytoplasm (nucleoplasmic ratio, or N/P) which was abnormally low in the fertilized egg, steadily increases. It generally reaches a

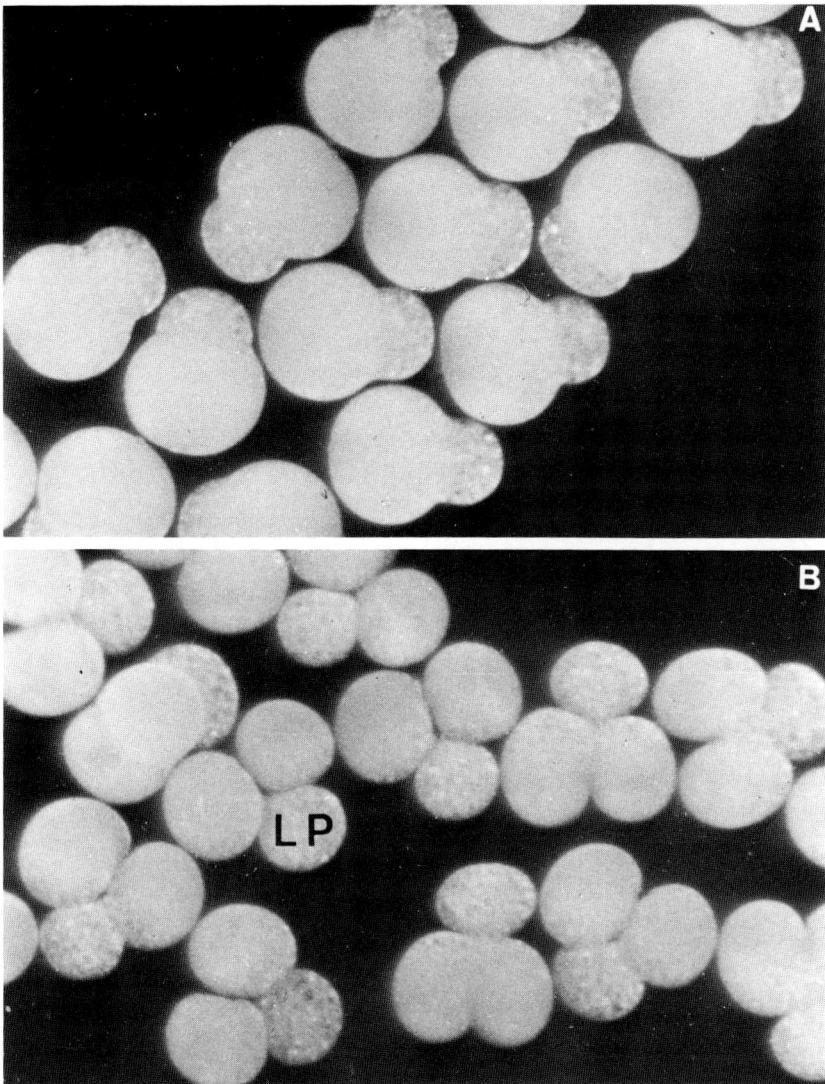

FIG. 57. Cleavage of *Ilyanassa* (mollusk) eggs. (a) Segregation of the first polar lobe, which contains heterogeneous material; (b) trefoil stage. LP, polar lobe. [Photographs of living eggs by Dr. Y. Gérin.]

constant value, characteristic of all cells in a given species, at the blastula stage, which marks the end of cleavage.

During cleavage, cytoplasmic material that plays an important role in development may be segregated into one or two blastomeres. Destruction of the

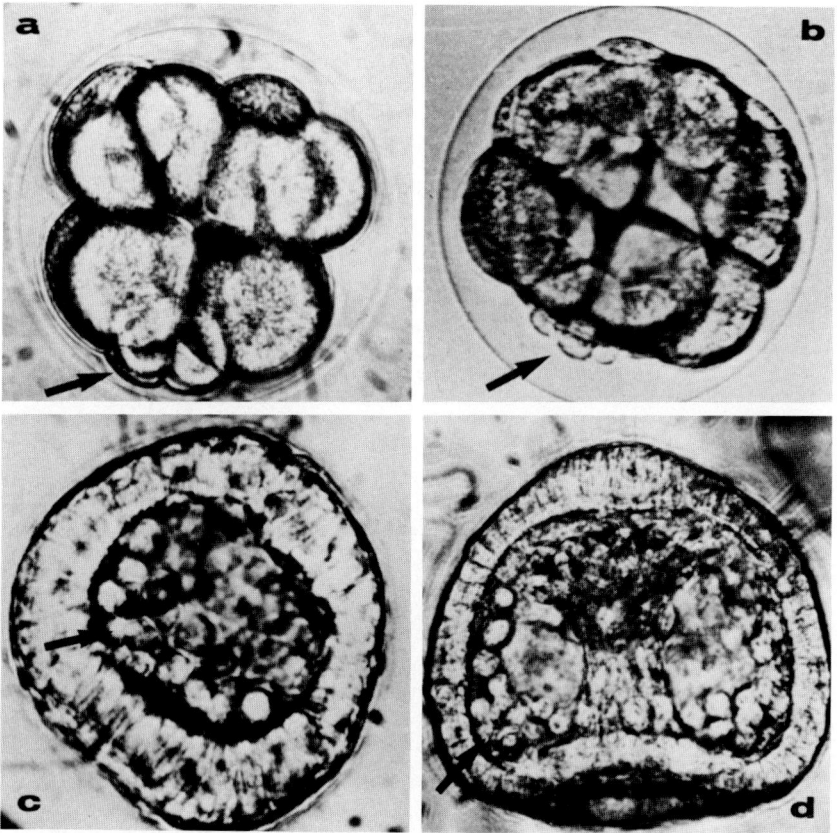

FIG. 58.   Cleavage of sea urchin eggs. (a) The 16-cell stage, with the micromeres at the vegetal pole (arrow); (b) the 32-cell stage, with the micromeres (arrow) at the vegetal pole; (c) late mesenchyme blastula; (d) late gastrula. In (c) and (d), the arrows point to the mesenchyme cells deriving from the micromeres. [Giudice, 1973.]

blastomeres that contain such specific cytoplasmic determinants (germinal localizations) leads to developmental deficiencies. Among these germinal localizations (to which we shall return later), special mention should be made of the gray crescent of amphibian eggs (a marker of the dorsal side of the future embryo and adult), the micromeres (Fig. 58) of the sea urchin eggs (which will give rise to

FIG. 59.   Development of an ascidian egg (*Cynthia*). (a) Side view of a fertilized egg showing the formation of the yellow crescent (cr, myoplasm) from the yellow hemisphere (yh); cp, clear cytoplasm (ectoplasm) above the yellow crescent. (b) Eight-cell stage. A, anterior and P, posterior sides of the embryo. pb, polar body marking the animal pole; cr, yellow crescent (myoplasm). (c) Young tadpole showing neural grove (n.p.), mesenchyme (m), and three rows of muscle cells (ms) deriving from the yellow cresent. [Conklin, 1905.]

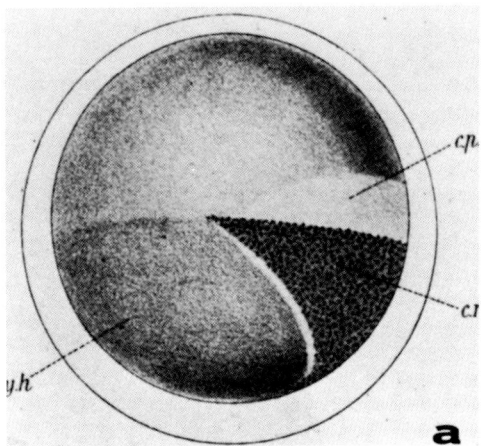

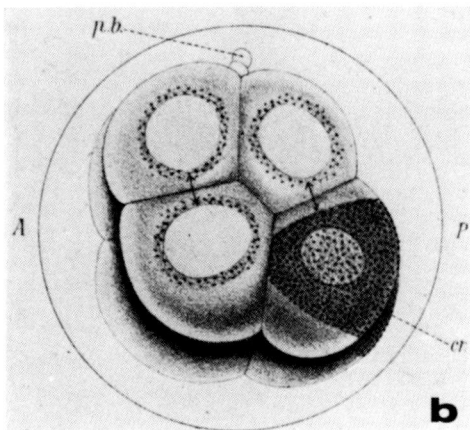

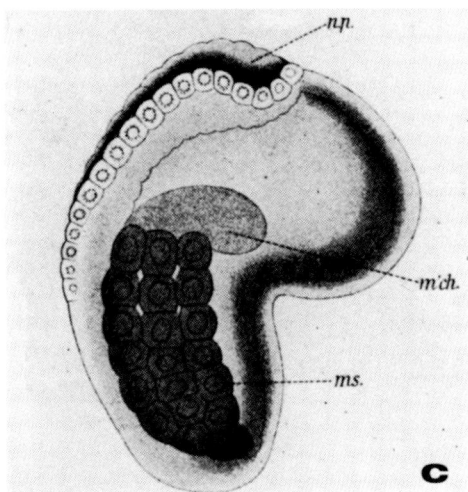

the larval skeleton), the mosaic of "plasms" present in ascidian eggs, and the polar lobe in molluskan eggs. As an example, Fig. 59 shows the localization, in an ascidian egg, of the myoplasm yellow crescent required for the differentiation of muscle cells in the tadpole larva. Of exceptional interest among the cytoplasmic determinants is the germ plasm of insects and amphibians, since it is absolutely necessary for the formation of the gonads, and thus for reproduction. This germ plasm is located at the posterior end of insect eggs (pole plasm) and at the vegetal pole of amphibian eggs (Fig. 60). Germ plasms in both insect and amphibian eggs are characterized by the presence of electron-dense RNP granules (Fig. 61); their role and precise chemical composition require further discussion. Destruction of the germ plasm, in both frog and *Drosophila* eggs, by localized uv irradiation does not prevent embryogenesis, but the adults become sterile. The same results have been obtained with the eggs of another insect,

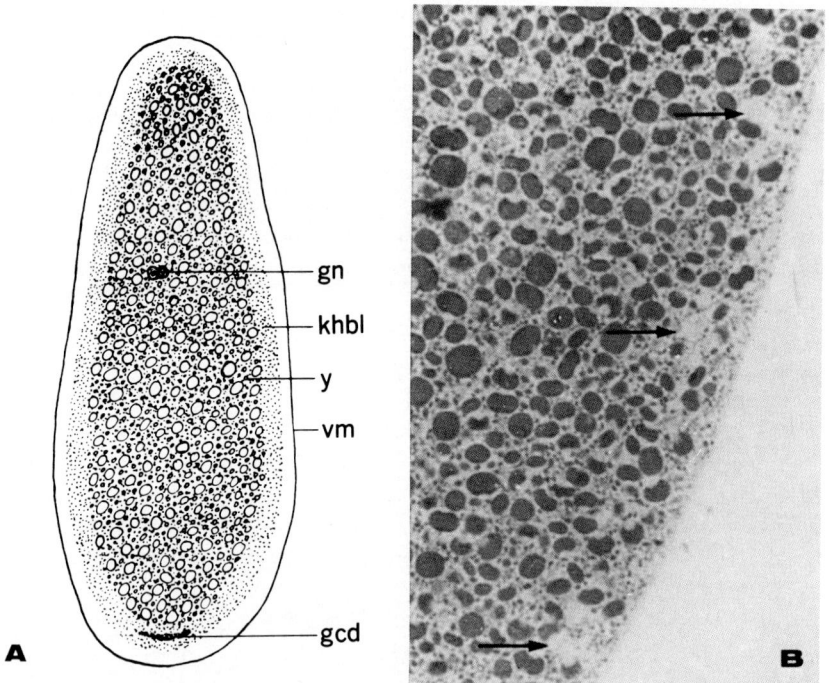

FIG. 60.   (A) Localization of the germ plasm (gcd) at the posterior end of an insect (*Calligrapha*) egg (4 hr after deposition, longitudinal section). gn, fusion of the two pronuclei; y, yolk; khbl, superficial clear cytoplasm; vm, vitelline membrane. (B) Semithin section in the vegetal hemisphere of a fertilized frog egg; arrows show germ plasm areas. [(A) Hegner, 1911; (B) Williams and Smith, 1971.]

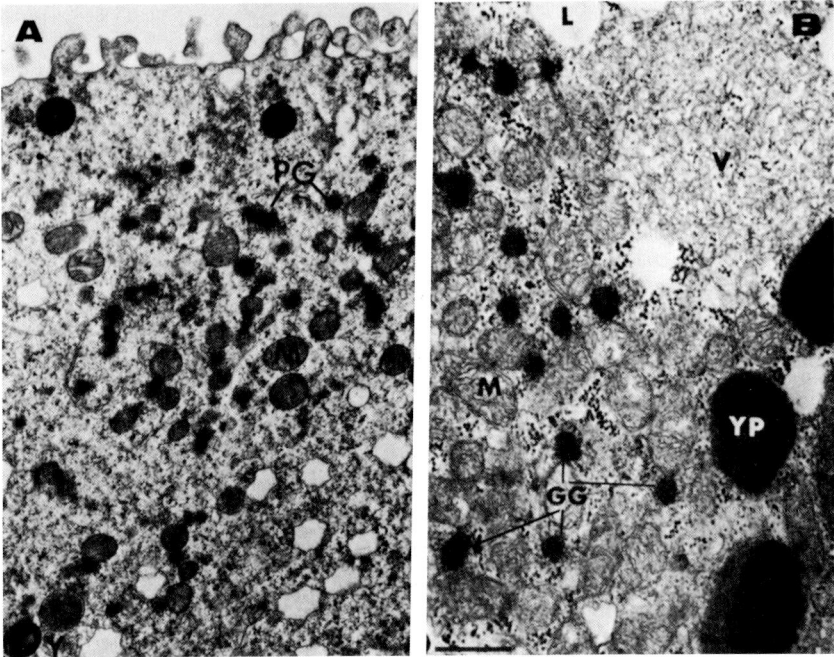

FIG. 61. Ultrastructure of germ plasms. (A) Posterior pole of an early cleavage *Drosophila* embryo (30 min old) containing polar granules (PG), yolk (Y), and mitochondria (M). (B) Vegetal pole of an unfertilized frog egg. GG, germinal granule; YP, yolk platelet; M, mitochondrion L, lipid droplet; V, vesicles. [(A) Illmensee and Mahowald, 1974; (B) Williams and Smith, 1971.]

Smittia, in which the action spectrum for the inhibition of pole cell formation shows a maximum at 260 nm, suggesting that a nucleic acid or a nucleoprotein is involved in this process (Brown and Kalthoff, 1983). In early embryos of the nematode *Caenorhabditis elegans,* segregation of the germ line granules results from the intervention of the MFs; the mitotic spindles and the microtubular network play no role in this process, as shown by the application of suitable inhibitors (Strome and Wood, 1983). If, as was done by Okada *et al.* (1974), uv-irradiated *Drosophila* eggs are injected with pole plasm material removed from a normal egg, normal adults are obtained. No less striking is the experiment of Illmensee and Mahowald (1974), who injected pole plasm material into the anterior part of a *Drosophila* egg and obtained germ cells in this area.

As one can see, the cytoplasmic determinants that accumulate in localized regions of the egg and are segregated by cleavage have considerable importance for embryogenesis in many species. Unfortunately, we know little about their molecular identity, which will be discussed later in this chapter.

Finally, we should recall here the unusual phenomenon of chromatin diminution, which takes place during early cleavage in the eggs of *Ascaris* and some insects (*Sciara,* for instance). As we saw in Chapter 5, Volume 1, elimination in the cytoplasm, followed by degradation, occurs for more than 20% of the total DNA during the first cleavage cycles in *Ascaris* eggs. In cecidomyids, centrifugation experiments have shown that retention or loss of chromosomes during cleavage depends on the presence or absence of a cytologically visible plasm, which thus shows a cytoplasmic determinant (Geyer-Duszyńska, 1959).

This brief outline clearly shows that egg cleavage involves more than mere cell division. Cleaving eggs reveal a diversity of patterns that is absent in cultured cells. For instance, why do the eggs of worms, mollusks, and sea urchins form micromeres at a given stage of cleavage? Micromere formation in such cases is not due to the unequal distribution of yolk, but to the localization in proximity to cell surface and to the orientation of the mitotic spindles present in given blastomeres. Such localization must depend on the composition of the surrounding cytoplasm, presumably the organization of the cytoskeleton. Clearly, what distinguishes the cleaving egg from dividing somatic cells is the bulk and heterogeneity of its cytoplasm, rather than specific properties of its nucleus. However, as we shall now see, nuclear division also has a number of characteristics that are not found in cultured cells.

B. Some Characteristics of Cleaving Eggs

The first demonstration that extensive DNA synthesis takes place during the cleavage of sea urchin eggs was given by the author in 1933. A paper by Parisi *et al.* (1978) contains a curve for DNA synthesis (Fig. 62) that is not very different from the one we obtained almost half a century ago. The work of Parisi *et al.* (1978) brought to light an interesting correlation between DNA synthesis and the mitotic index, which drops continuously during cleavage. While 60% of the cells are in mitosis in the 6-hr early blastula, there are only 11% at hatching and 4% at the mesenchyme blastula stage. Similar figures have been obtained by Brachet and De Petrocellis (1981), and it is clear that, in sea urchin as well as amphibian eggs, the mitotic rate falls continuously during cleavage. In the axolotl, for instance, cleavage is rapid and synchronous in all cells until the 10th mitotic cycle; in early blastulae it becomes asynchronous, while at gastrulation the mitotic index falls considerably.

Of great interest for embryologists is the fact that, from the 16-cell stage on, a gradient of mitotic activity, decreasing from the vegetal to the animal pole, can be detected in sea urchin eggs (Parisi *et al.,* 1978). This interest stems from the fact that the 16-cell stage is the time of embryonic determination. For instance, at this stage, micromeres are already committed to differentiate into primary mesenchyme; if they are isolated at the 16-cell stage, micromeres produce *in vitro* spicules (Okazaki, 1975). Curiously, treatment of sea urchin eggs with actinomycin D abolishes the mitotic gradient (Parisi *et al.,* 1979). It thus seems that

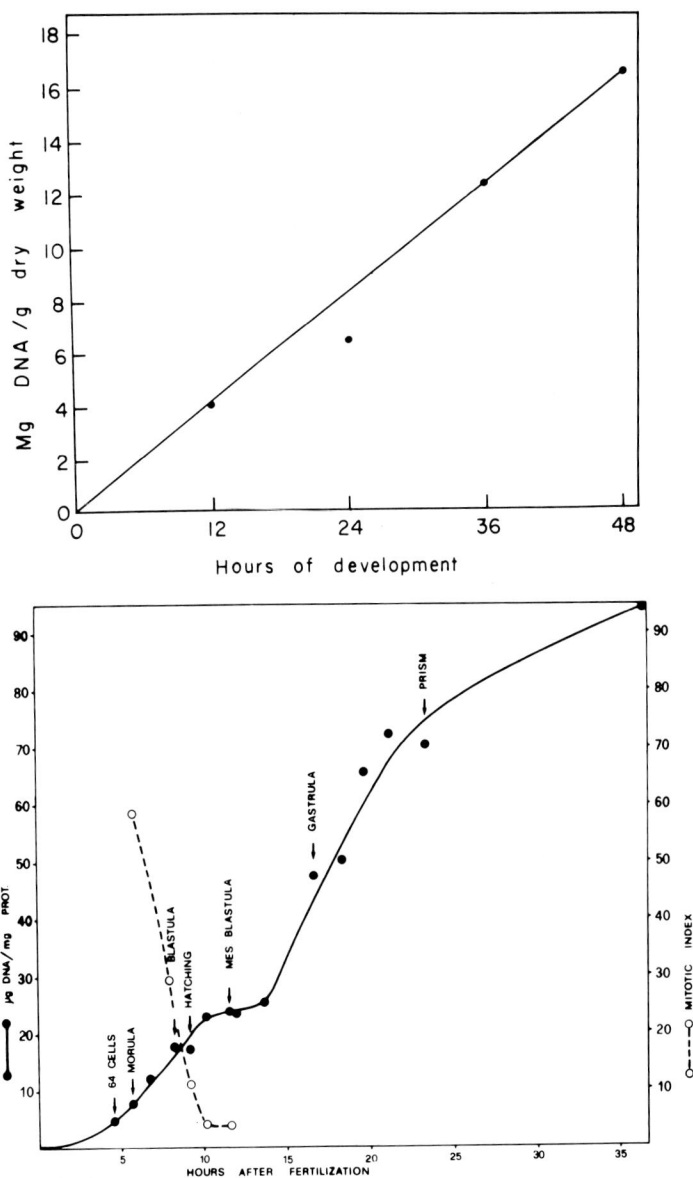

FIG. 62.   (a) DNA synthesis in developing sea urchin eggs. (b) Increase in the DNA content (solid line) and decrease in the mitotic index between the morula and mesenchyme blastula (dotted line) during sea urchin egg development. [(a) J. Brachet, 1933; (b) Parisi *et al.*, 1978.]

a transcriptional event in the micromeres controls the synchronization of the mitoses during cleavage. Interestingly, suppression of micromere formation by detergent treatment at the 4-cell stage leads to synchronization of the cleavage mitoses during several cycles. The micromeres look like coordinators of mitotic activity (Filosa *et al.*, 1985).

The cell cylce in cleaving eggs is, in general, very different from that of ordinary somatic cells. There is no G_1 phase, and DNA replication is exceedingly rapid (except in mammalian eggs, where cleavage is much slower than in most other animal species). For instance, in the large axolotl eggs, no G_1 phase can be detected. The S phase lasts for 8–10 min, the G_2 phase is confused with prophase, and the two together take about 25 min. The longest phase in the cell cycle is mitosis, which takes 80 min because the progression of the furrow from the animal to the vegetal pole is slow in large eggs. As mentioned in Chapter 5, Volume 1, the fastest known rate of DNA replication is in *Drosophila;* at the blastula stage, the whole genome is replicated in 3.4 min (Blumenthal *et al.*, 1973).

There is some evidence that the speed of the first cleavages in amphibian eggs, at least, is controlled by cytoplasmic factors and is not an autonomous property of the nuclei. This was shown by experiments in which Aimar *et al.* (1981) made reciprocal injections of cytoplasm between rapidly and slowly dividing eggs. Injection of cytoplasm from a rapidly cleaving egg (*Xenopus,* for instance) into an egg from a species that has a slower cleavage rate speeds up cleavage (and vice versa). According to Aimar *et al.* (1981), the cleavage timing system appears during maturation at the time of GVBD and does not depend upon MPF production. This cleavage cytoplasmic clock seems to be controlled by at least two different proteins (Aimar *et al.*, 1983).

The absence of a G_1 phase and the unusual speed of DNA replication during cleavage can be explained by the fact that fertilized eggs possess, since maturation, the machinery required for DNA replication. However, as already mentioned, there seems to be a need for increased production of deoxyribonucleotide precursors, since there is a marked increase in ribonucleotide reductase activity during early cleavage in both sea urchin (Noronha *et al.*, 1972; De Petrocellis and Rossi, 1976) and amphibian eggs (Tondeur-Six *et al.*, 1975). Inhibitors of this enzyme arrest cleavage, but not before the 8- to 16-cell stage in sea urchins and the middle blastula stage in amphibians (Brachet, 1967, 1968). This coincides with the time when the deoxyribonucleoside pool would be exhausted if it were not replenished by ribonucleotide reductase activity. Both the size of this pool and the enzymatic activity strongly decrease when, at the blastula stage, mitotic activity falls to a low rate (Mathews, 1975).

The key enzyme for DNA replication during early cleavage is, of course, DNA polymerase α. The amount of this enzyme present in unfertilized sea urchin eggs is sufficient for the production of 1000 nuclei (Morris and Rutter, 1976); the enzyme is believed to move from the cytoplasm into the nuclei when the chromosomes swell at telophase (Fansler and Loeb, 1969). Final proof that DNA polymerase-α activity is absolutely required for DNA synthesis and cleavage in sea urchin eggs comes from experiments by Ikegami et al. (1978, 1979). They found that aphidicolin arrests sea urchin egg development soon after amphimixy; the arrested eggs accumulate prophase asters (Brachet and De Petrocellis, 1981), but are unable to proceed further into cleavage (Fig. 63).

There is currently a good deal of interest in two processes that are closely related to the fast rate of DNA synthesis during cleavage: histone synthesis and nucleosome assembly. When DNA is replicated within a few minutes, is chromatin immediately assembled in nucleosomes? If so, where do the histones come from? We have already mentioned that unfertilized eggs possess a store of maternal histones. Nevertheless, fertilization results in strong synthesis of all types of histones due to intensive transcription of the repeated histone genes (which are repeated 400–1000 times in sea urchins and 50 times in *Xenopus,* as we have seen in Chapter 4, Volume 1). Close analysis of the histones synthesized by developing sea urchin embryos and of the corresponding mRNAs has shown unexpectedly that during cleavage there is a switch in histone synthesis: after maternal histones and their mRNAs are used, "early" or "cleavage" histones are synthesized. This is followed in the blastula by the synthesis of another set of "late embryonic" histones that are more closely related to those found in the adult. According to Kunkel and Weinberg (1978), new histone genes are switch-

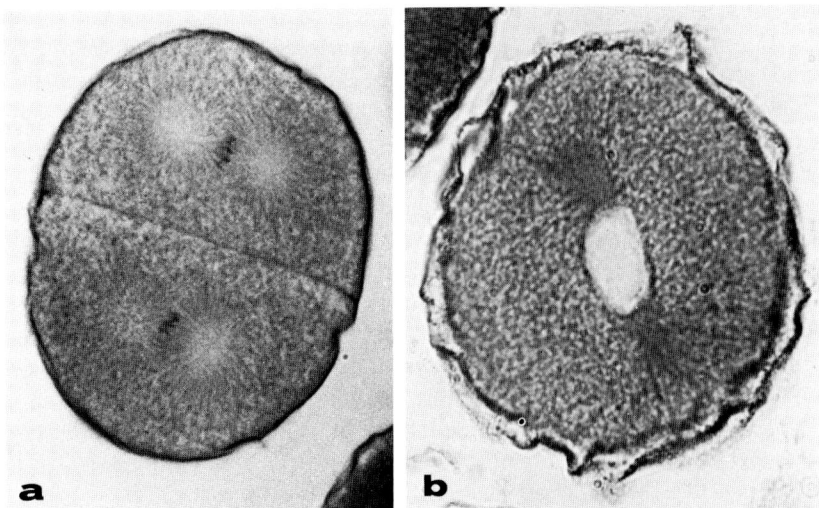

FIG. 63.   Effects of aphidicolin on fertilized sea urchin eggs. (a) After 2 hr, the controls display a normal mitotic activity. (b) The aphidicolin-treated eggs remain uncleaved. Aphidicolin inhibits DNA polymerase-α and DNA replication. [(a) H. Alexandre, unpublished; (b) Brachet and De Petrocellis, 1981.]

ed on (activated) in the hatching blastula. Spinelli *et al.* (1979) found similar electrophoretic patterns for histone mRNAs extracted from sea urchin oocytes, cleaving eggs, and morulae, but blastulae gave a very different pattern. However, according to Childs *et al.* (1979), the switch from genes coding for early histones to genes coding for late histones occurs progressively during cleavage of sea urchin eggs. During blastulation, which ends with the formation of "mesenchyme blastulae," the histone mRNA population undergoes more profound changes (Hieter *et al.*, 1979). A more recent analysis of early histone mRNAs during sea urchin egg cleavage has given the following information: the rate of early histone mRNA synthesis sharply increases at the 16-cell stage and reaches a peak at the 128-cell stage; it then decreases gradually until the 300-cell stage. At that time, late histone synthesis becomes more important. During cleavage, early histone gene transcription is intensive, occurring at least once every minute (Maxson and Wilt, 1981). It is worth pointing out here that the 16-cell stage really appears to be a turning point in sea urchin egg development. In fact, according to Senger and Gross (1978), quantitative changes in the pattern of histone synthesis can be detected, at this stage, between the micromeres and the other cells. The time course of histone synthesis in sea urchin eggs and embryos has been reinvestigated by Herlands *et al.* (1982). They found that cleavage stage (CS) histones are synthesized in the oocytes; at the morula stage, α variants of these histones become detectable; synthesis of late histones (β, γ, and δ) begins in the mesenchyme blastula. Both α and CS histones are stored in the egg, but unfertilized eggs synthesize only the CS histones. At fertilization, the CS and α mRNAs are both translated. Early histone synthesis is thus regulated at the translational level. Maxson and Wilt (1982) found that 7×10^6 molecules of core histone mRNAs and 2.5×10^5 molecules of histone H1 mRNA are accumulated between the 16- and 200-cell stages. The histone mRNAs then decay with a half-life of 1.5–2.0 hr. Sea urchin eggs are probably no exception, since it has been demonstrated that, when *Xenopus* eggs reach the blastula stage (1000–2000 cells), a new set of histone genes begins to operate (Woodland *et al.*, 1979).

Changes in chromatin conformation accompany the shifts in histone synthesis during sea urchin development. Spinelli *et al.* (1982) reported that, in spermatozoa, the histone genes are folded in nucleosomes and are resistant to DNase digestion. Their conformation changes at the 32- to 64-cell stage, where they become nuclease sensitive; the early histone genes are repressed again in mesenchyme blastulae, where the embryonic histone genes become active. Bryan *et al.* (1983) also found that, in early blastulae, the promoter regions of the five histone genes are highly sensitive to DNase digestion; this sensitivity is lost in hatched blastulae, where transcription of the histone genes stops.

We mentioned in Chapter 5, Volume 1 that the unusual speed of DNA replication during egg cleavage is correlated with the presence, on electron micrographs, of numerous "eyes" and clusters of "microbubbles" in spread chromatin (Wolstenholme, 1973; Baldari *et al.*, 1978; Buongiorno-Nardelli *et al.*,

1982). This means that a large proportion of total DNA is in single-stranded form during early cleavage. This fact, taken together with the presence of unusual species of histone molecules at this stage, raises the question of chromatin organization in dividing eggs. Is the chromatin in the form of nucleosomes during early cleavage? The answer seems to be yes. According to Shaw *et al.* (1981), nucleosomes already exist at the two-cell stage in sea urchins. Their core is composed of cleavage (early) histones, which persist in chromatin until the blastula stage. The nucleosome repeat length decreases when sea urchin spermatozoa and blastulae are compared (Savić *et al.*, 1981). That "linker" DNA is longer in sperm than in blastula chromatin had already been shown by Spadafora *et al.* (1976). According to Savić *et al.* (1981), the adjustment of the repeat length is dependent on the cell cycle and takes place when the egg begins its first cleavage; it is not due to the replacement of sperm-specific histone H1 by a cleavage-stage histone H1. Although the DNA fiber wrapped around the histone core of the nucleosomes has the same length at different stages of development, Simpson (1981) has reported that nucleosome core particles differ in their sensitivity to DNase I at various stages of sea urchin development. The core particles of blastulae are less stable under conditions of DNase I digestion, heating, and hypotonic treatments than those of plutei (which contain adult-type histones). Finally, Cognetti and Shaw (1981) found that chromatin extracted from micromeres is more resistant to DNase I treatment than that from the other blastomeres; in micromeres, chromatin seems to be more condensed than it is in macromeres and mesomeres, and would thus somewhat resemble sperm head chromatin. This is probably also true for blastula and pluteus chromatin, since in Feulgen-stained sections, the nuclei appear to be much more condensed in plutei than in blastulae (Fig. 64). However, it should be mentioned that, according to Keichline and Wasserman (1979), the organization of chromatin in nucleosomes does not change during sea urchin development. However, there is universal agreement that this organization is very different in sea urchin sperm and embryos. Spermatozoa possess sperm-specific histones that disappear soon after fertilization, and this affects chromatin structure.

It is too early to speculate about the embryological significance of the shift from cleavage to embryonic histones and about changes in nucleosome organization during development. It is likely, however, that they are responsible for chromatin condensation and that they limit the accessibility to DNA of the enzymes involved in its replication and transcription. Another factor that probably plays a role in the slowdown of DNA replication in blastulae is the decrease in "initiation factor" (IF) activity (Benbow *et al.*, 1975). This factor appears during maturation of *Xenopus* oocytes and is believed to be important for the initiation of DNA replication. Its activity steadily decreases, in parallel with mitotic activity, as cleavage proceeds in *Xenopus* eggs. Similar factors have been found in sea urchin eggs. Extracts of cleaving eggs stimulate DNA synthesis in nuclei isolated from sea urchin morulae; extracts from unfertilized eggs are

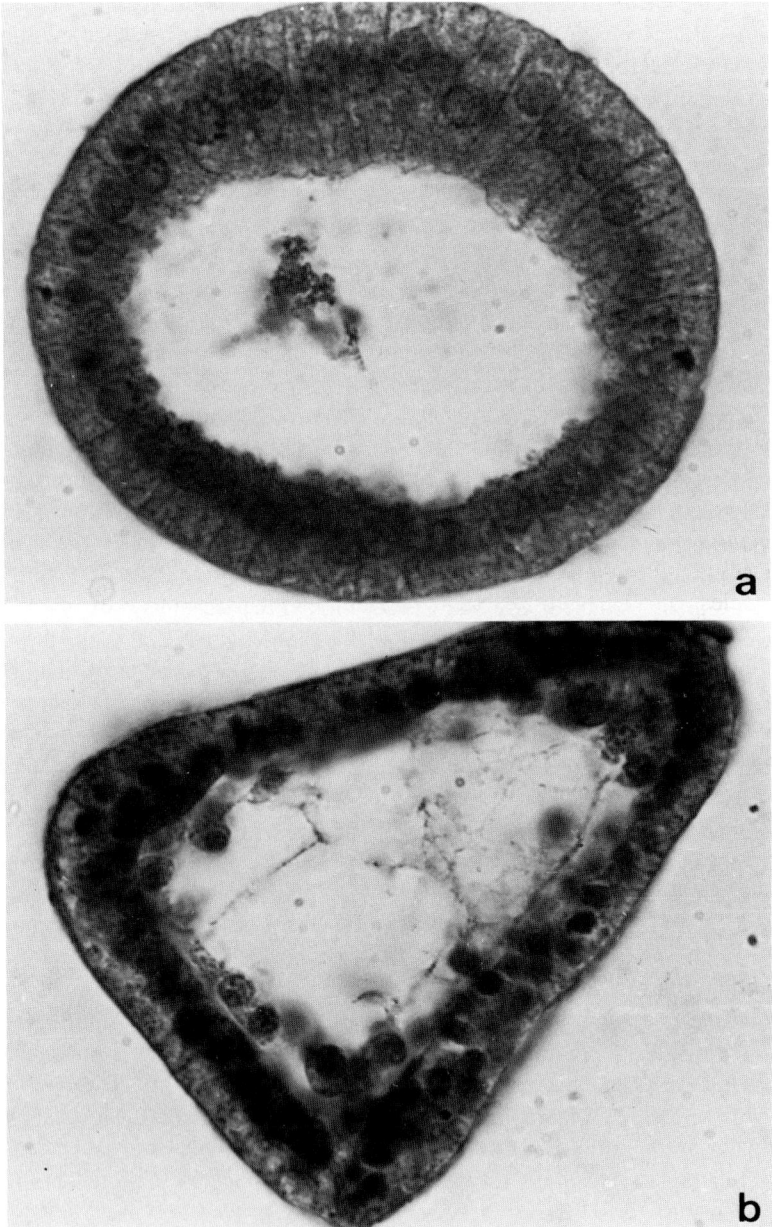

FIG. 64.   Sections of a sea urchin blastula (a) and a late (prism) gastrula (b). Note the increased condensation of the chromatin. The nuclei stain more strongly in (b) than in (a), although they have the same DNA content. [J. Brachet, unpublished.]

inactive. It is likely that the DNA synthesis-stimulating activity (which is due to three different proteins) decreases, as in amphibians, when mitotic activity slows down during gastrulation (Shimada, 1983).

The control of the cell cycle during cleavage of *Xenopus* eggs has been the subject of an important paper by Newport and Kirschner (1984). Injection of cytoplasm from unfertilized eggs (which contains the cytostatic factor CSF) arrests mitosis by blocking an endogenous cytoplasmic cell-cycle oscillator; this causes the stabilization of MPF. The effects of MPF are on the chromatin template, not on the replication machinery. This is demonstrated by experiments showing that DNA replication can be driven by addition and loss of MPF and that mitotic arrest is not mediated through nuclei and microtubules. The factors required for the inhibition of DNA synthesis at metaphase act stoichiometrically. Injection of large amounts of plasmid DNA reactivates the cell cycle in CSF-injected eggs. MPF activity of the cytoplasm is strong during mitosis, weak during the S-phase. CSF maintains MPF at high levels and this induces the arrest in metaphase. Ca^{2+} inactivates CSF, allowing the cell to progress in mitosis. MPF is probably synthesized periodically. In conclusion, the cell cycle in *Xenopus* cleaving eggs is controlled by addition and removal of MPF. This factor induces the phosphorylation of lamins A and C in the nuclear envelope and acts on nuclear membrance breakdown as a lamin kinase (Miyake-Lie and Kirschner, 1985). According to Stick and Hausen (1985), lamin C serves to the formation of the nuclear membrane during the first cleavages; lamins A and B are not synthesized before the blastula and gastrula stages.

We shall present only a brief discussion of RNA synthesis during cleavage, since this subject has been dealt with in great detail by E. Davidson in his excellent book (1976) and review (1982). In both sea urchin and amphibian eggs, rRNA synthesis is hardly detectable until the late blastula stage is reached. At the same time, typical basophilic, rRNA-containing nucleoli appear. Their formation is not merely a consequence of the slowdown of mitotic activity at the end of cleavage, since we found (Brachet *et al.*, 1972) that only "cleavage nucleoli" are present in the nuclei of sea urchin eggs arrested during early cleavage by treatment with inhibitors of DNA synthesis. These cleavage nucleoli (Fig. 65) which can be observed in normal sea urchin and amphibian eggs during cleavage, contain no RNA and do not incorporate labeled uridine; they have a fibrillar ultrastructure and are probably an accumulation of still unidentified nuclear proteins. The reason why rRNA synthesis is selectively repressed during cleavage (as it is in previtellogenic oocyte) is not known. There are some indications that cleaving *Xenopus* eggs contain a specific inhibitor of rRNA synthesis (Shiokawa and Yamana, 1979), but the evidence is far from compelling. Busby and Reeder (1983) injected into fertilized *Xenopus* eggs plasmids containing a repeating unit (with its spacer) of *Xenopus* rDNA. They observed no transcription before the late blastula stage; this shows that the control of transcription is

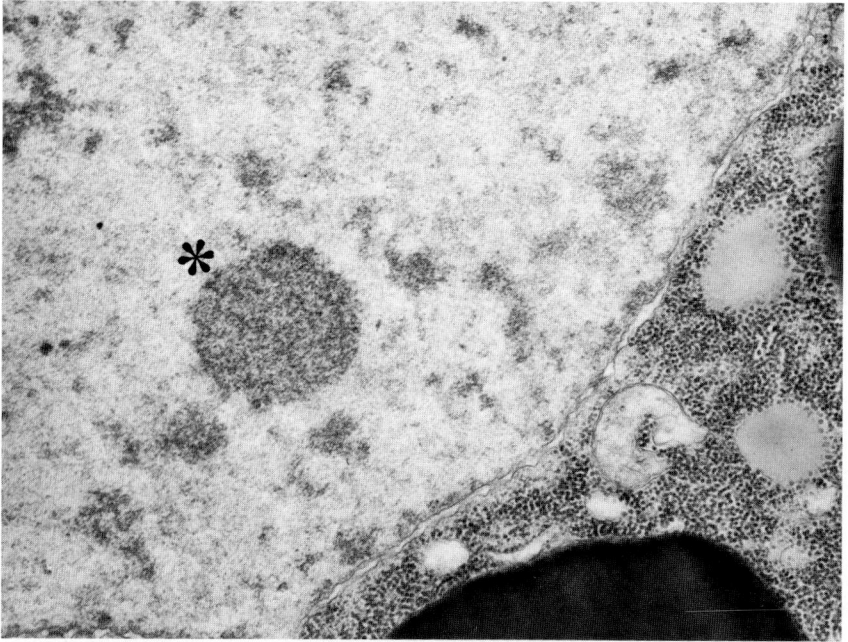

FIG. 65.   Cleavage nucleolus (asterisk) in an amphibian blastula seen under the electron microscope. It has a purely fibrillar structure and lacks the outer granular region. [Courtesy of Dr. M.
Geuskens.]

the same for endogenous and injected genes. These experiments showed, in
addition, that the rDNA transcription rate is strongly affected by spacer sequences far upstream from the promoter. It should be added that, in contrast to
sea urchin and amphibian eggs, rRNA synthesis is already present at the two-cell
stage in mouse eggs, where it precedes mRNA and tRNA synthesis, which
begins at the 16-cell stage (Young and Sweeney, 1979). As expected, typical
nucleoli are already present at the onset of cleavage in mouse eggs.

mRNA synthesis starts earlier than rRNA synthesis in amphibian and sea
urchin eggs. In *Xenopus,* mRNA synthesis can be detected at the 8- to 16-
blastomere stage and then steadily increases; it includes synthesis of poly(A)$^-$
histone mRNAs and various poly(A)$^+$ mRNA species. At the blastula stage,
according to Shiokawa *et al.* (1981), 85% of the overall protein synthesis is still
supported by maternal poly(A)$^+$ mRNAs, which are as efficient in translation as
the newly synthesized mRNAs. Later, when the embryos synthesize rRNAs and
make new ribosomes, the mRNA content increases six- to eightfold when rRNA
doubles. There is a particularly strong increase, at these relatively late stages of
embryogenesis, in the mRNAs that code for ribosomal proteins (Weiss *et al.,*
1981). Finally, synthesis of the tRNAs and of 5 S rRNA begins at the young
blastula stage, a few hours before 28 and 18 S RNA synthesis.

This picture can be completed because of Newport and Kirschner's (1982a,b) careful analysis of cleavage in *Xenopus* eggs (Kirschner *et al.*, 1985). After 12 rapid (30 min), synchronous cleavages, the embryo undergoes what they have called the "midblastula transition." The G_1 and G_2 phases of the cell cycle appear. If the cells are dissociated, they display motility, in contrast to the inertness of younger cells. Transcription of tRNAs, 5 S RNA, snRNAs, and a 7 S RNA that is transcribed on middle repetitive DNA sequences begins. Synthesis of rRNA, as we have seen, does not occur before a couple of hours after the midblastula transition. Experiments in which fertilized eggs were ligated with a hair, or centrifuged, or treated with cytochalasin have shown that the timing of the midblastula transition depends on a critical ratio between the volumes of the nucleus and the cytoplasm, and not on the number of cleavages, DNA replication cycles, or the time elapsed since fertilization. The midblastula transition is speeded up in polyspermic eggs, which led the authors to the interesting proposal that the nuclei titrate a cytoplasmic component. This hypothesis is supported by experiments of Newport and Kirschner (1982b), who injected into fertilized eggs a plasmid containing a yeast tRNA gene. They found that the injected DNA is transcribed transiently during early cleavage and that transcription is resumed when the midblastula transition occurs. There is no transient suppression of transcription if one injects an amount of plasmid DNA equal to the cellular DNA content after 12 cleavages (24 ng) in the blastula. It was suggested that DNA reacts with a cytoplasmic suppressor, which is progressively exhausted during the repeated cleavages; transcription would start when nuclear DNA is no longer saturated with this suppressor. Verification of this ingenious model will require the isolation and characterization of the still hypothetical cytoplasmic suppressor. Synthesis of the snRNA U1 has been studied in detail by Forbes *et al.* (1983b). Cleaving *Xenopus* eggs possess 8×10^8 molecules of this snRNA, an amount sufficient for 4000–8000 nuclei. Synthesis of U1 snRNA, catalyzed by RNA polymerase II, begins at the midblastula transition. There is a sevenfold increase in this particular RNA between cleavage and gastrulation.

In sea urchins, the sequence of events is basically the same. However, mRNA synthesis starts at the time of first cleavage and increases sharply at the 8- to 16-cell stage. Of the mRNAs synthesized during cleavage, 60% code for histones and (as usual) have no poly(A) terminal sequence; another 30% of the newly synthesized mRNAs are also devoid of a poly(A) tail, but code for other proteins. Thus, at the beginning of cleavage, only 10% of the newly synthesized mRNAs are of the poly(A)$^+$ type, but the proportion of the poly(A)$^+$ mRNAs synthesized during cleavage steadily increases and reaches 50% in the blastula (Nemer *et al.*, 1975). Unexpectedly, *Xenopus* and sea urchin eggs also possess maternal poly(A)$^+$ histone mRNAs. They are stored in RNP particles and are not found in the polyribosomes. As in other cells, only 10% of heterogeneous nuclear RNA (hnRNA) moves from the nucleus to the cytoplasm in cleaving sea urchin eggs. Among the mRNAs identified in sea urchin eggs is actin mRNA; its amount remains constant during the 8 hr following fertilization. Actin mRNA increases considerably (10–25 times in the cytoplasm, 15–40 times in the polyribosomes) when, after 18 hr of development, the eggs reach the mesenchyme blastula stage (Crain et al., 1981). A peculiarity of sea urchin eggs is the presence, in their cytoplasm, of very large mRNA molecules (Giudice et al., 1974) and of giant polyribosomes (Whiteley and Mizuno, 1981). At the four-cell stage, electron microscopy detects the presence in the cytoplasm of huge polyribosomes (13.6 μm long) composed of up to 277 ribosomes. Since the giant polyribosomes are also found in activated anucleate halves, their very large mRNA (6.5 $\times$ 10^4 nucleotides) must be of maternal origin.

Recent work by Cabrera et al. (1984) has shown that the turnover rate varies for different cytoplasmic mRNAs, and that there is no abrupt replacement of the maternal mRNAs by newly synthesized mRNAs during sea urchin development.

Recall at this point that cleavage, in both sea urchin and amphibian eggs, remains unaffected by actinomycin D. Normal blastulae are obtained in the presence of the drug, but gastrulation never begins. For this reason, it is believed that the RNAs synthesized during cleavage are required for later events in development than cleavage itself. A different situation prevails in the slowly dividing mammalian eggs in which actinomycin D quickly arrests cleavage. In contrast, cleavage of all eggs is quickly arrested by the addition of inhibitors of protein synthesis such as puromycin or emetin. In sea urchin eggs, the appearance of new proteins can be detected at the important 16-cell stage; many more are found in blastulae. According to a study by Bédard and Brandhorst (1983), sea urchin embryos synthesize about 900 different proteins during development; the rate of synthesis may increase 10 times and even 100 times for some of these proteins. There are few qualitative changes in the pattern of protein synthesis during cleavage; 60% of these changes occur between hatching and the beginning of gastrulation, which seems to be an important transition. Among the newly synthesized proteins, the histones are quantitatively predominant. There is little tubulin synthesis before ciliogenesis in hatching blastulae. Tubulin needed for spindle and aster edification at each mitotic cycle apparently arises from a preexisting pool of tubulin dimers. On the other hand, the activity of a tyrosine-specific protein kinase doubles at the time of first cleavage and has increased 20-fold by gastrulation. This enzyme phosphorylates tyrosine residues in nine membrane proteins; its role in sea urchin development is still unknown (Dasgupta and Garbers, 1983).

In mouse eggs, in addition to the already mentioned ''fertilization proteins,'' new proteins are already synthesized on maternal mRNAs after the two-cell stage (Brande et al., 1979; Bensaude et al., 1983).

There is no need to discuss in detail the energy requirements for cleavage, which were considered in Chapter 5, Volume 1, except to add that oxygen consumption shows a slow, but continuous, increase during cleavage in all eggs.

Anaerobiosis or KCN treatment immediately arrests cleavage in sea urchin eggs but has no immediate effect on frog eggs, which can reach the blastula stage even when respiration is almost completely suppressed (Brachet, 1934). This difference is due to the fact that under anaerobic conditions, ATP is utilized much more rapidly by sea urchin eggs than by frog eggs; in both cases, cell division is arrested when the ATP content has dropped by 50–60%. Whether there are small cyclic variations in the respiratory rate, and whether the increase in oxygen consumption during cleavage is connected to DNA synthesis, remains open to discussion.

Among the many factors, that regulate the synthesis of macromolecules during cleavage is the production of polyamines. One of the few enzymes that, like ribonucleotide reductase, are synthesized during cleavage is ornithine decarboxylase (ODC), the key enzyme for polyamine biosynthesis. This is the case in amphibians (Russell, 1971), sea urchins (Manen and Russell, 1973; Kusunoki and Yasumosu, 1976), and the nudibranch *Phestilla* (Manen *et al.,* 1977). According to Kusunoki and Yasumosu (1976, 1978), there are cyclic changes in ODC activity and polyamine content during sea urchin egg cleavage. The same authors (1978) found that an inhibitor of ODC (α-hydrazino-ornithine) slows cleavage and arrests development, at the early morula stage, in the Japanese sea urchins *Hemicentrotus* and *Anthocidaris*. These results are at variance with those we obtained (Brachet *et al.,* 1978, and unpublished results) on two Mediterranean sea urchin species (*Paracentrotus* and *Arbacia*). Three inhibitors of ODC (α-methylornithine, α-difluoromethylornithine, and α-hydrazino-ornithine) have no effect on sea urchin cleavage, but arrest development at the late blastula stage. Addition of putrescine or spermidine suppresses the inhibitory effects of the ornithine analogs, and normal plutei can be obtained under certain conditions (Fig. 66). DNA synthesis is affected more than RNA and protein synthesis by ODC inhibitors. This suggests that polyamine synthesis is not required for cleavage, but is necessary for the establishment of a regular cell cycle at the blastula stage. However, the three inhibitors affect a third Mediterranean species (*Sphaerechinus*) in a way similar to that of the Japanese species studied by Kusunoki and Yasumosu (1978). These species differences probably reflect differences in the pool size and the rate of polyamine synthesis in eggs of various sea urchin species. Unfertilized eggs of *Paracentrotus* and *Arbacia* already have a rather high polyamine content, in contrast to those of the Japanese species, in which the unfertilized eggs contain very little putrescine and spermidine. There is intense synthesis of polyamines during cleavage in the Japanese sea urchins and very little in the Mediterranean ones. Taken together, these observations suggest that polyamines are somehow involved in the control of DNA synthesis during early development.

Perhaps to the surprise of some readers, the actual process of mitosis during cleavage will not be discussed here; this is to avoid unnecessary repetition of

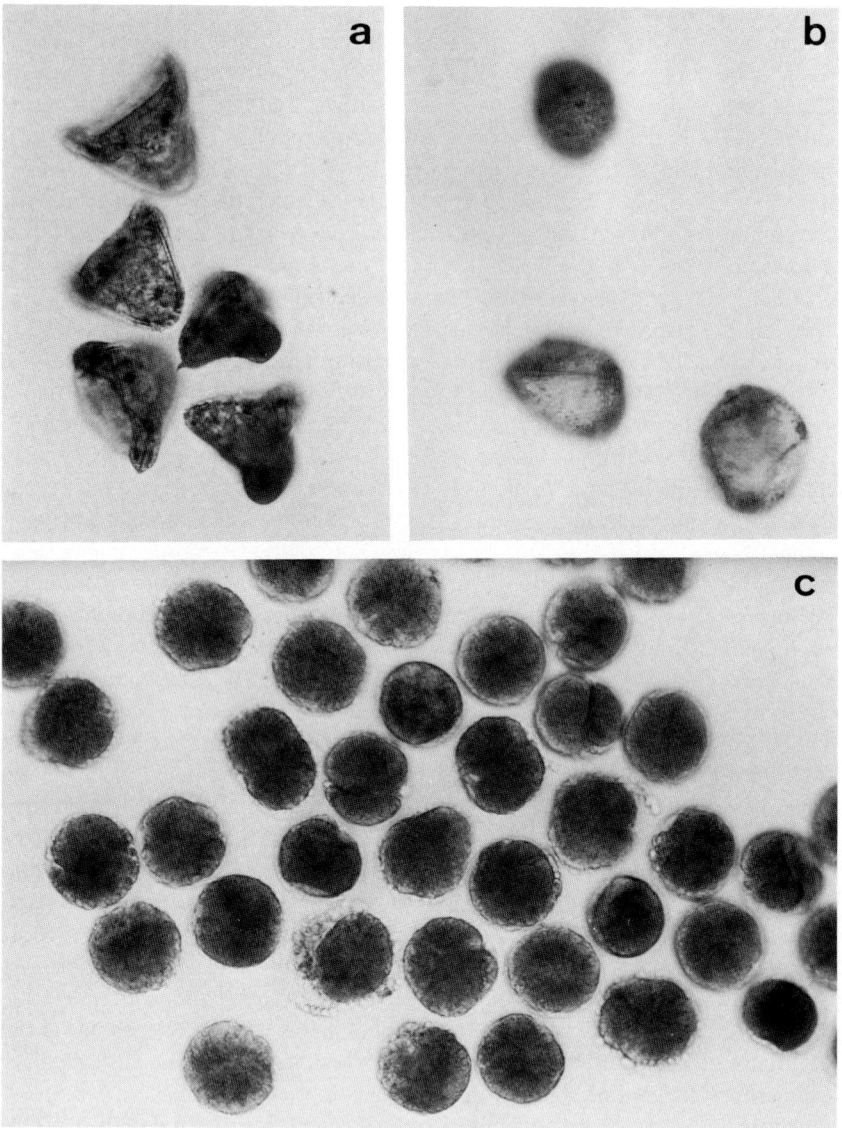

FIG. 66.   Effects of a polyamine synthesis inhibitor (hydroxyornithine) on sea urchin egg development. (a) Controls, young plutei; (b) treated with a mixture of hydroxyornithine and spermidine; (c) treatment with hydroxyornithine alone; incipient cytolysis. Spermidine in (b) has protected the embryos against cytolysis and has allowed development to proceed until the gastrula stage. [Original photographs by H. Alexandre.]

what has already been said in Chapter 5; after all, mitosis is basically the same in cleaving eggs and dividing cells. We wish to reemphasize that much of what we know about mitosis results from work done on sea urchin and amphibian eggs. The factor that induces maturation in *Xenopus* oocytes (MPF) is responsible for nuclear membrane breakdown and chromosome condensation in cleaving *Xenopus* eggs and in a number of somatic cells. Entry into prophase is apparently due to protein phosphorylation induced by MPF. Much of our knowledge about the organization of the spindle and asters and about chromosome movement at anaphase derives from the pioneer work of Mazia and Dan (1952), who isolated an intact mitotic apparatus from cleaving sea urchin eggs. Finally, the now generally accepted contractile ring theory of cytokinesis was first proposed as an explanation for furrowing during sea urchin egg cleavage.

C. THE BLASTULA STAGE: SEA URCHIN GASTRULATION

The end of cleavage is the poorly defined blastula stage (one speaks of early, middle, and late blastulae), which is characterized by the reduction of mitotic activity, the appearance of true nucleoli, and the synthesis of high-molecular-weight ribosomal RNA and many proteins. We shall limit ourselves here to sea urchin blastulae in which, after hatching and ciliation, gastrulation soon follows. As shown in Fig. 58c, the last stage of blastulation is the formation of mes-enchyme blastulae. The basal part of the larva has thickened in a basal plate. In this basal plate, the cells that come from the micromeres move into the blastocele, where they form the primary mesenchyme, which will give rise to the skeleton (spicules) of the pluteus larva. Gastrulation begins with an infolding (invagination) of the basal plate; the invaginated endoderm cells will later form the archenteron, or primitive gut. After this "primary" gastrulation, cells at the tip of the archenteron move freely in the blastocele cavity, where they form the secondary mesenchyme. Complete invagination of the archenteron ("second-ary" gastrulation) results (as shown by Gustafson and Wolpert, 1967) from the contraction of the secondary mesenchyme cells, which are firmly attached to the tip of the archenteron on one side and the anterior tip of the larvae on the other (Fig. 58d).

According to Gustafson and Wolpert (1967), primary gastrulation results from the autonomous activity of the basal plate cells, which display ameboid activity and lose contact with neighboring cells and with the hyaline layer. Secondary gastrulation is due to the contractility of MFs present in the secondary mes-enchyme cells.

Many changes in the pattern of protein synthesis occur at the blastula stage. Among the identified proteins that appear or increase at that stage is the "hatch-ing enzyme," a protease that is secreted by the blastula cells and digests the fertilization membrane. Another enzyme, which appears in the medium soon after hatching, is an arylsulfatase (Rapraeger and Epel, 1981), which is probably involved in the metabolism of sulfated proteoglycans and glycoproteins. These sulfated molecules are certainly involved in cell mobility and in the shape changes that are of fundamental importance during gastrulation. That glycosylation of proteins is absolutely necessary for successful gastrulation is shown by the fact that two specific inhibitors of protein glycosylation, tunicamycin and compactin, inhibit gastrulation in sea urchins (Schneider et al., 1978; Carson and Lennarz, 1981). Compactin inhibits the synthesis of dolichol, a polyisoprenoid lipid required for dolichyl phosphate–mediated glycosylation of proteins. The marked increase in actin mRNA in mesenchyme blastulae (Crain et al., 1981) probably plays some roll in cell contractility and motility, which are the prominent features of gastrulation. One phenomenon that is not yet understood is that a heat shock at 31°C induces the synthesis of a set of hsps in sea urchin gastrulae, but not in younger embryos. Synthesis of these hsps is prevented by actinomycin D, indicating that it results from the translation of newly synthesized mRNAs (Roccheri et al., 1981). This conclusion is reinforced by the finding that, during heat shocks, there is a parallel stimulation of the synthesis of the hsps and their mRNAs (Roccheri et al., 1982).

It is certain that, when sea urchin eggs undergo gastrulation, transcription greatly increases. Fregien et al. (1983) found that five different mRNAs become 3–47 times more abundant between the blastula and pluteus stages; they appear at hatching and, interestingly, have a specific localization in later embryos. Two of the newly synthesized mRNAs are limited to the ectoderm (Spec mRNAs) and another to the endoderm; the two remaining mRNAs are present in the two cell layers. Among the Spec mRNAs, one encodes a protein similar to the Ca^{2+}-binding protein troponin C (Carpenter et al., 1984). Also, in *Xenopus*, a new class of poly(A)$^+$ RNAs (called DG RNA) appears at gastrulation; these DG RNAs decrease at neurulation and disappear in tadpoles (Sargent and Dawid 1983). Newly synthesized mRNAs replace 30% of the maternal mRNAs at gastrulation and 100% during neurulation. Some of the DG RNAs are synthesized at a very rapid rate; it is likely that they code for proteins that are important for gastrulation itself.

Due to the production of ectoderm-, mesoderm-, and endoderm-specific mRNAs, the surface of the sea urchin gastula becomes a mosaic of antigens, which can be detected with monoclonal antibodies (McClay and Wessel, 1985). Some of these antigens are cell layer-specific.

Sea urchin blastulae are easily dissociated in Ca^{2+}-free seawater and therefore provide a very favorable material for the study of cell aggregation (Giudice, 1973). As mentioned in Chapter 3, Volume 1, Giudice and Mutolo (1970) found that dissociated blastula cells from *Paracentrotus* and *Arbacia* quickly form mixed aggregates but species-specific sorting out follows after a few hours. The *Arbacia* cells form red aggregated distinct from the white aggregates formed by *Paracentrotus* cells. Aggregation, without species specificity, of sea urchin blastula cells is promoted by proteins that can be extracted with butanol from membranes or the hyaline layer of dissociated blastula cells (Noll et al., 1981; McCarthy and Spiegel, 1983). Oppenheimer and Meyer (1982a,b) found that the supernatant of dissociated sea urchin blastulae favors reaggregation and stage specificity in both species; the specific aggregation factor is a lectin that binds cells together by reacting with D-galactoside–like and acetylglucosamine residues.

Similar experiments done at much earlier stages of development have provided us with a different type of information. Sea urchin eggs were dissociated, at the two-cell stage, in Ca^{2+}-free seawater and allowed to develop for a few hours. The supernatant (conditioned medium) was collected and, after readdition of $CaCl_2$, was used as a culture medium for fertilized eggs from another female. It was found that their development stopped at the blastula stage, showing that the conditioned medium contained factors that prevented gastrulation (Brachet and Aimi, 1972). It is likely that the inhibitory factor is mainly hyaline, the sulfated glycoprotein that surrounds the egg surface after fertilization. In fact, when sea urchin eggs are placed under adverse conditions (treatment with α-hydrazino-ornithine, for instance) they often form a hyaline layer of excessive thickness and fail to undergo gastrulation (Fig. 67). This confirms the importance of glycoproteins for gastrulation movements.

A last question might be: are cellularization and DNA replication required for gastrulation? It is obvious that cellularization provides the egg with a plasticity that allows the performance of the complex, massive cell movements that are required for its transformation in a tridermic gastrula. However, it has been often

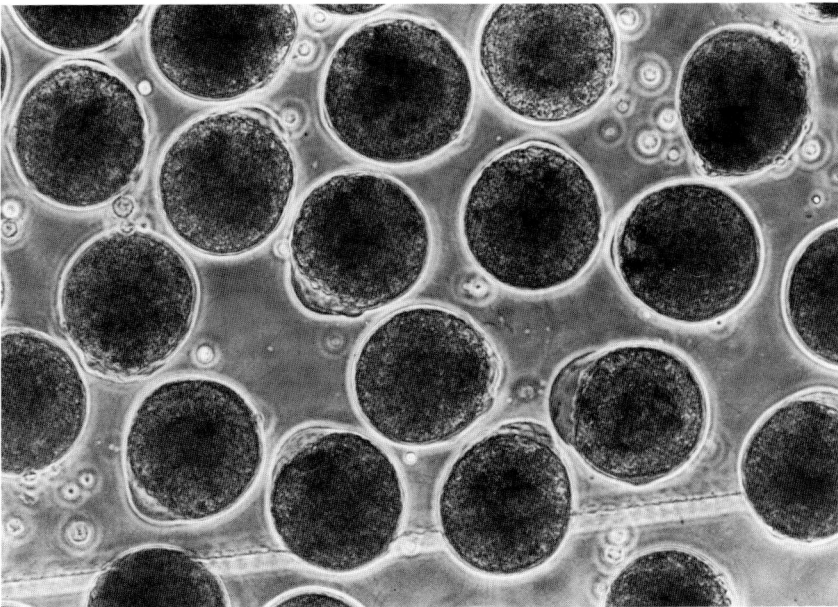

FIG. 67.   Sea urchin blastulae covered with an exceedingly thick hyaline layer after treatment with hydroxyornithine. [Original photograph by Dr. R. Tencer.]

reported (and we ourselves have observed it) that unfertilized amphibian eggs left for several days in their culture medium may display, near the vegetal pole, an unfolding that mimics gastrular invagination (Fig. 68). This occurs in the absence of DNA synthesis and mitotic activity, but does not lead to further devel-

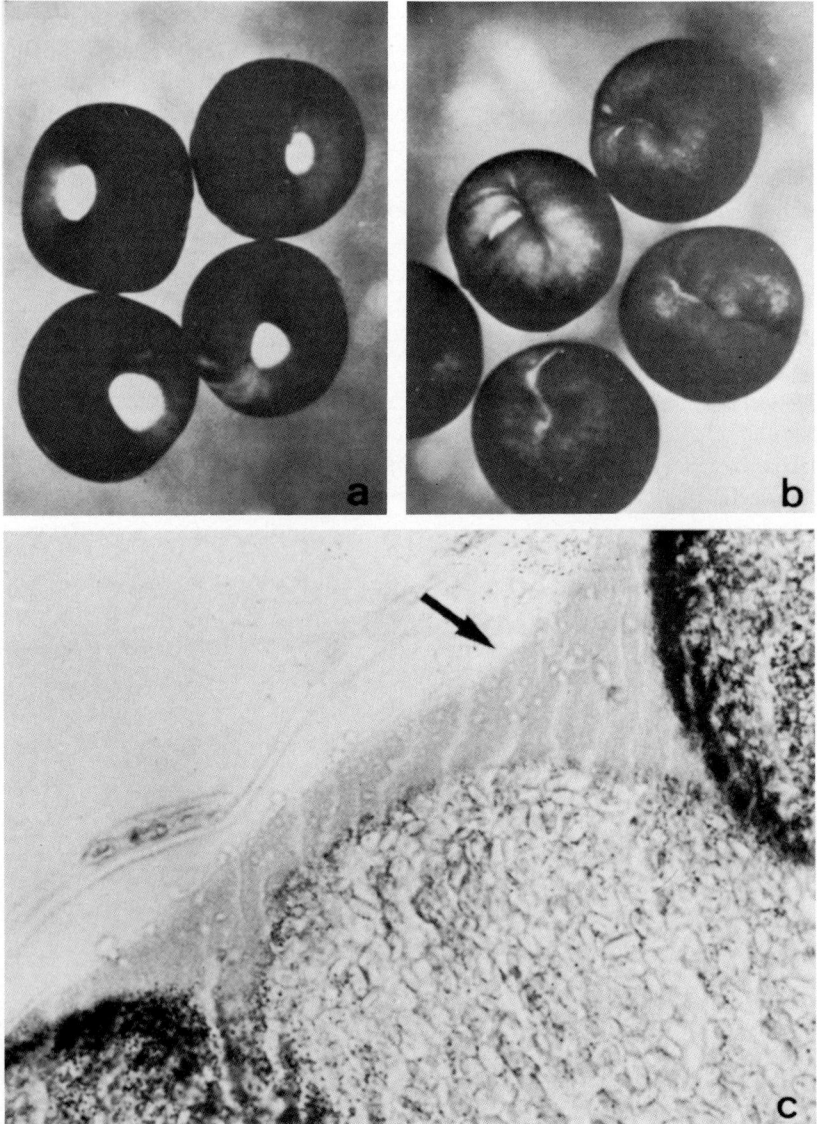

FIG. 68.   Pseudogastrulation in unfertilized frog eggs kept for many hours in Ringer's solution. Progression of the pseudoblastoporal lip from (A) to (B). (C) Section through a pseudo-yolk plug. [(A, B) Smith and Ecker, 1970; (C) Baltus *et al.*, 1973.]

opment. This "invagination" probably results from an autonomous activity of the cell cortex; it is a curiosity rather than a true morphogenetic event.

We became interested in the possible role of DNA synthesis in sea urchin gastrulation (Brachet and De Petrocellis, 1981). The addition of aphidicolin quickly arrested nuclear DNA synthesis at various stages of development. The results can be summarized as follows: at all cleavage stages up to the early blastula stage, suppression of DNA synthesis was followed by an immediate arrest of cell division and development. When swimming blastulae were treated with aphidicolin, only primary gastrulation took place. A basal plate was formed and a few cells moved into the blastocele cavity, but there was no invagination of the presumptive gut (Fig. 69). If, however, mesenchyme blastulae are treated with aphidicolin, gastrulation is completed. Thus nuclear DNA synthesis is required for cleavage and for preparation for gastrulation, but gastrulation itself (secondary gastrulation) no longer requires an increase in cell number and depends entirely on cell migration and contractility. During amphibian gastrulation, migration of the mesodermal cells is promoted by extracellular fibrils; they have a parallel organization and guide cell migration. These extracellular fibrils are absent or sparse in lethal hybrid (see Section 8 of this chapter) which are arrested at the late blastula or early gastrula stage (Nakatsuji and Johnson, 1984). An essential component of the extracellular fibrils in amphibian early gastrulae is fibronectin. Injection of antibodies against fibronectin inhibits gastrulation, but not neurulation (Boucaut et al., 1984a). Inhibition of amphibian gastrulation can also be obtained by treatment with a synthetic decapeptide which recognizes fibronectin-mediated interactions (Boucaut et al., 1984b). There is little doubt that similar mechanisms of "contact guidance" operate during sea urchin egg gastrulation. Primary mesenchyme cells leave the basal plate when they lose affinity for hyaline and acquire affinity for fibronectin (Fink and McClay, 1985).

D. DIFFERENTIATION WITHOUT CLEAVAGE

There is one famous case in which differentiation is said to take place in the absence of cellularization—the so-called differentiation without cleavage in the polychaete Chaetopterus eggs, discovered by F. R. Lillie (1902). Lillie found that unfertilized Chaetopterus eggs undergo maturation and can be activated to form a fertilization membrane by treatment with excess KCl (Fig. 70b,c). The activated eggs undergo pseudocleavage, forming polar lobes and appearing to divide into AB and CD blastomeres of unequal size, but the furrow that separates them immediately vanishes. After several abortive attempts to cleave, the eggs undergo extensive ameboid activity (Fig. 70a), leading to a more or less complete segregation of yolk and hyaloplasm (Fig. 70d,e). In some of the eggs (at best 20–40% only in the Mediterranean species of Chaetopterus), the clear cytoplasm completely surrounds the yolk mass (Fig. 70f), a process called "overflow" or "pseudogastrulation" by Lillie (1902). Where the overflow of clear cytoplasm is complete, eggs undergo hatching and ciliation. Finally, the

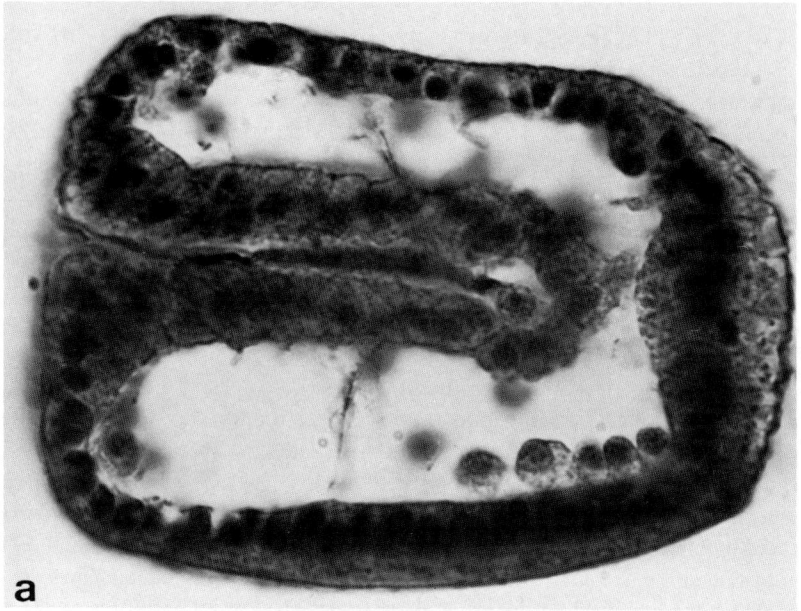

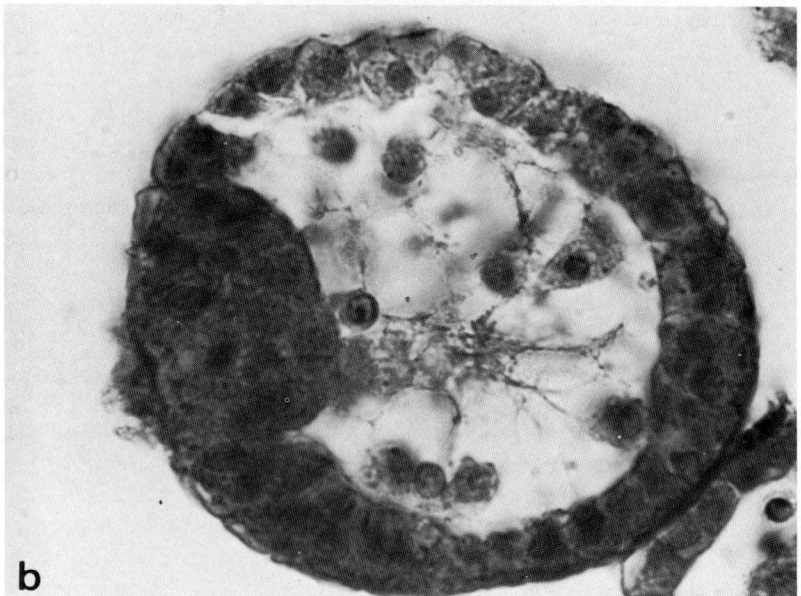

FIG. 69. Effects of aphidicolin on sea urchin blastulae. (a) Controls: complete invagination of the archenteron; (b) no invagination of the gut in the treated blastulae. However, if the aphidicolin treatment begins a few hours later (at the onset of gastrulation), there is no inhibition of archenteron invagination. The experiments show that DNA synthesis is required for the initiation of gastrulation, but not for gastrulation itself. [Brachet and De Petrocellis, 1981.]

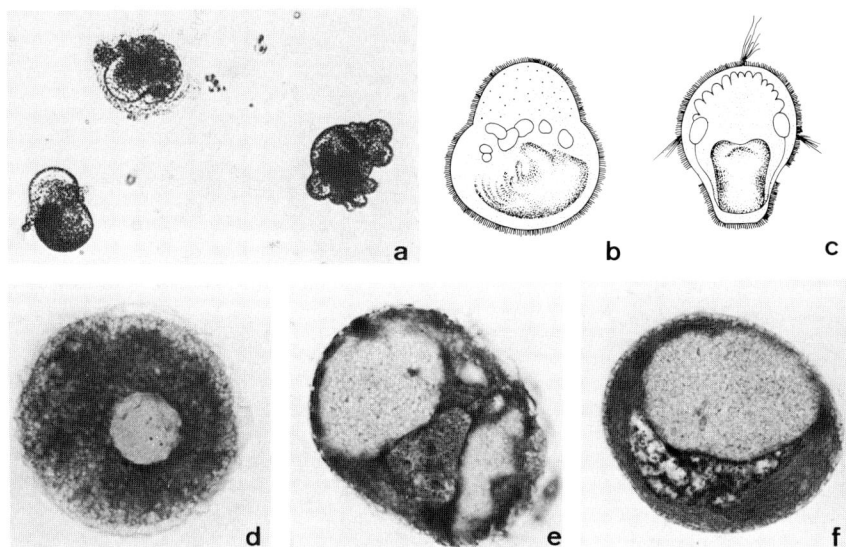

Fig. 70.   Differentiation without cleavage in KCl-activated *Chaetopterus* eggs. (a) Living eggs 12 hr after activation. Note the strong ameboid activity (right) and the segregation between the yolk and the clear cytoplasm (left). The embryo at the top is cytolyzing. (b) Differentiation without cleavage at its final stage and (c) normal larvae from fertilized *Chaetopterus* eggs. Note the absence of an apical tuft and intestine in (a). (d) Section through a *Chaetopterus* activated egg 4 hr after KCl treatment; the nucleus has enlarged, and there is no segregation between the yolk and the hyaloplasm. (e) Incomplete and anarchic segregation between the yolk (lightly stained) and the basophilic cytoplasm (16 hr after activation). (f) Same as (e), but complete segregation; the basophilic cytoplasm completely surrounds the yolk mass (pseudogastrulation). In both (e) and (f), the large nucleus lies between the basophilic cytoplasm and the yolk mass. [(a) Brachet and Donini-Denis, 1978; (b, c) redrawn from Lillie, 1902; (d–f) Alexandre *et al.*, 1982.]

unicellular swimming larvae digest their yolk reserves and become filled with vacuoles (Fig. 70b). Lillie (1902) was struck by the general similarities between normal trochophore and unicellular swimming larvae (Fig. 70b,c). However, as he correctly pointed out, larvae arising from differentiation without cleavage, in contrast to the multinucleated trochophores, lack the apical tuft of long cilia and the enteric cavity.

Lillie did not study the cytological aspects of differentiation without cleavage and was not much interested in the nucleus of the unicellular ciliated larvae. Pasteels (1934) filled this gap and showed that pseudocleavage corresponds to a succession of monasterial cycles of chromosome replication. These cycles lead to the formation of a highly polyploid nucleus that finally breaks down into a crown of aneuploid nuclei that vary greatly in size and DNA content (Brachet, 1937) (Fig. 71). As a result of these nuclear changes, the overall DNA content increases more than 300 times during differentiation without cleavage. The rate

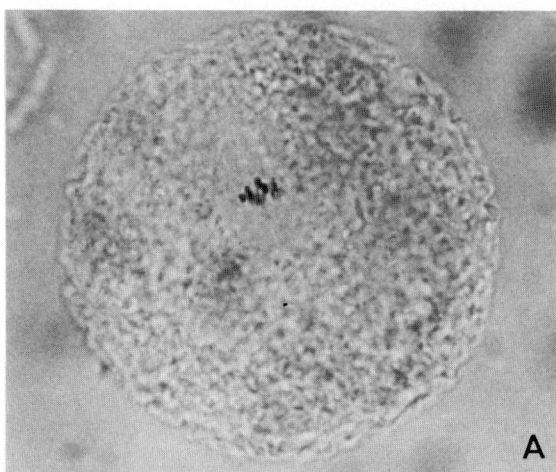
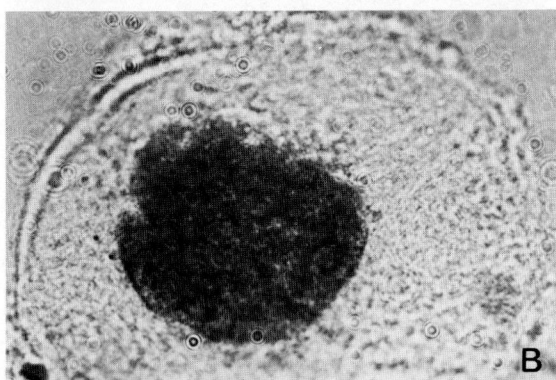
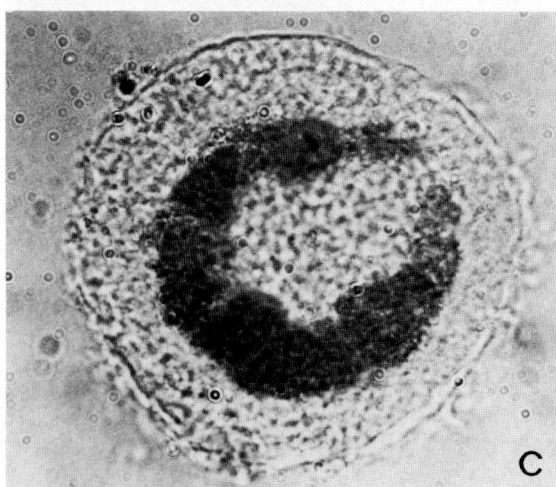

of DNA synthesis during this process, however, is much lower than that in fertilized diploid *Chaetopterus* eggs (Brachet 1938).

The fact that activation can be induced by KCl in *Chaetopterus* eggs may seem to contradict what we have said about the paramount importance of calcium ions at fertilization. It has been shown (Ikegami *et al.,* 1976; Brachet and Donini-Denis, 1978) that KCl is inefficient if *Chaetopterus* eggs are placed in Ca^{2+}-free seawater. In addition, the divalent ions ionophore A23187 induces activation, but it is not followed by differentiation without cleavage, which is probably due to toxicity of the ionophore. According to a report by Eckberg and Carroll (1982), A23187 induces maturation in a Ca^{2+}-free medium, provided that the ionophore is added at high concentrations, probably releasing sequestered calcium ions.

Microtubules are certainly involved in pseudocleavage and segregation since both are suppressed by colchicine. In contrast, cytochalasin B does not inhibit segregation between yolk and clear cytoplasm, but prevents ciliation (Brachet and Donini-Denis, 1978; Eckberg, 1981). An interesting observation was recently made by Eckberg (1981): if one inhibits the two first cleavages of fertilized *Chaetopterus* eggs with cytochalasin B they undergo differentiation without cleavage, similar to KCl-activated unfertilized eggs.

Segregation results from the unusually strong ameboid activity displayed by the eggs a few hours after KCl activation. The eggs appear to have been struggling and wriggling in order to achieve a new type of organization. If the struggle is successful and leads to a harmonious overflow of clear hyaloplasm around a central opaque yolk mass, hatching and ciliation will follow. This strong ameboid activity probably results from autonomous cortical activities. Cortical contractions may be so strong that the egg is fragmented into pieces; even anucleate "minicells" maintain a strong ameboid activity for many hours (Brachet and Donini-Denis, 1978).

Watching *Chaetopterus* eggs during differentiation without cleavage suggests that the ectoplasm (cortex and subcortical material) plays the active role and that the yolk-rich endoplasm is only passively displaced. This belief is strengthened by observations based on *in situ* hybridization (Jeffery and Wilson, 1983). It was found that, in *Chaetopterus,* the ectoplasm of the matured egg contains 15 times more poly(A)$^+$ RNAs (as well as histone and actin mRNAs) than the endoplasm and the clear plasm that corresponds to the nuclear sap of the GV after its breakdown at maturation; 90% of the mRNAs are accumulated in the cortex and,

FIG. 71. DNA synthesis and distribution during differentiation without cleavage in *Chaetopterus* (Feulgen staining). (A) Diploid set of chromosomes during the first monasterial cycle. (B) Strongly polyploid spherical nucleus in an 18-hr embryo in which segregation was only partial. (C) Same as (B), but complete segregation of the yolk and hyaloplasm; in the "crown" of daughter nuclei, a gradient in DNA content is visible. [(A) Original photograph by H. Alexandre; (B, C) Brachet and Donini-Denis, 1978.]

later, in the polar lobe. This cortical localization remains intact during cleavage of fertilized eggs.

We have become interested in the following questions. Is DNA synthesis required for differentiation without cleavage? If so, is "early" DNA synthesis during the first monasterian cycles sufficient to allow development up to the ciliated, swimming stage (Brachet *et al.,* 1981; Alexandre *et al.,* 1982)? In order to answer these questions, we submitted KCl-activated *Chaetopterus* eggs to aphidicolin pulses and then followed segregation and ciliation. Finally, we estimated the DNA content of the nuclei by Feulgen cytophotometry. The following results were obtained. If DNA synthesis is completely suppressed by continuous treatment with aphidicolin, pseudocleavage takes place, as in the controls, despite the fact that there is no chromosome replication. Repeated formation of a polar lobe and abortive cleavage furrow formation thus are probably intrinsic properties of the egg cytoplasm and are independent of nuclear DNA replication in activated eggs. In fertilized eggs, aphidicolin immediately stops cleavage, as in sea urchin eggs. Activated eggs continuously treated with aphidicolin never undergo segregation and, of course, one never sees a single hatched, ciliated unicellular larva. Pulse experiments with aphidicolin have clearly shown that early DNA synthesis, which occurs during the initial monasterial cycles of chromosome replication, is much more important for differentiation without cleavage than late DNA synthesis during that segregation period. It is possible to obtain embryos that have the same DNA content after treatment with aphidicolin during either the first hours of development or after 5–6 hours of development. In the former, there is little segregation and one never obtains swimming larvae; in the second, the percentage of swimming embryos is almost as high as in the controls. Thus, the capacity of the egg to differentiate without cleavage depends not on the quantity but on the quality of its DNA. Closer analysis has shown that the crucial event takes place at about the fifth chromosome replication cycle. Differentiation without cleavage no longer requires DNA synthesis (Alexandre *et al.,* 1982) after this cycle, just as sea urchin gastrulation can take place in the absence of DNA synthesis after the mesenchyme blastula stage.

Chaetopterus eggs raise other, still unanswered, questions. For instance, why are they unable to form an amphiaster? This is probably because it is impossible to assemble a second centriole. If so, how do they form hundreds of basal bodies in ciliated larvae? Why is cytokinesis abortive?

In summary, differentiation without cleavage only mimics the normal development of fertilized *Chaetopterus* eggs, since there is no real gastrulation. These eggs remain an excellent model for the study of cell differentiation, particularly for the analysis of cell motility and ciliogenesis. In addition, watching differentiation without cleavage under the microscope is a fascinating sight. An imaginative observer (like the author) can entertain the illusion that he is following the transformation of an ameba into a ciliate.

VI. CYTOPLASMIC DETERMINANTS (GERMINAL LOCALIZATIONS)

As pointed out at the beginning of this chapter, cytoplasmic heterogeneity has long been recognized as a major factor in embryonic development. A wide variety of experiments performed on fertilized or cleaving eggs of many animal species clearly showed that certain regions of the egg are absolutely necessary for normal morphogenesis. This led to the idea that eggs of many species are a mosaic of germinal localizations. These experiments were easy to perform. For instance, we have seen that moderate centrifugation of fertilized frog eggs leads to microcephaly (Fig. 2) because it modifies the animal–vegetal polarity gradient. Turning the frog egg upside down and compressing it slightly leads to twinning, while pricking the gray crescent of frog eggs with a needle suppresses, completely or partially, the differentiation of axial organs (nervous system, chorda). More delicate experiments involved cutting the eggs into pieces and following the fate of isolated blastomeres or groups of blastomeres. A considerable amount of patient and skillful work was the bases of the already discussed theory of T. H. Morgan (1934). During egg cleavage, genetically identical totipotent nuclei would be distributed, as the result of mitotic activity, in a mosaic of the chemically different territories (germinal localizations) that build up the egg cytoplasm. The heterogeneous cytoplasmic environment controls gene activity in the nuclei. Specific genes would be activated in a given germinal localization, while others would remain silent. Local gene activation would modify the surrounding cytoplasm, thus increasing its initial heterogeneity and leading to further gene activation (or repression). Repeated gene–cytoplasm interactions would ultimately lead to cell differentiation.

Currently, all too few investigators are still interested in germinal localizations. This is unfortunate, because cytoplasmic determinants will forever remain a cornerstone of developmental biology. We would like to know much more about these cytoplasmic determinants, which play a primordial role during the very early stages of ontogenesis that have such important consequences for the entire life of the individual. It must be remembered that if the egg from which one is derived had been centrifuged soon after fertilization, the result perhaps would be a microcephalic idiot instead of an intelligent molecular biologist cloning and sequencing genes—yet the idiot and the scientist would have exactly the same genes. The unfortunate consequence of this lack of interest in the cytoplasmic determinants of eggs is that we still know very little about their molecular identity.

Theoretically, two different molecular mechanisms can be conceived to explain how a given germinal localization gives rise to a given organ [how the myoplasm of ascidian eggs (Fig. 59) gives rise to muscle cells in tadpoles, for instance]. Either maternal mRNAs, coding for muscle-specific proteins, are accumulated in the myoplasm, or, in accordance with Morgan's theory, the myoplasm contains factors that selectively activate the genes coding for specific muscle proteins. We would be dealing with preformation in the first hypothesis and with epigenesis in the second.

There is a reason for selecting ascidian eggs as a model. As we shall see, these eggs are the only ones that give us a glimpse of the chemical nature of cytoplasmic determinants. We shall begin with other germinal localizations.

A. The Gray Cresent of Amphibian Eggs[7]

Figure 72 shows diagrammatically the gray region, which, about 1 hr after fertilization, becomes apparent on the dorsal side of frog eggs. This is due to a localized thinning out of the pigmented cortical layer.

As shown by Roux's (1903) classic experiments on localized fertilization, the gray crescent appears on the side opposite the entrance point of the spermatozoon. Destruction of the gray crescent with a hot needle (Roux, 1903; A. Brachet, 1904) is followed by a marked reduction or even absence of the dorsal organs; pricking of the opposite side of the egg (thus, of the ventral side) has almost no effect on development. However, remarkable experiments by Ancel and Vintemberger (1948) demonstrated that the dorsal side and, as a consequence, the plane of bilateral symmetry can be determined at will by modifying the orientation of freshly fertilized eggs (Fig. 72c). With proper design of the egg rotation conditions, the gray crescent can form on the side where the sperm entered the egg. This has been verified by Gerhart et al. (1981), who concluded that displacement of the egg content by gravity can determine the dorsoventral axis without any relationship to the entrance point of the spermatozoon and the position of the gray crescent; such displacements can induce twinning.

It seems, therefore, clear that the gray crescent is a marker of the dorsal side only if the egg has been fertilized under normal conditions. The role played by the spermatozoon is probably linked to the formation of the spermaster, which (like gravity) displaces the egg contents. Changes in the cortical layer of the egg are presumably responsible for the determination, at will, of the plane of symmetry in Ancel and Vintemberger's (1948) classic experiments. An important factor, as shown by Pasteels (1964), is that a thick mass of yolk (called by him the "vitelline wall") comes in close contact with the dorsal cortex. The most recent paper on the subject (Ubbels et al., 1983) gives the following account of gray crescent formation. Growth of the spermaster induces rearrangements of the animal pole material, which reacts with the egg cortex and the yolk to form the vitelline wall, which is a "vegetalizing-dorsalizing center." If the spermaster MTs are destroyed by vinblastine, the same rearrangements occur; they are not due to the spermaster, but to gravity. The sperm centriole, acting as an MTOC, structures the cytoskeleton and modifies the localization of the yolk. Rotation of

[7]Reviewed by Brachet (1977), Kirschner et al. (1980), and Gerhart et al. (1981).

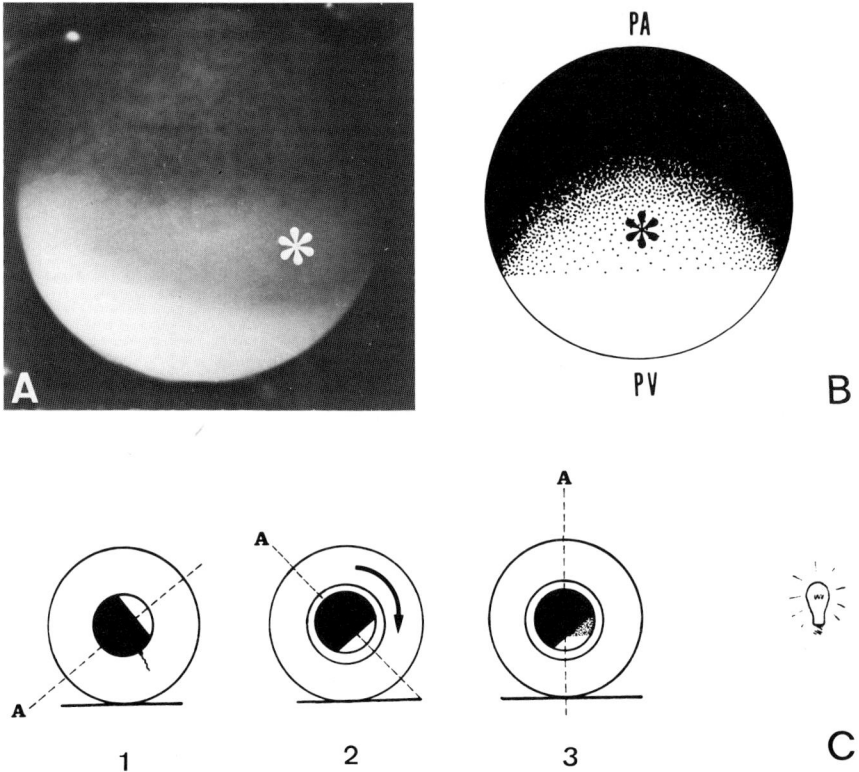

FIG. 72. The gray crescent of amphibian eggs. (A) Photograph of a gray crescent (asterisk) induced precociously by placing an axolotl egg at 35.5°C for 10 min. (B) Schematic representation of a frog egg about 2 hr after fertilization. PA, animal pole; PV, vegetal pole; asterisk, gray crescent. (C) A classic experiment by Ancel and Vintemberger (1948). (1) An unfertilized egg is placed in an oblique position and then fertilized (A, animal pole). (2) The egg rotates under the influence of gravity, after uplifting of the fertilization membrane, as indicated by the arrow. (3) The gray crescent always appears on the side facing the lamp. [(A) Courtesy of Dr. M. Namenwirth; (B, C) Brachet, 1974.]

the egg alters its cytoskeleton. Since neither the dorsal cytoplasm nor the gray crescent is affected by rotation of the egg, they are not the ultimate dorsal determinants, but only markers of its dorsal side.

The outcome of one particular experiment will be eagerly awaited by many embryologists: what would happen when amphibian eggs are fertilized, in the absence of gravity, in a Spacelab?

A gray crescent is also visible in the eggs of the axolotl. As shown by Bendford and Namenwirth (1974), formation of a gray crescent in these eggs is speeded up by a brief heat shock. More recently, Gautier and Beetschen (1983a) showed that inhibition of protein synthesis, by injection of diphtheria toxin, induces the formation of a gray crescent in progesterone-treated, maturing axolotl oocytes. This suggests that a protein that inhibits symmetrization is synthesized during maturation. Gautier and Beetschen (1983b) reported that enucleate maturing axolotl oocytes do not form a gray crescent after diphtheria toxin injection. If the oocyte has been enucleated before the addition of progesterone, symmetrization takes place if nuclear sap from an isolated GV is injected either before or after diphtheria toxin injection. This "correcting factor" is already present in the nuclei of small (500- to 600-μm) axolotl oocytes; it is present in *Pleurodeles* and *Xenopus* GVs, but is less active in the latter. The correcting nuclear factors acts only if cytoplasmic maturation took place in the recipient axolotl oocyte. Identification of its chemical nature would greatly increase our knowledge of the molecular mechanisms of symmetrization.

In all of these experiments, neither the genes nor the whole nucleus play any role in these early, but fundamental, morphogenetic events. Here a word of caution is necessary. We found (Brachet and Hubert, 1972) that slight injury to the dorsal cortex of *Xenopus* eggs is often followed by mitotic abnormalities—in particular, pluricentric mitoses (Fig. 37, Chapter 5, Volume 1). If the egg succeeds in cleaving, the result will be aneuploidy. The various blastomeres will contain a different number of chromosomes. Aneuploidy is lethal in amphibian eggs and often arrests development at the blastula stage. However, we do not think that these cytological findings seriously affect the conclusions drawn long ago by Roux (1903) and A. Brachet (1904), because they selected eggs that develop beyond the gastrula stage.

The unknown cytoplasmic determinants present in the gray crescent are uv sensitive (Grant and Youngdahl, 1974; Malacinski *et al.,* 1975). If the vegetal pole of an amphibian egg—or, better, its gray crescent—is uv irradiated, strongly microcephalic embryos (similar to those shown in Fig. 2) are obtained. Since uv light is particularly efficient at the wavelength where nucleic acids absorb uv most strongly (around 2600 Å), it is possible that the cytoplasmic determinants of fertilized amphibian eggs are RNP particles. Present evidence is much too scanty to allow us to draw safe conclusions.

B. INSECT EGGS

We have already mentioned the germ plasm of *Drosophila* eggs and observed that this uv-sensitive posterior part of early embryos is absolutely necessary for gonad formation. It has been reported that there is a localized synthesis of specific proteins, suggesting differential gene activation, in the pole plasm of *Drosophila* (Gutzeit and Gehring, 1979).

Ligation experiments by Seidel (1932) have shown that normal development of many insect eggs requires the activation by a nucleus of the cytoplasm located in the posterior half of the egg. After penetration of a nucleus, the posterior region becomes the formation center (*Bildungszentrum*). This region produces diffusible substances that induce the development of the anterior part of the egg.

It is immaterial whether the nucleus was originally present in the anterior or posterior part of the egg. Similar experiments by Spemann (1928) on partially ligated *Triturus* eggs had already established that development is normal if one of the nuclei of the dorsal side "colonizes" the uncleaved ventral half at the morula stage. These experiments of Spemann (1928) and Seidel (1932) completely disposed of the old Weissman–Roux theory, which held that the dorsal and ventral blastomeres contained qualitatively different "nuclear determinants" (i.e., genes) responsible for ulterior differentiation.

Unfortunately, nothing is known about the chemical nature of the cytoplasmic determinants present in the formation center of Seidel (1932). Somewhat more is known about cytoplasmic determinants in the egg of the midge *Smittia* due to the work of Kalthoff (1982). A number of experiments have shown that in *Smittia* the development of the head depends on cytoplasmic determinants located in the posterior half of the egg. These determinants are uv and ribonuclease sensitive. If one uv irradiates or injects with RNase the posterior part of the egg, one obtains "double-abdomen" larvae that possess abdominal structures instead of a head (Fig. 73). These experiments supported the view that the cytoplasmic determinants in insect eggs might be composed of RNP, or that maternal RNAs located at the posterior end of the egg were involved in the differentiation of the anterior part. However, these conclusions will remain unproved until the cytoplasmic determinants of insect eggs have been isolated and characterized. A good start in this direction is the demonstration, by Jäckle and Kalthoff (1980), that *Smittia* eggs synthesize a "posterior indicating" protein.

C. Sea Urchin Eggs[8]

Sea urchin eggs have been the prototype of "regulatory" eggs (in contrast to mosaic eggs) since the classic experiments of H. Driesch (1891). He showed that blastomeres separated from each other at the two-cell stage can give rise to complete plutei (Fig. 4). However, the equally classic studies of Hörstadius (1939) have demonstrated that two opposite gradients play a very important role in the early determination of sea urchin eggs. One of them decreases from the animal to the vegetal pole, while the other has its maximum at the vegetal pole, where the micromeres form at the 16-cell stage, and progressively decreases toward the animal pole (Fig. 74). Among other things, Hörstadius (1939) discovered that implantation of micromeres in the animal half of a morula induces the formation of a second archenteron, thus leading to the production of a secondary embryo. Isolated micromeres can differentiate into spicules; however, removal of the micromeres does not prevent the rest of the morula from developing in a normal pluteus larva.

Unfortunately, we still ignore the chemical nature of the animalizing and vegetalizing factors, and we have little information about the differences in

----

[8]Reviewed by Davidson (1982) and McClay and Wessel (1985).

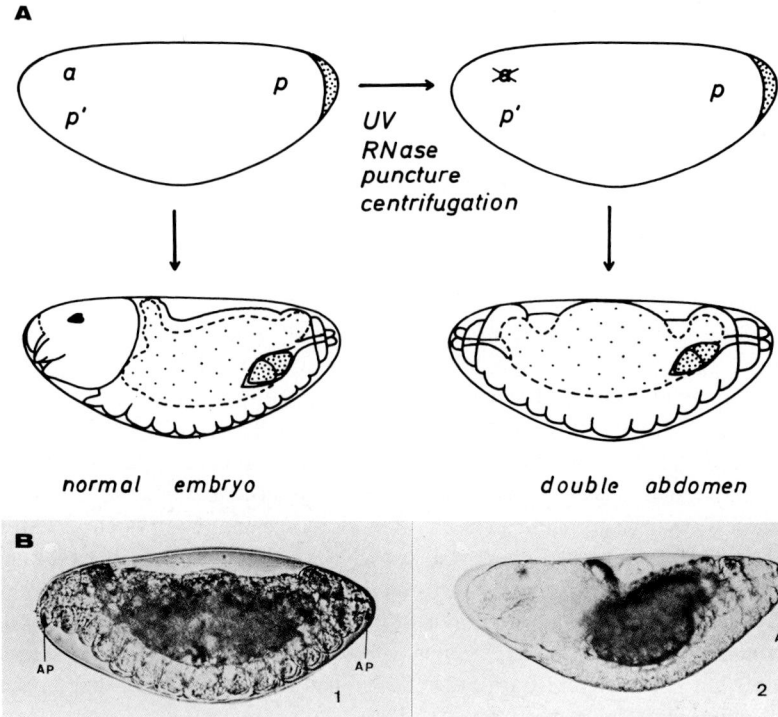

Fig. 73.   (A) Schematic representation of double abdomen induction in *Smittia* by various types of experimental interferences, all of which appear to inactivate or displace anterior determinants designated (a). They are thought to cooperate with other factors (p′) in the anterior half of the embryo so as to allow formation of the head and thorax upon inactivation or displacement of (a). p′ is assumed to cause abdomen formation in the anterior half; the formation of the posterior abdomen is ascribed to similar or identical factors (p). Germ cells (dotted) are present only in the posterior abdomen. (B) (1) *Smittia* embryo showing the aberrant pattern of the double abdomen. Anal papillae (AP) can be seen at both ends. Each abdomen consists of seven terminal posterior segments, fused in mirror image symmetry near the egg equator. (2) *Smittia* embryos with a normal segment pattern, which developed after application of RNAse to the posterior pole. [(A) Kalthoff, 1979; (B) Kandler-Singer and Kalthoff, 1976.]

molecular composition between micromeres and other blastomeres at the 16-cell stage. We have already mentioned the existence of an actinomycin D–sensitive mitotic gradient during the early cleavage of sea urchin eggs. It thus appears as if a transcriptional event originating from the micromeres (Filosa *et al.*, 1985) exerts control on the synchronization of mitoses during cleavage (Parisi *et al.*, 1979). We have also seen that there are quantitative differences in the pattern of histone synthesis in micromeres compared to other cells (Senger and Gross, 1978) and that chromatin from micromeres is more resistant to DNase digestion than chromatin isolated from the other blastomeres (Cognetti and Shaw, 1981).

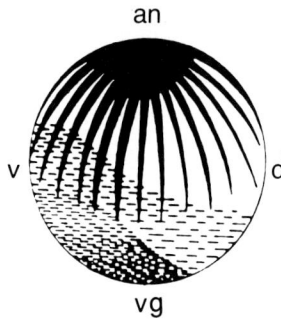

FIG. 74.   Schematic representation of the hypothetical double gradients in sea urchin eggs. an, animal pole; vg, vegetal pole. v, ventral side; d, dorsal side. The animal gradient is indicated by continuous lines, the vegetal gradient by horizontal broken lines. It is assumed that the vegetal gradient is displaced on the ventral side, so that its effects would be greater in the future ventral half. [Dalcq, 1957.]

This suggests that there are differences in the organization of chromatin between micromeres and other cells when the embryo undergoes determination. Such differences might, in theory, be reflected by quantitative or qualitative alterations in DNA replication or transcription. Surprisingly, the sequences complexity of micromere RNA is less than that of whole 16-cell embryos (Ernst *et al.*, 1980). However, it seems unlikely that gene expression is deeply modified in the micromeres, since studies by Tufaro and Brandhorst (1979) and Hutchins and Brandhorst (1979) show that the same proteins are synthesized in micromeres, macromeres, and mesomeres. Differences in gene expression between micromeres and macromeres, if they exist, are thus too subtle to be detected even by two-dimensional electrophoresis. However, that more subtle differences really exist seems likely in view of the observation of Spiegel and Spiegel (1978) that, in dissociated 16-cell stages, micromeres undergo selective aggregation. Their cell membrane thus differs in some respect form that of the other blastomeres.

One classic approach to the study of sea urchin embryogenesis is the induction of vegetalization or animalization. Treatment of the eggs with LiCl produces overdevelopment of the endoderm (vegetalization) and leads to exogastrulation; in contrast, a variety of chemicals induce animalization. Development stops at the blastula stage, and invagination of the endoderm to form the archenteron does not occur. So far, attempts to find clear-cut biochemical differences between normal, animalized, and vegetalized larvae have been frustrated. However, a promising result has been reported by Shepherd *et al.* (1983). In animalized larvae (obtained by treatment with $ZnCl_2$), the synthesis of two RNA species that increase 10 times during normal development is repressed; in contrast, another RNA increases considerably in animalized, but not in normal, larvae. These results should encourage embryologists to reinvestigate animalization and vegetalization from the molecular viewpoint.

D. The Polar Lobe of *Ilyanassa* Eggs

As already shown in Fig. 57, spiral cleavage is characterized by the formation, at the trefoil stage, of a purely cytoplasmic polar lobe that is particularly well developed in the egg of the mollusk *Ilyanassa*. Removal of the polar lobe at the trefoil stage results in characteristic developmental abnormalities (Clement, 1952), as shown in Fig. 3. Polar lobes isolated from *Ilyanassa* cleaving eggs are capable of protein synthesis (Clement and Tyler, 1967) and thus contain maternal mRNAs. Removal of the polar lobe affects the pattern of protein synthesis in the lobeless embryo, as shown by Collier (1961), Teitelman (1973), and Newrock and Raff (1975), as well as the rate of RNA (Koser and Collier, 1976; Collier, 1977) and DNA (Collier, 1975) synthesis. However, all of these changes in macromolecule synthesis appear relatively late in development and coincide with the first appearance of developmental abnormalities. In lobeless embryos, suppression of the polar lobe has no immediate effects on the synthesis of macromolecules, so far as we are currently aware.

More recently, protein synthesis in isolated polar lobes and lobeless eggs of *Ilyanassa* was analyzed by two-dimensional electrophoresis of [^{35}S]methionine-labeled proteins (Collier and McCarthy, 1981; Brandhorst and Newrock, 1981). The main results of these studies are that synthesis of polar lobe–specific proteins cannot be detected and that more than 98% of the proteins synthesized during early embryogenesis are translated from maternal mRNAs. Removal of the polar lobe does not affect the pattern of protein synthesis during early cleavage, but differences between normal and lobeless embryos become apparent at the mesentoblast stage. There is, therefore, no indication of a selective localization of translatable mRNAs in the polar lobe.

Some proteins are no longer synthesized, and new ones are synthesized between early cleavage and 24-hr embryos of *Ilyanassa;* curiously, some of these changes also take place in cultured, isolated polar lobes and thus result from a selective translation of maternal mRNAs (Brandhorst and Newrock, 1981). According to Collier and McCarthy (1981), the mRNA's are not differentially translated in lobeless eggs and isolated polar lobes, which contain the same set of mRNAs. What is clear is that the polar lobe does not control protein synthesis during cleavage, either quantitatively or qualitatively. Among the proteins synthesized at approximately the same rate by intact eggs, lobeless eggs and isolated polar lobes are all of the histones and the HMG-14 and -17 proteins (Collier and McCarthy, 1981).

Thus, *Ilyanassa* eggs are an excellent material for the analysis of post-transcriptional control in protein synthesis. Collier and McCarthy (1981) studied, among other things, the effects of actinomycin D on protein synthesis in *Ilyanassa* eggs. Although space limitations do not allow us to describe their interesting paper in detail, one of its main conclusions is that the cytoplasm of the polar lobe contains both repressors of translation and activators for unmasking mRNA from mRNP of ovarian origin. Actinomycin D has complex effects. It decreases by 41% the overall protein synthesis between the trefoil and mesentoblast stages, although, as we have seen, 98% of this protein synthesis results from the translation of preexisting maternal mRNAs. This unexpected result is believed to be due to some nonspecific effect of actinomycin D on protein synthesis. Fourteen proteins accumulate in mesentoblast-stage embryos after culture in the presence of actinomycin D. This is a case of what has been called "superinduction." This phenomenon was discovered by G. Tomkins (reviewed by Tomkins *et al.,* 1972), who found that in hepatoma cells actinomycin D increases the synthesis of the enzyme tyrosine aminotransferase (TAT) when it has been induced by glucocorticoids. We have seen that this also takes place in *Xenopus* oocytes treated with a mixture of actinomycin D and progesterone; the synthesis of a few proteins is enhanced. Several explanations have been proposed for the molecular mechanisms of superinduction. Collier and McCarthy (1981) think that, in *Ilyanassa* eggs, increases in protein synthesis in response to actinomycin D treatment result from transcriptionally dependent events that were repressed by the drug.

It would be interesting to see whether stimulation of the synthesis of certain proteins after actinomycin D treatment occurs in other anucleate systems. For instance, a comparison of actinomycin D–treated nucleate and anucleate halves of *Arbacia* eggs (or of normal and enucleated *Xenopus* oocytes) might be rewarding and give valuable information about the mechanisms of superinduction.

The most important conclusion drawn by Collier and McCarthy (1981) is that development up to the mesentoblast stage is not the result of differential gene transcription, as one would have expected, but of the translational control of precociously transcribed structural genes.

E. Ascidian Eggs[9]

We have already mentioned that ascidian eggs are composed of a mosaic of "plasms" (Conklin, 1905). One example, the localization of the myoplasm, is shown in Fig. 59. According to Ries (1957), who studied, with the indophenoloxidase reaction, the distribution of respiratory enzymes in the eggs of various invertebrates the myoplasm stains selectively. The same distribution was found for Janus green–staining granules by Reverberi (1956); there is thus no doubt that mitochondria are accumulated in the myoplasm. This has been confirmed by electron microscopy and by the fact that the myoplasm-containing posterior blastomeres possess, at the four-cell stage, 2.7 times more cytochrome oxidase than the anterior ones (Berg, 1956). However, the oxygen consumption of the anterior and posterior blastomeres is about the same (Holter and Zeuthen,

---

[9]Short review by Jeffery (1985).

1944), a fact that proves that local accumulation of mitochondria does not necessarily lead to an increased respiratory rate. In addition, Whittaker (1979) has demonstrated that the mitochondria are only markers of the myoplasm and are not the cause of muscle differentiation in the tadpole larva.

The chemical nature of these agents remains unknown, but due to the work of Whittaker (1974, reviewed in 1979), considerable progress has been made in elucidating the nature of the germinal localizations in ascidian eggs. The strategy used by Whittaker (1973, 1979) was to follow, using cytochemical methods, three "marker" enzymes in normal embryos, in eggs where cytokinesis had been blocked by cytochalasin B treatment, and in blastomeres isolated at the eight-cell stage. He also used actinomycin D to discover whether these enzymes were synthesized on preformed or newly synthesized mRNAs; he used puromycin to ascertain that they were synthesized *de novo*. The three marker enzymes were acetylcholinesterase for muscle differentiation, alkaline phosphatase for endoderm, and tyrosinase for brain melanocytes. Whittaker's experiments showed that the three enzymes became detectable during cleavage in well-localized areas of the egg and at a determined time after fertilization. Suppression of cytokinesis by cytochalasin B did not modify the general pattern of enzyme localization and appearance. This pattern was thus independent of cleavage into blastomeres. The experiments with puromycin showed that the three marker enzymes were newly synthesized. Thus, enzyme synthesis takes place in well-defined regions of the egg and at a given time of development, whether or not the eggs have undergone cleavage. Still more interesting are the results obtained by Whittaker (1979) with actinomycin D, since this drug allows a distinction between preformed and newly synthesized messengers. He found that alkaline phosphatase is synthesized in the endoderm on a maternal mRNA. Appearance of the same enzyme in the ectoderm (where it is present in much lower amounts) requires the synthesis of new mRNA molecules by the nuclei; the same was found for tyrosinase. Finally, cholinesterase mRNA is synthesized at gastrulation and is not a maternal mRNA (Meedel and Whittaker, 1983). The enzyme acetylcholinesterase is synthesized in the same amounts in controls and in eggs arrested at the four- or eight-cell stage. In such eggs, synthesis of cytochrome oxidase (the old marker of muscle differentiation in ascidians) is very small, while it doubles in the controls. Whittaker's (1983) conclusion was that a cytoplasmic determinant segregated during muscle lineage during cleavage releases genetic expression in muscle cells. This conclusion fits perfectly with Morgan's (1934) hypothesis, and is supported by recent experiments in which Deno and Satoh (1984) transplanted the myoplasm into anterior blastomeres. In a more recent paper, Meedel and Whittaker (1984) have separated manually the blastomeres of *Ciona* cleaving eggs and made estimations of acetylcholinesterase activity in the descendants of the different blastomeres. They also injected whole RNA extracted from the various blastomeres into *Xenopus* oocytes in order to detect acetylcholinesterase mRNA translation. These delicate experiments showed that only the muscle and mesenchyme lineages have functional acetylcholinesterase mRNA. In other words, the gene coding this enzyme is active only in muscle and mesenchyme tissues; its transcription does not begin before gastrulation, since no specific mRNA is translated in *Xenopus* oocytes before this stage. The experiments also showed that no cytoplasmic acetylcholinesterase mRNA is segregated in the egg. Curiously, typical myofilaments are present in ascidian eggs where cleavage was arrested with cytochalasin B until hatching time; their formation is inhibited by actinomycin D (Crowther and Whittaker, 1985).

The results discussed by Whittaker (1979) in his excellent review article show that there are two types of molecular mechanisms in the formation of germinal localizations. According to the marker enzymes, synthesis and localization depend either on a local accumulation of preformed maternal mRNAs or on the activation of specific nuclear genes by cytoplasmic determinants of still unknown nature as had been postulated by Morgan in 1934. Thus, preformation and epigenesis coexist in the same egg during early development; this conclusion is not very surprising.

Another approach has been used by Jeffery *et al.* (1983). Using *in situ* hybridization techniques, they found that 45% of the poly(A)$^+$ RNAs are localized in the ectoplasm and 50% in the endoplasm of fertilized eggs; only 5% of the polyadenylated RNAs are located in the myoplasm. A completely different distribution has been found for actin mRNA: 45% of the actin sequences are in the myoplasm, 40% in the ectoplasm, and only 15% in the endoplasm. In contrast, the histone mRNAs are ubiquitous. These observations demonstrate the heterogeneity at the molecular level of ascidian eggs. Jeffery (1984) reported that poly(A)$^+$ RNA, actin and histone mRNAs are all associated with the cytoskeletal framework in ascidian eggs.

The analysis has gone one step further with the work of Satoh and Ikegami (1981a,b) and Satoh (1982a,b), who used aphidicolin, a specific inhibitor of DNA polymerase-α, which arrests DNA replication as efficiently in ascidian eggs as in sea urchin eggs. Satoh and Ikegmai (1981a,b) first found that the time of first acetylcholinesterase appearance is the same in normal embryos and in embryos that have been arrested at the 32-cell stage by cytochalasin treatment; the same result was obtained by colchicine treatment. While in cytochalasin-treated embryos nuclear division continued normally, in the colchicine-arrested embryos only repeated cycles of nuclear breakdown and nuclear envelope reformation took place; there was no inhibition of DNA synthesis. In contrast, aphidicolin arrested both DNA replication and cleavage. Embryos that had been permanently arrested by treatment with aphidicolin in the cleavage stages up to the 64-cell stage did not develop acetylcholinesterase activity, while embryos treated with aphidicolin from the 76-cell stage on were capable of enzyme synthesis (at a low level, however). It was concluded that the eighth DNA replication cycle is of crucial importance for acetylcholinesterase synthesis in ascidian eggs, and that the DNA replication cycle is closely associated with a clock mechanism determining the time of initiation of enzyme synthesis. The work of Satoh (1982b) has extended these conclusions to two other marker enzymes. If one treats ascidian morulae with aphidicolin, there is no synthesis of tyrosinase (the marker enzyme for brain pigment cells); this enzyme is synthesized if DNA replication is halted at the gastrula stage. Endoderm alkaline phosphatase is synthesized on a maternal mRNA. The enzyme is not synthesized if aphidicolin is added at the 16-cell stage, but is synthesized if the inhibitor of DNA replication is added at the 32-cell stage. Clearly, several DNA replication cycles are required for histospecific enzyme synthesis, and the number of these cycles varies for each enzyme.

We have seen that the existence of a crucial mitotic cycle during cleavage is also very likely in *Chaetopterus* (Alexandre *et al.*, 1982) and in the mouse (Alexandre, 1982) eggs. Such crucial mitotic events have been called "quantal mitosis" by Holtzer *et al.* (1972), who postulated their existence to explain differentiation in cultures of embryonic cells (as we shall see in the next chapter). In a quantal mitosis, the two daughter cells are not identical, as in ordinary mitosis, and the difference between the two is transmitted from one cell generation to the next. We do not know the molecular mechanisms of these crucial mitotic events. A possibility (but no more than that) is that it is based on DNA demethylation by successive replication cycles. The intriguing question of the timing mechanisms in development has been reviewed by Satoh (1982b); he concludes that the early morphogenetic events are controlled by a cytoplasmic clock and that later events are associated with the DNA replicating cycles. In agreement with this view are the results of Petzoldt *et al.* (1983), who compared protein synthesis in normal and polyploid (4n to 64n) cytochalasin-treated mouse embryos. They found that stage-specific translation is temporally correlated with chromosome replication, not with cytokinesis or cell-to-cell interactions.

Finally, we should briefly mention the cellular mechanisms that lead to the segregation, after fertilization, of the various plasms. The accumulation of the myoplasm [Conklin's (1905) yellow crescent] and of numerous mitochondria at the vegetal pole is a consequence of fertilization in ascidian eggs. Zalokar (1974) has shown that this segregation of the myoplasm does not take place if ascidian eggs are treated with cytochalasin B. Jeffery (1982) studied the segregation of an orange crescent [homologous to Conklin's (1905) yellow crescent] in the eggs of *Boltenia*. This crescent, in addition to containing an orange pigment, is rich in mitochondria and is a marker for the differentiation of muscle cells in the tadpole larvae. Jeffery (1982) showed that the segregation of the orange crescent in *Boltenia* eggs can be induced at will by local application of the divalent ions ionophore A23187. He concluded that segregation of the yellow crescent results from a local release of calcium ions, which would activate a cytochalasin B–sensitive cortical MF system.

VII. NUCLEAR DETERMINANTS OF EARLY EMBRYONIC DEVELOPMENT

In the preceding section, we played the role of the "devil's advocate" in favoring the cytoplasm in the famous case of cytoplasmic determinants versus nuclear genes. Nuclear genes do not require an advocate, since their importance in development as well as in all cellular activities is generally acknowledged. Developmental genetics is today a respectable discipline taught in many universities. Summing up the case, the conclusions could be as follows: cytoplasm plays a leading role during early cleavage. After the blastula stage, the importance of the nucleus, i.e., of gene expression, becomes primordial. It is overwhelming when embryonic differentiation takes place.

In this section, we shall briefly summarize some of the evidence demonstrating that integrity of the nucleus is required to obtain full and normal morphogenesis. As we shall see, haploidy, aneuploidy, in some cases hybridization, and many mutations sooner or later prove lethal for the embryo. We shall focus on the cases in which lethality is an early event and stops development at gastrulation or neurulation; these cases demonstrate that even primary morphogenesis is under nuclear control.

A. HAPLOIDY, POLYPLOIDY, POLYSPERMY, AND ANEUPLOIDY

The following discussion will be limited to the amphibians and will summarize what we have said about them in greater detail in "Biochemical Cytology" (Brachet, 1957).

Haploidy can be obtained by a variety of means. In Bataillon's (1910) classic method of parthenogenesis, unfertilized frog eggs, which have been covered with frog blood, are pricked with a fine needle. Pricking of the egg results in activation (cortical reaction); to obtain development, a "second factor" (Bataillon, 1910) is required, which is provided by the introduction of a nucleate red blood cell into the egg. An aster forms around the inoculated red blood cell, and normal mitoses repeatedly follow. We now know that the second factor is not, as Bataillon believed in 1910, the nucleus of the red blood cell, but an MTOC—in general, a centriole and the surrounding osmiophilic material.

A much simpler method used to obtain haploid frog embryos by gynogenesis (in which the nucleus is provided by the egg) is fertilization with uv-irradiated sperm. The percentage of fertilized eggs obtained is as high as it is in controls fertilized with normal sperm; the uv-irradiated sperm always degenerates during one of the first cleavage mitotic cycles (Dalcq and Simon, 1932; Brachet, 1954).

Androgenesis is technically a little more difficult to achieve. The egg is fertilized normally, and the maturation spindle bearing the egg chromosomes is removed with a glass needle and a thin pipette; the haploid sperm nucleus alone will take part in development.

Whatever the method chosen to obtain haploids (gynogenesis or androgenesis) the results are the same. Cleavage is normal, but development is retarded from the gastrulation stage onward. Some of the haploids die during gastrulation, but the majority develop the haploid syndrome, including microcephaly, deficient blood circulation, reduction of the gills, edema and ascites, and, finally, death at an early larval stage. The reasons for the haploid syndrome are not known for certain, although it is probably not due to the presence of lethal genes that, in the haploid condition, would not be balanced by their normal alleles. As found by Bataillon (1910), if a monasterial cycle of chromosome duplication occurs before a second aster has formed around the inoculated red blood cell, the homozygous diploids (which should now contain two copies of the lethal genes) can reach the adult state. Synthesis of both RNA and DNA is slowed in haploids (Brachet, 1944, 1954), but this does not help to explain why haploids in frogs are always lethal. The fact that frog haploid cells can be maintained for many generations in tissue culture conditions (Mezger-Freed, 1953) suggests that haploids might be unable to produce enough of the necessary gene products, as was proposed by Moore (1955). These gene products might be provided by an artificial culture medium, but this issue remains highly speculative.

In amphibians, homozygous diploids are viable whether they result from gynogenesis or androgenesis. This does not seem to be true (despite claims to the contrary) for mouse eggs (reviewed by McLaren, 1984). Surani and Barton (1983) never obtained development beyond the 25-somite stage in mouse parthenogenetic diploids; they ascribed lethality to homozygosity. In other experiments, Surani et al. (1984) removed the male pronucleus from a fertilized mouse egg and replaced it by a female pronucleus. These eggs with two female pronuclei never developed until birth after their implantation into pseudogravid foster mothers. Similarly, McGrath and Solter (1984) reported that neither gynogenetic nor androgenetic diploid mouse embryos develop into adults. In mice, the two pronuclei (male and female) seem to be required for normal development.

This conclusion has been confirmed by Barton et al. (1984): mouse eggs with two female pronuclei do not develop further than the 25 somites stage, and eggs with two male pronuclei develop slowly and never come to term. It seems that the male genome is absolutely required for the development of the extra-embryonal tissues involved in the implantation of the embryo into the uterus, and the female genome for certain embryonic stages. Mann and Lovell-Badge (1984) have reported that parthenogenetic mouse embryos differentiate into a number of tissues if they are fused with normal embryos. If pronuclei from parthenogenetic diploid eggs are transferred into the cytoplasm of fertilized, enucleated eggs, death occurs shortly after implantation into the uterus. The reverse experiment (transfer of pronuclei from fertilized eggs into parthenogenetic enucleated eggs) yield viable youngs. The experiments show that death of the parthenogenetic mouse embryos is not due to a lack of cytoplasmic factors coming from the sperm.

Polyspermy has been studied in frog eggs by A. Brachet (1910) and Herlant

(1911). Only one of the nuclei of the supernumerary spermatozoa fuses with the egg nucleus; the other nuclei remain haploid, but they may undergo mitosis. If polyspermy is not too heavy (dispermy or trispermy), cleavage follows, giving rise to a mosaic of haploid and diploid cells. Although part of the embryo is diploid, polyspermy is, in general, more lethal than haploidy. The haploid half of a polyspermic embryo has small nuclei with only one nucleolus and is often underdeveloped; the haploid half has a lower RNA content than the diploid half (Brachet, 1944).

Polyploidy has been extensively studied by Fankhauser (1945, 1952) and Fischberg (1958). Triploidy is easily obtained by refrigerating or heating at 36°C freshly fertilized amphibian eggs; such temperature shocks suppress the elimination of the second polar body. Triploid females are generally fertile, and tetraploids and a few hexa- and heptaploids (Fankhauser, 1945) are found in their offspring. In polyploids, the size of the cells is approximately proportional to the number of chromosome sets; however, the size of the larvae remains normal. Another way to obtain polyploid embryos is to treat fertilized or cleaving eggs with cytochalasin. DNA replication continues in the absence of cytokinesis. We have seen that, in polyploid mouse embryos obtained this way, the synthesis of stage-specific proteins is temporarily correlated with chromosome replication (Petzoldt et al., 1983).

In contrast to polyploidy, aneuploidy, that is, unbalanced chromosome complements, has very serious consequences for development. Viability is greatly reduced, and morphogenesis is abnormal if only one or two chromosomes are added to the normal complement. If chromosome inbalance is marked, as a result of multipolar or monocentric mitoses during early cleavage, no development beyond the blastula stage is possible (Fankhauser, 1934). Strong aneuploidy is thus as lethal as the complete absence of a nucleus in amphibians.

These experiments clearly show that a balanced chromosomal constitution is necessary for satisfactory development beyond the blastula stage. However, the results obtained with amphibian eggs are not necessarily valid for all species. García-Bellido et al. (1983) have shown that, in Drosophila, deletions that may reach 12% of the genome are not lethal before a late stage of embryogenesis. Maternal gene products accumulated in the oocyte are sufficient to allow normal segmentation of the embryo and cuticular differentiation.

B. Lethal Hybrids

Addition of sperm from a foreign species to unfertilized eggs can lead to different types of results, depending on the taxonomic distance between the two species: (1) nothing happens: the foreign spermatozoon does not even contact the egg surface; (2) the foreign spermatozoon comes in contact with the egg surface, but does not penetrate the egg; this produces only activation (like pricking an unfertilized frog egg with a clean needle); (3) the foreign nucleus penetrates the egg, but sooner or later degenerates; this leads to gynogenesis, as in the case of fertilization with uv-irradiated sperm, the result of which is a "pseudohybrid"; (4) amphimixy and cleavage are normal, but development of the diploid embryos eventually stops (lethal hybrids); (5) finally, viable hybrids (sterile or fertile) can be obtained.

In the following discussion, we shall focus on early lethal hybrids in sea urchins and amphibians, limiting ourselves to the combinations in which attempts to discover the biochemical reasons for lethality have been made. A frequent difficulty in these experiments is that the mechanisms that block polyspermy may efficiently prevent the penetration of foreign sperm. In order to obtain enough hybrid embryos for biochemical work, the experimenter must often pretreat the eggs with agents that slow the cortical reaction.

One of these lethal combinations in echinoderms is *Paracentrotus* ♀ × *Arbacia* ♂ (Baltzer, 1910; Baltzer, *et al.*, 1954) in which cleavage is normal but gastrulation is fatal to the hybrids. A few of the latter, in which the sperm nucleus has presumably degenerated, escape lethality and develop into plutei of the maternal type. In this combination, paternal antigens could not be detected by the then available immunological methods 1 hr after fertilization, but were detectable at the blastula stage a short time before the arrest of development at gastrulation (Harding *et al.*, 1954, 1955). It would certainly be rewarding to repeat these experiments with modern immunofluorescence techniques in order to follow the intracellular localization of the paternal antigens. Using the same combination (*Paracentrotus* ♀ × *Arbacia* ♂), Denis and Brachet (1969a,b) studied, by molecular hybridization, the synthesis of paternal and maternal DNAs and RNAs, and found that the arrested early gastrulae contained about twice as much maternal as paternal DNA. This must result from a partial elimination of paternal chromosomes during the successive cleavage cycles. Both genomes are transcribed in the lethal hybrids; transcription of the paternal (foreign) genome is thus favored in this interspecific lethal hybrid.

Similar results have been obtained on lethal hybrids between sand dollar (*Dendraster*) eggs and sea urchin (*Stronglyocentrotus*) sperm (reviewed by Whiteley and Whiteley, 1975). Both the paternal and maternal genomes are transcribed as well in the lethal hybrid embryos as in the respective parental embryos (Lee and Whiteley, 1982); however, the proteins of the hybrid embryos are mostly maternal. This is the case for esterases (Ozaki, 1975), malate dehydrogenase (Ozaki and Whiteley, 1970), and the hatching enzyme (Barrett and Angelo, 1969; Showman and Whiteley, 1980). Immunological studies have also shown that, in these hybrids, most of the antigens are maternal (Badman and Brookbank, 1970), although there is a substantial synthesis of paternal histone H1 during the early development of these hybrids (Easton and Whiteley, 1979).

A gross underrepresentation of paternal proteins is also found in sea urchin hybrids that can reach a normal pluteus larval stage. Two-dimensional gel electrophoresis has shown that only a few proteins specified by the paternal genome can be detected; in these hybrid combinations, the paternal transcripts are also underrepresented (Tufaro and Brandhorst, 1982). This suggests that much of the mRNA that is translated into proteins is persistent maternal mRNA. However, in such sea urchin interspecies hybrids, both maternal and paternal histone mRNAs are already detectable during early cleavage (Easton and Whiteley, 1979; Maxson and Egrie, 1980).

Crain and Bushman (1983) have studied the actin-coding RNAs in two interspecific sea urchin hybrids that may reach the pluteus stage. They found that the transcripts of both paternal and maternal actin genes are present in hybrid blastulae. They proposed an interesting explanation for the variety of observations on paternal genome expression in hybrid sea urchin embryos. The paternal genes for universal (housekeeping), highly conserved proteins would always be expressed; this is the case for the histones and actin genes. On the other hand, paternal genes whose expression is regulated by interactions with specific factors in the embryo would be poorly (or not at all) expressed in the hybrids. The majority of the proteins synthesized in an embryo would fall into this second category; the greater complexity of their regulation would prevent their expression in hybrids. More work on a variety of messages and a large number of hybrid combinations is required before the validity of this hypothesis can be ascertained.

As one can see, sea urchin hybrids raise interesting and important questions regarding the control of transcription and translation during early embryogenesis. Particularly intriguing is the finding by Lee and Whiteley (1982) that many of the messages for paternal proteins present in the hybrids are not translated.

We now come to the work done with amphibian lethal hybrids [reviewed by Baltzer (1952) for crosses among urodeles and Moore (1955) for combinations between anuran species]. Among anurans, many lethal hybrids stop development at the late blastula or early gastrula stage. The introduction of a foreign nucleus is obviously a marked disadvantage compared to haploidy. In fact, the development of many lethal hybrids (*R. pipiens* ♀ × *R. sylvatica* ♂, *R. pipiens* ♀ × *R. catesbeiana* ♂, *R. esculenta* ♀ × *R. temporaria* ♂, *Bufo vulgaris* ♀ × *R. temporaria* ♀) is only slightly better than the cleavage that occurs in completely anucleate eggs. Gastrulation never proceeds further than the appearance of the dorsal lip, and in many instances, there is even no sign of gastrulation. The same situation is found in urodeles for the combination *T. palmatus* ♂ × *Salamandra atra* ♂ or *maculosa* ♂. In most interspecific combinations of urodeles, lethality does not begin before neurulation or is absent altogether. These combinations are truly diploid, since there is no elimination of paternal chromatin.

In F. Baltzer's laboratory, much work has been done, especially by E. Hadorn (1932, 1934), on "hybrid merogones," in which the female nucleus has been removed (by pricking) after fertilization with foreign sperm. Removal of the maternal nucleus in the combination *T. palmatus* ♀ × *T. cristatus* ♂ reduces, as one would expect, the capacity for development and viability (Fig. 75); while development of the diploid hybrid is normal, the haploid hybrid merogone dies

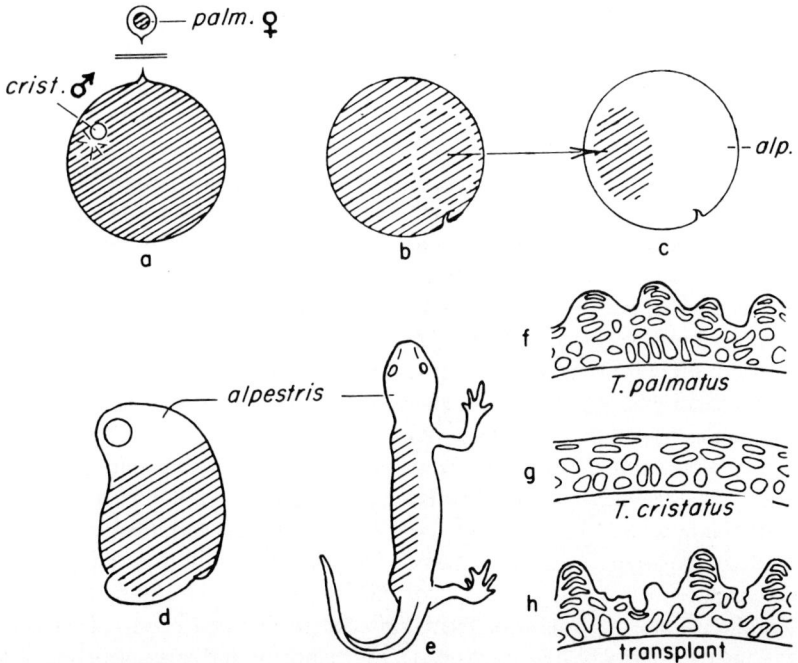

FIG. 75.   Diagrammatic representation of Hadorn's (1936) experiments. (a) Production of the hybrid merogone *T. palmatus* ♀ × *T. cristatus* ♂; (b, c) graft of a piece of hybrid ectoderm on a normal *T. alpestris* host; (d, e) further development of the host, with its graft; (f) section through the epidermis of *T. palmatus;* (g) section through the epidermis of *T. cristatus;* (h) section through the epidermis formed by the transplant, showing typical protuberances of *T. palmatus* epidermis. [After Fankhauser, 1955.]

after neurulation. Important experiments by Hadorn (1932, 1934) further showed that transplantation of a fragment from a lethal hybrid merogone gastrula to a normal host of the same age results in normal differentiation of the lethal fragment; the species of the host does not matter, indicating that the factors from the host that "save" the hybrid merogone fragment from lethality have no species specificity.

Similar results have been obtained with diploid lethal hybrids between anurans. If the dorsal lip of the blastoporus (organizer) of an arrested lethal *R. esculenta* ♀ × *R. temporaria* ♂ (early gastrula) is grafted into a normal *Triturus* gastrula, it does not undergo cytolysis, but instead induces a secondary nervous system in the *Triturus* host and differentiates into chorda and somites (Brachet, 1944). Similar results have been obtained by J. A. Moore (1947, 1948) and B. C. Moore (1954) for the comparable *R. pipiens* ♀ × *R. sylvatica* ♂ combination. Closer analysis of these experiments has shown that both the competence of the ectoderm and the inducing power of the organizer are reduced in this hybrid combination. (See Chapter 3 for details about induction by the organizer.) However, it seems that fragments of early lethal hybrids between anurans, where development is already arrested at the blastula instead of the young gastrula stage, cannot undergo morphogenesis after transplantation into a normal embryo (King and Briggs, 1953; Brachet, 1954).

The reason for the "revitalization" of fragments from lethal hybrids implanted in a normal host remains unknown. The simplest explanation is that lethal hybrids are incapable of synthesizing substances that are required for further morphogenesis; these substances would diffuse from the healthy host into the graft, allowing revitalization and further development. Another possibility is that lethal hybrids accumulate toxic substances that are destroyed by the host's normal cells. There are no experimental facts for or against these hypotheses, nor do we know why, in some combinations, development stops earlier than in others and why, in the former, revitalization by grafting does not seem to be possible. Interestingly, fusion of aneuploid mouse embryos with normal embryos at the 8- to 16-cell stage rescues the aneuploid cells from lethality (Epstein *et al.*, 1982). We are thus dealing with a widespread phenomenon.

Another ingenious and interesting approach for the study of nucleocytoplasmic interactions in lethal hybrids has been devised by King and Briggs (1953): analysis of lethality by nuclear transplantations. They worked with the combination *R. pipiens* ♀ × *R. catesbeiana* ♂, which stops development at the very beginning of gastrulation and does not undergo revitalization by transplantation into a normal host. The purpose of the experiments was to find out whether the irreversible block to embryogenesis involves irreversible changes in the nuclei. As shown schematically in Fig. 76, nuclei from diploid or haploid hybrid blastulae were transferred to enucleate *R. pipiens* unfertilized eggs. The nuclei were taken out of the hybrid blastulae either at the onset of developmental arrest (i.e., 26 hr after the addition of *R. catesbeiana* sperm to the *R. pipiens* eggs) or at various intervals after this time. The experiments showed that nuclei of hybrid gastrulae that had just stopped developing are still capable of reparticipating in the complete course of hybrid development. Development proceeds normally in the injected enucleated egg until the early gastrula stage and then stops abruptly. At 10–20 hr after the onset of gastrulation arrest, the nuclei of the gastrulae are no longer capable of provoking normal cleavage in the recipient eggs. At that time, cytological signs of nuclear alterations become conspicuous in the hybrid nuclei.

These experiments of King and Briggs (1953) show that the nuclei of the hybrids are still unchanged at the time when development stops and even for several hours thereafter. They are not revitalized by several mitotic cycles in normal cytoplasm. The arrest in development apparently results from inadequate interactions between the nucleus and the foreign cytoplasm rather than from a sudden, irreversible change at the beginning of gastrulation. More recent studies

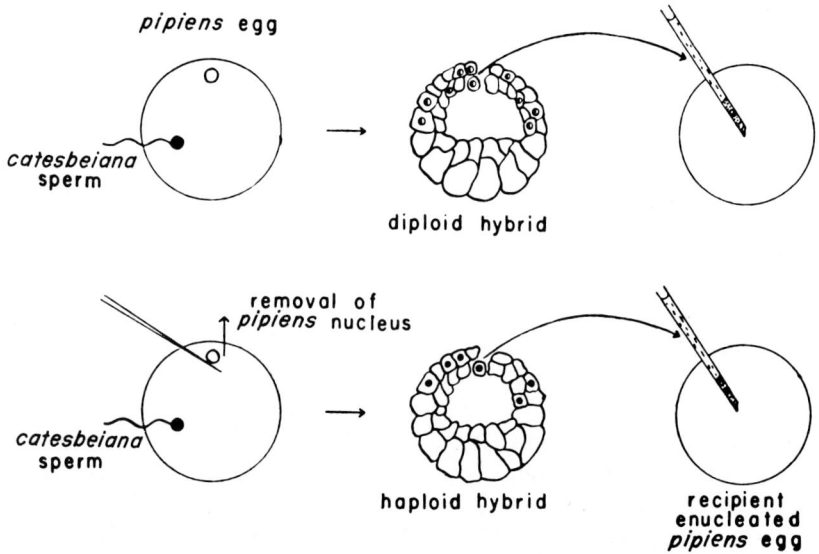

FIG. 76. Method of preparing hybrid blastulae and transplanting their nuclei into enucleated eggs. [King and Briggs, 1953.]

by Hennen (1974) with the same methodology have confirmed that there are no irreversible changes in the nuclei of merogone hybrids as long as they do not undergo prepycnotic degeneration.

Finally, Briggs and King (1955) investigated whether a nucleus removed from a *Triturus* gastrula would be capable of eliciting cleavage in enucleated eggs of *R. pipiens*. Cleavage was obtained in about 50% of the recipient eggs but led, in a majority of the cases, to the formation of only partial blastulae; most of the "cells" forming the partial blastulae were nonnucleated. Thus, the transplanted urodele nuclei had undergone degeneration after a few rounds of DNA replication in anuran egg cytoplasm.

Promising results obtained recently by Delarue and Aimar (1984) deserve mention. They found that grafting a diploid nucleus of the toad *Bufo bufo* into a fertilized egg of *Bufo calamita* is followed by developmental arrest during gastrulation. Injection of clear cytoplasm (hyaloplasm) taken out of centrifuged *B. calamita* eggs into fertilized *B. bufo* eggs also arrests development during gastrulation. Preliminary analysis of the hyaloplasm has shown that it contains at least two inhibitory protein factors. Identification of these proteins might throw some light on the molecular mechanisms that arrest the development of many amphibians hybrids at gastrulation.

As one can see, many important problems dealing with nucleocytoplasmic interactions during early development have been solved by the nuclear transfer technique devised and used by Robert Briggs and Tom King. Comparison of biochemical activities between normal and hybrid embryos, as we shall now see, has not yet added as much as one would have hoped to our knowledge in the field. This is probably due to the fact that the biochemical parameters so far studied (respiration, overall nucleic acid and protein synthesis) are crude. More valuable information would certainly be gained by the use of the much more specific immunological and nucleic acid hybridization methods now available.

It has been found by Barth (1946) that, in the combination *R. pipiens* ♀ × *R. sylvatica* ♂, in which development stops during gastrulation, oxygen consumption first increases normally, as in the controls, and then stops increasing (Fig. 77); a final rise is probably due to incipient cytolysis. As shown by Sze (1953), the metabolic block is generalized and does not specifically affect one region of the arrested gastrulae. All parts of a dissected blocked blastula respire at a much lower rate than their normal counterparts.

Barth's (1946) findings cannot be generalized to all amphibian lethal hybrids. Barth and Barth (1954) reported that in the *R. pipiens* ♀ × *R. clamitans* ♂ hybrid, in which development also ceases during gastrulation, a normal increase in oxygen consumption after the arrest of development is observed. In the *R. esculenta* ♀ × *R. temporaria* ♂ combination, we found (Brachet, 1954) that oxygen consumption at the blastula stage is slightly lower in the hybrids than in the *R. temporaria* controls. During gastrulation, which occurs only in the controls, the increase in the respiratory rate is practically the same in hybrids and controls. In the lethal combination *Bufo vulgaris* ♀ × *R. temporaria* ♂, which never proceeds further than the blastula stage, oxygen consumption of the hybrids increases at the same rate as in the controls for 1 day, although develop-

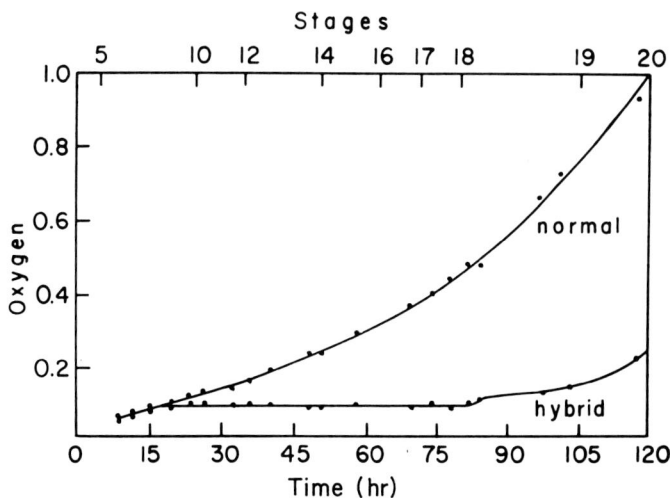

FIG. 77.   Oxygen consumption of normal and lethal (*R. pipiens* ♀ × *R. sylvatica* ♂) embryos. [Barth, 1946.]

ment has stopped (Brachet, 1954). In urodeles, Chen (1953) found that, in lethal hybrids between *Triturus* and *Salamandra* in which development stops at the early gastrula stage, respiration is slightly lower in hybrids than in controls. In both sets of eggs, it increases in the same fashion between the blastula and gastrula stages and then remains constant in the hybrids, while it increases considerably during and after gastrulation in the controls.

In the arrested *R. pipiens* ♀ × *R. sylvatica* ♂ gastrulae, the rate of glycogenolysis is markedly slowed (Gregg, 1948), suggesting a partial block of at least one stage in the glycolytic pathway. Probably related to this decrease in glycolytic activity is the fact that the blocked lethal hybrids show a decreased capacity for keeping ATP in its phosphorylated form under anaerobic conditions (Barth and Jaeger, 1947; Brachet, 1954). Since, as we have seen, NAD synthesis is often a function of the nucleus, a possible explanation of all of these results is that NAD synthesis is impeded when the nucleus is abnormal. This hypothesis has not yet been experimentally tested.

The cytology of arrested lethal hybrids has been studied in a few combinations. In the *R. esculenta* ♀ × *R. temporaria* ♂ lethal hybrids, mitotic activity and development stop together (Brachet, 1944, 1954); occasional expulsion of part of the chromatin occurs at that time. After staining for RNA detection using Unna, the nuclei form a mosaic. Some are normal, while others are characterized by an overload of RNA. In some cells, the nucleoli are too large and too numerous, and the chromatin itself stains strongly with pyronine. It appears as if the lethal hybrids synthesized nuclear RNAs at an abnormally rapid rate or, rather, that this RNA could not be properly processed and transferred to the cytoplasm. Only [³H]uridine pulses, followed by autoradiography and chemical analysis, could show whether this explanation is valid. If a fragment of the lethal hybrid is grafted into a normal gastrula in order to revitalize it, the nuclei become more normal and mitotic activity is resumed.

These effects of lethal hybridization can be mimicked by heat shocks (Brachet, 1948, 1949). Heating a frog gastrula at 36–37°C for 1 hr irreversibly arrests gastrulation movements and mitotic activity (Fig. 78a). If the dorsal lip of a heated gastrula is grafted into the blastocele of a normal embryo (whatever the species to which this host belongs), it is revitalized. The dorsal lip of the heated gastrula differentiates into chorda, somites, and archenteron, and induction of a secondary nervous system occurs in the host (Fig. 78b). As in the lethal hybrids, contact with the normal host returns the swollen nuclei to normal, and mitotic activity is resumed. It is likely that, in both cases, diffusion of substances (devoid of species specificity) that are required for further development occurs at the expense of the host and allows the resumption of DNA and probably RNA synthesis.

The use of heat shocks has become very popular because they induce the synthesis of a set of specific hsps and decrease the synthesis of the other proteins.

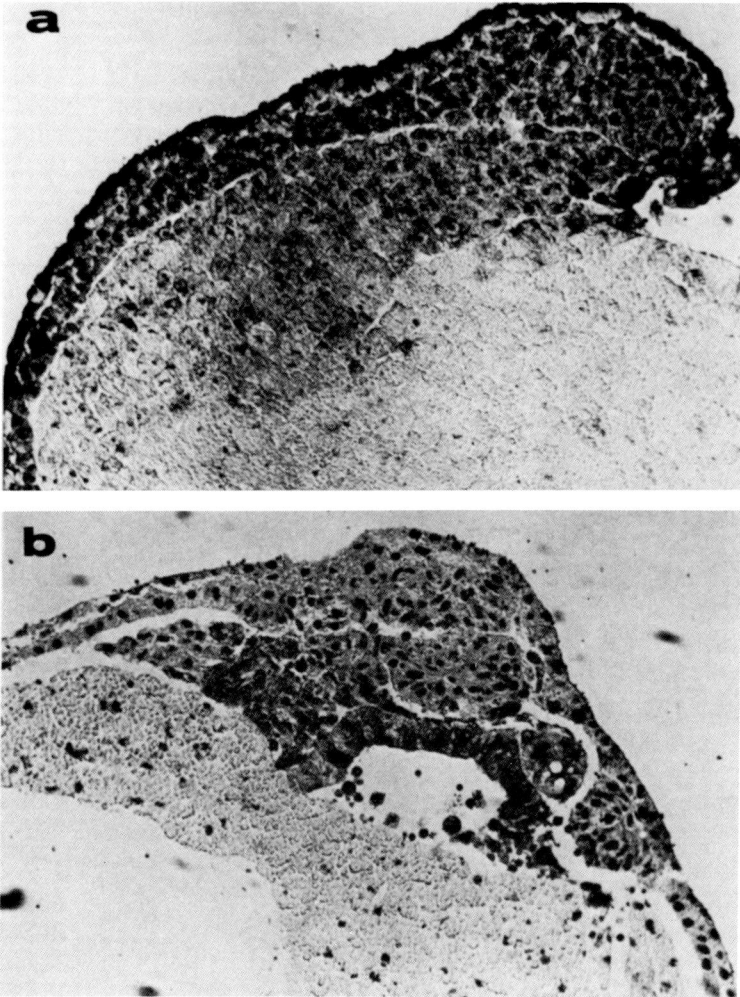

FIG. 78.   (a) Dorsal lip of a frog gastrula irreversibly arrested by a heat shock. (b) A dorsal lip of a frog embryo arrested as in (a) has been implanted in a normal axolotl gastrula; it has differentiated into chorda, somites, and archenteron (dark), and has induced several neural nodules in the host (less pigmented cells). [Brachet, 1957.]

It would thus be rewarding to study protein synthesis in heat-shocked amphibian embryos. Synthesis of hsp is due to selective activation of a few genes [reviewed by Bienz (1985) and Pelham (1985)]. It does not take place in sea urchin (Roccheri *et al.*, 1981) and in *Drosophila* (Dura, 1981) before gastrulation, and thus at a time when mRNA synthesis becomes important. In both cases, hsp synthesis is actinomycin D sensitive. All we know about the biochemical effects of heat shocks on amphibian gastrulae is that RNA synthesis is more sensitive to heating than DNA replication (Steinert, 1951; Hasyawa, 1955) and that contact of a fragment removed from a heated gastrula with a normal host allows the return of normal DNA and RNA synthesis. Whether the production (or lack of production) of hsp plays a role in the arrest of development in heat-shocked amphibian gastrulae remains to be determined.

In the *R. pipiens* ♀ × *R. sylvatica* ♂ cross, cytophotometric DNA measurements have shown that there is no difference in the DNA content of the nuclei between hybrids and controls until the blastula stage. Afterward the hybrid nuclei have a DNA content between the diploid and haploid values, while values between 2C and 4C are found in the controls (B. C. Moore, 1954). This suggests a discrete elimination of part of the DNA when development is arrested. Since we have obtained similar results in the sea urchin *Paracentrotus* ♀ × *Arbacia* ♂ lethal hybrid (Brachet *et al.*, 1962), discrete losses of genetic material in arrested lethal hybrids might be a frequent phenomenon.

In other lethal combinations, mitotic abnormalities are conspicuous. This is true for the *T. palmatus* ♀ × *S. atra* ♂ lethal hybrid, in which the blastula stage is a "critical phase." Development may proceed until the early gastrula stage, but in the blastula, mitotic abnormalities (especially the presence of sticky chromosomes) followed by pycnosis are frequently seen (Schönmann, 1938; Zeller, 1956).

Curiously, it seems that, despite a loss of DNA in individual nuclei, overall DNA synthesis continues in arrested embryos. For instance, in the previously mentioned *T. palmatus* ♀ × *S. atra* ♂ lethal hybrids, there is a reduced, but definite, DNA synthesis after development has been blocked (Chen, 1954). Still more impressive are the results obtained by Gregg and Løvtrup (1955) with *R. pipiens* ♀ × *R. sylvatica* ♂ hybrid embryos, which were obtained with an apparently specific microbiological technique for DNA estimation (Fig. 79). As shown in Fig. 78a, the DNA content of the hybrids, although they do not complete gastrulation, is identical to that of the controls until the latter reach the neural fold stage. Unfortunately, thymidine incorporation in lethal hybrids has not yet been studied, and it must be admitted that correct chemical estimation of the DNA content of amphibian egg is difficult due to the presence of "yolk DNA." Nevertheless, we can certainly dismiss the idea that development of the lethal hybrid stops when an initial store of yolk DNA has been exhausted; net synthesis of DNA is apparently still possible when development has stopped.

There is little to say about RNA synthesis, since the few data available deal with overall RNA synthesis. Quantitative estimations by Steinert (1951) on the RNA content of the lethal *R. esculenta* ♀ × *R. temporaria* ♂ hybrid have shown that lethal and normal embryos have the same RNA content when development stops in the lethals. Afterward RNA continues to be synthesized in the controls, but there is a decrease in the RNA content of the lethal hybrids, probably due to incipient cytolysis. Similar results have been obtained by Chen (1954), who worked with the *T. palmatus* ♀ × *Salamandra* ♂ combination. He found a normal RNA content in the hybrids during cleavage and a decrease when, at the critical stage, many cells undergo cytolysis in this combination.

With the demise of Fritz Baltzer, Ernst Hadorn, and Lester Barth, interest in amphibian lethal hybrids was lost. The only recent work done with the classic *R. pipiens* ♀ × *R. sylvatica* ♂ is that by Johnson on the properties of the cell membrane in control and arrested hybrid gastrulae. By studying disaggregation and reaggregation of dissociated gastrulae from hybrid and control embryos, he showed that there are definite differences between the two. While all the cells taken from a normal late frog gastrula show an active "circus movement" (rotation of a hyaline cytoplasmic blister), very few dissociated ectoderm cells display this movement in arrested hybrid embryos. In addition, cells from arrested hybrid embryos, in confirmation of our findings on another lethal hybrid combination in the frog (Brachet *et al.*, 1962), have a greatly reduced ability to reaggregate *in vitro* (Johnson and Adelman, 1981). Furthermore, Johnson (1981) demonstrated, using chemical methods, that the surface glycoproteins of hybrid and normal gastrulae are not identical. This is an important finding, since it suggests that, in lethal hybrids, the maternal and paternal genes coding for these glycoproteins might be expressed simultaneously (Wright and Subtelny, 1971).

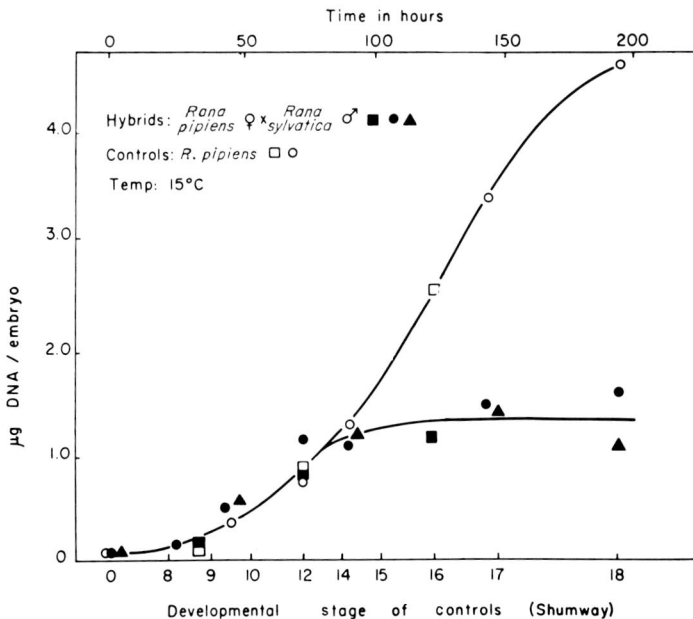

FIG. 79.   DNA synthesis in normal and hybrid (*R. pipiens* ♀ × *R. sylvatica* ♂) embryos. [Gregg and Løvtrup, 1955.]

It is possible that changes in surface glycoproteins and cell adherence play an important role in the developmental arrest of several hybrid combinations at the early gastrula stage, since it is known that treatment of cleaving *Xenopus* eggs with lectins prevents gastrulation (R. Tencer, unpublished results). One should also keep in mind the fact that the late blastula-early gastrula stage is the one in which new nuclear gene transcription takes place; in the hybrid, anarchic gene expression might lead to developmental arrest. I have mentioned that normal frog gastrulae possess extracellular fibrils which promote cell migration and that these fibrils are reduced in number or even absent in lethal hybrids (Nakatsuji and Johnson, 1984; Delarue *et al.*, 1985). The lack of extracellular fibrils is probably responsible for the arrest of development before or at the beginning of gastrulation in frog hybrid combinations. Since antibodies against fibronectin arrest gastrulation (Boucaut *et al.*, 1984a), it would be interesting to find out whether the fibronectin genes are expressed in the chordomesoderm of lethal hybrids.

Recent work by Mohun *et al.* (1984) and Gurdon *et al.* (1984) shows that such a study should be possible. Using radioactive cDNA probes, they found that transcription of the genes coding for α-skeletal and β-cardiac actins does not begin in *Xenopus* before the end of gastrulation. Transcription occurs exclusively in the mesoderm, which will differentiate later in embryonic muscle. Since it occurs several hours before the morphological appearance of the somites, transcription of skeletal and cardiac actins is a molecular model of early cell determination (Mohun *et al.*, 1984). In other experiments Gurdon *et al.* (1984) injected genetically marked nuclei of larval muscles into enucleated *Xenopus* eggs. They found that the α-actin genes are inactive at the blastula stage and active in early gastrulae. α-Actin RNA is synthesized at the right time after nuclear transplantation, but only in the equatorial (mesodermal) regions of the embryo. Dissociation and reaggregation experiments done on animal, equatorial, and vegetal regions of eggs at various stages of development have shown that regulation of the α-actin gene transcription is not affected by the lack of cell contacts. Finally, elegant experiments by Gurdon *et al.* (1985) have demonstrated that the cytoplasmic constituents required for the activation of the actin genes at gastrulation are already localized in the sub-equatorial region (probably the gray crescent) of recently fertilized *Xenopus* eggs. This is what Morgan would have expected in 1934. It would be of great interest to perform similar experiments on lethal hybrids between amphibians and sea urchins. In normal sea urchin embryos, several lineage-specific genes are expressed before overt morphological differentiation of the skeleton, gut and aboral ectodermal wall of the late embryo. The subject has been recently reviewed by Angerer and Davidson (1984), who point out that the transcripts of these genes accumulate at given stages of development and that they are not represented in maternal mRNA. A study of these genes in echinoderm hybrids might be rewarding.

It must be admitted that the reasons for developmental arrest and morphological abnormalities after the introduction of sperm from a different species into an egg remain almost unknown; lethal hybridization raises many questions that could be answered using current methods. How do the two genomes "cohabit" in the same nucleus? Are the paternal and maternal DNA molecules separated in the nucleus? Or do they undergo recombinational events by illegitimate crossing-over? Are there sister chromatid exchanges between paternal and maternal DNAs? How would spread chromatin isolated from arrested gastrulae look under the electron microscope—is it organized in nucleosomes, and would one see transcription units? What is the pattern of nuclear RNA synthesis? Are the processing and migration of the rRNA and mRNA precursors normal? Is the pattern of protein synthesis—in particular, histone synthesis—different in lethal and normal embryos? What are the molecular mechanisms of revitalization? So long as these questions (and many others) are unanswered, the reasons for the block in the development of lethal hybrids will remain a mystery.

C. A FEW EARLY LETHAL MUTATIONS

The field of developmental genetics has become so large that all we can do here is to say a few words about certain mutations that arrest development at an early stage (blastula or gastrula) and therefore have particular interest for cell biologists. The ideal materials for these studies are, of course, *Drosophila* and mouse eggs, since we know so much about the genetics of these two organisms. *Drosophila* eggs are opaque and surrounded by a thick chorion (which can be permeabilized by brief treatment with octane); in order to see what happens during the cleavage stages, the eggs must be sectioned and stained. Precursors for nucleic acid and protein synthesis must be injected into the eggs. Mouse eggs do not have these disadvantages, but few eggs (compared to amphibians) can be obtained from a single mouse, and they do not cleave and reach the blastocyst (blastula) stage unless they are cultured *in vitro*. Amphibian eggs are a more favorable material for biochemical studies, but we know of only a few early lethal mutants in amphibians that are of interest to cell biologists.

The first evidence for a role of the genes in early morphogenesis came from the work of Poulson (1940, 1945) on *Drosophila*. He showed that the complete loss of the X chromosomes (*nullo-X* condition) results in problems early in development. The migration of the cleavage nuclei toward the egg periphery in order to form the blastoderm becomes abnormal, and development stops; the nuclei accumulate at the two poles of the egg instead of being evenly distributed (Fig. 80). Loss of one-half of the X chromosome produces the same effects, while loss of the other half is less lethal. Cleavage and migration of the nuclei proceed normally, but gastrulation of the blastoderm soon becomes abnormal and stops. Much more limited chromosomal deficiencies may lead to serious characteristic abnormalities; development of the nervous system, mesoderm, and gut may be profoundly affected. Ede (1956) and Counce (1956) found that point mutations or small deletions have the same effect on the migration of the nuclei

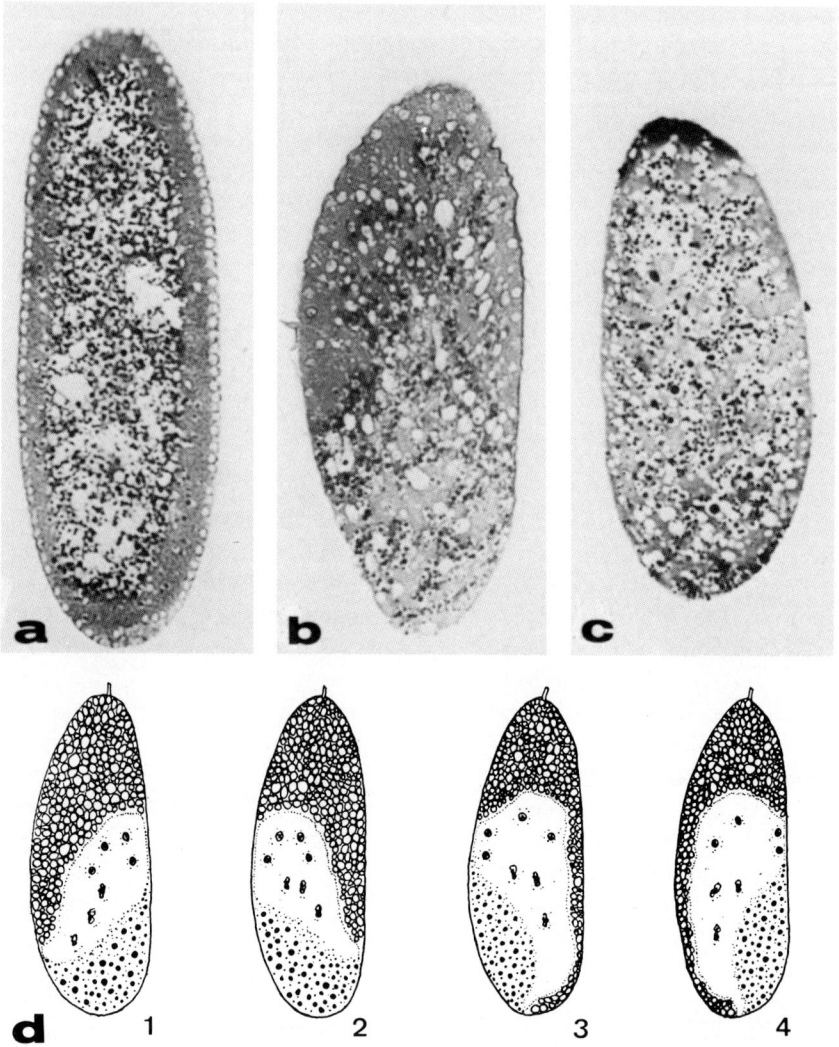

Fig. 80.   Effects of X chromosome deficiency (*nullo-X*) on the development of *Drosophila*. (a) Sagittal section of a normal egg before completion of the blastoderm. (b) Nearly sagittal section of a *nullo-X* egg at the close of nuclear migration. Nuclei and cytoplasm around the anterior end. Only a few nuclei in the posterior half of the egg. (c) Another nearly sagittal section of a *nullo-X* egg. Nuclei in the anterior half of the right side. Breakdown of the cortical layer at the posterior end. (d) Schematic representation of various types of distribution of materials encountered in *nullo-X* eggs at later stages. Types 1 and 2 are most commonly found. Only the cell outlines are indicated; nuclei lying in the central mass of cytoplasm are shown. [Poulson, 1940.]

during cleavage as the absence of whole X chromosomes. These genetic studies led to the conclusion that individual genes control embryogenesis at very early stages. However, a more likely explanation is that these genes control the organization and chemical composition of the egg cytoplasm during oogenesis. If this organization is abnormal in the unfertilized egg, it is not surprising that the cleavage nuclei migrate toward the poles instead of the egg periphery. Such early mutants are sometimes called "cytoplasmic" mutants because the mutated genes affect the cytoplasm of the oocyte and the egg. This expression is misleading because we are thinking here of nuclear genes and not of cytoplasmic (mitochondrial or chloroplastic) genes; it is better to speak of maternal effect genes.

All we know about the biochemistry of the *nullo-X* lethal embryos is that their initial oxygen consumption is normal and that it increases, just as in the controls, until development stops (Fig. 81). At that time, the respiration of the lethal embryos stops increasing, almost completely, in contrast to that of the normally developing controls. The reasons for this block in respiration, which we found in *R. pipiens* ♀ × *R. sylvatica* ♂ hybrids, are totally unknown.

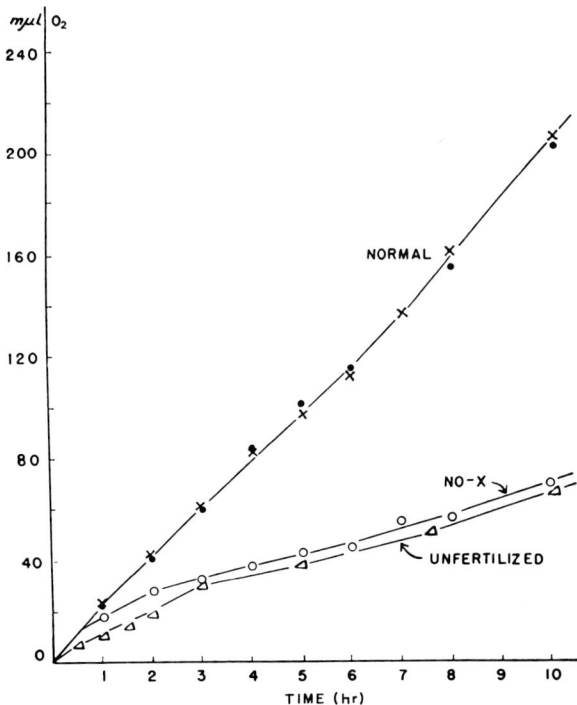

FIG. 81.   Oxygen uptake of normal, unfertilized, and *nullo-X* (*NO-X*) single eggs of *Drosophila*. [Poulson, 1945.]

Another interesting early lethal mutation in *Drosophila* is *deep orange* (*dor*), which has been studied by Garen and Gehring (1972). In the homozygous condition, (*dor/dor*) development stops at the gastrulation stage. The interesting point is that injection into such *dor/dor* eggs of cytoplasm from a normal, wild-type egg suppresses the lethal effects of the mutation. Development of part of the injected eggs can proceed until an advanced stage of embryogenesis. This experiment shows that lethality is due to the absence in the egg cytoplasm of the *dor* mutants of a substance (or substances) required for embryogenesis beyond the blastoderm stage.

Comparable results have been reported by Santamaria and Nüsslein-Volhard (1983), who studied the morphogenetic effects of the *dorsal* locus in *Drosophila* embryos. By a maternal effect, this locus induces dorsalization of the embryos. The normal dorsoventral pattern can be partially restored by injecting cytoplasm from wild-type embryos into dorsal mutants at the syncytial stage of cleavage. Ventral cytoplasm is more effective than cytoplasm taken from other regions of the wild-type donor embryos. This suggests that, in *Drosophila* eggs, the dorsoventral pattern depends on cytoplasmic determinants synthesized and distributed during oogenesis under the control of the *dorsal* locus. Continuation of these studies by Anderson and Nüsslein-Volhard (1984) disclosed that mutations with maternal effects at 10 different loci give completely dorsalized *Drosophila* embryos. Injection of RNA from wild type embryos into the eggs of mutants at 6 loci partially reverse the dorsoventral polarity. In one of the mutants (called *shake*), injection of poly(A)$^+$ RNA completely restored the normal dorsoventral pattern. The information for this pattern is thus stored in maternal RNAs. Almost identical results have been reported by Preiss *et al.* for the *Krüppel* mutation (a deletion of two segments in *Drosophila* larvae): injection of a cloned DNA from a blastoderm-specific RNA induces the phenotypic rescue of the mutants (Preiss *et al.,* 1985). In addition, injection of antisense mRNA (see Chapter 4, Volume 1) of *Krüppel* into normal embryos induces phenocopies of the *Krüppel* mutation (Rosenberg *et al.,* 1985).

Drosophila is an important material for developmental geneticists because mutations in the so-called *homeotic* genes affect morphogenesis. Bender *et al.* (1983) have studied homeotic mutations in the *bithorax* (*bx*) complex (located on chromosome III); they transform segments in other segments. An extreme case is the *bx*^{3} mutation, which transforms halters in wings. The result is the production of adult flies with four wings instead of two. As shown by Bender *et al.* (1983), mutations in the *bithorax* complex are due to DNA rearrangements and not to point mutations. Most of the mutants have insertions of a particular genetic mobile element called *gypsy* by the authors; it affects the functioning of DNA sequences distant from the insertion. In revertants, *gypsies* remain inserted at the same site.

McGinnis *et al.* (1984a) in W. Gehring's laboratory, recently reported that a repetitive DNA sequence is specifically localized in the homeotic *bithorax* and *antennapedia* complexes which are required for correct segmental development. This sequence, which was called "homeo box," is arising considerable interest today, as well as mutations of a newly discovered homeotic gene (*fushi tarazu, ftz*), which reduces the number of segments. Mutations at the *ftz* locus give embryos with only half the normal number of body segments (Laughton and Scott, 1984). The interest in the homeo box is because of the fact that this DNA sequence is not found only in the 3' portion of the *antennapedia, fushi tarazu* and *ultrabithorax* genes of *Drosophila;* it is present in many invertebrates and vertebrates, including man (McGinnis *et al.,* 1984b). It codes a highly conserved, very basic sequence of 60 amino acids, which probably binds to DNA. *Xenopus* possesses a gene homologous to the *Drosophila* homeo box. It is expressed during gastrulation and gives three different transcripts under strict temporal control. It also codes for a basic, chromatin binding polypeptide of 60 amino acids which presumably controls the activity of a battery of genes (Carrasco *et al.,* 1984). A second *Xenopus* gene, containing a homeo domain, has been recently isolated by Müller *et al.* (1984). In contrast to the first gene, it is abundantly transcribed in full-grown oocytes. It might play a role in the early determination of embryonic cell types. In mice, a homeo box is located on chromosome 6, which contains several genes which, if mutated, cause developmental abnormalities (McGinnis *et al.,* 1984c). In man (reviewed by Ruddle *et al.,* 1985), DNA sequences which display more than 90% homology with the homeo box of the homeotic *Drosophila* genes code for 61 amino acids. It is generally believed that homeo boxes play an essential role in the all metameric organisms (in the separation of the somites, for instance).

Fushi tarazu homozygotes, which have no alternate segments, are lethal before hatching. The *ftz* gene codes for 1.9 kb RNA, which is expressed during early cleavage; it is no longer expressed after gastrulation (Kuroiwa *et al.,* 1984). Elegant hybridization experiments by Hafen *et al.* (1984) have given interesting informations about the localization of the *ftz* transcripts. They are detectable during cleavage; at the blastoderm stage, they are restricted to seven evenly spaced bands of cells, and they disappear at later stages. All this speaks for a key role of *ftz* in the segmentation pattern of the embryo. A similar pattern of periodic expression has been found for the *engrailed* locus, which has a homeo box sequence and is involved in the segmentation of the larva (Fjose *et al.,* 1985).

We mentioned briefly, in Chapter 4, Volume 1, the important work of Spradling and Rubin (1982) on another *Drosophila* transposable element, the so-called P element. This 3-kb transposable element can be injected into the posterior pole of *Drosophila* eggs and can serve as a vector for DNA-mediated gene transfer. In this way, it is possible to induce mutations in strains that lack P elements. Part of the DNA injected into the germ plasm enters the germline cell nuclei, where it is transcribed. The transcripts are translated in the cytoplasm; their product, a transposase, enters the nuclei, where it catalyzes the insertion of P elements into the chromosomes. This increases mutability and produces a syndrome called "hybrid dysgenesis" by *Drosophila* geneticists. It is characterized by sterility, induction of mutations, male recombination, segregation distortion, and nondisjunction. If a gene is inserted into a P element, it is possible to obtain genetic transformation in *Drosophila* via this transposable element vector. Rubin and Spradling (1982), who discovered this approach, used it to introduce the *rosy* gene into strains lacking this gene. Insertion of transposable P elements into the gene coding for RNA polymerase II is lethal for *Drosophila,* according to Searles *et al.* (1982).

The principle of insertion mutagenesis can be extended to organisms other than *Drosophila* by using viruses instead of P elements. Schnieke *et al.* (1983) found that insertion of a retrovirus (RNA virus) into the germline gene coding for α_1 (I) collagen results in a lethal mutation. Mouse embryos die on the 12th day of development. According to Harbers *et al.* (1984), introduction of the retrovirus into the first intron (upstream from the mRNA starting point) of the α_1 (I) collagen gene produces this lethal mutation. The embryos are unable to synthesize collagen, and the intron apparently plays a role in tissue-specific collagen expression. Death on the 12th day of development of mouse embryos homozygous for the mutation resulting from the insertion of a retrovirus in the collagen I gene results from the death of erythropoietic and mesenchymous cells (Löhler *et al.,* 1984). At the molecular level, the mutation is associated with an altered chromatin structure as shown by DNase I-digestion experiments. Insertion of the virus would prevent the developmentally regulated appearance of a DNase-hypersensitive site and activation of the collagen I gene during development.

However, for mouse eggs, interest has been mainly focused on the complex *t* locus, where many different alleles are known (reviewed by Frischauf, 1985). This *t* locus is the equivalent, for mouse embryos, of the *H-2* locus, which, in the adult mouse, controls the synthesis of the histocompatibility antigens responsible for the success or failure of organ transplantation (graft rejection or acceptance). Many *t* mutants are known and have been studied from both the embryological and genetic viewpoints. Of special interest for our purposes are the t^{12}/t^{12} mutants, which stop developing at the morula stage (Smith, 1956). Since mouse eggs cleave slowly, the synthesis of new kinds of mRNAs and proteins has already occurred during early cleavage and is immediately arrested by actinomycin D treatment. It is likely that the t^{12}/t^{12} mutation is a classic, probably affecting the expression of a gene required for early development, already active during cleavage, rather than a gene affecting the composition of the egg cytoplasm during oogenesis.

A mutation of great interest for cell biologists has been recently described by Magnusson *et al.* (1984). It was known that a mutation called *Os* causes oligodactily (reduction of the number of digits) or syndactily (fusion of the digits) in mice: homozygous *Os/Os* embryos are lethal. Magnusson *et al.* (1984) found that these homozygous mutants form blastocyst outgrowths where mitoses are accumulated. At first, they look like colchicine-treated embryos. But the spindles of the arrested mitotic figures are normal and it is clear that, in homozygous *Os* embryos, chromosome separation at metaphase is prevented. Since, as we have seen in Chapter 5, Volume 1, the molecular basis of chromosome separation and migration remains poorly understood, the availability of the *Os* mutants should throw a new light on this exceedingly important process.

We have already discussed the amphibian nucleolus in Chapter 4, Volume 1, particularly the famous *anucleolate* mutation (*nu-o*) of *Xenopus*. The deletion of the nucleolar organizer in homozygous *nu-o/nu-o* eggs and embryos results in an inability to synthesize 28 and 18 S rRNAs and to accumulate new ribosomes after fertilization. Anucleolate mutants (*nu-o/nu-o*) can hatch, but die at the time when the controls begin to feed. Heterozygotes (*nu-o/+*) are viable, despite the fact that their cells have only one nucleolus instead of two; they have a normal rRNA content and a normal load of ribosomes. Regulatory mechanisms allow the ribosomal genes in the uninucleolate embryos to synthesize as much 28 and 18 S rRNA as normal controls.

Also of interest is the *nc* (*no cleavage*) mutant of the axolotl (Raff *et al.*, 1976). The eggs can be fertilized, but they do not cleave because they are unable to assemble their large tubulin store into spindle and aster MTs. The mutation can be corrected (i.e., repeated cleavages do occur) after injection of basal bodies into the cytoplasm of fertilized eggs.

Finally, we should mention the *o* mutation, which affects the cytoplasm of axolotl eggs. As shown by Briggs and Cassens (1966) and by Briggs (1972), *o/o* homozygotes develop normally until the blastula stage, at which time developmental ceases and they are unable to undergo gastrulation. If one injects nuclear sap removed from the GV of an o^+/o^+ or o^+/o oocyte into one of the blastomeres at the two-cell stage, the injected blastomere develops much further than the uninjected one (Fig. 82). The GV (in contrast to the oocyte cytoplasm) thus contains a gene product necessary for embryonic development beyond gastrulation. This product has no species specificity and is already present in the GVs of young oocytes. RNA synthesis is very low in *o* mutants, and it is thus possible that the missing gene product plays a role in RNA synthesis. Brothers (1976) has reviewed the subject and included the results of her own nuclear transplantation experiments. She found that nuclei from *o* morulae support full development after transfer into anucleate unfertilized eggs. This capacity to support development is lost in *o* blastula nuclei, showing that irreversible alterations of the nuclei occur while the development of the mutants proceeds from the morula to the blastula stage. This is similar to what was said above about the transfer of nuclei from lethal hybrids (diploid or haploid) into enucleate unfertilized frog eggs (King and Briggs, 1953; Hennen, 1974).

D. Transgenic Mice: A Breakthrough in Genetic Research

To introduce a gene into an egg and to witness its expression, not only in the adult but also in its offspring, and to create new strains with predetermined genetic changes have been old dreams for all biologists. Lamarck believed, at the beginning of the last century, that this had been achieved during evolution by

o/o ♀   x o^+/o ♂

mutant eggs

injection
of normal
nucleoplasm

gastrular
arrest

normal
development

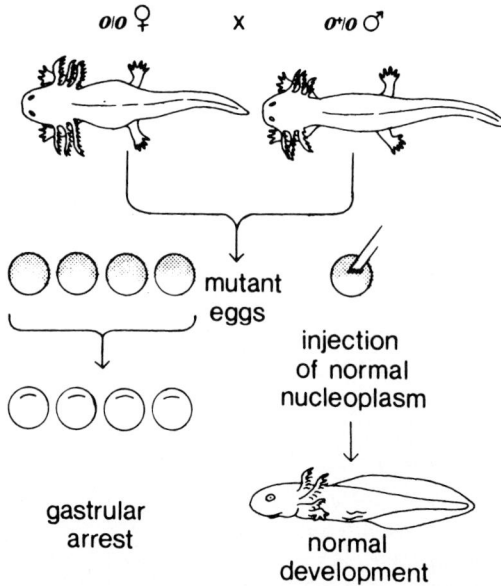

FIG. 82. A maternal effect mutation (o, for ova deficient) in the axolotl. Females homozygous for o produce eggs that always arrest at gastrulation. This gastrular arrest can be corrected by the injection of o^+ substance. This o^+ substance is produced during oogenesis under the direction of the normal allele of the o gene. The substance is found in the GV and in the cytoplasm of normal full-grown oocytes. [Reprinted by permission from *Nature* **260**, 112. Copyright © 1976 Macmillan Journals Limited.]

changes in the external environmental conditions of animals and plants; this led to the tomfoolery of Lyssenkism–Mitchourinism in the late 1940s. The possibility of inducing mutations by x rays and later by chemical mutagens remained unsatisfactory to many scientists because the induced mutations appeared at random and could not be directed at will by the experimenter. Thus, one could only select "interesting" mutants among a host of blindly produced mutants.

In the past, attempts were made to inject DNA into eggs, but these early experiments were done with whole unfractionated DNA. It was found in our laboratory (Sempinska and Ledoux, 1967) that, if tritiated *E. coli* DNA was injected into fertilized amphibian eggs, it was partly degraded and partly incorporated into the nuclei. It has been shown in Gurdon's laboratory that any DNA injected into an unfertilized *Xenopus* egg is replicated (Gurdon *et al.*, 1969). More recently, Rusconi and Schaffner (1981) went one step further with the advent of recombinant DNA technology. They injected into fertilized *Xenopus* eggs cloned rabbit β-globin genes. Using specific probes for the detection of globin genes, they found that the injected DNA was first replicated in the cytoplasm, with the formation of concatamers, and then largely degraded. However, part of the injected DNA was integrated into the *Xenopus* genome, and 3–

10 copies of rabbit β-globin genes could be detected in each cell of the adult. The rabbit globin gene is transcribed and correctly spliced in all *Xenopus* cells.

The major breakthrough—a true revolution—came from work done on mouse eggs. The basic idea was to introduce pure cloned genes into one of the pronuclei of fertilized mouse eggs by direct microinjection. All that is required for such experiments, besides a pure preparation of a given gene in sufficient amounts is a good micromanipulator, skill, patience, and some good luck. Luck is certainly required because the injected eggs, if they have not been injured by the experimental procedure, must be cultured *in vitro* until the blastocyst stage and then implanted in a pseudogravid foster mother. The blastocyst must then establish intimate connections with the uterus; then, with luck and good work, fetuses and even adults may be obtained.

In 1980, Gordon *et al.* injected into one of the pronuclei of fertilized mouse eggs the thymidine kinase *(TK)* gene of herpes simplex virus (HSV) coupled to SV40 (simian virus 40) DNA, which provides an origin for DNA replication. Out of 78 newborn mice, 2 possessed the donor DNA in all cells. These experiments were extended by Gordon and Ruddle (1981), who injected various genetic materials (including the interferon gene inserted in a plasmid) into mouse pronuclei and found that the injected genes were transferred to some of the newborn mice. They found two different patterns of evolution for the injected genetic material: it may or may not be integrated into the host genome. If it is integrated (and this is where luck still plays a role), the foreign gene will be maintained until adulthood. The injected DNA is covalently associated with the host sequences. Most important of all is that these linked sequences are transmitted, according to Mendel's laws, to the succeeding generations of progeny. It has thus been possible not only to introduce a given gene into an egg and an adult, but even to transmit it to the progeny. These "transgenic mice" (reviewed by Petri, 1982; Brinster and Palmiter, 1982; Gordon and Ruddle, 1985) fulfill the old dream of directing mutations and of manipulating heredity at will.

Similar experiments have been done, at the same time, by E. F. Wagner *et al.* (1981), and T. E. Wagner *et al.* (1981), and Costantini and Lacy (1981). T. E. Wagner *et al.* (1981) first injected cloned rabbit β-globin genes and *TK* gene from HSV and found that 15% of advanced fetuses possessed the two genes in their DNA. There were 3–50 copies per cell, apparently integrated into host cell DNA. E. F. Wagner *et al.* (1981) injected rabbit β-globin genes (intact or fragmented) into the male pronucleus of mouse eggs at the time its chromatin underwent decondensation resulting from the loss of the sperm-specific basic proteins. Not only adult mice originating from the gene-injected mice, but also their offspring, produced rabbit β-globin, demonstrating that the foreign gene had been expressed. The injected gene fragments possessed the TATA and CAAT boxes, which, as we have seen, are required for accurate transcription by RNA polymerase II. The input was either 20,000 gene fragments or 20,000 gene-containing plasmids per male pronucleus. The authors point out that, in addition to the remarkable fact that the rabbit globin genes are transmitted from the injected mouse to her offspring, the yield of successful gene transfers is much higher than that obtained by the classic techniques of gene transfer in cultured cells (where only $1:10^5 - 1:10^7$ cells are transformed, which thus requires selection of the transformed cells on selection media). In almost identical experiments, Costantini and Lacy (1981) also obtained transmission of the rabbit β-globin gene through the germ line to the offspring. They showed that, in the descendants of the gene-injected mice, *all* cells possessed the rabbit gene. Similar results were obtained by Stewart *et al.* (1982), who injected human β-globin genes into the male pronucleus of mouse fertilized eggs. They found that the injected genes are conserved in the adult and transmitted to the offspring in Mendelian fashion; this proves that the human genes were stably integrated into the mouse genome. Interferon genes injected into a mouse pronucleus are also transmitted as Mendelian characters to the offspring (Gordon, 1983).

A new step forward was made when Palmiter *et al.* (1982) injected into the mouse pronucleus fused eukaryotic genes. They fused the powerful promoter of the metallothionein I gene with the structural gene of rat growth hormone prior to injection. The outcome was that out of 21 mice that reached the adult stage, 6 were giants. More recently, Palmiter *et al.* (1983) fused the metallothionein I promoter with the gene coding for human growth hormone and injected the fused genes into mouse pronuclei. Seventy percent of the mice resulting from this experiment were larger than the control mice. Treatment with heavy metals (cadmium, zinc), which induces the expression of the metallothionein I gene, increases the synthesis of growth hormone. The serum of the transgenic mice contains growth hormone and insulin-like growth factor I (which is synthesized in response to growth hormone). The pituitaries of the transgenic mice are histologically abnormal, with a marked decrease in acidophilic cells. The fusion genes are expressed in all tissues of the transgenic mice, but in varying degrees. It seems that their expression is influenced by the integration site and by the surrounding tissues.

Is there a preferential expression of the injected genes in certain organs of transgenic mice? This question arouses great interest today. Lacy *et al.* (1983) injected rabbit γ-globin genes into the pronuclei of fertilized mouse eggs and showed, by *in situ* hybridization of metaphase chromosomes, that they are incorporated into one or two chromosomal loci. Each locus possesses 3–40 copies of the injected genes, arranged in tandem arrays. These sequences are stably inherited as Mendelian markers. But no rabbit globin could be detected in the hematopoietic cells, while the foreign protein was present, at a low level, in skeletal muscle and testis. It was concluded that the injected gene is expressed in inappropriate tissues in transgenic mice. However, different results have been obtained for other genes. McKnight *et al.* (1983) injected the chicken transferrin (a hepatic protein) gene into mouse pronuclei; 15–30% of the descendants possessed the gene, which, again, was integrated in multiple copies in a tandem arrangement. In five out of seven transgenic mice, there was a 5- to 10-fold preferential expression of the transferrin gene in the liver compared to other tissues. Chicken transferrin was secreted by the liver in the serum, and the chicken gene was also expressed by the descendants. Finally, in contrast to the results of Lacy *et al.* (1983), Brinster *et al.* (1983) obtained tissue-specific expression of an injected, rearranged, and functional immunoglobulin gene. This gene was expressed in the spleen, but not in the liver of the mice that had inherited the injected gene. The reasons why injected rabbit γ-globin genes are expressed in inappropriate tissues in transgenic mice, and why the immunoglobin genes are expressed in a tissue-specific way, are still unknown. Recent work has confirmed that immunoglobin genes are expressed in a selective way in transgenic mice. Storb *et al.* (1984) found that all transgenic mice express the rearranged mouse kappa *Ig* gene only in the spleen. There is a correlation with B-lymphocytes, not with T-lymphocytes. Introduction of the kappa *Ig* gene into the pronucleus of a fertilized mouse egg is followed by its selective expression in lymphoid cells and by synthesis of functional antibodies (Grossschedl *et al.* 1984). Similar findings have been made for the gene coding a specific pancreatic enzyme, elastase I; after injection of this gene into a pronucleus, four out of five transgenic mice expressed selectively and at a high rate (10,000 molecules per cell) elastase I mRNA in the pancreas (Swift *et al.*, 1984; Ornitz *et al.*, 1985). Many experiments in which transgenic mice expressed the foreign injected gene in a tissue-specific manner have been published. Predominant, if not exclusive, expression of foreign β-globin genes in the erythroid lineage (Townes *et al.*, 1985; Magram *et al.*, 1985); the myosin light chain gene in skeletal muscle (Shani, 1985); the insulin gene in endocrine pancreas (Hanahan, 1985); and the E_α

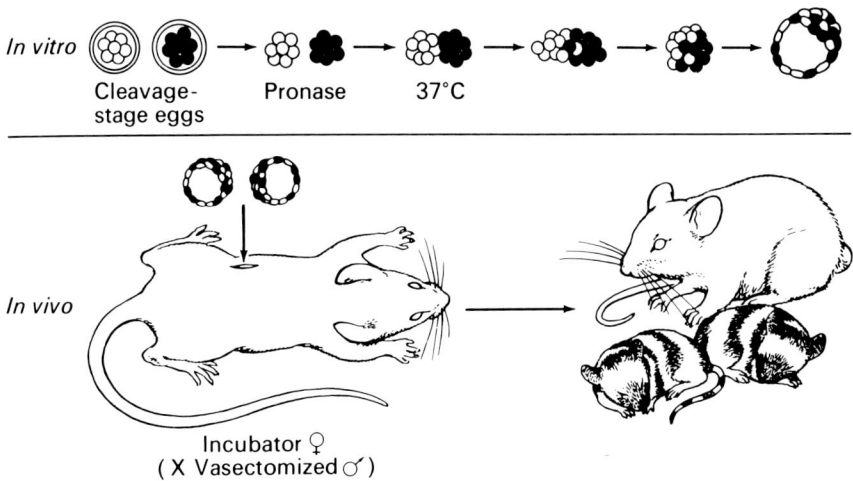

FIG. 83. Diagram of procedures for obtaining allophenic mice from aggregated morulae of different genotypes for fur color. Aggregates of the 8- to 12-cell stage quickly form a sphere that can be implanted in a genetically marked host. [From "Methods in Mammalian Embryology" ed. by J. L. Daniel, Jr. W. H. Freeman and Company. Copyright © 1971.]

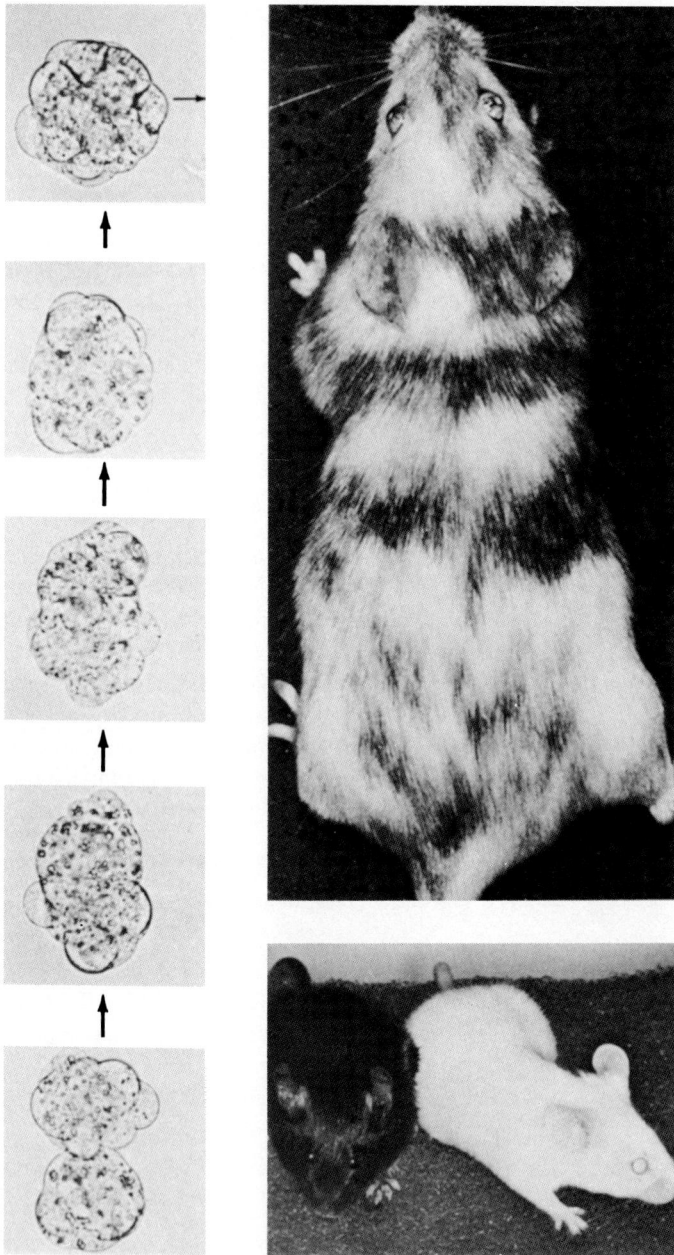

FIG. 84.   Obtaining allophenic mice of multiembryo origin. Lower left: two living cleavage-stage embryos of pigmented (CC) and albino (cc) genotypes aggregated *in vitro*. Successive photographs show the formation of one spherical embryo (CC ↔ cc) from all of the blastomeres of two eggs. One of the allophenic mice from these paired genotypes is at the upper right. Note the transverse clones of black and white in the coat and the radiating clones in the eyes. At the lower right are two of the offspring of this allophenic female mouse after mating with a pure-strain albino male.

gene (which controls the immune response) in macrophages (Le Meur *et al.*, 1985) has been reported. These findings should lead to a better understanding of the tissue-specific factors that selectively activate one or a few genes during cell differentiation (see Chapter 3).

More and more experiments will certainly be done on transgenic mice and will probably be extended to other animal species. Using this methodology, it might be possible to produce giant cows, pigs, and sheep for the benefit of agriculture; this has already been tried by Hammer *et al.* (1985). Giant men would be of little use, except perhaps as basketball players. The dangers of extending the transgenic methodology to humans are great from both an ethical and a biological viewpoint. After *in vitro* fertilization and reimplantation of human eggs, 80% of failures are observed; two-thirds of the early embryos are cytologically abnormal, and 20% are haploid (Angell *et al.*, 1983). Still worse results would be obtained if one of the pronuclei of an artificially fertilized egg received an injection before culture and reimplantation.

This chapter would be incomplete without mentioning the outstanding work of Beatrice Mintz, since it has important implications for both cell differentiation and cell malignancy, which will be discussed in the next chapter. The discussion here will be limited to a summary of B. Mintz's more striking experiments. A first attempt to combine experimental embryology and genetics was the production of allophenic (tetraparental) mice (Tarkowski, 1961; Mintz, 1962). Two mouse morulae possessing different genetic markers (skin color, isozyme pattern) were fused together (Fig. 83). After *in vitro* culture and implantation in a foster mother of known genotype, "mosaic" mice were obtained. An example is the famous "zebra" mouse shown in Fig. 84, in which the fur coat has alternating pigmented and nonpigmented streaks. These mice result from the differentiation of two hemimorulae of different genotypes. Careful analysis of such allophenic mice has allowed Mintz (1974) to increase considerably our understanding of embryonic differentiation.

Fusion of cleaving eggs from different species is possible. Embryo manipulation (fusion at the four- to eight-cell stage) has allowed the production of sheep–goat chimeras (Fehilly *et al.*, 1984) and the birth of a goat kid from a sheep mother (Meinecke-Tillmann and Meinecke, 1984). The day will perhaps come where the centaurs and sphinxes of mythology will be produced at will by embryologists.

Other studies of Mintz deal with mouse teratocarcinoma, a tumor originating from the ovary or the testis (reviewed by F. Jacob, 1978). Some of its cells differentiate into a monstrous embryo, while others remain undifferentiated and are malignant. If these stem cells (embryonal carcinoma cells) are injected into a mouse, the animal is killed by a fast-growing tumor. Malignant and differentiat-

---

One comes from a genetically pigmented germ cell and the other from an albino germ cell. [Mintz, 1971.]

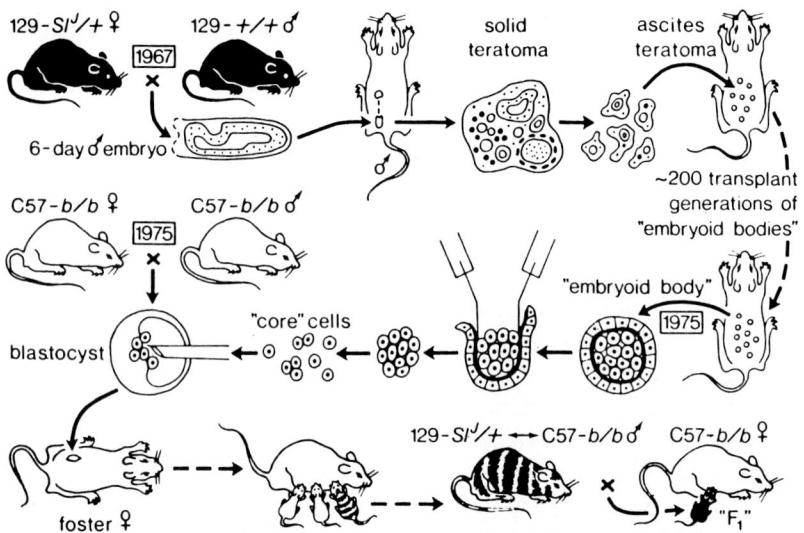

FIG. 85.   Normal genetically mosaic mice produced from malignant teratocarcinoma cells. In 1967, a teratoma was produced from a 6-day (chromosomally male, XY) embryo of the *steel* coat genotype placed under a testis capsule. The primary tumor was minced and transplanted intraperitoneally. It became an ascitis tumor of "embryoid bodies" with teratocarcinoma (embryonal carcinoma) cores. In 1975, the malignant core cells were injected into blastocysts from parents of the non-agouti brown strain. The blastocysts were transferred to the uterus of a pseudopregnant foster mother (mated with a sterile vasectomized male). Normal mice were born; some had coat color mosaicism (striped) and/or internal tissue contributions of the tumor strain. A mosaic male was test mated to C57 b/b females. Production of F_1-like offspring proved that he had normal sperm derived from the teratocarcinoma cells. [Mintz and Illmensee, 1975.]

ing cell lines can be separated and cultured under *in vitro* conditions. A remarkable result obtained by Brinster (1974), Papaioannou *et al.* (1975), and Mintz and Illmensee (1975) is that if a small number (three to five) of malignant carcinoma cells are implanted into a normal mouse blastocyst (blastula), they are cured of their malignancy and can participate normally in embryogenesis (Fig. 85). Undifferentiated teratocarcinoma cells, like eggs, are totipotent and are thus capable, in a suitable environment, of differentiating into all kinds of tissues. More recently, Stewart and Mintz (1981) went one step further. After injecting three to five euploid teratocarcinoma cells into mouse blastocysts, they obtained mosaic, zebra-striped adult mice (Fig. 86). In this case, the genetic marker was *agouti,* a fur color marker. The important new finding is that the totipotent teratocarcinoma cells can give rise to oocytes in adult females. If one crosses them with males of another strain (GC 57), one obtains, in F_1, *agouti* mice that in F_2 give rise to new strains.

Finally, Wagner and Mintz (1982) used teratocarcinoma cells as vectors for gene transfer in mice. Human β-globin genes and *TK* genes of HSV were introduced into undifferentiated totipotent teratocarcinoma cells; the transformed cells were grafted into normal blastocysts. The outcome, again, was the produc-

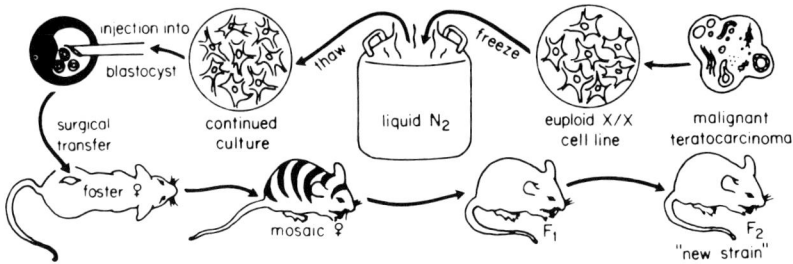

FIG. 86. Successive generations of mice produced from cultured euploid teratocarcinoma cells. Diagrammatic summary, starting at the upper right, of the experiment. The malignant teratocarcinoma was obtained as shown in Fig. 85, and euploid teratocarcinoma stem cells (chromosomally female, XX) were cultured; they were frozen and stored in liquid nitrogen. After thawing, propagation in culture continued. Blastocysts of C57 BL/6 (black) strain were microinjected with cells from the culture, and the embryos were transferred to the uterus of a pseudopregnant foster mother for development to term. One of the two-color (striped) mosaic females that was born had tumor-derived normal cells in addition to embryo-derived cells in her germ line (and other tissues). When she was mated with a C57 male, she produced some F_1 progeny with the predominant *agouti* color of the 129 strain. A subsequent mating between an F_1 female and an F_1 male (both heterozygous for marker genes) yielded F_2 offspring among which were segregants homozygous for the tumor-strain alleles. These animals are wild-type models of a "new strain." [Stewart and Mintz, 1981.]

tion of new strains, but with predetermined genetic changes due to the introduction of foreign genes into the blastocysts.

Thanks to the tremendous progress made in molecular biology with the advent of DNA recombinant technology, and thanks to the intelligent combination of genetics and experimental embryology, a new era has begun in general biology. Since eggs are the only cells that can give rise to adults capable of producing an offspring, and since we can now manipulate, at will, the genetic background of fertilized eggs, interest in the egg cell should intensify in the years to come.

REFERENCES

Adams, R. L. P., Burdon, R. H., Gibb, S., and McKay, E. L. (1981). *Biochim. Biophys. Acta* **655**, 329.

Adamson, E. D., and Woodland, H. R. (1977). *Dev. Biol.* **57**, 136.

Afzelius, B. A. (1957). *Z. Zellforsch. Mitrosk. Anat.* **45**, 660.

Aimar, C., Delarue, M., and Vilain, C. (1981). *J. Embryol. Exp. Morphol.* **64**, 259.

Aimar, C., Vilain, C., and Delarue, M. (1983). *Cell Differ.* **13**, 293.

Aimi, J. (1974). *Experientia* **30**, 837.

Alexandre, H. (1982). *C. R. Hebd. Séances Acad. Sci.* **294**, 1001.

Alexandre, H., De Petrocellis, B., and Brachet, J. (1982). *Differentiation* **32**, 132.

Ancel, P., and Vintemberger, P. (1948). *Biol. Bull. (Woods Hole, Mass.)* **31**, Suppl. 1.

Anderson, D. M., Galau, G. A., Britten, R. J., and Davidson, E. H. (1976). *Dev. Biol.* **51**, 138.

Anderson, D. M., Richter, J. D., Chamberlin, M. E., Price, D. H., Britten, R. J., Smith, L. D., and Davidson, E. H. (1982). *J. Mol. Biol.* **155**, 281.

Anderson, K. V., and Nüsslein-Volhard, C. (1984). *Nature* **311**, 223.

Angell, R. R., Aitken, R. J., Van Look, P. F. A., Lumsdem, M. A., and Templeton, A. A. (1983). *Nature (London)* **303**, 336.

Angerer, L. M., and Angerer, R. C. (1981). *Nucleic Acids Res.* **9**, 2819.
Angerer, R. C., and Davidson, E. (1984). *Science* **226**, 1153.
Arad, G., and Beebee, T. J. C. (1981). *FEBS Lett.* **136**, 247.
Arceci, R. J., and Gross, P. R. (1977). *Proc. Natl. Acad. Sci. U.S.A.* **74**, 5016.
Attardi, D. G., De Paolis, A., and Tocchini-Valentini, G. P. (1981). *J. Biol. Chem.* **256**, 3654.
Bachvarova, R., and De Leon, V. (1977). *Dev. Biol.* **58**, 248.
Badman, W. S., and Brookbank, J. W. (1970). *Dev. Biol.* **21**, 243.
Bakken, A., Morgan, G., Sollner-Webb, B., Roan, J., Busby, S., and Reeder, R. H. (1982). *Proc. Natl. Acad. Sci. U.S.A.* **79**, 56.
Bałakier, H. (1978). *Exp. Cell Res.* **112**, 137.
Bałakier, H., and Czołowska, R. (1977). *Exp. Cell Res.* **110**, 466.
Baldari, C. T., Amaldi, F., and Buonguorno-Nardelli, M. (1978). *Cell (Cambridge, Mass.)* **15**, 1095.
Ballentine, R. (1939). *Biol. Bull (Woods Hole, Mass.)* **77**, 328.
Ballinger, D. G., and Hunt, T. (1981). *Dev. Biol.* **87**, 277.
Baltus, E., and Hanocq-Quertier, J. (1985). *Cell Differentiation* **16**, 161.
Baltus, E., Quertier, J., Ficq, A., and Brachet, J. (1965). *Biochim. Biophys. Acta* **95**, 408.
Baltus, E., Brachet, J., Hanocq-Quertier, J., and Hubert, E. (1973). *Differentiation* **1**, 127.
Baltus, E., Hanocq-Quertier, J., Pays, A., and Brachet, J. (1977). *Proc. Natl. Acad. Sci. U.S.A.* **74**, 3461.
Baltus, E., Hanocq-Quertier, J., and Guyaux, M. (1981). *FEBS Lett.* **123**, 37.
Baltzer, F. (1910). *Arch. Zellforsch.* **5**, 497.
Baltzer, F. (1952). *Symp. Soc. Exp. Zool.* **6**, 230.
Baltzer, F., Harding, C., Lehman, H. E., and Boff, P. (1954). *Rev. Suisse Zool.* **61**, 402.
Barnard, E. A., Miledi, R., and Sumikawa, K. (1982). *Proc. R. Soc. London, Ser. B* **215**, 241.
Barrett, D., and Angelo, G. M. (1969). *Exp. Cell Res.* **57**, 159.
Barrett, P., Johnson, R. M., and Sommerville, J. (1984). *Exp. Cell Res.* **153**, 299.
Barth, L. G. (1946). *J. Exp. Zool.* **103**, 463.
Barth, L. G., and Barth, L. J. (1954). "The Energetics of Development." Columbia Univ. Press, New York.
Barth, L. G., and Jaeger, L. (1947). *Physiol. Zool.* **20**, 133.
Barton, S. C., Surani, M. A. H., and Norris, M. L. (1984). *Nature* **311**, 374.
Bassüner, R., Huth, A., Manteuffel, R., and Rappoport, T. A. (1983). *Eur. J. Biochem.* **133**, 321.
Bataillon, E. (1910). *Arch. Zool. Exp. Géné.* **56**, 101.
Baulieu, E. E., Godeau, F., Schorderet-Slatkine, S., and Schorderet, M. (1978). *Nature (London)* **275**, 593.
Baud, C., and Barisch, M. E. (1984). *Dev. Biol.* **105**, 423.
Baur, R., Wohlert, H., and Kroger, H. (1978). *Hoppe-Seyler's Z. Physiol. Chem.* **359**, 274.
Bédard, P. A., and Brandhorst, B. P. (1983). *Dev. Biol.* **96**, 74.
Begg, D. A., Rebhun, L. I., and Hyatt, H. (1982). *J. Cell Biol.* **93**, 24.
Bellé, R., Boyer, J., and Ozon, R. (1978). *Biol. Cell.* **32**, 97.
Bellé, R., Boyer, J., and Ozon, R. (1982). *Dev. Biol.* **90**, 315.
Bellé, R., Boyer, J., and Ozon, R. (1984). *J. Exp. Zool.* **231**, 131.
Benavente, R., Krohne, G., Stick, R., and Franke, W. W. (1984). *Exp. Cell Res.* **151**, 224.
Benbow, R. M., Pestell, R. Q., and Ford, C. C. (1975). *Dev. Biol.* **43**, 159.
Bender, W., Akam, M., Karch, F., Beachy, P. A., Peifer, M., Spierer, P., Lewis, E. B., and Hogness, D. S. (1983). *Science* **221**, 23.
Benford, H. H., and Namenwirth, M. (1974). *Dev. Biol.* **39**, 172.
Bensaude, O., Babinet, C., Morange, M., and Jacob, F. (1983). *Nature (London)* **305**, 331.
Berg, W. E. (1956). *Biol. Bull. (Woods Hole, Mass.)* **110**, 1.
Berridge, M. J. (1984). *Biochem. J.* **220**, 345.

Bibring, T., and Baxandall, J. (1981). *Dev. Biol.* **83,** 122.

Bienz, M. (1984). *Proc. Natl. Acad. U.S.A.* **81,** 3138.

Bienz, M. (1985). *Trends Biochem. Sci.* **10,** 157.

Bienz, M., and Gurdon, J. B. (1982). *Cell (Cambridge, Mass.)* **29,** 811.

Bienz, M., Kubli, E., Kohli, J., de Henau, S., and Grosjean, H. (1980). *Nucleic Acids Res.* **8,** 5169.

Bienz, M., Kubli, E., Kohli, J., de Henau, S., Huez, G., Marbaix, G., and Grosjean, H. (1981). *Nucleic Acids Res.* **9,** 3835.

Billett, F. S., and Adam, E. (1976). *J. Embryol. Exp. Morphol.* **36,** 697.

Birchmeier, C., Grosschedl, R., and Birnstiel, M. L. (1982). *Cell (Cambridge, Mass.)* **28,** 739.

Bird, A. P. (1977). *Cold Spring Harbor Symp. Quant. Biol.* **42,** 1179.

Bird, A. P., and Southern, E. M. (1980). *J. Mol. Biol.* **118,** 27.

Bird, A. P., Taggart, M. H., and Smith, B. A. (1979). *Cell (Cambridge, Mass.)* **17,** 889.

Bird, A. P., Taggart, M., and MacLeod, D. (1981). *Cell (Cambridge, Mass.)* **26,** 381.

Birnstiel, M., Speirs, J., Purdom, I., Jones, K., and Loening, U. E. (1969). *Nature (London)* **219,** 454.

Blondeau, J. P., and Baulieu, E. E. (1984). *Biochem. J.* **219,** 785.

Blondeau, J. P., and Baulieu, E. E. (1985). *J. Biol. Chem.* **260,** 3617.

Blumenthal, A. B., Kriegstein, H. J., and Hogness, D. S. (1973). *Cold Spring Harbor Symp. Quant. Biol.* **38,** 205.

Bogenhagen, D. F., Wormington, W. M., and Brown, D. D. (1982). *Cell (Cambridge, Mass.)* **28,** 413.

Boldt, J., Schuel, H., Schuel, R., Dandekar, P. V., and Troll, W. (1981). *Gamete Res.* **4,** 365.

Boloukhère, M. (1984). *J. Cell Sci.* **65,** 73.

Bona, M., Scheer, U., and Bautz, E. K. F. (1981). *J. Mol. Biol.* **151,** 81.

Bonner, W. M. (1975a). *J. Cell Biol.* **64,** 421.

Bonner, W. M. (1975b). *J. Cell Biol.* **64,** 431.

Boucaut, J. C., Darribère, T., Boulekbache, H., and Thiéry, J. P. (1984a). *Nature* **307,** 364.

Boucaut, J. C., Darribère, T., Poole, J. J., Aoyama, H., Yamada, K. M., and Thiéry, J. P. (1984b). *J. Cell Biol.* **99,** 1822.

Bower, D. J., Errington, L. H., Cooper, D. N., Morris, S., and Clayton, R. M. (1983). *Nucleic Acids Res.* **9,** 2513.

Bozzoni, I., Beccari, E., Luo, Z. X., Amaldi, F., Pierandrei-Amaldi, P., and Campioni, N. (1981). *Nucleic Acids Res.* **9,** 1069.

Brachet, A. (1904). *Arch. Biol.* **21,** 103.

Brachet, A. (1910). *Arch. Zool. Exp. Gen. [5]* **6,** 1.

Brachet, A. (1922). *Arch. Biol.* **32,** 205.

Brachet, A. (1930). "L'oeuf et les facteurs de l'ontogénèse." Doin, Paris.

Brachet, J. (1933). *Arch. Biol.* **44,** 519.

Brachet, J. (1934). *Arch. Biol.* **45,** 611.

Brachet, J. (1937). *Arch. Biol.* **48,** 561.

Brachet, J. (1938). *Biol. Bull. (Woods Hole, Mass.)* **74,** 93.

Brachet, J. (1939). *Arch. Exp. Zellforsch. Besonders Gewebezuecht.* **22,** 541.

Brachet, J. (1940). *Arch. Biol.* **51,** 151.

Brachet, J. (1942). *Arch. Biol.* **53,** 207.

Brachet, J. (1944). *Ann. Soc. R. Zool. Belg.* **75,** 49.

Brachet, J. (1948). *Experientia* **4,** 353.

Brachet, J. (1949). *Bull. Soc. Chim. Biol.* **31,** 724.

Brachet, J. (1950). *Experientia* **6,** 56.

Brachet, J. (1954). *Arch. Biol.* **65,** 1.

Brachet, J. (1957). "Biochemical Cytology." Academic Press, New York.

Brachet, J. (1965a). *C. R. Hebd. Seances Acad. Sci.* **261,** 1092.

Brachet, J. (1965b). *Nature (London)* **208,** 596.
Brachet, J. (1967). *Nature (London)* **214,** 1132.
Brachet, J. (1968). *Curr. Mod. Biol.* **1,** 314.
Brachet, J. (1974). "Introduction to Molecular Embryology." Springer-Verlag, Berlin and New York.
Brachet, J. (1976). *In* "Progress in Differentiation Research" (M. Müller-Bérat, ed.), p. 422. North-Holland Publ., Amsterdam.
Brachet, J. (1977). *Curr. Top. Dev. Biol.* **11,** 133.
Brachet, J., and Aimi, J. (1972). *Exp. Cell Res.* **72,** 46.
Brachet, J., and De Petrocellis, B. (1981). *Exp. Cell Res.* **135,** 179.
Brachet, J., and Donini-Denis, S. (1978). *Differentiation* **11,** 19.
Brachet, J., and Hubert, E. (1972). *J. Embryol. Exp. Morphol.* **27,** 121.
Brachet, J., and Ledoux, L. (1955). *Exp. Cell Res.* **3,** 27.
Brachet, J., Bieliavsky, N., and Tencer, R. (1962). *Bull. Cl. Sci., Acad. R. Belg.* **48,** 255.
Brachet, J., Ficq, A., and Tencer, R. (1963). *Exp. Cell Res.* **32,** 168.
Brachet, J., Hanocq, F., and Van Gansen, P. (1970). *Dev. Biol.* **21,** 157.
Brachet, J., O'Dell, D., Steinert, G., and Tencer, R. (1972). *Exp. Cell Res.* **73,** 463.
Brachet, J., Baltus, E., De Schutter, A., Hanocq, F., Hanocq-Quertier, J., Hubert, E., Iacobelli, S., and Steinert, G. (1974). *Mol. Cell Biochem.* **3,** 189.
Brachet, J., Pays-De Schutter, A., and Hubert, E. (1975a). *Differentiation* **3,** 3.
Brachet, J., Baltus, E., De Schutter-Pays, A., Hanocq-Quertier, J., Hubert, E., and Steinert, G. (1975b). *Proc. Natl. Acad. Sci. U.S.A.* **72,** 1574.
Brachet, J., Baltus, E., and Hanocq-Quertier, J. (1976). *C. R. Hebd. Séances Acad. Sci.* **283,** 263.
Brachet, J., Mamont, P., Boloukhère, M., Baltus, E., and Hanocq-Quertier, J. (1978). *C. R. Hebd. Séances Acad. Sci.* **287,** 1289.
Brachet, J., De Petrocellis, B., and Alexandre, H. (1981). *Differentiation* **19,** 47.
Brandhorst, B. P., and Hewrock, K. M. (1981). *Dev. Biol.* **83,** 250.
Brande, P., Pelhma, H., Flach, G., and Lobattto, R. (1979). *Nature (London)* **282,** 102.
Bravo, R., Otero, C., Allende, C. C., and Allende, J. E. (1978). *Proc. Natl. Acad. Sci. U.S.A.* **75,** 1242.
Breindl, M., Harbers, K., and Jaenisch, R. (1984). *Cell* **38,** 9.
Bresch, H. (1978). *Exp. Cell Res.* **111,** 205.
Briggs, R. (1972). *J. Exp. Zool.* **181,** 271.
Briggs, R. (1979). *Int. Rev. Cytol., Suppl.* **9,** 107.
Briggs, R., and Cassens, G. (1966). *Proc. Natl. Acad. Sci. U.S.A.* **55,** 1103.
Briggs, R., and King, T. J. (1952). *Proc. Natl. Acad. Sci. U.S.A.* **38,** 455.
Briggs, R., and King, T. J. (1953). *J. Exp. Zool.* **122,** 485.
Briggs, R., and King, T. J. (1955). *In* "Biological Specificity and Growth" (E. Butler, ed.), p. 207. Princeton Univ. Press, Princeton, New Jersey.
Briggs, R., Green, E. U., and King, T. J. (1951). *J. Exp. Zool.* **116,** 455.
Brinster, R. L. (1974) *J. Exp. Med.* **140,** 1049.
Brinster, R. L., and Palmiter, R. D. (1982). *Trends Biochem. Sci.* **7,** 438.
Brinster, R. L., Chen, H. Y., and Trumbauer, M. E. (1981). *Science* **211,** 396.
Brinster, R. L., Ritchie, K. A., Hammer, R. E., O'Brien, R. L., Arp, B., and Storb, U. (1983). *Nature (London)* **306,** 332.
Brothers, A. J. (1976). *Nature (London)* **260,** 112.
Brown, D. D. (1981). *Science* **211,** 667.
Brown, D. D. (1984). *Cell* **37,** 359.
Brown, D. D., and Blackler, A. W. (1972). *J. Mol. Biol.* **63,** 75.
Brown, D. D., and Dawid, I. B. (1968). *Science* **160,** 272.
Brown, D. D., and Gurdon, J. B. (1964). *Proc. Natl. Acad. Sci. U.S.A.* **51,** 139.

Brown, D. D., and Gurdon, J. B. (1977). *Proc. Natl. Acad. Sci. U.S.A.* **74,** 2064.

Brown, D. D., Carroll, D., and Brown, R. D. (1977). *Cell (Cambridge, Mass.)* **12,** 1045.

Brown, P. M., and Kalthoff, K. (1983). *Dev. Biol.* **97,** 113.

Brown, R. D., and Brown, D. D. (1976). *J. Mol. Biol.* **102,** 1.

Bryan, P. N., Olah, J., and Birnstiel, M. L. (1983). *Cell (Cambridge, Mass.)* **33,** 843.

Buongiorno-Nardelli, M., Micheli, G., Carri, M. T., and Marilley, M. (1982). *Nature (London)* **298,** 100.

Burch, J. B. E., and Weintraub, H. (1983). *Cell (Cambridge, Mass.)* **33,** 65.

Burdon, R. H., and Adams, R. L. P. (1980). *Trends Biochem. Sci.* **5,** 294.

Busa, W. B., and Nuccitelli, R. (1985). *Science* **228,** 1325.

Busby, S., and Reeder, R. H. (1981). *J. Cell Biol.* **91,** 128a.

Busby, S. J., and Reeder, R. H. (1982). *Dev. Biol.* **91,** 458.

Busby, S. J., and Reeder, R. H. (1983). *Cell (Cambridge, Mass.)* **34,** 989.

Cabrera, C. V., Lee, J. J., Ellison, J. W., Britten, R. J., and Davidson, E. H. (1984). *J. Mol. Biol.* **174,** 85.

Callan, H. G. (1963). *Int. Rev. Cytol.* **15,** 1.

Callan, H. G. (1972). *Proc. R. Soc. London, Ser. B* **181,** 19.

Callan, H. G. (1982). *Proc. R. Soc. London, Ser. B* **214,** 417.

Callen, J. C., and Mounolou, J. C. (1978). *Biol. Cell* **33,** 5.

Callen, J. C., Dennebouy, N., and Mounolou, J. C. (1980). *J. Cell Sci.* **41,** 307.

Cameron, I. L., Hunter, K. E., and Cragoe, E. J., Jr. (1982). *Exp. Cell Res.* **139,** 455.

Capco, D. G. (1982). *J. Exp. Zool.* **219,** 147.

Capco, D. G., and Jäckle, H. (1982). *Dev. Biol.* **94,** 41.

Capco, D. G., and Jeffery, W. R. (1981). *Nature (London)* **294,** 255.

Capco, D. G., and Jeffery, W. R. (1982). *Dev. Biol.* **89,** 1.

Carpenter, C. D., Bruskin, A. M., Hardin, P. E., Keast, M. J., Anstrom, J., Tyner, A. L., Brandhorst, B. P., and Klein, W. H. (1984). *Cell (Cambridge, Mass.)* **36,** 663.

Carrasco, A. E., McGinnis, W., Gehring, W. J., and De Robertis, E. M. (1984). *Cell* **37,** 309.

Carroll, A. G., and Ozaki, H. (1979). *Exp. Cell Res.* **119,** 307.

Carroll, D. (1983). *Proc. Natl. Acad. Sci. U.S.A.* **80,** 6902.

Carroll, D., and Brown, D. D. (1976). *Cell (Cambridge, Mass.)* **7,** 467.

Carson, D. D., and Lennarz, W. J. (1981). *J. Biol. Chem.* **256,** 4679.

Cartaud, A., Huchon, D., Marot, J., Ozon, R., and Demaille, J. G. (1981). *Cell Differ.* **10,** 357.

Cascio, S. M., and Wassarman, P. M. (1982). *Dev. Biol.* **89,** 397.

Chamberlain, J. T., and Metz, C. B. (1972). *J. Mol. Biol.* **64,** 593.

Chandler, D. E., and Heuser, J. (1979). *J. Cell Biol.* **83,** 94, 104, 106.

Chen, P. S. (1953). *Exp. Cell Res.* **5,** 275.

Chen, P. S. (1954). *Experientia* **10,** 212.

Childs, G., Maxson, R. E., and Kedes, L. H. (1979). *Dev. Biol.* **73,** 153.

Cho, W. K., Stern, S., and Biggers, J. D. (1974). *J. Exp. Zool.* **187,** 383.

Ciampi, M. S., Melton, D. A., and Cortese, R. (1982). *Proc. Natl. Acad. Sci. U.S.A.* **79,** 1388.

Cicirelli, M. F., and Smith, L. D. (1985). *Dev. Biol.* **107,** 254.

Cicirelli, M. F., Robinson, K. R., and Smith, L. D. (1983). *Dev. Biol.* **100,** 133.

Ciliberto, G., Castagnoli, L., Melton, D. A., and Cortese, R. (1982). *Proc. Natl. Acad. Sci. U.S.A.* **79,** 1195.

Clark, T. G., and Merriam, R. W. (1977). *Cell (Cambridge, Mass.)* **12,** 883.

Clark, T. G., and Rosenbaum, J. C. (1979). *Cell (Cambridge, Mass.)* **18,** 1101.

Clement, A. C. (1952). *J. Exp. Zool.* **121,** 593.

Clement, A. C., and Tyler, A. (1967). *Science* **158,** 1457.

Coburn, M., Schuel, H., and Troll, W. (1981). *Dev. Biol.* **84,** 235.

Cognetti, G., and Shaw, B. R. (1981). *Nucleic Acids Res.* **9,** 5609.

Collier, J. R. (1961). *Acta Embryol. Morphol. Exp.* **4,** 70.

Collier, J. R. (1975). *Exp. Cell Res.* **95,** 254.

Collier, J. R. (1977). *Exp. Cell Res.* **106,** 390.

Collier, J. R., and McCarthy, M. E. (1981). *Differentiation* **19,** 31.

Colman, A. (1982). *Trends Biochem. Sci.* **7,** 435.

Colman, A., Lane, C. D., Craig, R., Boulton, A., Mohun, T., and Morser, J. (1981a). *Eur. J. Biochem.* **113,** 339.

Colman, A., Morser, J., Lane, C., Besley, J., Wylie, C., and Valle, G. (1981b). *J. Cell Biol.* **91,** 770.

Colman, A., Besley, J., and Valle, G. (1982). *J. Mol. Biol.* **160,** 459.

Colombo, R., Benedusi, P., and Valle, G. (1981). *Differentiation* **20,** 45.

Conklin, E. G. (1905). *J. Acad. Nat. Sci. Philadelphia* **13,** 1.

Costantinides, P. G., Jones, P. A., and Gevers, W. (1977). *Nature (London)* **67,** 364.

Costantini, F., and Lacy, E. (1981). *Nature (London)* **294,** 92.

Counce, S. J. (1956). *Z. Indukt. Abstamm. Vererbungsl.* **87,** 443, 462, 482.

Crain, W. R., Jr., and Bushman, F. D. (1983). *Dev. Biol.* **100,** 190.

Crain, W. R., Jr., Durica, D. S., and Van Doren, K. (1981). *Mol. Cell Biol.* **1,** 711.

Crampton, J. M., and Woodland, H. R. (1979). *Dev. Biol.* **70,** 453.

Crowther, R. J., and Whittaker, J. R. (1985). *Wilhelm Roux's Arch. Dev. Biol.* **194,** 87.

Cutler, D., Lane, C., and Colman, A. (1981). *J. Mol. Biol.* **153,** 917.

Dabauvalle, M. C., and Franke, W. W. (1984). *Exp. Cell Res.* **153,** 308.

Dalcq, A. (1928). "Les bases physiologiques de la maturation." Presses Universitaires de France, Paris.

Dalcq, A., and Simon, S. (1932). *Protoplasma* **14,** 497.

Dalcq, A. (1957). "Introduction to General Embryology." Oxford Univ. Press, London and New York.

Dale, B., de Santis, A., and Hoshi, M. (1979). *Nature (London)* **282,** 87.

Danilchik, M. V., and Hille, M. B. (1981). *Dev. Biol.* **84,** 291.

Dasgupta, J. D., and Garbers, D. L. (1983). *J. Biol. Chem.* **258,** 6174.

Davidson, E. H. (1976). "Gene Activity in Early Development," 2nd ed. Academic Press, New York.

Davidson, E. H., and Britten, R. J. (1979). *Science* **204,** 1052.

Davidson, E. H., and Hough, B. R. (1971). *J. Mol. Biol.* **56,** 491.

Davidson, E. H., and Posakony, J. W. (1982). *Nature (London)* **297,** 633.

Davidson, E. H., Hough-Evans, B. R., and Britten, R. J. (1982). *Science* **217,** 17.

Dawid, I. B. (1966). *Proc. Natl. Acad. Sci. U.S.A.* **56,** 269.

Dawid, I. B., Brown, D. D., and Reeder, R. H. (1970). *J. Mol. Biol.* **51,** 341.

Delarue, M., and Aimar, C. (1984). *Dev., Growth Differ.* **26,** 1.

Delarue, M., Darribere, T., Aimar, C., and Boucaut, J. C. (1985). *Wilhelm Roux's Arch. Dev. Biol.* **194,** 275.

De Leon, D. V., Cox, K. H., Angerer, L. M., and Angerer, R. C. (1983). *Dev. Biol.* **100,** 197.

Denis, H., and Brachet, J. (1969a). *Proc. Natl. Acad. Sci. U.S.A.* **62,** 194.

Denis, H., and Brachet, J. (1969b). *Proc. Natl. Acad. Sci. U.S.A.* **62,** 438.

Denis, H., and Mairy, M. (1972). *Eur. J. Biochem.* **25,** 524.

Denis, H., and Wegnez, M. (1977). *Dev. Biol.* **58,** 212.

Denny, P., and Tyler, A. (1964). *Biochem. Biophys. Res. Commun.* **14,** 245.

Deno, T., and Satoh, N. (1984). *Dev., Growth Differ.* **26,** 43.

De Petrocellis, B., and Rossi, M. (1976). *Dev. Biol.* **48,** 250.

De Robertis, E. M. (1983). *Cell (Cambridge, Mass.)* **32,** 1021.

De Robertis, E. M., and Gurdon, J. B. (1977). *Proc. Natl. Acad. Sci. U.S.A.* **74,** 2470.

De Robertis, E. M., and Mertz, J. E. (1977). *Cell (Cambridge, Mass.)* **12,** 175.

De Robertis, E. M., and Olson, M. V. (1979). *Nature (London)* **278**, 137.
De Robertis, E. M., Partington, G. A., Longthorne, R. F., and Gurdon, J. B. (1977). *J. Embryol. Exp. Morphol.* **40**, 199.
De Robertis, E. M., Longthorne, R. F., and Gurdon, J. B. (1978). *Nature (London)* **272**, 254.
De Robertis, E. M., Black, P., and Nishikura, K. (1981). *Cell (Cambridge, Mass.)* **23**, 89.
De Robertis, E. M., Lienhard, S., and Parisot, R. F. (1982). *Nature (London)* **295**, 572.
Dettlaff, T. A. (1966). *J. Embryol. Exp. Morphol.* **16**, 183.
Di Berardino, M. A. (1979). *Int. Rev. Cytol., Suppl.* **9**, 129.
Di Berardino, M. A. (1980). *Differentiation* **17**, 17.
Di Berardino, M. A., and Hoffner, N. (1970). *Dev. Biol.* **23**, 185.
Di Berardino, M. A., Hoffner, N. J., and Etkin, L. D. (1984). *Science* **224**, 946.
Dickinson, D. G., and Baker, R. F. (1978). *Proc. Natl. Acad. Sci. U.S.A.* **75**, 5627.
Dickinson, D. G., and Baker, R. F. (1979). *Science* **205**, 816.
Dictus, W. J. A. G., Van Zoelen, E. J. J., Tetteroo, P. A. T., Tertoolen, L. G. J., De Laat, S. W., and Bluemink, J. G. (1984). *Dev. Biol.* **101**, 201.
Dingwall, C., Sharnick, S. V., and Laskey, R. A. (1982). *Cell (Cambridge, Mass.)* **30**, 449.
Dingwall, G. (1985). *Trends Biochem. Sci.* **10**, 64.
Dorée, M., Kishimoto, T., Le Peuch, C. J., Demaille, J. C., and Kanatani, H. (1981). *Exp. Cell Res.* **135**, 237.
Dorée, M., Picard, A., Cavadore, J. C., Le Peuch, C., and Demaille, J. G. (1982). *Exp. Cell Res.* **139**, 135.
Dreyer, C., and Hausen, P. (1983). *Dev. Biol.* **100**, 412.
Driesch, H. (1891). *Z. Wiss. Zool.* **53**, 160.
Duncan, R., and Humphreys, T. (1983). *Dev. Biol.* **96**, 258.
Duncan, R., and McConkey, E. H. (1984). *Exp. Cell Res.* **152**, 520.
Dura, J. M. (1981). *Mol. Gen. Genet.* **184**, 381.
Earnshaw, W. C., Rekvig, O. P., and Hannestad, K. (1982). *J. Cell Biol.* **92**, 871.
Easton, D. P., and Whiteley, A. H. (1979). *Differentiation* **12**, 127.
Ebert, K. M., and Brinster, R. L. (1983). *J. Embryol. Exp. Morphol.* **74**, 159.
Eckberg, W. R. (1981). *Differentiation* **19**, 55.
Eckberg, W. R., and Carroll, A. G. (1982). *Cell Differ.* **11**, 155.
Ede, W. (1956). *Wilhelm Roux' Arch. Entwicklungsmech. Org.* **148**, 416, 437.
Ehrenfeld, E. (1982). *Cell (Cambridge, Mass.)* **28**, 435.
Eisen, A., Kiehart, D. P., Wieland, S. J., and Reynolds, G. T. (1984). *J. Cell Biol.* **99**, 1647.
Engelke, D. R., Ng., S.-Y., Shastry, B. S., and Roeder, R. G. (1980). *Cell (Cambridge, Mass.)* **19**, 717.
Epel, D. (1964). *Biochem. Biophys. Res. Commun.* **17**, 62.
Epel, D. (1977). *Sci. Am.* **237**, 131.
Epel, D. (1978). *Curr. Top. Dev. Biol.* **12**, 185.
Epel, D., Patton, C., Wallace, R. W., and Cheung, W. Y. (1981). *Cell (Cambridge, Mass.)* **23**, 543.
Eppig, J. J., and Downs, S. M. (1984). *Biol. Reprod.* **30**, 1.
Epstein, C. J., Smith, S. A., Zamora, T., Sawicki, J. A., Magnuson, T. R., and Cox, D. R. (1982). *Proc. Natl. Acad. Sci. U.S.A.* **79**, 4376.
Ernst, S. G., Hough-Evans, B. R., Britten, R. J., and Davidson, E. H. (1980). *Dev. Biol.* **79**, 119.
Etkin, L. D. (1982). *Differentiation* **21**, 149.
Evans, T., Rosenthal, E. T., Youngblom, J., Distel, D., and Hunt, T. (1983). *Cell (Cambridge, Mass.)* **33**, 389.
Eysen, A., and Reynolds, G. T. (1985). *J. Cell Biol.* **100**, 1522.
Fankhauser, G. (1934). *J. Exp. Zool.* **67**, 349.
Fankhauser, G. (1945). *Q. Rev. Biol.* **20**, 20.

Fankhauser, G. (1952). *Int. Rev. Cytol.* **1**, 165.

Fankhauser, G. (1955). *In* "Analysis of Development" (B. H. Willier, P. A. Weiss, and V. Hamburger, eds.), p. 126. Saunders, Philadelphia, Pennsylvania.

Fansler, B., and Loeb, L. A. (1969). *Exp. Cell Res.* **57**, 305.

Fehilly, C. B., Willadsen, S. M., and Tucker, E. M. (1984). *Nature (London)* **307**, 634.

Feldherr, C. M., and Ogburn, J. A. (1980) *J. Cell Biol.* **87**, 589.

Feldherr, C. M., Cohen, R. J., and Ogburn, J. A. (1983). *J. Cell Biol.* **96**, 1486.

Feldherr, C. M., Kallenbach, E., and Schultz, N. (1985). *J. Cell Biol.* **99**, 2216.

Ficq, A. (1970). *Exp. Cell Res.* **63**, 453.

Filosa, S., Andreuccetti, P., Parisi, E., and Monroy, A. (1985). *Dev., Growth Differ.* **27**, 29.

Finidori-Lepicard, J., Schorderet-Slatkine, S., Hanoune, J., and Baulieu, E. E. (1981). *Nature (London)* **292**, 255.

Fink, R. D., and McClay, D. R. (1985). *Dev. Biol.* **107**, 66.

Fischberg, M. (1958). *J. Embryol. Exp. Morphol.* **6**, 393.

Fisher, G. W., and Rebhun, L. I. (1983). *Dev. Biol.* **99**, 456.

Fjose, A., McGinnis, W. J., and Gehring, W. J. (1985). *Nature (London)* **313**, 284.

Flemming, W. (1882). "Zellsubstanz, Kern und Zelltheilung," p. 424. Vogel, Leipzig.

Flynn, J. M., and Woodland, H. R. (1980). *Dev. Biol.* **75**, 222.

Foerder, C. A., Klebanoff, S. J., and Shapiro, B. M. (1978). *Proc. Natl. Acad. Sci. U.S.A.* **75**, 3183.

Folger, K., Anderson, J. N., Hayward, M. A., and Shapiro, D. J. (1983). *J. Biol. Chem.* **258**, 8908.

Forbes, D. J., Kirschner, M. W., and Newport, J. W. (1983a). *Cell (Cambridge, Mass.)* **34**, 13.

Forbes, D. J., Kornberg, T. B., and Kirschner, M. W. (1983b). *J. Cell Biol.* **97**, 62.

Forbes, D. J., Kirschner, M. W., Caput, D., Dahlberg, J. E., and Lund, E. (1984). *Cell (Cambridge, Mass.)* **38**, 681.

Ford, C. C., Wall, V. M., and Smith, C. (1983). *Cell Biol. Int. Rep.* **7**, 545.

Ford, P. J., Mathieson, T., and Rosbach, M. (1977). *Dev. Biol.* **77**, 417.

Forer, A., Masui, Y., and Zimmerman, A. M. (1977). *Exp. Cell Res.* **106**, 430.

Franke, W. W., Rathke, P. C., Seib, E., Trendelenburg, M. F., Osborn, M., and Weber, K. (1976). *Cytobiologie* **14**, 111.

Franz, J. K., Gall, L., Williams, M. A., Picheral, B., and Franke, W. W. (1983). *Proc. Natl. Acad. Sci. U.S.A.* **80**, 6254.

Freeman, G., and Lundelius, J. W. (1982). *Wilhelm Roux's Arch. Dev. Biol.* **191**, 69.

Fregien, N., Dolecki, G. J., Mandel, M., and Humphreys, T. (1983). *Mol. Cell. Biol.* **3**, 1021.

Frischauf, A. M. (1985). *Trends Genet.* **1**, 100.

Fritz, A., Parisot, R., Newmeyer, D., and De Robertis, E. M. (1984). *J. Mol. Biol.* **178**, 273.

Fruscoloni, P., Al-Atia, G. R., and Jacobs-Lorena, M. (1983). *Proc. Natl. Acad. Sci. U.S.A.* **80**, 3359.

Galau, G. A., Klein, W. H., Davis, M. M., Wold, B. J., Britten, R. J., and Davidson, E. H. (1976). *Cell (Cambridge, Mass.)* **7**, 487.

Gall, J. G. (1956). *J. Biophys. Biochem. Cytol., Suppl.* **2**, 393.

Gall, J. G. (1968). *Proc. Natl. Acad. Sci. U.S.A.* **60**, 553.

Gall, J. G., and Callan, H. G. (1962). *Proc. Natl. Acad. Sci. U.S.A.* **48**, 562.

Gall, J. G., and Pardue, M. L. (1969). *Proc. Natl. Acad. Sci. U.S.A.* **63**, 378.

Gall, L., Picheral, B., and Gounon, P. (1983). *Biol. Cell.* **47**, 331.

Galli, G., Hofstetter, H., Stunnenberg, H. G., and Birnstiel, M. L. (1983). *Cell (Cambridge, Mass.)* **34**, 823.

García-Bellido, A., Moscoso del Prado, J., and Botas, J. (1983). *Mol. Gen. Genet.* **192**, 253.

Garen, A., and Gehring, W. (1972). *Proc. Natl. Acad. Sci. U.S.A.* **69**, 2982.

Gargiulo, G., Razvi, F., and Worcel, A. (1984). *Cell (Cambridge, Mass.)* **38**, 511.

Gautier, J., and Beetschen, J. C. (1983a). *Wilhelm Roux's Arch. Dev. Biol.* **192**, 196.
Gautier, J., and Beetschen, J. C. (1983b). *C.R. Hebd. Séances Acad. Sci.* **296**, 815.
Gelfand, R. A., and Smith, L. D. (1983). *Dev. Biol.* **99**, 427.
Gerber-Huber, S., May, F. E. B., Westley, B. R., Felber, B. K., Hosbach, H. A., Andres, A. C., and Ryffel, G. U. (1983). *Cell (Cambridge, Mass.)* **33**, 43.
Gerhart, J., Ubbels, G., Black, S., Hara, K., and Kirschner, M. (1981). *Nature (London)* **292**, 511.
Gerhart, J., Wu, M., and Kirschner, M. (1984). *J. Cell Biol.* **98**, 1247.
Geuskens, M. (1965). *Exp. Cell Res.* **39**, 413.
Geyer-Duszyńska, I. (1959). *J. Exp. Zool.* **141**, 391.
Gillies, R. J., Rosenberg, M. D., and Deamer, D. W. (1981). *J. Cell Physiol.* **108**, 115.
Giudice, G. (1973). "Developmental Biology of the Sea Urchin Embryo." Academic Press, New York.
Giudice, G., and Mutolo, V. (1970). *Adv. Morphog.* **8**, 115.
Giudice, G., Sconzo, G., Albanese, I., Ortolani, G., and Cammarata, M. (1974). *Cell Differ.* **3**, 287.
Godeau, F., Schorderet-Slatkine, S., and Baulieu, E. E. (1978). *Biol. Cell* **32**, 83.
Godsave, S. F., Wylie, C. C., Lane, E. R., and Anderton, B. H. (1984a). *J. Embryol. Exp. Morphol.* **83**, 157.
Godsave, S. F., Anderton, B. H., Heasman, J., and Wylie, G. C. (1984b). *J. Embryol. Exp. Morphol.* **83**, 169.
Goldberg, M. L., Paro, R., and Gehring, W. J. (1982). *EMBO J.* **1**, 93.
Golden, L., Schafer, U., and Rosbach, M. (1980). *Cell (Cambridge, Mass.)* **22**, 835.
Goodhart, M., Ferry, N., Buscaglia, M., Baulieu, E. E., and Hanoune, J. (1984). *EMBO J.* **3**, 2653.
Gordon, J. W., and Ruddle, F. H. (1981). *Science* **214**, 1244.
Gordon, J. W., and Ruddle, F. H. (1985). *Gene* **33**, 121.
Gordon, J. W., Scangos, G. A., Plotkin, D. J., Barbosa, J. A., and Ruddle, F. H. (1980). *Proc. Natl. Acad. Sci. U.S.A.* **77**, 7380.
Gottesfeld, J., and Bloomer, L. S. (1982). *Cell (Cambridge, Mass.)* **28**, 781.
Gounon, P., and Karsenti, E. (1981). *J. Cell Biol.* **88**, 410.
Grainger, J. L., Winkler, M. M., Shen, S. S., and Steinhardt, R. A. (1979). *Dev. Biol.* **68**, 396.
Grainger, R. M., Hazard-Leonards, R. M., Samaha, F., Hougan, L. M., Lesk, M. R., and Thomsen, G. H. (1983). *Nature (London)* **306**, 88.
Grant, P., and Youngdahl, P. (1974). *J. Exp. Zool.* **190**, 289.
Green, M. R., Maniatis, T., and Melton, D. A. (1983). *Cell (Cambridge, Mass.)* **32**, 681.
Gregg, J. R. (1948). *J. Exp. Zool.* **109**, 119.
Gregg, J. R., and Løvtrup, S. (1955). *Biol. Bull. (Woods Hole, Mass.)* **108**, 29.
Grey, R. D., Bastiani, M. J., Webb, D. J., and Schertel, E. R. (1982). *Dev. Biol.* **89**, 475.
Griffith, J. K., Griffith, B. B., and Humphreys, T. (1981). *Dev. Biol.* **87**, 220.
Grippo, P., and Lo Scavo, A. (1972). *Biochem. Biophys. Res. Commun.* **48**, 280.
Grippo, P., Locorotondo, G., and Taddei, C. (1977). *J. Exp. Zool.* **200**, 143.
Gross, K. W., Jacobs-Lorena, M., Baglioni, C., and Gross, P. R. (1973). *Proc. Natl. Acad. Sci. U.S.A.* **70**, 2614.
Gross, P., and Cousineau, G. (1964). *Exp. Cell Res.* **33**, 368.
Grosschedl, R., and Birnstiel, M. L. (1980). *Proc. Natl. Acad. Sci. U.S.A.* **77**, 7102.
Grosschedl, R., Wasylyk, B., Chambon, P., and Birnstiel, M. L. (1981). *Nature (London)* **294**, 178.
Grosschedl, R., Weaver, D., Baltimore, D., and Constantini, F. (1984). *Cell* **38**, 647.
Guerrier, P., Meijer, L., Moreau, M., and Longo, F. J. (1983). *J. Exp. Zool.* **226**, 303.
Gundersen, C. B., Miledi, R., and Parker, I. (1984). *Nature (London)* **308**, 421.
Gurdon, J. B. (1968). *J. Embryol. Exp. Morphol.* **20**, 401.
Gurdon, J. B. (1970). *Proc. R. Soc. London, Ser. B* **176**, 303.

Gurdon, J. B. (1974). "The Control of Gene Expression in Animal Development." Harvard Univ. Press, Cambridge, Massachusetts.

Gurdon, J. B. (1977). *Proc. R. Soc. London, Ser. B* **198,** 211.

Gurdon, J. B., and Laskey, R. A. (1970). *J. Embryol. Exp. Morphol.* **24,** 227.

Gurdon, J. B., and Woodland, H. R. (1970). *Curr. Top. Dev. Biol.* **5,** 39.

Gurdon, J. B., Elsdale, T. R., and Fischberg, M. (1958). *Nature (London)* **182,** 64.

Gurdon, J. B., Lane, C. D., Woodland, H. R., and Marbaix, G. (1971). *Nature (London)* **233,** 177.

Gurdon, J. B., Lingrel, J. B., and Marbaix, G. (1973). *J. Mol. Biol.* **51,** 539.

Gurdon, J. B., Laskey, R. A., and Reeves, O. R. (1975). *J. Embryol. Exp. Morphol.* **34,** 93.

Gurdon, J. B., De Robertis, E. M., and Partington, G. (1976). *Nature (London)* **260,** 116.

Gurdon, J. B., Brennan, S., Fairman, S., and Mohun, T. J. (1984). *Cell* **38,** 691.

Gurdon, J. B., Muhun, T. J., Fairman, S., and Brennan, S. (1985). *Proc. Natl. Acad. Sci. U.S.A.* **82,** 139.

Gustafson, T., and Wolpert, L. (1967). *Biol. Rev. Cambridge Philos. Soc.* **42,** 442.

Gutzeit, H. O., and Gehring, W. J. (1979). *Wilhelm Roux's Arch Dev. Biol.* **187,** 151.

Hadorn, E. (1932). *Wilhelm Roux' Arch. Entwicklungsmech. Org.* **125,** 496.

Hadorn, E. (1934). *Wilhelm Roux' Arch. Entwicklungsmech. Org.* **131,** 238.

Hafen, E., Kuroiwa, A., and Gehring, W. J. (1984). *Cell* **37,** 833.

Haigh, L. S., Owens, B. B., Hellewell, S., and Ingram, V. M. (1982). *Proc. Natl. Acad. Sci. U.S.A.* **79,** 5332.

Hamkalo, B. A., and Miller, O. L., Jr. (1973). *Annu. Rev. Biochem.* **42,** 379.

Hammer, R. E., Pursel, V. G., Rexroad, C. E., Jr., Wall, R. J., Bolt, D. J., Ebert, K. M., Palmiter, R. D., and Brinster, R. L. (1985). *Nature (London)* **315,** 680.

Hanahan, D. (1985). *Nature (London)* **315,** 680.

Hanas, J. S., Bogenhagen, D. F., and Wu, C. W. (1983). *Proc. Natl. Acad. Sci. U.S.A.* **80,** 2142.

Hanocq, F., Kirsch-Volders, M., Hanocq-Quertier, J., Baltus, E., and Steinert, G. (1972). *Proc. Natl. Acad. Sci. U.S.A.* **69,** 1322.

Hanocq, F., De Schutter, A., Hubert, E., and Brachet, J. (1974). *Differentiation* **2,** 75.

Hanocq-Quertier, J., and Baltus, E. (1981). *Eur. J. Biochem.* **120,** 351.

Hanocq-Quertier, J., Baltus, E., and Brachet, J. (1976). *Proc. Natl. Acad. Sci. U.S.A.* **73,** 2028.

Hanocq-Quertier, J., Baltus, E., and Brachet, J. (1978). *Biol. Cell.* **32,** 103.

Hara, K., Tydeman, P., and Kirschner, M. (1980). *Proc. Natl. Acad. Sci. U.S.A.* **77,** 462.

Harbers, K., Kuehn, M., Delius, H., and Jaenisch, R. (1984). *Proc. Natl. Acad. Sci. U.S.A.* **81,** 1504.

Harding, C. V., Harding, D., and Perlmann, P. (1954). *Exp. Cell Res.* **6,** 202.

Harding, C. V., Harding, D., and Bamberger, J. W. (1955). *Exp. Cell Res., Suppl.* **3,** 181.

Harris, P. (1969). *In* "The Cell Cycle" (G. M. Padilla, G. L. Whitson, and I. L. Cameron, eds.), p. 315. Academic Press, New York.

Harris, P., Osborn, M., and Weber, K. (1980). *Exp. Cell Res.* **126,** 19.

Harsa-King, M. L. (1982). *J. Cell Biol. (Woods Hole, Mass.)* **95,** 153a.

Harvey, E. B. (1933). *Biol. Bull. (Woods Hole, Mass.)* **64,** 125.

Harvey, E. B. (1936). *Biol. Bull. (Woods Hole, Mass.)* **71,** 101.

Hasyawa, H. (1955). *Nature (London)* **175,** 1031.

Heasman, J., Quarmby, J., and Wylie, C. C. (1984). *Dev. Biol.* **105,** 458.

Hegner, R. W. (1911). *Biol. Bull. (Woods Hole, Mass.)* **20,** 237.

Heidemann, S. R., and Gallas, P. T. (1980). *Dev. Biol.* **80,** 489.

Heidemann, S. R., and Kirschner, M. W. (1975). *J. Cell Biol.* **67,** 105.

Heidemann, S. R., and Kirschner, M. W. (1978). *J. Exp. Zool.* **204,** 431.

Heidemann, S. R., Sander, G., and Kirschner, M. W. (1977). *Cell (Cambridge, Mass.)* **10,** 337.

Hennen, S. (1974). *Dev. Biol.* **36,** 447.

Herlands, L., Allfrey, V. G., and Poccia, D. (1982). *J. Cell Biol.* **94,** 219.

Herlant, M. (1911). *Arch. Biol.* **26,** 173.
Hermann, J., Mulner, O., Bellé, R., Marot, J., Tso, J., and Ozon, R. (1984). *Proc. Natl. Acad. Sci. U.S.A.* **81,** 5150.
Hieter, P. A., Hendricks, M. B., Hemminki, K., and Weinberg, E. S. (1979). *Biochemistry* **18,** 2707.
Hill, R. S., and MacGregor, H. C. (1980). *J. Cell Sci.* **44,** 87.
Hirai, S., Nagahama, Y., Kishimoto, T., and Kanatani, H. (1981). *Dev., Growth Differ.* **23,** 465.
Hirano, K. I. (1982). *Dev., Growth Differ.* **24,** 273.
Hogan, P., and Gross, P. R. (1971). *J. Cell Biol.* **49,** 692.
Holland, L. Z., and Gould-Somero, M. (1982). *Dev. Biol.* **92,** 549.
Holter, H., and Zeuthen, E. (1944). *C. R. Trav. Lab. Carlsberg, Ser. Chim.* **25,** 33.
Holtzer, H., Weintraub, H., Mayne, R., and Mochan, B. (1972). *Curr. Top. Dev. Biol.* **7,** 229.
Honda, B. M., and Roeder, R. G. (1980). *Cell (Cambridge, Mass.)* **22,** 119.
Hörstadius, S. (1939). *Biol. Rev. Cambridge Philos. Soc.* **18,** 132.
Houamed, K. M., Bilbe, G., Smart, T. G., Constanti, A., Brown, D. A., Barnard, E. A., and Richards, B. M. (1984). *Nature* **310,** 318.
Hough-Evans, B. R., and Anderson, D. M. (1981). *Biochem. Cell. Regul.* **3,** 83.
Hough-Evans, B. R., Ernst, S. G., Britten, R. J., and Davidson, E. H. (1979). *Dev. Biol.* **69,** 258.
Hourcade, D., Dressler, D., and Wolfson, J. (1973). *Proc. Natl. Acad. Sci. U.S.A.* **70,** 2926.
Huchon, D., Ozon, R., Fischer, E. H., and Demaille, J. G. (1981a). *Mol. Cell Endocrinol.* **22,** 211.
Huchon, D., Ozon, R., and Demaille, J. G. (1981b). *Nature (London)* **294,** 358.
Huchon, D., Crozet, N., Cantenot, N., and Ozon, R. (1981c). *Reprod. Nutr. Dev.* **21,** 135.
Huchon, D., Jessus, C., and Ozon, R. (1985). *C. R. Acad. Sci. Paris* **300,** 463.
Huez, G., Marbaix, G., Hubert, E., Leclercq, M., Nudel, V., Soreq, H., Salomon, R., Lebleu, B., Revel, M., and Littauer, V. J. (1974). *Proc. Natl. Acad. Sci. U.S.A.* **71,** 3143.
Hülser, D., and Schatten, G. (1982). *Gamete Res.* **5,** 363.
Hultin, T. (1961). *Experientia* **17,** 410.
Hultin, T. (1964). *Exp. Cell Res.* **34,** 608.
Humphreys, T. (1971). *Dev. Biol.* **26,** 201.
Hurkman, W. J., Smith, L. D., Richter, J., and Larkins, B. A. (1981). *J. Cell Biol.* **89,** 292.
Hutchins, R., and Brandhorst, B. R. (1979). *Wilhelm Roux's Arch. Dev. Biol.* **186,** 95.
Ikegami, S., Okada, T. S., and Koida, S. S. (1976). *Dev., Growth Differ.* **18,** 33.
Ikegami, S., Taguchi, T., and Ohashi, M. (1978). *Nature (London)* **275,** 458.
Ikegami, S., Amemiya, S., Oguro, M., Nagano, H., and Mano, Y. (1979). *J. Cell Physiol.* **100,** 439.
Illmensee, K., and Mahowald, A. P. (1974). *Proc. Natl. Acad. Sci. U.S.A.* **71,** 1016.
Imoh, H., Okamoto, M., and Eguchi, G. (1983). *Dev., Growth Differ.* **25,** 1.
Imperato-McGinley, J., and Peterson, R. E. (1976). *Am. J. Med.* **61,** 264.
Iwamatsu, T., and Ohta, T. (1974). *J. Exp. Zool.* **187,** 3.
Iwamatsu, T., Miki-Nomura, T., and Ohta, T. (1976). *J. Exp. Zool.* **195,** 97.
Iwao, Y., and Katagiri, C. (1984). *J. Exp. Zool.* **230,** 115.
Jäckle, H., and Kalthoff, K. (1980). *Proc. Natl. Acad. Sci. U.S.A.* **77,** 6700.
Jacob, F. (1978). *Proc. R. Soc. London Ser. B* **201,** 249.
Jacob, F., and Monod, J. (1963). "Cytodifferentiation and Macromolecular Synthesis" (M. Locke, ed.), p. 30. Academic Press, New York.
Jacq, C., Miller, J. R., and Brownlee, G. G. (1977). *Cell (Cambridge, Mass.)* **12,** 109.
Jaenisch, R., and Jähner, D. (1984). *Biochim. Biophys. Acta* **782,** 1.
Jaffe, L. A., Sharp, A. P., and Wolf, D. P. (1983). *Dev. Biol.* **96,** 317.
Jamrich, M., Warrior, R., Steele, R., and Gall, J. G. (1983). *Proc. Natl. Acad. Sci. U.S.A.* **80,** 3364.
Jeffery, W. R. (1982). *Science* **216,** 545.

Jeffrey, W. R. (1984). *Dev. Biol.* **103**, 482.
Jeffrey, W. R. (1985). *Cell (Cambridge, Mass.)* **41**, 11.
Jeffery, W. R., and Wilson, L. T. (1983). *J. Embryol. Exp. Morphol.* **75**, 225.
Jeffery, W. R., Tomlinson, C. R., and Brodeur, R. D. (1983). *Dev. Biol.* **99**, 408.
Jenkins, N. A., Kaumeyer, J. F., Young, E. M., and Raff, R. A. (1978). *Dev. Biol.* **63**, 279.
Johnson, C. H., and Epel, D. (1981a). *J. Cell Biol.* **89**, 284.
Johnson, C. H., and Epel, D. (1981b). *J. Cell Biol.* **91**, 180a.
Johnson, C. H., and Epel, D. (1983). *Dev. Biol.* **92**, 461.
Johnson, J. D., Epel, D., and Paul, M. (1976). *Nature (London)* **262**, 661.
Johnson, K. E. (1981). *Cell Differ.* **10**, 47.
Johnson, K. E., and Adelman, M. R. (1981). *J. Cell Sci.* **49**, 205.
Jones, P. A., and Taylor, S. M. (1981). *Nucleic Acids Res.* **9**, 2933.
Jones, R. E., De Feo, D., and Piatigorsky, J. (1981). *J. Biol. Chem.* **256**, 8172.
Jordana, X., Allende, C. C., and Allende, J. E. (1982). *FEBS Lett.* **143**, 124.
Kado, R. T., Marcher, K., and Ozon, R. (1981). *Dev. Biol.* **84**, 471.
Kallenbach, R. J. (1982). *Cell Biol. Int. Rep.* **6**, 1025.
Kallenbach, R. J. (1983). *Eur. J. Cell Biol.* **30**, 159.
Kallenbach, R. J., and Mazia, D. (1982). *Eur. J. Cell Biol.* **28**, 68.
Kalthoff, H. (1979). *In* "Determinants of Spatial Organization" (S. Subtelny and I. R. Konigsberg, eds.), p. 102. Academic Press, New York.
Kalthoff, H., and Richter, D. (1982). *Biochemistry* **21**, 741.
Kalthoff, H., Darmer, D., Towbin, J., Gordon, J., Amons, R., Möller, W., and Richter, D. (1982). *Eur. J. Biochem.* **122**, 439.
Kanatani, H., Shirai, H., Nakanishi, K., and Kurukova, T. (1969). *Nature (London)* **221**, 273.
Kandler-Singer, I., and Kalthoff, H. (1976). *Proc. Natl. Acad. Sci. U.S.A.* **73**, 3740.
Karsenti, E., and Gounon, P. (1979). *Biol. Cell* **34**, 91.
Karsenti, E., Gounon, P., and Bornens, M. (1978). *Biol. Cell* **31**, 219.
Karsenti, E., Newport, J., Hubble, R., and Kirschner, M. (1984). *J. Cell Biol.* **98**, 1730.
Katagiri, C., and Moriya, M. (1976). *Dev. Biol.* **50**, 235.
Kato, K. H., and Sugiyama, M. (1971). *Dev., Growth Differ.* **13**, 359.
Kaumeyer, J. F., Jenkins, N. A., and Raff, R. A. (1978). *Dev. Biol.* **63**, 266.
Kaye, J. S., Bellard, M., Dretzen, G., Bellard, F., and Chambon, P. (1984). *EMBO J.* **3**, 1137.
Keichline, L. D., and Wasserman, P. H. (1979). *Biochemistry* **18**, 214.
Keller, H., and Bessis, M. (1975). *Nature (London)* **258**, 723.
Kessel, R. G., and Subtelny, S. (1981). *J. Exp. Zool.* **217**, 119.
Killian, C. E., Bland, C. E., Kuzawa, J. M. and Nishioka, D. (1985). *Exp. Cell Res.* **158**, 519.
King, T. J., and Briggs, R. (1953). *J. Exp. Zool.* **123**, 61.
King, T. J., and Briggs, R. (1954). *J. Embryol. Exp. Morphol.* **2**, 73.
King, W. J., and Greene, G. L. (1984). *Nature (London)* **307**, 745.
Kinsey, W. H. (1984). *Dev. Biol.* **105**, 137.
Kinsey, W. H., and Lennarz, W. J. (1981). *J. Cell Biol.* **91**, 325.
Kirschner, M., Gerhart, J. C., Hara, K., and Ubbels, G. A. (1980). *In* "The Cell Surface, Mediator of Developmental Processes" (S. Subtelny and N. K. Wessels, eds.), p. 187. Academic Press, New York.
Kirschner, M., Newport, J., and Gerhart, J. (1985). *Trends Genet.* **1**, 41.
Kishimoto, T., and Kanatani, H. (1973). *Exp. Cell Res.* **82**, 296.
Kishimoto, T., Hirai, S., and Kanatani, H. (1981). *Dev. Biol.* **81**, 177.
Kishimoto, T., Kuriyama, R., Kondo, H., and Kanatani, H. (1982). *Exp. Cell Res.* **137**, 121.
Kleinschmidt, J. A., and Franke, W. W. (1982). *Cell (Cambridge, Mass.)* **29**, 799.
Kleinschmidt, J. A., Scheer, U., Dabauvalle, M. C., Bustin, M., and Franke, W. W. (1983). *J. Cell Biol.* **97**, 838.

Kofoid, E. C., Knauber, D. C., and Allende, J. E. (1979). *Dev. Biol.* **72,** 374.

Korn, L. J., and Brown, D. D. (1978). *Cell (Cambridge, Mass.)* **15,** 1145.

Korn, L. J., and Gurdon, J. B. (1981). *Nature (London)* **289,** 461.

Korn, L. J., Gurdon, J. B., and Price, J. (1982). *Nature (London)* **300,** 354.

Koser, R., and Collier, J. R. (1976). *Differentiation* **6,** 47.

Kostellow, A. B., Weinstein, S. P., and Morrill, G. A. (1982). *Biochim. Biophys. Acta* **720,** 356.

Krohne, G. K., and Franke, W. W. (1980a). *Exp. Cell Res.* **129,** 167.

Krohne, G., and Franke, W. W. (1980b). *Proc. Natl. Acad. Sci. U.S.A.* **77,** 1034.

Krohne, G., Franke, W. W., and Scheer, U. (1978). *Exp. Cell Res.* **116,** 85.

Krohne, G., Dabauvalle, M. C., and Franke, W. W. (1981). *J. Mol. Biol.* **151,** 121.

Krohne, G., Stick, R., Kleinschmidt, J. A., Moll, R., Franke, W. W., and Hausen, P. (1982). *J. Cell Biol.* **94,** 749.

Krystal, G., and Poccia, D. (1979). *Exp. Cell Res.* **123,** 207.

Krystal, G., and Poccia, D. (1981). *Exp. Cell Res.* **134,** 41.

Kunkel, N. S., and Weinberg, E. S. (1978). *Cell (Cambridge, Mass.)* **14,** 313.

Kunnath, L., and Locker, J. (1983). *EMBO J.* **2,** 317.

Kuriyama, R., and Borisy, G. G. (1983). *J. Cell Sci.* **61,** 175.

Kuroiwa, A., Hafen, E., and Gehring, W. J. (1984). *Cell* **37,** 825.

Kusunoki, S., and Yasumasu, I. (1976). *Biochem. Biophys. Res. Commun.* **68,** 881.

Kusunoki, S., and Yasumasu, I. (1978). *Dev. Biol.* **67,** 336.

Labhart, P., and Oller, T. (1982). *Cell (Cambridge, Mass.)* **28,** 279.

Lacy, E., Roberts, S., Evans, E. P., Burtenshaw, M. D., and Costantini, F. D. (1983). *Cell (Cambridge, Mass.)* **34,** 343.

Lane, C. D. (1981). *Cell (Cambridge, Mass.)* **24,** 281.

Lane, C. D., Marbaix, G., and Gurdon, J. B. (1971). *J. Mol. Biol.* **61,** 73.

Lane, C. D., Colman, A., Mohun, T., Morser, J., Champion, J., Kourides, I., Craig, R., Higgins, S., James, T. C., Applebaum, S. W., Ohlson, R. I., Paucha, E., Houghton, M., Matthews, J., and Miflin, B. J. (1980). *Eur. J. Biochem.* **111,** 225.

Lane, C. D., Champion, J., Colman, A., James, T. C., and Applebaum, S. W. (1983a). *Eur. J. Biochem.* **130,** 529.

Lane, C. D., Champion, J., and Craig, R. (1983b). *Eur. J. Biochem.* **136,** 141.

Laskey, R. A., and Earnshaw, W. C. (1980). *Nature (London)* **286,** 763.

Laskey, R. A., and Gurdon, J. B. (1970). *Nature (London)* **228,** 1332.

Laughton, A., and Scott, M. P. (1984). *Nature* **310,** 25.

La Volpe, A., Taggart, M., McStay, B., and Bird, A. (1983). *Nucleic Acids Res.* **11,** 5361.

Lee, Y. R., and Whiteley, A. H. (1982). *Cell Differ.* **11,** 311.

Le Meur, M., Gerlinger, P., Benoist, C., and Mathis, D. (1985). *Nature (London)* **316,** 38.

Levenson, R., and Marcu, K. (1976). *Cell (Cambridge, Mass.)* **9,** 311.

Levine, M., Garen, A., Lepesant, J. A., and Lepesant-Kejzlarova, J. (1981). *Proc. Natl. Acad. Sci. U.S.A.* **78,** 2417.

Levine, M., Rubin, G. M., and Tijan, R. (1984). *Cell* **38,** 667.

Lillie, F. R. (1902). *Arch. Entwicklungsmech. Org.* **14,** 477.

Lindahl, P. E., and Holter, H. (1941). *C.R. Trav. Lab. Carlsberg, Ser. Chim.* **24,** 49.

Littauer, U. Z., and Soreq, H. (1982). *Prog. Nucleic Acid Res. Mol. Biol.* **27,** 53.

Liu, F. T., and Orida, N. (1984). *J. Biol. Chem.* **259,** 10649.

Lohka, M. J., and Masui, Y. (1983a). *Science* **220,** 719.

Lohka, M. J., and Masui, Y. (1983b). *Exp. Cell Res.* **148,** 481.

Lohka, M. J., and Masui, Y. (1984). *J. Cell Biol.* **98,** 1222.

Löhler, J., Timpl, R., and Jaenisch, R. (1984). *Cell* **38,** 597.

Lorch, I. J., Danielli, J. F., and Hörstadius, S. (1953). *Exp. Cell Res.* **4,** 253.

McCarthy, R. A., and Spiegel, M. (1983). *Cell Differ.* **13,** 93.

McCarthy, R. A., and Spiegel, M. (1983). *Cell Differ.* **13**, 93.
McClay, D. R., and Wessel, G. M. (1985). *Trends Genet.* **1**, 12.
McCrae, M. A., and Woodland, H. R. (1981). *Eur. J. Biochem.* **116**, 467.
McGrath, J. M., and Solter, D. (1984). *Science* **226**, 1317.
McGinnis, W., Levine, M. S., Hafen, E., Kuroiwa, A., and Gehring, W. J. (1984a). *Nature (London)* **308**, 428.
McGinnis, W., Garber, R. L., Wirz, J., Kuroiwa, A., and Gehring, W. J. (1984b). *Cell* **37**, 403.
McGinnis, W., Hart, C. P., Gehring, W. J., and Ruddle, F. H. (1984c). *Cell* **38**, 675.
McGrath, J., and Solter, D. (1984). *Cell (Cambridge, Mass.)* **37**, 179.
McKnight, G. S., Hammer, R. E., Kuenzel, E. A., and Brinster, R. L. (1983). *Cell (Cambridge, Mass.)* **34**, 335.
McKnight, S. L., and Kingsbury, R. (1982). *Science* **217**, 316.
McLaren, A. (1984). *Nature (London)* **309**, 671.
MacLeod, D., and Bird, A. (1983). *Nature (London)* **306**, 200.
Magnuson, T., and Epstein, C. J. (1984). *Cell (Cambridge, Mass.)* **38**, 823.
Magram, J., Chada, K., and Costantini, F. (1985). *Nature (London)* **315**, 338.
Mairy, M., and Denis, H. (1971). *Dev. Biol.* **24**, 143.
Malacinski, G. M., Benford, H., and Chung, H. M. (1975). *J. Exp. Zool.* **191**, 97.
Maller, J., Poccia, D., Nishioka, D., Kidd, P., Gerhart, J., and Hartman, H. (1976). *Exp. Cell Res.* **99**, 285.
Maller, J., Wu, M., and Gerhart, J. C. (1977). *Dev. Biol.* **58**, 295.
Maller, J. L., and Krebs, E. G. (1977). *J. Biol. Chem.* **252**, 1712.
Maller, J. L., Butcher, F. R., and Krebs, E. G. (1979). *J. Biol. Chem.* **254**, 579.
Manen, C. A., and Russell, D. H. (1973). *J. Embryol. Exp. Morphol.* **29**, 331.
Manen, C. A., Hadfield, M. G., and Russell, D. H. (1977). *Dev. Biol.* **57**, 454.
Manes, C., and Menzel, P. (1981). *Nature (London)* **293**, 589.
Mann, J. R., and Lovell-Badge, R. A. (1984). *Nature* **310**, 66.
Marbaix, G., Huez, G., Burny, A., Cleuter, Y., Hubert, E., Leclercq, M., Chantrenne, H., Soreq, H., Nudel, U., and Littauer, U. Z. (1975). *Proc. Natl. Acad. U.S.A.* **72**, 3065.
Marinos, E., and Billett, F. S. (1981). *J. Embryol. Exp. Morphol.* **62**, 395.
Marot, J., Tso, J., Huchon, D., Mulner, O., and Ozon, R. (1984). *Gamete Res.* **9**, 339.
Martin, K. A., and Miller, O. L., Jr. (1983). *Dev. Biol.* **98**, 338.
Martin, R. P., Sibler, A. P., Dirheimer, G., de Henau, S., and Grosjean, H. (1981). *Nature (London)* **293**, 235.
Martindale, M. Q., and Brandhorst, B. P. (1984). *Dev. Biol.* **101**, 512.
Martini, G., Tato, F., Gandini Attardi, D., and Tocchini-Valentini, G. P. (1976). *Biochem. Biophys. Res. Commun.* **72**, 875.
Masui, Y. (1974). *J. Exp. Zool.* **187**, 141.
Masui, Y., and Clarke, H. J. (1979). *Int. Rev. Cytol.* **57**, 185.
Masui, Y., and Markert, C. L. (1971). *J. Exp. Zool.* **177**, 129.
Masui, Y., Forer, A., and Zimmerman, A. M. (1978). *J. Cell Sci.* **31**, 117.
Mathews, C. K. (1975). *Exp. Cell Res.* **92**, 47.
Matsumoto, L., Kasamatsu, H., Piko, L., and Vinograd, J. (1974). *J. Cell Biol.* **63**, 146.
Mattaj, I. W., Lienhard, S., Zeller, R., and De Robertis, E. M. (1983). *J. Cell Biol.* **97**, 1261.
Maundrell, K. (1975). *J. Cell Sci.* **17**, 579.
Maxson, R. E., Jr., and Egrie, J. C. (1980). *Dev. Biol.* **74**, 335.
Maxson, R. E., Jr., and Wilt, F. H. (1981). *Dev. Biol.* **83**, 380.
Maxson, R. E., Jr., and Wilt, F. H. (1982). *Dev. Biol.* **94**, 435.
Mazia, D., and Dan, K. (1952). *Proc. Natl. Acad. Sci. U.S.A.* **38**, 826.
Mazia, D., and Ruby, A. (1974). *Exp. Cell Res.* **85**, 167.

Mazia, D., Schatten, G., and Steinhardt, R. (1975). *Proc. Natl. Acad. Sci. U.S.A.* **72,** 4469.
Mazzei, G., Meijer, L., Moreau, M., and Guerrier, P. (1981). *Cell Differ.* **10,** 139.
Méchali, M., Méchali, F., and Laskey, R. A. (1983). *Cell (Cambridge, Mass.)* **35,** 63.
Meedel, T. H., and Whittaker, J. R. (1983). *Proc. Natl. Acad. Sci. U.S.A.* **80,** 4761.
Meedel, T. H., and Whittaker, J. R. (1984). *Dev. Biol.* **105,** 479.
Meijer, L., and Guerrier, P. (1981). *Dev. Biol.* **88,** 318.
Meijer, L., and Guerrier, P. (1984). *Int. Rev. Cytol.* **86,** 129.
Meijer, L., Paul, M., and Epel, D. (1982). *Dev. Biol.* **94,** 62.
Meinecke-Tillmann, S., and Meinecke, B. (1984). *Nature (London)* **307,** 637.
Melton, D. A., De Robertis, E. M., and Cortese, R. (1980). *Nature (London)* **284,** 143.
Merriam, R. W., Sauterer, R. A., and Christensen, K. (1983). *Dev. Biol.* **95,** 439.
Mertz, J. E., and Gurdon, J. B. (1977). *Proc. Natl. Acad. Sci. U.S.A.* **74,** 1502.
Mertz, J. E., and Miller, T. J. (1983). *Mol. Cell. Biol.* **3,** 126.
Meyerhof, P. G., and Masui, Y. (1977). *Dev. Biol.* **61,** 214.
Meyerhof, P. G., and Masui, Y. (1979). *Exp. Cell Res.* **123,** 345.
Mezger-Freed, L. (1953). *J. Cell. Comp. Physiol.* **41,** 493.
Miake-Lye, R., Newport, J., and Kirschner, M. (1983). *J. Cell Biol.* **97,** 81.
Miki-Noumura, T. (1977). *J. Cell Sci.* **24,** 203.
Miledi, R., Parker, I., and Sumikawa, K. (1982). *EMBO J.* **1,** 1307.
Miller, J. R., and Melton, D. A. (1981). *Cell (Cambridge, Mass.)* **24,** 829.
Miller, O. L., and Beatty, B. R. (1969). *Science* **164,** 955.
Mills, A. D., Laskey, R. A., Black, P., and De Robertis, E. M. (1980). *J. Mol. Biol.* **139,** 561.
Mintz, B. (1962). *Am. Zool.* **2,** 432.
Mintz, B. (1971). *In* "Methods in Mammalian Embryology" (J. L. Daniel, Jr., ed.), p. 191. Freeman, San Francisco, California.
Mintz, B. (1974). *Annu. Rev. Genet.* **8,** 411.
Mintz, B., and Illmensee, K. (1975). *Proc. Natl. Acad. Sci. U.S.A.* **72,** 3585.
Miyake-Lie, R., and Kirschner, M. (1985). *Cell* **41,** 165.
Mohandas, T., Sparkes, R. S., and Shapiro, L. J. (1981). *Science* **211,** 393.
Mohun, T. J., Lane, C. D., Colman, A., and Wylie, C. C. (1981). *J. Embryol. Exp. Morphol.* **61,** 367.
Mohun, T. J., Brennan, S., Dathan, N., Fairman, S., and Gurdon, J. B. (1984). *Nature (London)* **311,** 716.
Monroy, A. (1960). *Experientia* **16,** 114.
Monroy, A., and Rosati, F. (1983). *Gamete Res.* **7,** 85.
Monroy, A., Maggio, R., and Rinaldi, A. M. (1965). *Proc. Natl. Acad. Sci. U.S.A.* **54,** 107.
Moon, R. T., Danilchik, M. V., and Hille, M. B. (1982). *J. Cell Biol.* **93,** 389.
Moore, B. C. (1954). *Science* **120,** 786.
Moore, J. A. (1947). *J. Exp. Zool.* **105,** 349.
Moore, J. A. (1948). *J. Exp. Zool.* **108,** 127.
Moore, J. A. (1955). *Adv. Genet.* **7,** 139.
Moreau, M., Guerrier, P., and Dorée, M. (1976). *J. Exp. Zool.* **197,** 435.
Moreau, M., Guerrier, P., Dorée, M., and Ashley, C. C. (1978). *Nature (London)* **272,** 251.
Morgan, G. T., MacGregor, H. C., and Colman, A. (1980). *Chromosoma* **80,** 309.
Morgan, G. T., Bakken, A. H., and Reeder, R. H. (1982). *Dev. Biol.* **93,** 471.
Morgan, T. H. (1934). "Embryology and Genetics." Columbia Univ. Press, New York.
Moriya, M., and Katagiri, C. (1976). *Dev., Growth Differ.* **18,** 349
Morrill, G. A., and Murphy, J. B. (1972). *Nature (London)* **238,** 282.
Morrill, G. A., Schatz, F., and Zabrenetzky, V. S. (1975). *Differentiation* **4,** 143.
Morrill, G. A., Kostelow, A. B., Mahajan, S., and Gupta, R. K. (1984). *Biochim. Biophys. Acta* **804,** 107.

Morris, P. W., and Rutter, W. J. (1976). *Biochemistry* **15**, 3106.

Moss, T. (1983). *Nature (London)* **302**, 223.

Moy, G. W., Brandriff, B., and Vacquier, V. D. (1977). *J. Cell Biol.* **73**, 788.

Müller, M. M., Carrasco, A. E., and De Robertis, E. M. (1984). *Cell (Cambridge, Mass.)* **39**, 157.

Mulner, O., Huchon, D., Thibier, C., and Ozon, R. (1979). *Biochim. Biophys. Acta* **582**, 179.

Mulner, O., Cartaud, A., and Ozon, R. (1980). *Differentiation* **16**, 31.

Murphy, J. T., Burgess, R. R., Dahlberg, J. E., and Lund, E. (1982). *Cell (Cambridge, Mass.)* **29**, 265.

Nakatsuji, N., and Johnson, K. E. (1984). *J. Cell Sci.* **68**, 49.

Neff. A. W., Wakahara, M., Jurand, A., and Malacinski, G. M. (1984). *J. Embryol. Exp. Morph.* **80**, 197.

Nemer, M., Dubroff, L. M., and Graham, M. (1975). *Cell (Cambridge, Mass.)* **6**, 171.

Newport, J., and Kirschner, M. (1982a). *Cell (Cambridge, Mass.)* **30**, 675.

Newport, J., and Kirschner, M. (1982b). *Cell (Cambridge, Mass.)* **30**, 687.

Newport, J. W., and Kirschner, M. W. (1984). *Cell (Cambridge, Mass.)* **37**, 731.

Newrock, K. M., and Raff, R. A. (1975). *Dev. Biol.* **42**, 242.

Nielsen, R. J., Thomas, G., and Maller, J. L. (1982). *Proc. Natl. Acad. Sci. U.S.A.* **79**, 2937.

Nojima, H., and Kornberg, R. D. (1983). *J. Biol. Chem.* **258**, 8151.

Noll, H., Metranga, V., Palma, P., Cutrono, F., and Vittorelli, L. (1981). *Dev. Biol.* **87**, 229.

Noronha, J. M., Sheys, G. H., and Buchanan, M. (1972). *Proc. Natl. Acad. Sci. U.S.A.* **69**, 2006.

O'Connor, C. M., and Smith, L. D. (1976). *Dev. Biol.* **52**, 318.

Ohta, T., and Iwamatsu, T. (1980). *J. Exp. Zool.* **214**, 93.

Okada, M., Kleinman, I. A., and Schneiderman, H. A. (1974). *Dev. Biol.* **37**, 43.

Okada, A., Shin, T., Dworkin-Rastl, E., Dworkin, M. B., and Zubay, G. (1985). *Differentiation* **29**, 14.

Okazaki, K. (1975). *Am. Zool.* **15**, 567.

Old, R. W., Callan, H. G., and Gross, K. W. (1977). *J. Cell Sci.* **27**, 57.

Oppenheimer, S. B., and Meyer, J. T. (1982a). *Exp. Cell Res.* **137**, 472.

Oppenheimer, S. B., and Meyer, J. T. (1982b). *Exp. Cell Res.* **139**, 451.

Opresko, L., Wiley, H. S., and Wallace, R. A. (1979). *J. Exp. Zool.* **209**, 367.

Orgel, L. E., and Crick, F. H. C. (1980). *Nature (London)* **284**, 604.

Ornitz, D. M., Palmiter, R. D., Hammer, R. E., Brinster, R. L., Swift, G. H., and MacDonald, R. J. (1985). *Nature (London)* **313**, 600.

Ozaki, H. (1975). *In* "Isozymes" (C. L. Markert, ed.), Vol. 3, p. 543. Academic Press, New York.

Ozaki, H., and Whiteley, A. H. (1970). *Dev. Biol.* **21**, 196.

Palaćek, J., Habrová, V., Nedvidek, J., and Romanovsky, A. (1985). *J. Embry. Exp. Morph.* **87**, 75.

Palmiter, R. D., Brinster, R. L., Hammer, R. E., Trumbauer, M. E., Rosenfeld, M. G., Birnberg, N. C., and Evans, R. M. (1982). *Nature (London)* **300**, 611.

Palmiter, R. D., Norstedt, G., Gelinas, R. E., Hammer, R. E., and Brinster, R. L. (1983). *Science* **222**, 809.

Papaioannou, V. E., McBurney, M. W., Gardner, R. L., and Evans, M. J. (1975). *Nature (London)* **258**, 70.

Pardue, M. L., Brown, D. D., and Birnstiel, M. L. (1973). *Chromosoma* **42**, 191.

Parisi, E., Filosa, S., De Petrocellis, B., and Monroy, A. (1978). *Dev. Biol.* **65**, 38.

Parisi, E., Filosa, S., and Monroy, A. (1979). *Dev. Biol.* **72**, 167.

Pasteels, J. J. (1934). *Arch. Anat. Microsc. Morphol. Exp.* **30**, 161.

Pasteels, J. J. (1940). *Arch. Biol.* **49**, 629.

Pasteels, J. J. (1964). *Adv. Morphog.* **3**, 363.

Pasteels, J. J., Castiaux, P., and Vandermeersche, G. (1958). *Arch. Biol.* **69**, 627.

Pays, A., Hubert, E., and Brachet, J. (1977). *Differentiation* **8**, 79.

Pelham, H. (1985). *Trends Genet.* **1**, 31.

Pelham, H. R., and Brown, D. D. (1980). *Proc. Natl. Acad. Sci. U.S.A.* **77**, 4170.

Pennock, D. G., and Reeder, R. H. (1984). *Nucleic Acids Res.* **12**, 2225.

Perlman, S., and Rosbash, M. (1978). *Dev. Biol.* **63**, 197.

Perry, G., and Epel, D. (1981). *Exp. Cell Res.* **134**, 65.

Pestell, R. G. (1975). *Biochem. J.* **145**, 527.

Petri, W. (1982). *Nature (London)* **299**, 399.

Petzoldt, U., Burki, K., Illmensee, G. R., and Illmensee, K. (1983). *Wilhelm Roux's Arch. Dev. Biol.* **192**, 138.

Phillips, C. R. (1982). *J. Exp. Zool.* **223**, 265.

Picard, A., and Dorée, M. (1983). *Exp. Cell Res.* **145**, 325.

Picard, A., Giraud, F., Le Bouffant, F., Sladeczek, F., Le Peuch, C., and Dorée, M. (1984). *FEBS Lett.* **182**, 446.

Picard, B., and Wegnez, M. (1979). *Proc. Natl. Acad. Sci. U.S.A.* **76**, 241.

Pollock, J. M., Swihart, M., and Taylor, J. H. (1978). *Nucleic Acids Res.* **5**, 4855.

Posakony, I. W., Flytzanis, C. N., Britten, R. J., and Davidson, E. H. (1983). *J. Mol. Biol.* **167**, 361.

Poulson, D. F. (1940). *J. Exp. Zool.* **83**, 271.

Poulson, D. F. (1945). *Am. Nat.* **79**, 340.

Preiss, A., Rosenberg, U. B., Kienlin, A., Seifert, E., and Jäckle, H. (1985). *Nature (London)* **313**, 27.

Pruitt, S. C., and Grainger, R. M. (1981). *Cell (Cambridge, Mass.)* **23**, 711.

Pure, E., Luster, A. D., and Unkeless, J. C. (1984). *J. Exp. Med.* **160**, 606.

Raff, R. A., Greenhouse, G., Gross, K. W., and Gross, P. R. (1971). *J. Cell Biol.* **50**, 516.

Raff, R. A., Colot, H. V., Selvig, S. E., and Gross, P. R. (1972). *Nature (London)* **235**, 211.

Raff, R. A., Brandis, J. W., Huffman, C. J., Koch, A. L., and Leister, D. E. (1981). *Dev. Biol.* **86**, 265.

Raff, E. C., Brothers, A. J., and Raff, R. A. (1976). *Nature (London)* **260**, 615.

Rapoport, T. A. (1981). *Eur. J. Biochem.* **115**, 665.

Rapraeger, A. C., and Epel, D. (1981). *Dev. Biol.* **88**, 269.

Razin, A., and Riggs, A. D. (1980). *Science* **210**, 604.

Razin, A., Webb, C., Szyf, M., Yisraeli, J., Rosenthal, A., Naveh-Many, T., Sciaky-Gallili, N., and Cedar, H. (1984). *Proc. Natl. Acad. Sci. U.S.A.* **81**, 2275.

Reeder, R. H., and Roan, J. G. (1984). *Cell (Cambridge, Mass.)* **38**, 39.

Reeder, R. H., Brown, D. D., Wellauer, P. K., and Dawid, I. B. (1976). *J. Mol. Biol.* **105**, 507.

Reverberi, G. (1956). *Experientia* **12**, 55.

Richter, J. D., and Smith, L. D. (1981). *Cell (Cambridge, Mass.)* **27**, 183.

Richter, J. D., and Smith, L. D. (1984). *Nature (London)* **309**, 378.

Richter, J. D., Jones, N. C., and Smith, L. D. (1982a). *Proc. Natl. Acad. Sci. U.S.A.* **79**, 3789.

Richter, J. D., Wasserman, W. J., and Smith, L. D. (1982b). *Dev. Biol.* **89**, 159.

Richter, J. D., Smith, L. D., Anderson, D. M., and Davidson, E. H. (1984). *J. Mol. Biol.* **173**, 227.

Ries, E. (1937). *Pubbl. Staz. Zool. Napoli* **16**, 363.

Rinaldi, A. M., and Parente, A. (1976). *Dev. Biol.* **49**, 260.

Rinaldi, A. M., Storage, A., Arzone, A., and Mutolo, V. (1977). *Cell Biol. Int. Rep.* **1**, 249.

Rinaldi, A. M., De Leo, G., Arzone, A., Salcher, I., Storace, A., and Mutolo, V. (1979a). *Proc. Natl. Acad. Sci. U.S.A.* **76**, 1916.

Rinaldi, A. M., Salcher-Cillari, I., and Mutolo, V. (1979b). *Cell Biol. Int. Rep.* **3**, 179.

Rinaldi, A. M., Salcher-Cillari, I., and Valenti, A. M. (1981). *Cell Biol. Int. Rep.* **5**, 987.

Rinaldi, A. M., Carra, E., Salcher-Cillari, I., and Oliva, A. O. (1983). *Cell Biol. Int. Rep.* **7**, 211.

Robinson, K. R. (1979). *Proc. Natl. Acad. Sci. U.S.A.* **76**, 837.

Roccheri, M. C., Di Bernardo, M. G., and Giudice, G. (1981). *Dev. Biol.* **83**, 173.

Roccheri, M. C., Sconzo, G., Di Carlo, M., Di Bernardo, M. G., Pirrone, A., Bambino, R., and Giudice, G. (1982). *Differentiation* **22**, 175.

Roeder, R. G. (1983). *J. Biol. Chem.* **258**, 1932.

Rosbash, M., and Ford, P. J. (1974). *J. Mol. Biol.* **85**, 87.

Rosbash, M., Ford, P. J., and Bishop, J. V. (1974). *Proc. Natl. Acad. Sci. U.S.A.* **71**, 3746.

Rosenberg, M. P., and Lee, H. H. (1981). *J. Exp. Zool.* **217**, 389.

Rosenberg, U. B., Preiss, A., Seifert, E., Jäckle, H., and Knippel, D. C. (1985). *Nature (London)* **313**, 703.

Rosenthal, E. T., Tansey, T. R., and Ruderman, J. V. (1983). *J. Mol. Biol.* **166**, 309.

Roux, W. (1903). *Anat. Anz.* **23**, 65, 113, 161.

Rubin, G. M., and Spradling, A. C. (1982). *Science* **218**, 348.

Ruddell, A. and Jacobs-Lorena, P. N. (1985). *Proc. Natl. Acad. Sci. U.S.A.* **82**, 3316.

Ruddle, F. H., Hart, C. P., and McGinnis, W. (1985). *Trends Genet.* **1**, 48.

Ruderman, J. V., and Pardue, M. L. (1977). *Dev. Biol.* **60**, 48.

Ruderman, J. V., Woodland, H. R., and Sturgess, E. A. (1979). *Dev. Biol.* **71**, 71.

Rungger, D., Rungger-Brändle, E., Chaponnier, C., and Gabbiani, G. (1979). *Nature (London)* **282**, 320.

Runnström, J. (1928). *Protoplasma* **4**, 388.

Rusconi, S., and Schaffner, W. (1981). *Proc. Natl. Acad. Sci. U.S.A.* **78**, 5051.

Russell, D. H. (1971). *Proc. Natl. Acad. Sci. U.S.A.* **68**, 523.

Sadler, S. E., and Maller, J. L. (1981). *J. Biol. Chem.* **256**, 6368.

Sadler, S. E., and Maller, J. L. (1982). *J. Biol. Chem.* **257**, 355.

Sadler, S. E., and Maller, J. L. (1983). *Dev. Biol.* **98**, 165.

Sakai, M., and Kubota, H. Y. (1981). *Dev., Growth Differ.* **23**, 41.

Sakai, M., and Shinagawa, A. (1983). *J. Cell Sci.* **63**, 69.

Salmon, E. D., Leslie, R. J., Saxton, W. M., Karow, M. L., and McIntosh, J. R. (1984). *J. Cell Biol.* **99**, 2165.

Santamaria, P., and Nüsslein-Volhard, C. (1983). *EMBO J.* **2**, 1695.

Sargent, T. D., and Dawid, I. B. (1983). *Science* **222**, 135.

Sato, E., and Koide, S. S. (1984). *J. Exp. Zool.* **230**, 135.

Satoh, N. (1982a). *Differentiation* **21**, 37.

Satoh, N. (1982b). *Differentiation* **22**, 156.

Satoh, N., and Ikegami, S. (1981a). *J. Embryol. Exp. Morphol.* **61**, 1.

Satoh, N., and Ikegami, S. (1981b). *J. Embryol. Exp. Morphol.* **64**, 61.

Savić, A., Richman, P., Williamson, P., and Poccia, D. (1981). *Proc. Natl. Acad. Sci. U.S.A.* **78**, 3706.

Sawai, T. (1979). *J. Embryol. Exp. Morphol.* **51**, 183.

Schackmann, R. W., Christen, R., and Shapiro, B. M. (1981). *Proc. Natl. Acad. Sci. U.S.A.* **78**, 6066.

Schatten, G. (1984). *Subcell Biochem.* **10**, 359.

Schatten, G., and Schatten, H. (1981). *Exp. Cell Res.* **135**, 311.

Schatten, G., Schatten, H., Bestor, T., and Balczon, R. (1981). *J. Cell Biol.* **91**, 185.

Scheer, U. (1978). *Cell (Cambridge, Mass.)* **13**, 535.

Scheer, U. (1981). *J. Cell Biol.* **88**, 599.

Scheer, U. (1982). *Biol. Cell* **44**, 213.

Scheer, U., and Somerville, J. (1982). *Exp. Cell Res.* **139**, 410.

Scheer, U., Franke, W. W., Trendelenburg, M. F., and Spring, H. (1976a). *J. Cell Sci.* **22**, 503.

Scheer, U., Trendelenburg, M. F., and Franke, W. W. (1976b). *J. Cell Biol.* **69**, 465.

Scheer, U., Sommerville, J., and Bustin, M. (1979). *J. Cell Sci.* **40**, 1.

Scheer, U., Sommerville, J., and Müller, U. (1980). *Exp. Cell Res.* **129**, 115.

Scheer, U., Hinssen, H., Franke, W. W., and Jockusch, B. M. (1984). *Cell (Cambridge, Mass.)* **39**, 111.

Scherer, G., Telford, J., Baldari, C., and Pirrotta, V. (1981). *Dev. Biol.* **86**, 438.

Schmidt, T., Patton, C., and Epel, D. (1982). *Dev. Biol.* **90**, 284.

Schneider, E. G., Nguyen, H. T., and Lennarz, W. J. (1978). *J. Biol. Chem.* **253**, 2348.

Schnieke, A., Harbers, K., and Jaenisch, R. (1983). *Nature (London)* **304**, 315.

Schönmann, W. (1938). *Wilhelm Roux' Arch. Entwicklungsmech. Org.* **138**, 345.

Schorderet-Slatkine, S., Schorderet, M., and Baulieu, E. E. (1976). *Nature (London)* **262**, 289.

Schorderet-Slatkine, S., Schorderet, M., and Baulieu, E. E. (1977). *Differentiation* **9**, 67.

Schorderet-Slatkine, S., Schorderet, M., Boquet, P., Godeau, F., and Baulieu, E. E. (1978). *Cell (Cambridge, Mass.)* **15**, 1269.

Schorderet-Slatkine, S., Finidori-Lepicard, J., Hanoune, J., and Baulieu, E. E. (1981). *Biochem. Biophys. Res. Commun.* **100**, 544.

Schorderet-Slatkine, S., Schorderet, M., and Baulieu, E. E. (1982). *Proc. Natl. Acad. Sci. U.S.A.* **79**, 850.

Schroeder, T. (1979). *Dev. Biol.* **70**, 311.

Schultz, R. M., Montgomery, R. R., and Belanoff, J. R. (1983). *Dev. Biol.* **97**, 264.

Searles, L. L., Jokerst, R. S., Bingham, P. M., Voelker, R. A., and Greenleaf, A. L. (1982). *Cell (Cambridge, Mass.)* **31**, 585.

Seidel, F. (1932). *Wilhelm Roux' Arch. Entwicklungsmech. Org.* **126**, 213.

Selvig, S. A., Greenhouse, G. E., and Gross, P. R. (1972). *Cell Differ.* **1**, 5.

Sempinska, E., and Ledoux, L. (1967). *Biochim. Biophys. Acta* **138**, 570.

Senger, D. R., and Gross, P. R. (1978). *Dev. Biol.* **65**, 404.

Severinsson, L., and Peterson, P. A. (1984). *J. Cell Biol.* **99**, 226.

Shani, M. (1985). *Nature (London)* **314**, 283.

Shapiro, H. (1935). *J. Cell Comp. Physiol.* **6**, 101.

Shaw, B. R., Cognetti, G., Sholes, W. M., and Richards, R. G. (1981). *Biochemistry* **20**, 4971.

Shen, S. S., and Steinhardt, R. A. (1978). *Nature (London)* **272**, 253.

Shen, S. S., and Steinhardt, R. A. (1980). *Exp. Cell Res.* **125**, 55.

Shepherd, G. W., Rondinelli, E., and Nemer, M. (1983). *Dev. Biol.* **96**, 520.

Shimada, H. (1983). *Dev. Biol.* **97**, 454.

Shinagawa, A. (1983). *J. Cell Sci.* **64**, 147.

Shinagawa, A. (1985). *J. Embry. Exp. Morph.* **85**, 33.

Shiokawa, K., and Yamana, K. (1979). *Dev., Growth Differ.* **21**, 501.

Shiokawa, K., Misumi, Y., and Yamana, K. (1981). *Wilhelm Roux's Arch. Dev. Biol.* **190**, 103.

Showman, R. M., and Whiteley, A. H. (1980). *Dev., Growth Differ.* **22**, 305.

Showman, R. M., Wells, D. E., Anstrom, J., Hursh, D. A., and Raff, R. A. (1982). *Proc. Natl. Acad. Sci. U.S.A.* **79**, 5944.

Signoret, J., David, J. C., Lefresne, J., and Houillon, C. (1983). *Proc. Natl. Acad. Sci. U.S.A.* **80**, 3368.

Signoret, J., Lefresne, J., and David, J. C. (1984). *Differentiation* **26**, 235.

Simpson, R. T. (1981). *Proc. Natl. Acad. Sci. U.S.A.* **78**, 6803.

Skoblina, M. N. (1976). *J. Embryol. Exp. Morphol.* **36**, 67.

Skoultchi, S., and Gross, P. R. (1973). *Proc. Natl. Acad. Sci. U.S.A.* **70**, 2840.

Slater, D. W., Slater, I., and Bollum, F. J. (1978). *Dev. Biol.* **63**, 94.

Slater, I., and Slater, D. W. (1974). *Proc. Natl. Acad. Sci. U.S.A.* **71**, 1103.

Slater, I., and Slater, D. W. (1979). *Differentiation* **13**, 109.

Smith, L. D., and Ecker, R. E. (1970). *Curr. Top. Dev. Biol.* **5**, 1.

Smith, L. J. (1956). *J. Exp. Zool.* **132,** 51.

Solari, A., and Deutscher, M. P. (1982). *Nucleic Acids Res.* **10,** 4397.

Solari, A., and Deutscher, M. P. (1983). *Mol. Cell. Biol.* **3,** 1711.

Sollner-Webb, B., and McKnight, S. L. (1982). *Nucleic Acids Res.* **10,** 3391.

Sommerville, J., and Malcolm, D. B. (1976). *Chromosoma* **55,** 183.

Sommerville, J., and Scheer, U. (1982). *Chromosoma* **86,** 95.

Sommerville, J., Malcolm, D. B., and Callan, H. G. (1978). *Philos. Trans. R. Soc. London, Ser. B* **283,** 359.

Soreq, H., Sagar, A. D., and Sehgal, P. B. (1981). *Proc. Natl. Acad. Sci. U.S.A.* **78,** 1741.

Spadafora, C., Bellard, M., Compton, J. L., and Chambon, P. (1976). *FEBS Lett.* **69,** 281.

Spemann, H. (1928). *Z. Wiss. Zool.* **132,** 105.

Spiegel, M., and Spiegel, E. (1978). *Exp. Cell Res.* **117,** 269.

Spieth, J., and Whiteley, A. H. (1981). *Wilhelm Roux's Arch. Dev. Biol.* **190,** 111.

Spinelli, G., Gianguzza, F., Casano, C., Acierno, P., and Burckhardt, J. (1979). *Nucleic Acids Res.* **6,** 545.

Spinelli, G., Albanese, I., Anello, L., Ciaccio, M., and Di Liegro, I. (1982). *Nucleic Acids Res.* **10,** 7977.

Spradling, A. C., and Rubin, G. M. (1982). *Science* **218,** 341.

Stauffer, E. (1945). *Rev. Suisse Zool.* **52,** 231.

Steinert, G., Thomas, C., and Brachet, J. (1976). *Proc. Natl. Acad. Sci. U.S.A.* **73,** 833.

Steinert, G., Felsani, A., Kettmann, R., and Brachet, J. (1984). *Exp. Cell Res.* **154,** 203.

Steinert, M. (1951). *Bull. Soc. Chim. Biol.* **33,** 549.

Steinhardt, R. A., and Alderton, J. M. (1982). *Nature (London)* **295,** 154.

Steinhardt, R. A., and Epel, D. (1974). *Proc. Natl. Acad. Sci. U.S.A.* **71,** 1915.

Steinhardt, R. A., Shen, S., and Mazia, D. (1972). *Exp. Cell Res.* **72,** 195.

Steinhardt, R. A. *et al.* (1977). *Dev. Biol.* **58,** 189.

Stewart, T. A., and Mintz, B. (1981). *Proc. Natl. Acad. Sci. U.S.A.* **78,** 6314.

Stewart, T. A., Wagner, E. F., and Mintz, B. (1982). *Science* **217,** 1046.

Stick, R., and Hausen, P. (1985). *Cell* **41,** 191.

Storb, U., O'Brien, R. L., McMullen, M. D., Gollahon, K. A., and Brinster, R. M. (1984). *Nature* **310,** 238.

Strome, S., and Wood, W. B. (1983). *Cell (Cambridge, Mass.)* **35,** 15.

Stunnenberg, H. G., and Birnstiel, M. L. (1982). *Proc. Natl. Acad. Sci. U.S.A.* **79,** 620.

Sugiyama, M. (1951). *Biol. Bull. (Woods Hole, Mass.)* **101,** 335.

Sumikawa, K., Houghton, M., Emtage, J. S., Richards, B. M., and Barnard, E. A. (1981). *Nature (London)* **292,** 862.

Sumikawa, K., Parker, I., Amano, T., and Miledi, R. (1984). *EMBO J.* **3,** 2291.

Sunkara, P. S., Wright, D. A., and Rao, P. N. (1979). *Proc. Natl. Acad. Sci. U.S.A.* **76,** 2799.

Sunkara, P. S., Wright, D. A., and Nishioka, K. (1981). *Dev. Biol.* **87,** 351.

Surani, M. A. H., and Barton, S. C. (1983). *Science* **222,** 1034.

Surani, M. A. H., Barton, S. C., and Norris, M. L. (1984). *Nature (London)* **308,** 548.

Swanson, R. F., and Dawid, I. B. (1970). *Proc. Natl. Acad. Sci. U.S.A.* **66,** 117.

Swift, G. H., Hammer, R. E., MacDonald, R. J., and Brinster, R. L. (1984). *Cell (Cambridge, Mass.)* **38,** 639.

Sze, L. C. (1953). *Science* **117,** 479.

Takeshima, K., and Nakano, E. (1983). *Eur. J. Biochem.* **137,** 437.

Taniguchi, T., Pang, R. H. L., Yip, Y. K., Henriksen, D., and Vilcek, J. (1981). *Proc. Natl. Acad. Sci. U.S.A.* **78,** 3469.

Tarkowski, A. (1961). *Nature (London)* **190,** 857.

Tarkowski, A. K., and Bałakier, H. (1980). *J. Embryol. Exp. Morphol.* **55,** 319.

Teitelman, G. (1973). *J. Embryol. Exp. Morphol.* **29,** 267.

Thomas, C. (1970). *Biochim. Biophys. Acta* **224,** 99.

Thomas, C. (1974). *Dev. Biol.* **39,** 191.

Thomas, C., and Schram, A. (1977). *Biol. Cell.* **30,** 50.

Thomas, C., Hanocq, F., and Heilporn, V. (1977). *Dev. Biol.* **57,** 226.

Thomas, T. L., Posakony, J. W., Anderson, D. M., Britten, R. J., and Davidson, E. H. (1981). *Chromosoma* **84,** 319.

Tomkins, G. M., Levinson, B. B., Baster, J. D., and Dethlefsen, L. (1972). *Nature (London), New Biol.* **239,** 9.

Tondeur-Six, N., Tencer, R., and Brachet, J. (1975). *Biochim. Biophys. Acta* **395,** 41.

Tonegawa, S. (1979–1980). *Harvey Lect.* **75,** 61.

Tourte, M., Mignotte, F., and Mounolou, J. C. (1984). *Eur. J. Cell Biol.* **34,** 171.

Townes, T. M., Lingrel, J. B., Chen, H. Y., Brinster, R. L., and Palmiter, R. D. (1985). *EMBO J.* **4,** 1715.

Trendelenburg, M. F. (1981). *Biol. Cell.* **42,** 1.

Trendelenburg, M. F., and Gurdon, J. B. (1978). *Nature (London)* **276,** 292.

Tso, J., Thibier, C., Mulner, O., and Ozon, R. (1982). *Proc. Natl. Acad. Sci. U.S.A.* **79,** 5552.

Tucciarone, L. M., and Lanclos, K. D. (1981). *Biochem. Biophys. Res. Commun.* **99,** 221.

Tufaro, F., and Brandhorst, B. P. (1979). *Dev. Biol.* **72,** 390.

Tufaro, F., and Brandhorst, B. P. (1982). *Dev. Biol.* **92,** 209.

Ubbels, G. A., Hara, K., Koster, C. H., and Kirschner, M. W. (1983). *J. Embryol. Exp. Morphol.* **77,** 15.

Vacquier, V. (1981). *Dev. Biol.* **84,** 1.

Vacquier, V., and Brandriff, B. (1975). *Dev. Biol.* **47,** 12.

Valle, G., Besley, J., and Colman, A. (1981). *Nature (London)* **291,** 338.

Van Assel, S., and Brachet, J. (1968). *J. Embryol. Exp. Morphol.* **19,** 261.

Vandekerckhove, J., Franke, W. W., and Weber, K. (1981). *J. Mol. Biol.* **152,** 413.

Van Dongen, W. M. A. M., Zaal, R., Moorman, A. F. M., and Destrée, O. H. J. (1981). *Dev. Biol.* **86,** 303.

Van Gansen, P., and Schram, A. (1972). *J. Cell Sci.* **10,** 339.

Vardimon, L., Kressmann, A., Cedar, H., Maechler, M., and Doerfler, W. (1982). *Proc. Natl. Acad. Sci. U.S.A.* **79,** 1073.

Varley, J. M., MacGregor, H. C., Nardi, I., Andrews, C., and Erba, H. P. (1980). *Chromosoma* **80,** 289.

Venezky, D. L., Angerer, L. M., and Angerer, R. C. (1981). *Cell (Cambridge, Mass.)* **24,** 385.

Vitto, A., Jr., and Wallace, R. A. (1976). *Exp. Cell Res.* **97,** 56.

Vlad, M., and MacGregor, H. C. (1975). *Chromosoma* **50,** 327.

Wadsworth, P., and Sloboda, R. D. (1983). *J. Cell Biol.* **97,** 1249.

Wagner, E. F., and Mintz, B. (1982). *Mol. Cell. Biol.* **2,** 190.

Wagner, E. F., Stewart, T. A., and Mintz, B. (1981). *Proc. Natl. Acad. Sci. U.S.A.* **78,** 5016.

Wagner, T. E., Hoppe, P. C., Jollick, J. D., Scholl, D. R., Hodinka, R. L., and Gault, J. B. (1981). *Proc. Natl. Acad. Sci. U.S.A.* **78,** 6376.

Wahli, W., Dawid, I. B., Ryffel, G. V., and Weber, R. (1981). *Science* **212,** 298.

Wakefield, L., and Gurdon, J. B. (1983). *EMBO J.* **2,** 1613.

Wakefield, L., Ackerman, E., and Gurdon, J. B. (1983). *Dev. Biol.* **95,** 468.

Waksmundzka, M., Krysiak, E., Karasiewicz, J., Czołowska, R., and Tarkowski, A. L. (1984). *J. Embryol. Exp. Morphol.* **79,** 77.

Wallace, R. A., and Jared, D. W. (1976). *J. Cell Biol.* **69,** 345.

Wallace, R. A., and Misulovin, Z. (1978). *Proc. Natl. Acad. Sci. U.S.A.* **75,** 5534.

Wallace, R. A., Jared, D. W., Dumont, J. N., and Sega, M. W. (1973). *J. Exp. Zool.* **184,** 321.

Wallace, R. A., Misulovin, Z., and Etkin, L. D. (1981). *Proc. Natl. Acad. Sci. U.S.A.* **78,** 3078.

Warburg, O. (1908). *Hoppe-Seyler's Z. Physiol. Chem.* **57,** 1.

Ward, G. E., and Vacquier, V. D. (1983). *Proc. Natl. Acad. Sci. U.S.A.* **80,** 5578.
Ward, G. E., Vacquier, V. D., and Michel, S. (1983). *Dev. Biol.* **95,** 360.
Wasserman, W. J., and Houle, J. G. (1984). *Dev. Biol.* **101,** 436.
Wasserman, W. J., and Masui, Y. (1976). *Science* **191,** 1266.
Wasserman, W. J., and Smith, L. D. (1978). *J. Cell Biol.* **78,** R15–R22.
Wasserman, W. J., and Smith, L. D. (1981). *J. Cell Biol.* **89,** 389.
Wasserman, W. J., Pinto, L. H., O'Connor, C. M., and Smith, L. D. (1980). *Proc. Natl. Acad. Sci. U.S.A.* **77,** 1534.
Wasserman, W. J., Richter, J. D., and Smith, L. D. (1982). *Dev. Biol.* **89,** 152.
Wasserman, W. J., Houle, J. G., and Samuel, D. (1984). *Dev. Biol.* **105,** 315.
Webb, A. C., and Camp, C. J. (1979). *Exp. Cell Res.* **119,** 414.
Webb, D. J., and Nuccitelli, R. (1981). *J. Cell Biol.* **91,** 562.
Weintraub, H. (1983). *Cell (Cambridge, Mass.)* **32,** 1191.
Weisbrod, S., Wickens, M. P., Whytock, S., and Gurdon, J. B. (1982). *Dev. Biol.* **94,** 216.
Weiss, Y. C., Vaslet, C. A., and Rosbash, M. (1981). *Dev. Biol.* **87,** 330.
Wellauer, P. K., Dawid, I. B., Brown, D. D., and Reeder, R. H. (1976). *J. Mol. Biol.* **105,** 461.
Welshons, W. V., Lieberman, M. E., and Gorski, J. (1984). *Nature (London)* **307,** 747.
Westley, B., Wyler, T., Ryffel, G., and Weber, R. (1981). *Nucleic Acids Res.* **9,** 3557.
Whitaker, D. M. (1933). *J. Gen Physiol.* **16,** 479.
Whitaker, M., and Irvine, R. F. (1985). *Nature (London)* **312,** 636.
Whitaker, M. J., and Steinhardt, R. A. (1981). *Cell (Cambridge, Mass.)* **25,** 95.
Whitaker, M. J., and Steinhardt, R. A. (1982). *Q. Rev. Biophys.* **15,** 593.
Whiteley, A. H., and Mizuno, S. (1981). *Wilhelm Roux's Arch. Dev. Biol.* **190,** 73.
Whiteley, H. R., and Whiteley, A. H. (1975). *Curr. Top. Dev. Biol.* **9,** 39.
Whittaker, J. R. (1973). *Proc. Natl. Acad. Sci. U.S.A.* **70,** 2096.
Whittaker, J. R. (1979). *In* "Determinants of Spatial Organization" (S. Subtelny and I. R. Konigsberg, eds.), p. 29. Academic Press, New York.
Wiblet, M., Baltus, E., and Brachet, J. (1975). *C.R. Hebd. Séances Acad. Sci.* **281,** 1891.
Wickens, M. P., Woo, S., O'Malley, B. W., and Gurdon, J. B. (1980). *Nature (London)* **285,** 628.
Wiley, H. S., and Wallace, R. A. (1981). *J. Biol. Chem.* **256,** 8626.
Williams, M. A., and Smith, L. D. (1971). *Dev. Biol.* **25,** 57.
Williams, M. A., Trendelenburg, M. F., and Franke, W. W. (1981). *Differentiation* **20,** 36.
Williams, M. A., Kleinschmidt, J. A., Krohne, G., and Franke, W. W. (1982). *Exp. Cell Res.* **137,** 341.
Wilt, F. H. (1973). *Proc. Natl. Acad. Sci. U.S.A.* **70,** 2345.
Wilt, F. H., and Mazia, D. (1974). *Dev. Biol.* **37,** 422.
Wolf, S. F., and Migeon, B. R. (1982). *Nature (London)* **295,** 667.
Wolstenholme, R. (1973). *Chromosoma* **43,** 1.
Woodland, H. R. (1974). *Dev. Biol.* **40,** 90.
Woodland, H. R. (1982). *Biosci. Rep.* **2,** 471.
Woodland, H. R., and Pestell, R. Q. W. (1972). *Biochem. J.* **127,** 597.
Woodland, H. R., Flynn, J. M., and Wyllie, A. J. (1979). *Cell (Cambridge, Mass.)* **18,** 165.
Wormington, W. M., and Brown, D. D. (1983). *Dev. Biol.* **99,** 248.
Wright, D. A., and Subtelny, S. (1971). *Dev. Biol.* **24,** 119.
Wu, C. (1984). *Nature (London)* **309,** 229.
Wu, M., and Gerhart, J. C. (1980). *Dev. Biol.* **79,** 465.
Yamamoto, K., and Yoneda, M. (1983). *Dev. Biol.* **96,** 166.
Yoneda, M., Ikeda, M., and Washitani, S. (1978). *Dev., Growth Differ.* **20,** 329.
Young, R. J., and Sweeney, K. (1979). *J. Embryol. Exp. Morphol.* **49,** 139.
Younglai, E. V., Godeau, F., and Baulieu, E. E. (1981). *FEBS Lett.* **127,** 233.

Younglai, E. V., Godeau, F., Mulvihill, B., and Baulieu, E. E. (1982). *Dev. Biol.* **91,** 36.

Zalokar, M. (1971). *Proc. Natl. Acad. Sci. U.S.A.* **68,** 1539.

Zalokar, M. (1974). *Wilhelm Roux' Arch Entwicklungsmech. Org.* **175,** 243.

Zampetti-Bosseler, F., Huez, G., and Brachet, J. (1973). *Exp. Cell Res.* **78,** 383.

Zasloff, M. (1983). *Proc. Natl. Acad. Sci. U.S.A.* **80,** 6436.

Zasloff, M., Rosenberg, M. and Santos, T. (1982). *Nature (London)* **300,** 81.

Zeller, C. (1956). *Wilhelm Roux' Arch. Entwicklungsmech. Org.* **148,** 311.

Zentgraf, H., Trendelenburg, M. F., Spring, H., Scheer, U., Franke, W. W., Müller, U., Drury, K. C., and Rungger, D. (1979). *Exp. Cell Res.* **122,** 363.

Ziegler, D., and Masui, Y. (1973). *Dev. Biol.* **35,** 283.

Ziegler, D., and Morrill, G. A. (1977). *Dev. Biol.* **60,** 318.

Zimmerman, W., and Weissbach, A. (1981). *Mol. Cell. Biol.* **1,** 680.

CHAPTER 3

CELL DIFFERENTIATION, CELL
TRANSFORMATION,
AND CELL AGING

Why include cell differentiation, malignant transformation, and cell aging in the same chapter? At first, the three processes seem to have little in common; yet, as we shall see, there is a common background. Cell differentiation, which is at the root of histology, physiology, and biochemistry, is the normal outcome of embryonic development; it eventually follows the initial stages of ontogenesis that were discussed in the preceding chapter, and provides diversity among cells of common origin. In a broad sense, cancer cells are unable to undergo differentiation and, instead, continue to divide; in many respects, tumor cells are more similar to early embryonic cells than to the differentiated cells of late embryos and adults. In culture, cancer cells display the transformed phenotype that can easily be recognized under the microscope from the well-differentiated cell types that characterize muscles, bones, cartilages, nervous system, etc. More important, cancer cells are endowed with tumorigenicity. After they are injected into a suitable animal host (the best available is the nude mouse, which, having no thymus, lacks immunological mechanisms for graft rejection), tumors appear; if these tumors are malignant, their metastases invade the host and eventually kill it. Finally, cells undergo several morphological and biochemical changes when they become senescent, either *in vivo* or in culture. Aging may lead to new phenotypes that are typical of senescent cells; this particular type of cell differentiation can only result in cell death. For an embryologist, aging and death are nothing more than the penultimate and ultimate stages of development.

I. CELL DIFFERENTIATION

A. GENERAL BACKGROUND

Except for a brief description of sea urchin gastrulation, our discussion of the eggs in the preceding chapter ended at the end of their cleavage period (late blastula stage). The cell movements characteristic of gastrulation (the so-called morphogenetic movements) are followed by organogenesis. Figure 1 shows the topography of the axial organs in an amphibian late neurula (nervous system, notochord, somites). As is generally known, in vertebrates the entire nervous system is a result of morphogenetic induction by the underlying chordomesoblast. Despite its importance, relatively little will be said about the biochemical

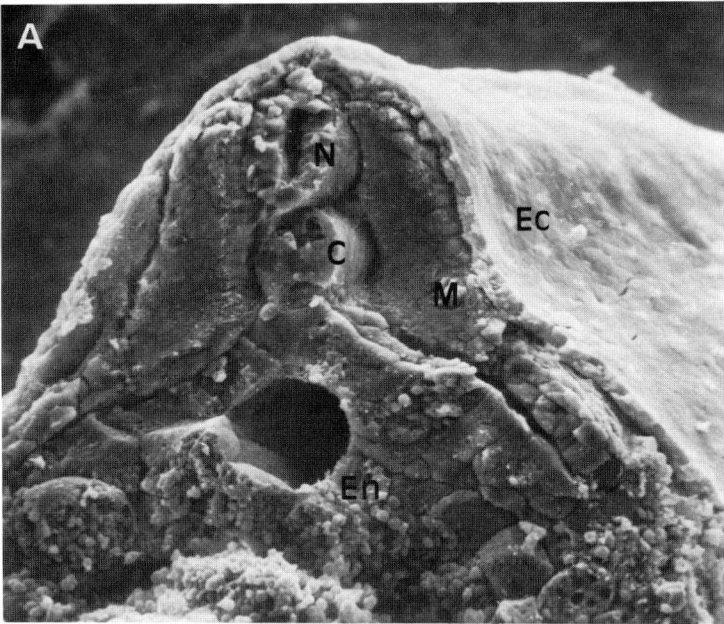

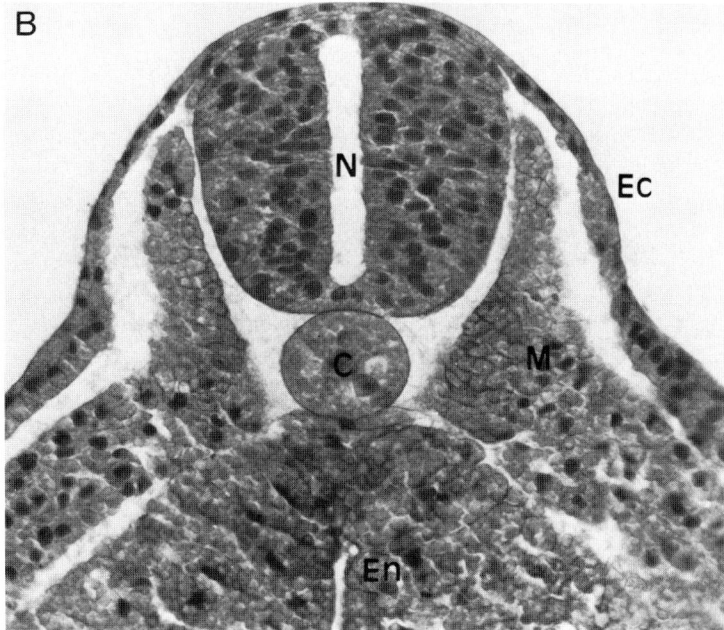

FIG. 1. Dorsal region of a *Xenopus neurula*. (A) Seen with the scanning electron microscope; view of an embryo broken into two pieces. (B) Seen with the light microscope (histological section). C, chorda; Ep, epiderm; En, endoderm; M, mesoderm; N, neural tube. [(A) Courtesy of Dr. L. Devos; (B) courtesy of Dr. H. Alexandre.]

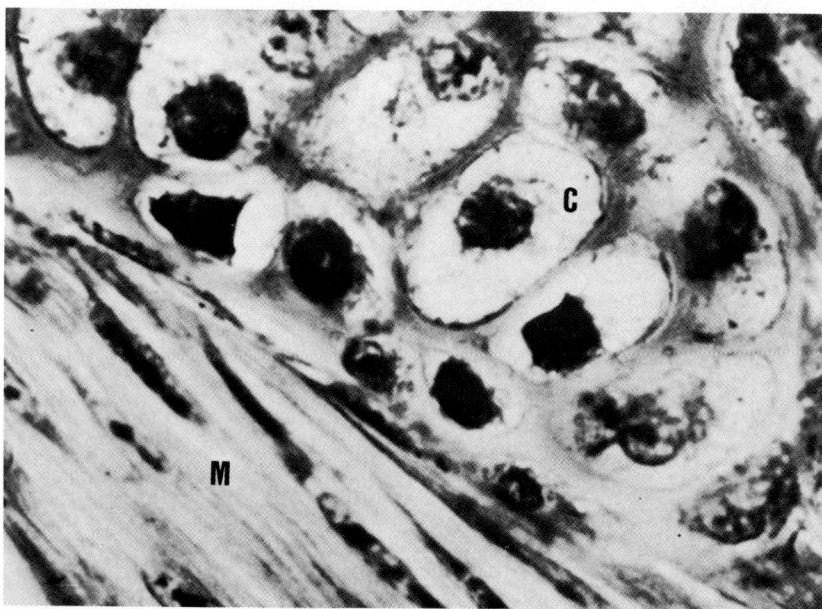

FIG. 2.   Cell differentiation in an amphibian embryo. C, Chrondrocytes; M, myocytes.

mechanisms of this primary (neural) induction, because little progress has been made in this field in the last few years. One should bear in mind that if neural induction had failed to take place during our embryonic life, we would have neither brains nor eyes nor nerves, and there would be no authors to write books or readers to read them.

In the neural tube of the neurula depicted in Fig. 1, all cells are still apparently undifferentiated. There are no morphologically, physiologically or biochemi-cally recognizable neurons at this early stage of nervous system formation. Both *in situ* and in explants of the neural tube, cell differentiation follows, the majority of the cells becoming typical neurons. Other cells will differentiate into pigment cells (melanocytes) that will migrate together with part of the neurons, from the neural crest, to distant but specific parts of the body (Le Douarin, 1980). In the neurula, all somitic cells look alike; at a slightly more advanced stage of develop-ment, part of each somite differentiates into muscle, while the remainder differ-entiates into cartilage. As shown in Fig. 2, muscle cells (myocytes) and cartilage cells (chondrocytes) are morphologically very different despite their apparently common origin. The two tissues are easy to recognize under the microscope and are sharply separated; there are no intermediary cell types at the border between myocytes and chondrocytes. Cell differentiation thus appears to be an all-or-none process. Chondrocytes and myocytes also differ profoundly in their chemical

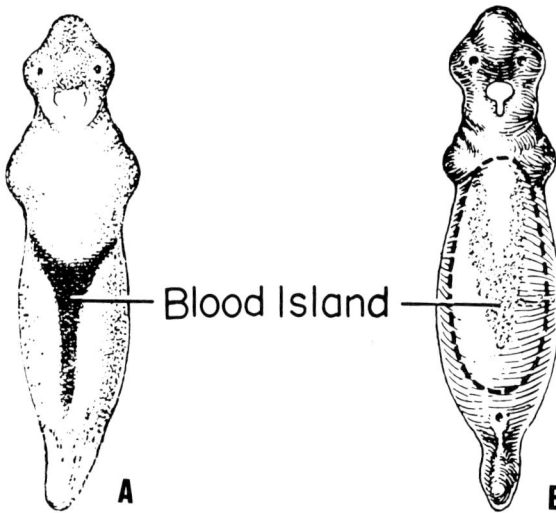

FIG. 3. (A) Ventral view of an axolotl embryo stage 31, showing the blood island (dark) stained by benzidine. (B) Ambystoma punctatum, stage 32, showing the blood island (stippled) and an area excised in the experimental production of bloodless embryos [Copenhaver, 1955.]

composition. Contractile proteins (muscle, α-actin, and myosin) and muscle creatine kinase are typical biochemical markers of muscle differentiation. In contrast, chondroproteins (sulfated proteoglycans, particularly chondroitin sulfate) are specific markers of differentiated cartilage cells.

As already mentioned, the term *luxury proteins* was coined by Holtzer *et al.* (1972, 1975) for proteins specific to the differentiated state. These are in contrast to *housekeeping proteins,* which are present in all cells and which are required for their survival. For instance, our cells can survive without hemoglobin, but not without energy-producing enzymes. Inhibition of hemoglobin synthesis during red blood cell differentiation (erythropoiesis) does not kill the cell; instead, it suppresses the most obvious marker of differentiation. Thus, hemoglobin is a typical example of a luxury protein. Needless to say, complete suppression of hemoglobin synthesis would be lethal for the organism as a whole.

Cell differentiation takes place at a given stage of embryonic development and in well-defined regions of the embryo. For example, erythrocyte differentiation and, as a corollary, hemoglobin synthesis occurs only in the mesodermic blood islands (Fig. 3). How apparently undifferentiated cells become "committed" to differentiate according to a given developmental program remains one of the major unsolved problems in biology. Many believe that a "clock" mechanism (for instance, a given number of DNA replication cycles) marks the time when a cell chooses its differentiation program, but such ideas will remain vague and unsatisfactory as long as we do not know the molecular nature of the clock and

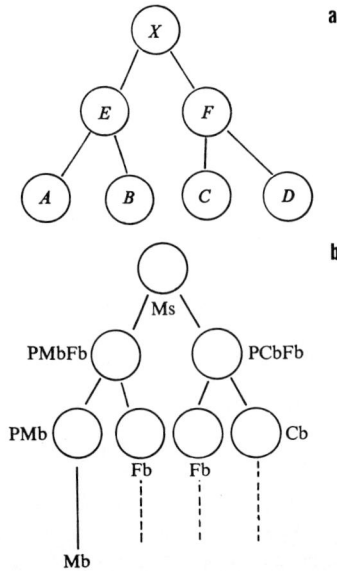

FIG. 4. Models of cell diversification. (a) Model ascribing a primary role of DNA synthesis and/or cell division in generating daughters with metabolic options different from those of their mother cells. The metabolic repertoire of any single cell would be a function of that cell's previous history of quantal, not proliferative, cell cycles. (b) Model assuming that a given mesenchyme cell (Ms) cannot be transformed directly into a myoblast (Mb), a fibroblast (Fb), or a chondroblast (Cb). A minimum of two quantal cell cycles is required for an Ms cell to yield a Cb, and a minimum of three quantal cell cycles is required to yield an Mb. The model predicts that fibroblasts may have more than one origin. PmbFB, presumptive myoblast and fibroblast cell; PCbFB, presumptive chondroblast and fibroblast cell; PMb, presumptive myoblast cell. [After Holtzer *et al.*, 1975, *Quat. Rev. Biophys.* **8**, 523–557 "Lineages, quantel cell cycles, and the generation of cell diversity." Cambridge University Press.]

the biochemical mechanisms that induce the cell to follow one genetic program or another. Words like *commitment* and *developmental program* are useful to describe the facts, but only mask our present ignorance of the underlying mechanisms.

One of the hypotheses proposed to explain cell differentiation is that of quantal mitoses (Holtzer *et al.*, 1972, 1975). As already mentioned, a quantal mitosis (Fig. 4) is a cell division that is believed to give rise to two nonidentical cells, one of which remains an undifferentiated stem cell, and the other of which is committed to differentiate according to a given program. The committed cell, by successive divisions, will give rise to a clone of differentiated cells. This clonal theory of embryonic differentiation is logical; however, it should be pointed out that the work of B. Mintz (1971) on allophenic (tetraparental) mice has established that adult tissues may be more than a single clone of cells. For instance, all the muscles of a mouse derive from a small number of cells committed to become myocytes in the somites; in other words, they derive from a few independent clones of cells (Gearhart and Mintz, 1972).

It is now customary to speak of terminal differentiation. The adjective *terminal* somehow implies that cell differentiation is an irreversible process. Is this necessarily true? Can terminally differentiated cells dedifferentiate and, if so, can they change their developmental program and differentiate along a different pathway than the preceding one? These questions remain the subject of strong debate. There is little doubt that there is at least one firmly established case of transdifferentiation (reviewed by Okada, 1980). During lens regeneration in urodeles (to which we shall return later) and in cultures of retina cells, the pigmented retina cells first lose their melanin granules (dedifferentiation) and then differentiate into lens cells (Eguchi and Okada, 1973; Moscona and Degenstein, 1981) as shown in Fig. 5. Transdifferentiation of cultured brain cells from young chick embryos (Nomura, 1982) and of glial cells from the neural retina of 13- to 16-day-old chick embryos (Moscona *et al.,* 1983) has also been reported. In both cases, some of the cells differentiated in small lenses (lentoids). According to Scarpelli and Rao (1981), a single dose of a chemical carcinogen transdifferentiates regenerating pancreatic cells into hepatocytes. Transdifferentiation is probably a rare event, but it exists and has considerably theoretical interest, since it demonstrates that a differentiated cell can switch from one genetic program to another. During this switch, genes that were active in retina cells are shut off, while others, which were previously silent in the same cells, are activated and direct the synthesis of lens-specific proteins. In the preceding chapter, it was stated that such reprogramming has been demonstrated by Gurdon's experiments on *Xenopus* oocytes injected with adult nuclei. That a similar shift in the genetic program may still occur at a much later stage of embryogenesis should not be a complete surprise.

Can cell differentiation be *partial,* that is, can committed cells stop their differentiation program before it ends? This possibility exists, at least in artificial *in vitro* systems. Dlugosz *et al.* (1983) reported that myoblasts (presumptive muscle cells) undergo partial differentiation after treatment with a phorbol ester. Muscle-specific proteins (desmin, meromyosin) are synthesized, but the later stages of myogenesis (cell fusion, formation of myofibrils) fail to take place.

It is generally assumed that there is a form of antagonism between cell proliferation and cell differentiation, and it is true that mitotic activity generally tends to decrease when a differentiated state is attained. In fact, one of the definitions of commitment is "irreversible withdrawal from the cell cycle." This definition will not satisfy embryologists, who know that in sea urchin eggs, micromeres are already committed at the 16-cell stage to differentiate into spicules; yet, they will divide many times before they actually become spicules. This is why many embryologists still prefer to speak of determination, an old and vague concept, rather than of commitment.

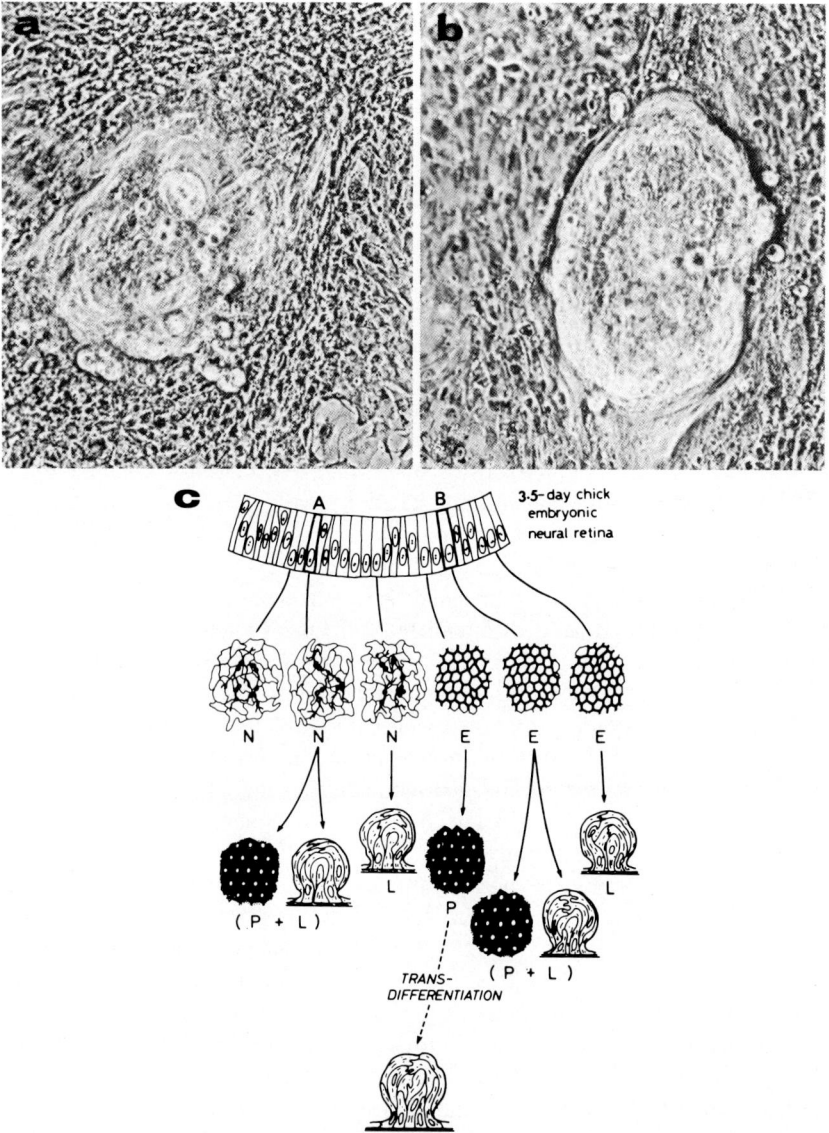

FIG. 5. (a) Lentoid body that appeared in a culture derived from a large number of retina pigment cells and (b) in a culture of a clone deriving from a pigment cell. (c) Schematic representation of the multiple pathways in differentiation starting from embryonic neural retina. Some of the pigmented cells can transdifferentiate into lentoid bodies. E, epithelial cells; N, neuronal cells; P, pigmented cells; L, lentoid body. [(a,b) Okada, 1973; (c) Okada, 1980.]

As mentioned in the preceding chapter, a certain number of cell division cycles are usually required before cells undergo differentiation (although this is not always true, as shown by morphogenesis in the absence of the nucleus in *Acetabularia*). This requirement for a given number of cell divisions before cell commitment is a strong argument for Holtzer *et al.*'s (1972, 1975) theory of quantal mitoses. We have seen that demethylation of DNA sequences during successive DNA replication cycles is one of the possible molecular explanations for the necessity of a given number of mitoses before cells differentiate (Holliday and Pugh, 1975). We shall return to this point later when we discuss the molecular mechanisms of cell differentiation, but the important point here is the following: mitosis can occur in cells that are morphologically and biochemically differentiated (as shown by the fact that they produce a specific luxury protein). As an example, Franke and Keenan (1979) have demonstrated, by electron microscopy, that in lactating mammary glands, dividing cells are fully differentiated. They are filled with casein and fat granules. Similar observations have been made on differentiating muscle and cartilage cells. However, the rule seems to be that agents that stimulate cell proliferation tend to inhibit cell differentiation, and vice versa. For instance, Taketani and Oka (1983) showed that growth factors that stimulate cell divisions [epidermal growth factor (EGF), phorbol esters] inhibit the differentiation of mouse mammary epithelial cells; this was shown by the inhibition of casein and α-lactalbumin synthesis. There is no contradiction between these results and those of Franke and Keenan (1979). Stimulation of cell proliferation prevents the commitment of mammary cells to differentiate; if they have started to differentiate, they may do so even if they are dividing. But there may be exceptions to this rule. Although 6-bromo cAMP is a mitogen for epidermal cells, it induces their differentiation into basal cells, as shown by an increase in keratin synthesis (Tong and Marcelo, 1983).

Scott *et al.* (1982a,b) proposed that in cells that will differentiate, a distinct state in the G_1 phase of the cell cycle precedes differentiation. After this G_0 stage of G_1 has been reached, cells will differentiate without further DNA synthesis. In the system studied by Scott *et al.* (1982b) (the differentiation of proadipocytes into adipocytes filled with a large lipidic droplet), five events follow each other: arrest of growth at the G_0 stage of G_1; nonterminal differentiation; terminal differentiation; loss of the differentiated phenotype; and reinitiation of cell proliferation. At the critical G_0 stage, addition of inhibitors of cAMP-phosphodiesterase induces proliferation. Thus, it remains to be seen whether the G_0 concept can be generalized to other systems when differentiation is morphologically and biochemically more complex than fat accumulation (myogenesis, for instance).

A stochastic model of cell differentiation has been proposed by Bennett (1983) on the basis of his observations on melanoma cell differentiation (accumulation of the brown pigment melanin). Initiation of melanin production occurs at highly variable times in different cells, although these cells do not differ genetically. No correlation with cell divisions could be found in this system. But the differentiation times are highly correlated in sister cells; dedifferentiation (loss of pigment) may occur in the cultures, and it is associated with cell proliferation. It was proposed that the functions associated with differentiation might switch on and off, and that an inhibitor of the off transition builds up on the on stage.

The existence of a variety of hypotheses to explain cell differentiation proves that we do not yet understand its mechanisms. Particularly difficult to explain are recent findings of Suda *et al.* (1984). When two daughter cells of hematopoietic blast cell progenitors were cultured separately, their descendants were different from the primitive lineage in 68 out of 387 cases. This highly variable differentiation of paired hematopoietic progenitors suggests that their intrinsic potentialities are different.

Two main models can be used for the study of cell differentiation: developing embryos and cultures of embryonic cells. The former have the advantage of representing an integrated system in which differentiation is a normal process; the second are easier to work with and are, therefore, more frequently used by students of cell differentiation. The results given by these two different approaches are, in general, similar and have led to the conclusion that cell differentiation results, in most cases, from the activation and expression of a limited number of genes coding for the so-called luxury proteins. As we know, the molecular mechanisms that control gene activity are not yet fully understood, although considerable progress has been made in this exciting field.

Although both whole embryos and cultures of embryonic cells are adequate systems for the analysis of cell differentiation, differences between the two can be found; results obtained on cultured embryonic cells do not always apply to whole embryos. This is true, for instance, in regard to the effects of bromo deoxyuridine (BrdUrd) on cell differentiation. As we shall see, this thymidine analog suppresses differentiation in cultures of many embryonic cells but exerts only marginal, unspecific effects on embryonic differentiation *in vivo*. This cannot be explained by a lack of permeability of intact embryos to the inhibitor, since autoradiography shows that ^{3}H-labeled BrdUrd is heavily incorporated into the nuclei of the treated embryos. We think that there are other, less trivial reasons for such differences between whole embryos and cultured embryonic cells. In embryos, in contrast to cultured cells, cellular movements, cellular affinities (positive or negative) between cell layers, morphogenetic gradients, and inductions play a fundamental role. When embryos are dissociated into individual cells, these cells lose their precise localization in a morphogenetic gradient; the result of this is the loss of their "positional information" (Wolpert, 1971). The preexisting cell-to-cell communications are broken down as soon as the embryos have been dissociated. New haphazard communications will then be established between a given associated cell and its neighbors. Clearly, the two systems are valid, but one should keep in mind that what is true for cultured cells is not necessarily true for a developing embryo. Ephrussi (1956), a pioneer of studies on *in vitro* cell differentiation, was correct when he concluded, like Lederberg (1952), that "embryology will ultimately have to be studied in embryos."

It is impossible to discuss all of the systems that have been used for the study of cell differentiation. We will therefore restrict ourselves to a limited number of selected examples. After a brief survey of primary (neural) and some secondary (lens, pancreas, etc.) inductions in embryos, we shall consider the results obtained on cultures of embryonic cells (myoblasts, chondroblasts, erythroblasts, melanoblasts, etc.). We shall then proceed to the experimental analysis of cell differentiation, using such tools as cell fusion or chemical additions (BrdUrd, phorbol esters, retinoids), which selectively inhibit or promote *in vitro* cell differentiation. Systems such as teratocarcinoma, neuroblastoma, and Friend mouse erythroleukemic cells provide a bridge between cell differentiation and cell transformation. The section will conclude with a discussion of the molecular mechanisms of cell differentiation.

B. Primary (Neural) Induction

The famous work of H. Spemann (1938) demonstrated that the dorsal lip of the blastoporus (chordoblast) from a young amphibian gastrula (Fig. 6) is capable of inducing a secondary embryo after implantation into the blastocele of another early gastrula. Because of this induction capacity, the dorsal lip of the blastoporus was called the *organizer* by Spemann. The easiest way to test the inducing capacity of a tissue is to work with explants from early amphibian gastrulae. As shown in Fig. 7, ectoblast removed from the animal pole region of a late blastula or early gastrula produces, after 2–3 days of explanation, only an undifferentiated cuboid epidermis; if such an ectoblast fragment is cultivated in contact with a piece of the dorsal lip of the blastoporus (chordomesoblast), a neural tube is induced. Thus, the ectoblast cells that were in direct contact with the inducing chordoblast have been transformed into neural cells and will eventually differentiate into neurons. If the same experiment is performed with ectoblast isolated from a late gastrula, there is no response to the inducing stimulus of the organizer; during the course of gastrulation, the epiblast cells have lost their competence to react to the inducer. This restriction in the potentialities of the ectoblast shows that aging starts much earlier than is generally believed.

There have been numerous attempts (1930–1960) to identify the chemical nature of the inducing agent and to study its distribution in various tissues. This work (in which the author was directly involved) now has only historical interest. After a period of great excitement, this elation turned to disappointment when we found (Waddington *et al.,* 1936) that dyes such a methylene blue or neutral red were capable of inducing neural structures in a newt ectoblast. Clearly, these dyes acted in an indirect way by releasing a "masked" inducing substance

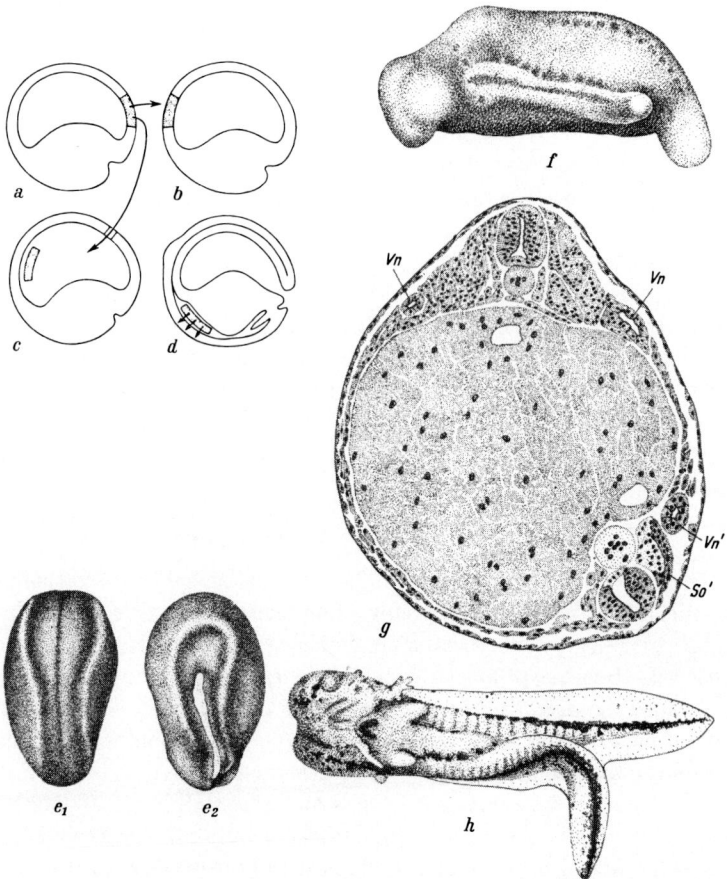

FIG. 6.   Transplantation of the dorsal lip of the blastoporus in amphibians. (a,b) Into ventral ectoderm; (a–c) Into blastocoel, (d) Position of (c) implant next to ectoderm (inductive effect). (e–h) *Triturus taeniatus* embryos with secondary embryonic anlagen. (e) Neurula. (e_1) Dorsal view of the primary neural plate. (e_2) Ventral view of the secondary neural plate (pale, longitudinally stretched, implanted strip). (f) Embryo at the tailbud stage with a secondary embryonic anlage (otic vesicles, neural tube, somite rows, and tailbud). (g) Cross section through (f); (h) Late-stage embryo with extremely well developed secondary twin. [(a–g) After Spemann and Mangold (1924); (h) after Holtfreter (1933). Kühn (1971).]

already present in the ectoblast. Holtfreter (1945) showed that a short acid (pH 3–4) or alkaline (pH 9) shock was sufficient to cause spontaneous neuralization of an ectoblast fragment in the complete absence of any inducing agent. In fact, blowing into the saline solution where an ectoblast piece was cultured was enough to induce its transformation into a neural tube. Both CO_2 and ammonia are efficient inducers of spontaneous neuralization.

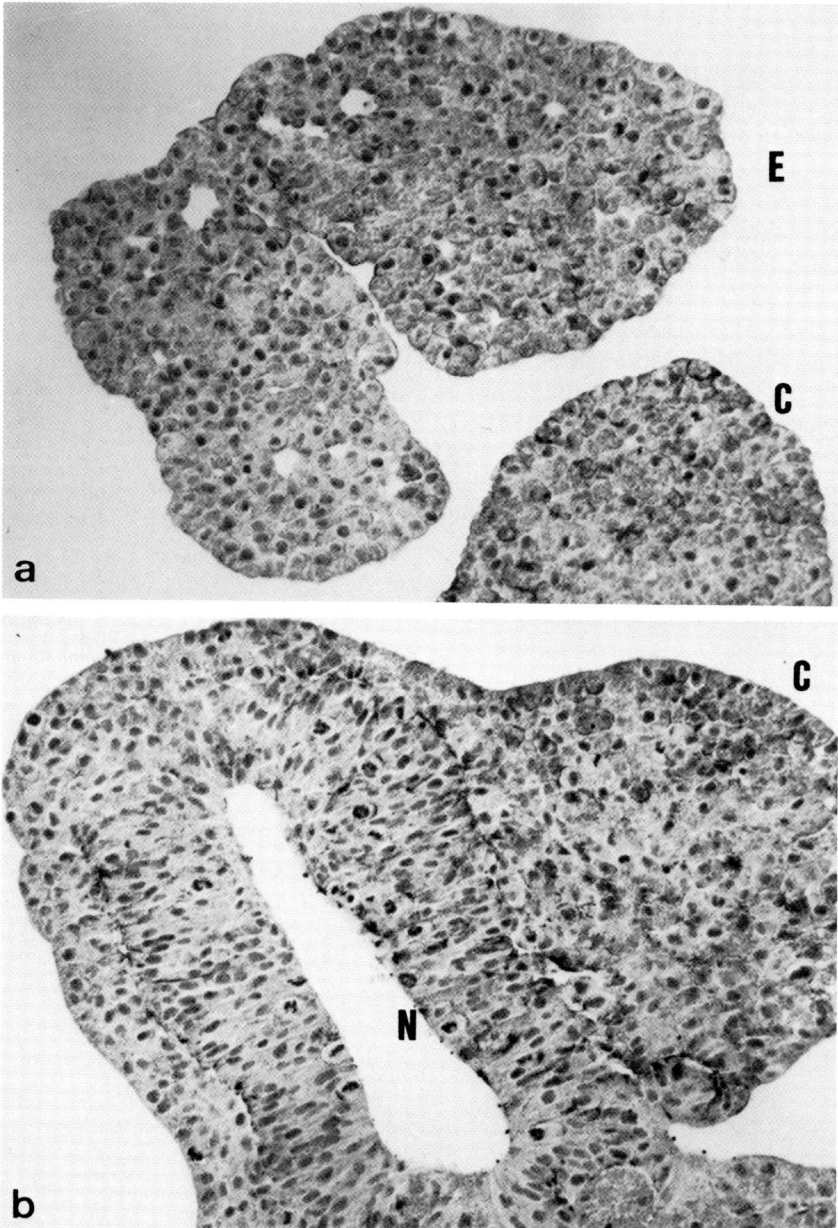

Fig. 7. Neural induction in cultivated amphibian cells (*Ambystoma mexicanum*) (a) Ectoblast isolated from a late blastula; it produces only ectoderm in culture. (b) Same as (a), but E is cultured in contact with a fragment of the dorsal lip of the blastoporus (chordoblast organizer). After contact with the organizer, the ectoblast cells form a neural tube. C, chordoblast; E, ectoderm; N, neural tube. [Courtesy of Dr. R. Tencer.]

Definite progress was made by the experiments of Toivonen (1950) and Yamada (1950). Pieces of various tissues grafted into early newt gastrulae showed at least two different types of inducing agents. As shown in Fig. 8, implantation of liver is followed by induction of cephalic structures (brain, eyes); in contrast, bone marrow induces caudal structures in which mesoderm derivatives (muscle, mesenchyme) are preponderant. There are thus distinct neural and mesodermic inductions, presumably due to different chemical agents.

Serious attempts to purify the two inducing factors were made by Tiedemann (1968). He isolated from chick embryos two active proteins: a neuralizing factor, which is bound to microsomes and ribosomes, and a soluble mesodermizing factor, which has been purified to near homogeneity. This mesodermizing factor produces an overdevelopment of chorda, muscles, mesenchyme, and gut; for this reason, Tiedemann (1968) called it a *vegetalizing* factor. Its activity is inhibited by a glycoprotein also isolated in Tiedemann's laboratory. At the cellular level, the vegetalizing factor inhibits the differentiation of ectoderm cells (formation of cilia and microvilli); it first stops mitotic activity and then induces a wave of cell divisions. It seems that, to be efficient, the vegetalizing protein must penetrate the cells, while the neuralizing factor acts on the cell surface. There is no doubt that Tiedemann's strenuous efforts have given us a very useful tool (the vegetalizing factor) for the analysis of primary morphogenesis. For instance, Grunz (1983) has shown that brief treatments of ectoderm fragments from young *Xenopus* gastrulae transform ectoderm into endoderm and derivatives of ventral mesoderm (red blood cells, pronephros). Longer treatments lead to the formation of more dorsal structures (pronephros, somites, and notochord). No nervous system is formed unless a large proportion of the explant has been transformed into somites and chorda, which, as has been known since Spemann's time, induce neural structures in ectoderm. It should be pointed out, however, that we do not know whether the chicken vegetalizing factor of Tiedemann is present in early amphibian gastrulae, or if the danger of spontaneous neuralization or mesodermization, even if every attempt is made to minimize it, can ever be completely excluded. However, recent work from Tiedemann's laboratory shows that *Xenopus* unfertilized eggs and gastrulae both contain neural inducing factors: ribonucleoprotein (RNP) particles isolated from eggs or embryos induce neural - archencephalic structures (brain, eyes); the proteins isolated from the RNP particles are active, but less than the intact RNP particles (which contain maternal mRNAs in addition to proteins). In contrast, ribosomal subunits have no inducing activity (Janeczek *et al.*, 1984a). Acidic proteins isolated from RNP particles isolated from *Xenopus* eggs, gastrulae, and neurulae all induce archencephalis structures; they are associated with ER vesicles (Janeczek *et al.*, 1984b).

Figure 9, Chapter 1 represents schematically the gradients that can be detected by cytochemical methods and by autoradiography in developing amphibian eggs. In the fertilized eggs, the ribosomes are distributed along a decreasing animal–vegetal gradient; this gradient does not seem to be modified during cleavage. At

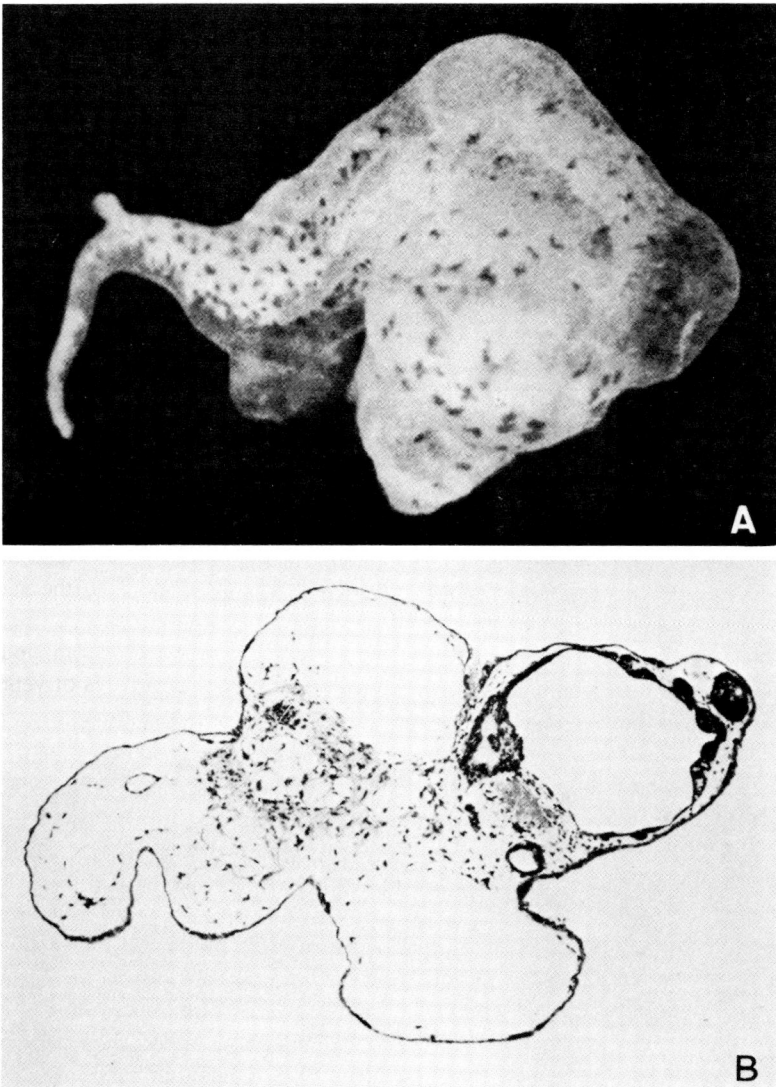

FIG. 8.   An external view (A) and a section (B) of an explant induced by liver + bone marrow tissues, resulting in a complete "embryo." A tail-like formation with myotomes and notochord is seen on the left (bone marrow effect) and, on the right, around the liver implant, forebrain structures with an eye can be identified. [Toivonen and Saxén, 1955; Saxén and Toivonen, 1962.]

the beginning of gastrulation, nuclear RNA and protein synthesis becomes more active on the dorsal than on the ventral side. Qualitative differences in the pattern of the proteins synthesized in the dorsal and ventral sides become detectable at gastrulation (Smith and Knowland, 1984). The synthesis of all macromolecules

(DNA, RNAs, and proteins) markedly increases during neurulation. At this stage, cephalocaudal and dorsoventral gradients of RNA and protein synthesis become more conspicuous.

The work of Denis (1964), Tiedemann (1968), and others (reviewed by Brachet and Malpoix, 1971) has shown that contact between inducing chordoblast and ectoblast is soon followed by DNA replication (as shown by a strong stimulation of mitotic activity) and by the synthesis of new mRNA and protein species. Obviously, then, gene activation is closely linked to neural induction. Unfortunately, there have been no serious attempts to identify, even with the powerful methods now available, the mRNAs and proteins that are synthesized as the result of neural induction. Injection of specific anti-sense mRNAs (Melton, 1985) should help us in reaching that goal.

That RNA and protein synthesis are required for neural induction is shown by experiments with actinomycin D, puromycin, and cycloheximide (Brachet *et al.*, 1964). Treatment of fertilized frog eggs with these inhibitors results in inhibition of gastrulation, while cleavage remains unaffected. After treatment of young gastrulae with the same inhibitors, microcephalic embryos are obtained (Brachet *et al.*, 1964). More interesting is the fact that actinomycin D treatment suppresses the competence of the reacting ectoblast, while the inducing power of the organizer remains unaffected (Denis, 1964). This shows that the response of the ectoblast to an inducer requires RNA synthesis. In contrast, such synthesis is not needed for the acquisition of inducing capacities in the organizer. Another interesting fact is that cycloheximide extends the duration of the ectoblast's competence toward Tiedemann's vegetalizing protein; this protein displays vegetalizing activity for many hours when it is tested on ectoblast fragments that have been treated with cycloheximide for several hours and then washed. In contrast, the vegetalizing protein has no effect on ectoblast from an advanced untreated gastrula (Grunz, 1972). It thus appears that the arrest of protein synthesis had prevented the aging of the ectoblast.

The already mentioned classic experiments of Holtfreter (1939) have shown that, in amphibian gastrulae, the different territories have positive or negative tissue affinities between each other (Fig. 18, Chapter 3, Volume 1). Thus, if one joins, side by side, fragments of ectoblast and entoblast, after a short period of fusion, sorting out takes place (negative affinity). On the other hand, mesoblast has a positive affinity for both ectoblast and entoblast. Treatment with Tiedemann's vegetalizing protein or with actinomycin D modifies the ectoblast's affinity. It now adheres to the entoblast and finally surrounds it. This "wrapping up" of the entoblast by the ectoblast does not occur if the ectoblast has been pretreated with cycloheximide (Grunz, 1972). Synthesis of macromolecules thus seems to be involved in the acquisition or maintenance of Holtfreter's tissue affinities, which presumably result from interactions between receptors and ligands.

Since the inducing chordomesoblast displays, during gastrulation, a very strong positive tissue affinity for the overlying ectoblast, it was of interest to find out whether substances are transferred from the inducer to the reacting ectoblast. The experiments designed to test this possibility have shown that a transfer of large molecules may, indeed, occur, but there is no evidence that this transfer is unidirectional, i.e., that it takes place only from the inducer to the reacting tissue (Brachet, 1950). Other experiments (reviewed by Toivonen, 1979) attempted to determine whether the inducing stimulus can cross a membrane interposed between the inducer and the reacting system. The author found that a cellophane membrane completely stopped neural induction, which is, therefore, not due to production of small molecules such as CO_2 or NH_4OH (Brachet, 1950). On the other hand, Millipore filters do not arrest the inducing stimulus, as shown by Toivonen. It thus seems likely that induction does not require intimate contact between inducing and reacting cells, and that it is mediated by nondialyzable molecules (presumably proteins or glycoproteins) that are secreted by the inducing cells. However, the cell surface plays an important role in neural induction, since this process is inhibited by lectins (Duprat et al., 1982).

Edelman (1983) and Edelman et al. (1983) have shown that cell adhesion molecules (CAMs; see Chapter 3, Volume 1) play an important role during early embryonic events, particularly in the organization and differentiation of the nervous system in chick embryos. CAMs have a marked tissue specificity. They are different in neurons (N-CAM), glial cells, liver (L-CAM), and striated muscles. During neurulation, N-CAM increases and L-CAM decreases in the chick embryo; N-CAM accumulates in the neural plate, the neural tube, and cardiac mesoderm; in contrast, L-CAM appears in all of the budding endodermal structures (liver, pancreas, etc.). This distribution suggests that CAMs play an important role in neural induction.

Recent work in Edelman's laboratory (Grumet et al., 1984; Grumet and Edelman, 1984) has demonstrated that a specific CAM (Ng-CAM) mediates adhesion between neuronal and glia cells in the chick embryo brain. Like N-CAM, Ng-CAM is located on the cell surface of the neurons; the same neuron can thus possess at least two different CAMs. N-CAM and Ng-CAM have common antigenic determinants, but their peptide maps differ. Both act by Ca^{2+}-independent cell-to-cell adhesion mechanisms; it is believed that Ng-CAM interacts with a small unidentified CAM present on the cell surface of the glia cells.

N-CAMs are present in the brains of all vertebrates; they allow adhesion between neural cells of different vertebrate species. Their conservation during evolution speaks for an important role in nervous system morphogenesis (Hoffman et al., 1984). This role has been recently reviewed by Rutishauser (1984).

These studies led Edelman (1984) to propose the *regulator hypothesis* of morphogenesis. CAM genes would be expressed before and independently of cytodifferentiation. The morphogenetic movements necessary for inductions would be controlled by CAMs; they would result from cell motility and CAM expression. The control of CAM genes by regulatory genes would determine the fate maps that are so familiar to all embryologists. Edelman's hypothesis certainly contains a good deal of truth, but we think that it neglects a few important facts. Cytoplasmic heterogeneity, morphogenetic gradients, positional information, and other factors should be taken into account in any general theory of morphogenesis. One might imagine that differential activation of the regulatory genes that control CAM gene activity results from cytoplasmic differences between the various presumptive territories of the very early embryo; if so, we would fall back on the already discussed theory of Morgan.

The importance of cell movements for gastrulation and neurulation, in particular for chordomesoderm migration (which is an absolute prerequisite for neural induction), hardly needs to be stressed. Recent work, already mentioned in Chapter 2, by Boucaut *et al.* (1984) and Nakatsuji and Johnson (1984) suggests that in amphibians, fibronectin, which is present in the dense fibrillar matrix underlying the blastocele roof, is an essential support for gastrulation movements. Antibodies directed against fibronectin inhibit invagination of the chordomesoderm, but have no effects on neurulation. Migration of the chordomesoderm takes place by "contact guidance" on an aligned network of extracellular fibers on the fibronectin-containing ectodermal layer. If one isolates this layer and orients it by mechanical tension, the cells move along the tension axis.

Finally, studies on the distribution of proteins in amphibian gastrulae and neurulae are beginning. It was found (Slack, 1984a) that the neural epidermis of axolotl neurulae contains abundant proteins that are not present elsewhere; some of them are cytokeratins typical of desmosomes. Specific glycoproteins appear in chorda, mesoderm, and epidermis. Slack (1984b) has also studied protein synthesis in explants of ectoderm isolated from the animal pole of young axolotl gastrulae. When the controls undergo neurulation, the explants express epidermal markers, not the mesodermal ones (ultrastructure, newly synthesized proteins, glycoproteins, and glycolipids). Cytokeratins appear as in whole embryos, but the chorda-specific proteins are absent. In addition to providing useful biochemical markers for further studies, continuation of this work might lead to substantial progress in our understanding of the molecular mechanisms of neural induction.

C. SECONDARY INDUCTIONS (LENS, PANCREAS, RETINA, KIDNEY)

Lens differentiation is of particular interest becuse it is closely linked to the synthesis of lens-specific proteins—α-, β- and γ-*crystallines*. These proteins can be isolated in pure form, allowing the use of immunological methods for their cytological detection or quantitative estimation. Crystalline mRNAs have been isolated and titrated with cDNA probes, and work is now proceeding on the sequence organization of cloned crystalline genes (reviewed by Piatigrosky,

1981, 1984). In addition, the lens of amphibians can easily be surgically removed and its regeneration followed with these available methods.

The embryonic lens was one of the first materials where immunological methods were used successfully in embryology (Woerdeman, 1955). These early experiments have shown that the crystallines became detectable at about the time when the lens was induced by the eye cup; they greatly increased in amount during lens differentiation, when the lens fibers became microscopically visible. This increase was due to an enormous increase in the synthesis of the crystalline mRNAs. A major effect of induction by the eye cup is, thus, the selective activation of the previously silent crystalline genes in the differentiating lens (see the review by Piatigorsky, 1981, for details). This selective activation does not result from changes in the methylation pattern of the γ-crystalline genes (Errington et al., 1983; Bower et al., 1983).

As for amphibian primary induction (Brachet, 1950), lens induction ceases when a cellophane membrane is interposed between the optic cup and the lens anlage (McKeehan, 1951). Activation of the varous crystalline genes (which takes place in a sequential order) thus results from the diffusion of a soluble, nondialyzable factor that is produced by the embryonic neural retina.

The so-called Wolffian regeneration of the lens has been studied in detail by Yamada (1967). If, as shown in Fig. 9, one removes the lens of an adult urodele (or of an embryo from any amphibian species), a new lens regenerates at the expense of the iris. Analysis of this process shows that a series of steps is involved. First, the iris cells lose their pigment and thus undergo dedifferentation; the cells apparently cast out their melanin granules before regeneration starts. The next step is intensive DNA replication, a phenomenon that, as we know, is, in general, a prerequisite for cell differentiation; it leads to an increase in the number of cells and, as a consequence, to an increase in the size of the regenerate. The following stage of lens regeneration is characterized by RNA synthesis; the number of both rRNA and mRNA molecules increases in the regenerating cells. Since regeneration is stopped by actinomycin D treatment, it is clear that this accumulation of ribosomes and messengers is absolutely required for the success of lens regeneration. Finally, immunofluorescence allows the detection of the α-, β-, and γ-crystallines that appear in a well-defined order; at the time of lens fiber differentiation, the crystallines can be seen in both the fibers and the cell nuclei. Finally, the nuclei degenerate and disappear. There is some evidence in favor of the view that a soluble factor produced by the retina is involved in lens regeneration. In a recent paper on lens regeneration, Yamada and McDevitt (1984) pointed out that dedifferentiation of iris epithelial cells must be preceded by a certain number of mitotic divisions. The retina factor is probably a mitogen for these cells. Cell-type conversion is controlled by the number of cell cycles traversed.

Lens Wolffian regeneration is of interest in that it is a clear case of transdeter-

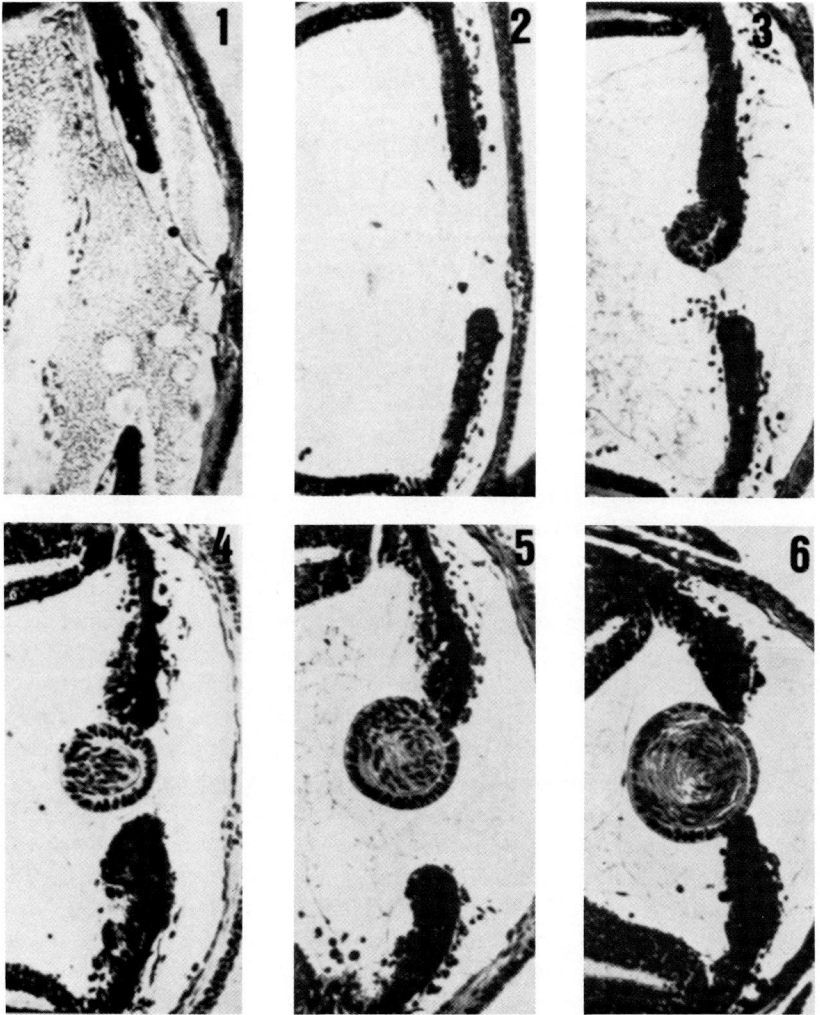

FIG. 9. Wolffian regeneration of the lens from the upper iris border. (1–6), successive stages (Sato, 1930). Spemann, 1968.

mination, a phenomenon that was discovered by E. Hadorn in transplanted imaginal disks of *Drosophila* larvae (Hadorn, 1968). It is essentially similar to the transdifferentiation studied by Eguchi and Okada (1973) and reviewed by Okada in 1980. As we have seen, their experiments were done on neural retina cells in culture. As in Yamada's (1967) experiments on whole animals, the pigmented cells lose their melanin granules and redifferentiate into lens cells, with the formation of isolated small lenses; these lentoids contain the typical crystallines. In both conditions, *in vivo* and *in vitro,* inactive crystalline genes are activated. The difference between transdetermination and transdifferentiation is that the former occurs in still undifferentiated embryos. It thus takes place at an earlier stage than transdifferentiation (just as embryonic regulation precedes regeneration of an already existing organ).

Lens formation and regeneration provide opportunities for analysis of cell differentiation by immunological methods. A different approach has been used successfully in the study of pancreas differentiation: the enzymological approach used by W. Rutter and his colleagues. Differentiation of the acini, which secrete the enzymes synthesized by the exocrine part of the pancreas, results from an induction by the surrounding mesenchyme. This induction, which is inhibited by treatment with cytochalasin B and thus requires the integrity of the actin microfilament (MF) network, can take place through a Millipore filter. One can recover from the filter a glycoprotein that acts on the surface of the pancreatic cells, where it induces DNA synthesis and mitotic activity (Levine *et al.,* 1973). This is followed, as in the lens, by rRNA, mRNA, and, finally, pancreatic enzyme synthesis. As with the crystallines, the various pancreatic enzymes (trypsin, chymotrypsin, amylase, carboxypeptidase, etc.) are synthesized sequentially in a rigorous and constant order. It is clear that the structural genes coding for all of these enzymes are activated in a sequential and individual manner, since there is a concomitant increase in their mRNAs. For instance, the content of amylase mRNA in the pancreatic cells increases as much as 400-fold during cell differentiation (Harding and Rutter, 1978). More recently, Van Nest *et al.* (1983) have followed the pattern of protein synthesis before and during differentiation of the embryonic pancreas. About 200 proteins are synthesized by the undifferentiated tissue; their synthesis decreases very strongly at the beginning of differentiation, while the expression of amylase and chymotrypsin greatly increases. The other pancreatic enzymes are not synthesized until a later stage of differentiation. Treatment with BrdUrd strongly modifies both morphogenesis and the pattern of protein synthesis. It alters, in a specific way, the complex developmental program of the pancreas.

The enzymological approach has also proved very useful in the studies of A. A. Moscona (1957) and M. Moscona *et al.* (1972) on the differentiation of neural retina cells in chick embryos. He found that the enzyme glutamine synthetase (a marker for glia cells) is synthesized when the neural retina cells undergo differentiation. Enzyme synthesis occurs 2–3 days earlier when the cells are dissociated and cultured *in vitro,* suggesting that cell-to-cell communications exert some kind of control on glutamine synthetase synthesis. Addition of corticosteroids to the medium speeds up enzyme synthesis in cultured retina cells. Interestingly, addition of actinomycin D and even, under given experimental conditions, cycloheximide produces a superinduction of enzyme synthesis, with a paradoxical increase in enzymatic activity. Proflavine and ethidium bromide, which are believed to act mainly on mitochondrial DNA synthesis, markedly reduce glutamine synthetase synthesis when it is induced by glucocorticoids. This synthesis is also inhibited by cordycepin, an inhibitor of poly(A) addition to mRNAs. As one can see, the neural retina system is complex, and it is likely that glutamine synthetase synthesis is regulated at both the transcriptional and translational levels.

Inductive processes are still at work at relatively late stages of embryogenesis. This is the case, for instance, in the development of kidney tubules in the metanephric blastema. Tubule formation in the blastema is triggered by the epithelial bud of the Wolffian duct (Grobstein, 1953); the inductor does not require cell-to-cell contact and is not arrested by a Millipore filter (Grobstein and Dalton, 1957). However, accurate electron microscopic (EM) studies by Wartiovaara *et al.* (1972) have shown that one should be careful in the interpretation of transfilter experiments. The cells on both sides of the filter may extend very long filopodia that could establish contact inside the pores. As in the previously examined systems, induction of kidney tubules is actinomycin D sensitive and goes hand in hand with increases in DNA replication and in RNA and protein synthesis (reviewed by Saxén *et al.*, 1968).

It is beyond the scope of this book to discuss all the complexities of sexual differentiation in gonads.[1] It is known that sex is determined by the heterochromosomes in a great majority of animal species. In mammals, the females are homozygous (XX chromosomes) and the males are heterozygous (XY chromosomes). It has been known for many years that the sexual phenotype can be reversed by addition of appropriate steroid hormones (for instance, masculinization can be induced by addition of testosterone to fertilized amphibian eggs). In mammals, in order to become a male, the fetus requires a müllerian duct–inhibiting substance (MIS), in addition to testosterone. This factor, which, like testosterone, is synthesized in the male gonad, prevents the formation of the uterus and the Fallopian tubes. MIS is a large glycoprotein that is responsible for the regression of müllerian ducts in males (Budzick *et al.*, 1983). There are indications that the biological effects of MIS might be mediated by protein dephosphorylation. A strong nucleotide pyrophosphatase activity is localized around the regressing ducts (Fallat *et al.*, 1983), and the phosphatase inhibitor sodium fluoride induces the regression of the ducts in the absence of MIS (Hutson *et al.*, 1984). Testosterone, like the other steroid hormones, must bind to specific receptors present only in the target organs in order to be active. In addition, more recent work has indicated the probable role of a male-specific antigen, called the *H-Y antigen*, in male sex differentiation (Wachtel *et al.*, 1975, reviewed by Silvers *et al.*, 1982). Experiments have shown that this antigen (which is coded by a single gene located on the Y heterochromosomes)

---

[1]See the supplement to *Differentiation* **23** (1983).

can induce the masculinization of bovine XX undifferentiated gonads; testosterone and MIS then appear. Much work currently is being done in many laboratories in order to establish more precisely the role of the H-Y antigen in the primary determination of sex.

D. DIFFERENTIATION OF CULTURED CELLS

The work on embryos that has just been presented leads to the conclusion that the key event in embryonic differentiation is the selective activation and expression of specific genes. This conclusion also emerges overwhelmingly from the innumerable studies on embryonic blast cells cultured under *in vitro* conditions. These studies, which were almost nonexistent when the author wrote "Biochemical Cytology" (Brachet, 1957), are now so numerous that it is impossible to list them here; cell differentiation, which has been an almost neglected subject for many years, is now very popular. This is largely due to the fact that advances in tissue culture technology now allow us to follow *in vitro* the differentiation of almost any type of stem cell.

One of the favorite systems for the study of cell differentiation in culture are the myoblasts (reviewed by Wakelam, 1985). These undifferentiated mesodermic cells fuse together in culture and the resulting myotubes differentiate into contractile fibers, which have the characteristic morphology of striated muscles. An advantage for molecular biologists interested in cell differentiation is that myogenesis is closely linked to the production of several well-defined proteins. Alpha-actin, muscle myosin and creatine kinase, α- and β-tropomyosin, and acetylcholine receptors are excellent biochemical markers of myogenesis. Once one has isolated the polyribosomes that are synthesizing one of these typical muscle proteins with a specific antibody, it is at least theoretically easy to obtain the corresponding mRNA, to synthesize its cDNA with reverse transcriptase, and to sequence it.

Early work on muscle differentiation *in vitro* was well summarized in 1969 by Yaffe, who emphasized that DNA synthesis stops at fusion and that, judging from actinomycin D sensitivity, myoblasts synthesize unstable mRNA molecules. When the differentiated stage is reached after fusion, the muscle cells contain stable mRNAs, which is shown by the fact that actinomycin D no longer exerts strong effects. More recent work by Buckingham *et al.* (1976) has shown that, in myoblasts, myosin mRNA is mainly localized in mRNP particles that are smaller than ribosomes. When cells fuse to form myotubes, myosin mRNA is shifted from the free mRNP particles to the polyribosomes, where it is translated. This is reminiscent of the process of sea urchin egg fertilization and might well be a fairly general phenomenon. However, it is not certain that the mRNAs stored in untranslatable form in ribonucleoprotein (RNP) complexes always serve as precursors for the polyribosomal mRNAs. Saidapet *et al.* (1982) showed that, during the transformation of myoblasts into myocytes in living chick embryos, the mRNAs for the muscle contractile proteins increase 45 times in the polyribosomes and only 3 times in the free mRNP particles. Further analysis led them to conclude that these particles do not store mRNA for the polyribosomes. If so, their function in the cell remains enigmatic. Both myoblasts and myotubes possess a large population of about 17,000 different mRNA species (Paterson and Bishop, 1977). The complexity of this population does not vary greatly during myogenesis, suggesting that differentiation requires the accumulation of numerous copies of a small number of tissue-specific gene transcripts. More recently, Schwartz and Rothblum (1981) studied the regulation of the actin genes during muscle differentiation. Myoblasts contain the mRNAs of β- and γ-actin, the ubiquitous cytoplasmic actins. In myotubes, these mRNAs decrease in amount, while the mRNA of the muscle-specific α-actin progressively increases. The synthesis of myosin heavy chain mRNA has been followed by Medford *et al.* (1983). As expected, it greatly increases (500 times in 6 days) during myoblast differentiation, with no change in its stability; its transcription rate increases as much as 100 times. Since withdrawal from the cell cycle increases the effective mRNA stability four to five times. Medford *et al.* (1983) suggest that a cell cycle–mediated regulation controls mRNA accumulation during cell differentiation.

A new marker of myocyte differentiation has been found by Nelson and Lazarides (1983). Primary chicken myoblasts express α,γ-spectrin (γ-spectrin is also called *fodrin*), and there is a switch to α,β-spectrin during their terminal differentiation. This switch is due to the synthesis of β-spectrin during myogenesis.

There are indirect indications, based on the use of inhibitors, of a role of polyamines and calmodulin in muscle differentiation. Erwin *et al.* (1983) have reported that α-difluoromethylornithine (DFMO), a specific inhibitor of ornithine decarboxylase, inhibits the insulin-induced differentiation of myoblasts; addition of putrescine to the culture medium reverses the inhibition. Fusion of myoblasts is inhibited by trifluoperazine, the classic inhibitor of calmodulin; this suggests that fusion is a Ca^{2+}-calmodulin-dependent process (Bar-Sagi and Prives, 1983). However, the mechanism of myoblast fusion is not yet completely known. A study by Parfett *et al.* (1981) shows that there is neither fusion nor differentiation of the myoblasts in the presence of concanavalin A, suggesting a role for mannosylated glycoproteins in myoblast differentiation and confirming that fusion is a necessary step in myogenesis. Curiously, ethidium bromide inhibits the differentiation of myoblasts into myotubes (Brunk, 1979). We have obtained comparable results by continuous treatment of fertilized amphibian eggs with ethidium bromide; this inhibitor of mitochondrial DNA synthesis has no effect on development until the tadpole stage. Outwardly, the treated embryos look normal, but they do not contract when they are touched with the tip of a needle. This paralysis is due to the fact that the somites have not differentiated into muscles and have remained filled with yolk platelets. Biochemical studies on the effects of ethidium bromide on *in vitro* and *in vivo* myogenesis might be rewarding.

In the embryo, cartilage is induced in the somites by a glycoprotein released by the sheath that surrounds the notochord. The chondroblasts differentiate into chondrocytes (Fig. 2) that incorporate sulfate and synthesize a metachromatic (i.e., stained in red with toluidine blue) acellular matrix; sulfate is incorporated into sulfated proteoglycans, in particular, chondroitin sulfate, which is the classic marker of cartilage differentiation. It has been reported that even dividing chondrocytes synthesize the cartilage matrix chemical constituents (Coon and Cahn, 1966) and that chondroitin sulfate production precedes the formation of a microscopically visible cartilage matrix. Differentiation of embryonic chondroblasts into cartilage has been extensively studied *in vitro* (Holtzer *et al.,* 1972; Lash, 1969; Levitt and Dorfman, 1974). Of interest is the fact that, according to Okayama *et al.* (1976), the proteoglycans synthesized by chondroblasts and differentiated cartilage are not the same; slightly different proteoglycans are found in the matrix of young adult and old cartilage cells. This shows that a complex developmental program is followed during cartilage differentiation. Soluble factors released by *in vitro* growing chondrocytes stimulate their differentiation into cartilage and selectively increase the synthesis of chondroproteins and α_1-collagen (Solursh and Meier, 1973). More recently, it has been shown that cartilage differentiation in living chicken embryos is accompanied by the disappearance of two non-histone proteins; one of them, called PCP 35.5, strongly binds to DNA sequences close to active (DNase-sensitive) chromatin and might thus play a role in the control of the genes coding for the synthesis of tissue-specific cartilage proteins (Perle *et al.,* 1982). However, one should not forget that changes in culture conditions strongly modify the chemical composition of the matrix. Benya and Shaffer (1982) have shown that while differentiated chondrocytes synthesize type II collagen and cartilage-specific proteoglycans, they "differentiate" after serial culture. This phenotype is replaced by another one in which the cells synthesize type I collagen and very few proteoglycans; culture in agarose gel restores the initial phenotype. The authors conclude that the phenotype is modulated by the cell shape, but their experimental results may also suggest that continuous interactions between the cell and the matrix play a role in chondrocyte differentiation and that cell shape is affected by the chemical composition of the matrix. The proliferation of mammalian chondrocytes is directly initiated by the pituitary growth hormone, according to Madsen *et al.* (1983). Curiously, a phorbol ester (PMA) induces the dedifferentiation of chondrocytes in fibroblast-like cells; the expression of type I and type II collagen mRNAs decreases 10 times in the treated cells (Finer *et al.,* 1985).

We shall briefly mention again the differentiation of some fibroblast strains into adipocytes. Treatment with insulin triggers the accumulation of triglycerides into the fibroblasts, which, therefore, acquire the phenotype characteristic of fat cells. Differentiation of fibroblasts into adipocytes is arrested by treatment with interferon, which inhibits the hexose monophosphate shunt required for lipid synthesis (Saneto and Johnson, 1982). During the accumulation of fat vacuoles into the cells characteristic of the transition from fibroblast to preadipocyte, cytoskeletal organization breaks down; simultaneously, the content of the cells in tubulin and actin mRNAs strongly decreases (Spiegelman and Farmer, 1982), while several other mRNAs accumulate up to 150 times (Spiegelman *et al.*, 1983). Krawisz and Scott (1982) have obtained preadipocyte–adipocyte differentiation by treatment with a heparinized medium containing human plasma without hormone addition. In fact, according to Steinberg and Brownstein (1982), insulin is not the real inducer of commitment in preadipose fibroblasts; it merely increases the synthesis and accumulation of lipids in already committed cells.

Much work has been done on red blood cells differentiation (erythropoiesis), since it is characterized by the appearance of hemoglobin, a protein that is easily detectable cytochemically by the bendizine peroxidase test (reviewed by Till and McCulloch, 1980, on the differentiation of hematopoietic stem cells). Hematopoiesis can be studied in the blood islands of early chick embryos, in the embryonic liver of mammals, and in the bone marrow of adults. It can also be studied *in vitro* on hematopoietic cell lines. Addition to erythroblasts of hematopoietin, a protein hormone present mainly in the kidney, induces a burst of erythropoiesis (stem cell proliferation followed by cell differentiation). One reason for the current interest in erythropoiesis is that we have an intimate knowledge of α- and β-globin gene organization (see Chapter 4, Volume 1) and that we can follow, at the molecular level, the shift from embryonic to fetal and, finally, adult hemoglobins. We know the chromosomal localization of the corresponding genes and the fact that an erythrocyte expresses only one type of hemoglobin (either fetal or adult).

Another reason for our present interest in erythropoiesis is the existence of many forms of thalassemias—hereditary diseases due to an imbalance in the synthesis of the α- and β-chains (see the recent brief review by Weatherall and Clegg, 1982). The molecular diversity of the α-thalassemias (excess of the α-chain over the β-chain) is truly amazing; this diversity can be due to gene deletion, incomplete or complete lack of β-gene expression, etc. The β-thalassemias result mainly from mutations that interfere with mRNA processing, translation, or stability, and may even be due, in some cases, to an instability of the synthesized protein. A study of thalassemia may ultimately lead to a genetic cure (by transfer of the normal gene into the patient's deficient genome) of the various forms of thalassemia.

Finally, we should mention Friend murine erythroleukemic (MEL) cells. Synthesis of hemoglobin and other markers of erythroid differentiation can be induced in these "pro-erythroblastoid" cells, which have been transformed by Friend leukemia virus, by treatment with such simple chemical agents as dimethyl sulfoxide or butyrate. We shall return later to MEL cells, since they provide, together with teratocarcinoma and neuroblastoma cells, an important bridge between cell differentiation and malignant transformation. The essential point, for our present purpose, is the following: whether one follows normal or chemically induced erythropoiesis, there is a close parallel between the number of hemoglobin mRNA molecules present in a cell and the amount of hemoglobin produced by the cell. In other words, differentiation of red blood cells is controlled at the transcriptional level. All cells, whether or not they will later express hemoglobin, have the same number of globin genes. In almost all cells, these genes are silent or expressed at a very low level. Only in cells that are committed to follow the erythropoietic pathway will the globin genes be activated (and transcribed); only in these cells will the globin mRNAs accumulate and be translated with the production of large amounts of hemoglobin. The same is true for the other red blood cell–specific proteins such as spectrin (a constituent of the plasma membrane underlying the cytoskeleton) and carbonic anhydrase. Thus, during erythropoiesis, a small number of genes are activated in a sequential fashion.

Many other systems are amenable to the analysis of *in vitro* cell differentiation. One of them is the formation of *melanocytes,* in which the presence of brown pigment granules is a convenient morphological marker. At the biochemical level, melanocyte differentiation can be followed by measuring the activity of enzymes involved in melanin synthesis (tyrosine hydroxylase, commonly called *tyrosinase,* and *DOPA-oxidase*). The pigmented phenotype of the melanocytes is often labile. Isolated melanocytes may lose their pigment and regain it when sparse cultures become confluent. When embryonic fibroblasts differentiate *in situ* or *in vitro,* they synthesize one of the types of collagen.

Particularly interesting is the differentiation of nerve cells. Neurons possess a number of morphological, physiological, and biochemical markers. They form neurites (axons and dendrites), possess typical electrophysiological properties when excited, and contain a number of nervous system–specific proteins (S100 protein, acetylcholinesterase, choline acetyltransferase, etc.). It is possible to specifically stimulate the outgrowth of neurites by addition of the nerve growth factor (NGF), a protein that has been isolated from the male rat submaxillary gland and extensively characterized (Levi-Montalcini and Angeletti, 1968). Since neurons no longer divide after birth, work on cultured nerve cells is usually done on embryonic neuroblasts or on cell lines derived from tumors (neuroblastomas, neurogliomas). Another system that might soon become amenable to biochemical experimentation is bone growth and differentiation. Farley and Baylink (1982) have isolated a 83,000-dalton protein from demineralized human bone (skeletal growth factor) that selectively stimulates DNA synthesis in the bones of chick embryos. Urist *et al.* (1984) isolated a bovine bone morphogenetic hormone ($M_r = 18,500$) that induces the differentiation of mesenchyme cells in cartilage and bone. Transformation of mesenchyme cells deriving from embryonic muscle can also be achieved by implantation of demineralized bone matrix (Sampath *et al.,* 1984).

The differentiation of skin keratinocytes can be followed by observing keratin synthesis and accumulation; differentiation can be obtained by increasing the Ca^{2+} content of the medium above 0.1 mM. This suffices to induce stratification and cornification of the cells, as well as the appearance of several marker proteins (Stanley and Yuspa, 1983).

This brief and incomplete overview of *in vitro* differentiation of embryonal cells shows the multiplicity of the systems. We shall now see how they can be used by experimenters.

E. EXPERIMENTAL ANALYSIS OF CELL DIFFERENTIATION

1. Cell Fusion[2]

B. Ephrussi, using inactivated Sendai virus, showed that fusion of a differentiated, pigmented melanoma cell with an undifferentiated embryonic fibroblast suppressed the differentiated phenotype. The resulting hybrid was unpigmented (Ephrussi *et al.*, 1964; Ephrussi and Weiss, 1965; Davidson *et al.*, 1966), and the enzymes responsible for melanin production could no longer be detected. Fusion of the differentiated cell with an undifferentiated cell thus "extinguished" the differentiated phenotype. Thus, it appears that the cytoplasm of the undifferentiated cell exerted a negative control, through a repressor, on melanocyte differentiation. Further experiments showed that, in this system, a gene dosage effect is in operation. If one fuses cells from a tetraploid melanoma with a diploid fibroblast, the hybrid cell retains its pigment (Fougère *et al.*, 1972). We have seen that, in hybrid cell lines, the chromosomal constitution is unstable. Chromosomes are generally lost during the successive cell divisions. The importance of this phenomenon for cell differentiation was quickly established by Sparkes and Weiss (1973). In a cell hybrid between a hepatoma cell and a normal epithelial cell, a typical hepatoma biochemical marker (alanine aminotransferase) is extinguished after cell fusion; this marker reappears after a few cell generations, when some of the chromosomes have been lost. Extinction and reexpression of the differentiated phenotype after a number of cell generations, in hybrids between differentiated and undifferentiated cells, is a general rule. Extinction seems to be due to a diffusible cytoplasmic factor and not, as was generally believed, to chromosome loss (Mével-Ninio and Weiss, 1981).

A step forward has been made by Killary and Fournier (1984), who studied the extinction of several hepatic markers in hybrids between rat hepatoma cells and mouse fibroblasts. Loss of all the markers is correlated with retention of mouse chromosomes 8, 9, 10, 11, and 13. In hepatoma × microcell (with a single mouse chromosome) hybrids, retention of chromosome 11 extinguishes the expression of tyrosine aminotransferase (a liver enzyme), but not that of the other hepatic

---

[2]Reviewed by Davidson (1971), book by Ephrussi (1972).

Ephrussi's work on the transplantation of eye disks isolated from *Drosophila* eye color mutants into larvae of another genotype was at the root of the one gene–one enzyme theory; he also discovered the mitochondrial *petite* mutations in yeast. Yet he never received the Nobel Prize and nobody—except perhaps in Stockholm—knows why.

markers. These experiments strongly suggest that extinction results from the production of repressors by chromosomes that have been retained in the hybrids, rather than to chromosome loss.

It is important to note that all of these results and conclusions have been obtained with hybrid *cell lines*. A few experiments recently done on hetero-karyons (isolated shortly after fusion) led to a different picture. In heterokaryons between chick myocytes and mouse adrenal cells, there is no extinction of the differentiated phenotypes. The heterokaryons retain their muscular functions and synthesize steroids (Wright, 1984a). In heterokaryons between chick myocytes and rat brain cells, activity of the muscle genes is even induced in the neuronal cells (Wright, 1984b). On the other hand, in heterokaryons between myocard cells and KB cells (cells originating from a human epidermoid carcinoma), muscle contractility disappears 2–4 hr after fusion (Goshima *et al.*, 1984). The results of more extensive biochemical studies on heterokaryons will be awaited with interest. In the following section, we shall deal with hybrid cell lines unless stated otherwise, since they are the classic material for the studies on cell differentiation.

Cell fusion has been extensively used for the analysis of cell differentiation since Ephrussi's pioneer work. For instance, Benda and Davidson (1971) fused glial cells with embryonal undifferentiated cells and observed the disappearance, in the resulting hybrid, of the S100 protein, which is a specific biochemical marker of nerve cells. In the classic experiments of H. Harris (see Chapter 1) in which a chicken red blood cell was fused with a mouse fibroblast or an HeLa cell, hemoglobin soon disappears from the hybrid. This was not surprising, at first sight, since the adult chicken red blood cell no longer synthesized hemo-globin. However, fusion reactivated the chick erythrocyte nucleus, which swelled and synthesized RNA; the globin genes were not reactivated. Deisseroth *et al.* (1975, 1976) and Conscience *et al.* (1977) carried the analysis one step further with Friend (MEL) cells induced by treatment with dimethyl sulfoxide (DMSO) to synthesize erythrocyte marker proteins. Fusion of such an induced Friend cell with a fibroblast suppressed the transcription of the globin genes, as shown by a strong decrease in globin mRNA in the hybrids. In MEL cells, DMSO increased the activity of cholinesterase and carbonic anhydrase, as well as the globin mRNA content; these increases were suppressed when an induced MEL cell was fused with a fibroblast or a hepatoma cell. Obviously, treatment of erythroleukemia cells with DMSO selectively stimulated the transcription of a few genes involved in erythroid differentiation; fusion with an undifferentiated cell suppressed the transcription of these specific differentiation genes.

Linder *et al.* (1981) extended the early findings of Davis and Harris (1975) to heterokaryons between chick erythrocytes and rat myoblasts. Although such hybrids do not synthesize hemoglobin, they can produce globin mRNAs, but are apparently unable to synthesize heme. Interestingly, Linder *et al.* (1981) found that the globin mRNAs present in these hybrids were encoding only the adult globin; there was no formation of embryonic globin mRNAs. Since only the adult globin genes were reactivated by fusion, it is clear that there was no "reprogramming," which would have occurred if embryonic chicken globin genes had been reactivated by the rat cytoplasm before the adult genes. Thus, only adult genes, which had been inactivated when the erythrocyte nucleus underwent its final condensation, could be reactivated by fusion with an active cell.

Fusion of an antibody-producing myeloma cell with a fibroblast (but not with a lymphocyte) suppresses immunoglobin synthesis (Coffino *et al.*, 1971). Hybridomas are the products of a fusion between a myeloma cell and a spleen cell taken from a mouse immunized against a given antibody. Hybridomas secrete monoclonal antibodies specific for this antigen (Köhler and Milstein, 1975) and, for this reason, are widely used by immunologists.

The hepatocyte–hepatoma system is also a favorable one because normal liver cells synthesize many enzymes and secrete serum proteins, in particular, serum albumin; in contrast, hepatoma cells, like the embryonic liver, produce α-fetoprotein. The genes coding for these two proteins display strong sequence homology and can be considered members of the same gene family, presumably deriving from the duplication of an ancestral gene (Jacodzinski *et al.*, 1981). Work done in the laboratory of M. Weiss has shown that fusion of a hepatocyte with a non-hepatocyte cell is, as usual, followed by extinction of the synthesis of liver-specific proteins (Weiss and Chaplain, 1971); in subclones from such extinguished hybrids, one or another of the many functions is reexpressed after the loss of given chromosomes. Interestingly, in hybrids between hepatoma and melanoma cells, there is mutual exclusion in the reexpression of the phenotypic markers. They reexpress either the hepatoma proteins or melanin, but never the two together (Fougère and Weiss, 1978). However, this is not an absolute rule, since, according to Allan and Harrison (1980), hybrids between erythroleukemic Friend cells and myeloma cells can produce hemoglobin and immunoglobin simultaneously. In the hepatocyte–hepatoma system, Szpirer *et al.* (1980) obtained "hepatocyte hybridomas" by fusing rat hepatocytes with mouse hepatoma cells. Such hybrids segregate rat chromosomes, and the number of their chromosomes (rat and mouse) can easily be counted. It was found that most of the hepatocyte hybridomas continued to synthesize one or more rat serum proteins. In this system, the corresponding structural genes remained active; however, the hybrids did not produce rat α-fetoprotein, although mouse α-fetoprotein synthesis was maintained. This suggests that silent rat (hepatocyte) α-fetoprotein genes coexist in the hybrid nuclei with active mouse (hepatoma) α-fetoprotein genes. More recently, Levilliers and Weiss (1983) made crosses between various strains of dedifferentiated hepatoma cells. None of the hybrid cell lines reverted to the original differentiation. Since there was no reexpression of hepatic functions, it is clear that no complementation between the two genomes took place. Complementation has been a powerful tool for the analysis of blocks in enzymatic pathways; the experiments of Levilliers and Weiss (1983) show that cell differentiation is more complex. We should also mention the work of Dickson *et al.* (1983), who made somatic hybrids between mouse neuroblastoma cells and human neurons, and found that human neuronal antigens were expressed; in this case, there was no extinction of the differentiated phenotype. These examples show the complexity of the interactions that take place when two different genomes are placed together in a common cytoplasm. It also shows that the extinction rule may have exceptions (like many laws promulgated by legislators).

In order to get some insight into the role played by the nuclear genes and possibly by the cytoplasm in the phenotypic expression of cell differentiation, Ringertz *et al.* (1978) fused together the cytoplasm of a fibroblast (thus, a cytoplast) with a mini-cell (a cytoplasmic fragment of a dividing cell) isolated from a myoblast. There was no inhibition of myogenesis. More straightforward are the results of Kahn *et al.* (1981) on cybrids obtained by fusing anucleate mouse fibroblasts and rat hepatoma cells. Extinction of serum protein synthesis lasted no longer than 12–20 hr. It was restored 48 hr after fusion, showing that a cytoplasmic factor exerts a negative control on gene expression, but this factor has a short life and is not renewed. It is probable that the nucleus itself contains or produces factors that inhibit cell differentiation. As shown by Liebermann and Sachs (1978), enucleation induces the differentiation (neurite formation) of nondifferentiated neuroblastoma cells. However, these observations remain isolated and much more work is needed before we can truly assess the respective roles of nuclear and cytoplasmic factors in cell differentiation.

It is not surprising that cell hybridization, especially when cells from different species are used as parents, gives results that are not easily amenable to a simple interpretation. We know next to nothing of the molecular interactions that take place in somatic hybrids. However, a promising start has been made by Sperling and Weiss (1980). As already mentioned, they found that the repeat length of chromatin varies in different cell types and has nothing to do with the generation time or the number of chromosomes. More important for our present purpose is that this repeat length changes when the differentiated phenotype is lost in somatic hybrids. This suggests that extinction of this phenotype is correlated with changes in the molecular organization of chromatin; such changes could affect the transcription of given genomic DNA sequences.

2. Effects of BrdUrd on Cell Differentiation[3]

We have already mentioned that BrdUrd is, in almost all instances, a powerful inhibitor of cell differentiation. Its remarkable negative effects on myogenesis,

[3]Reviewed by Rutter *et al.* (1973).

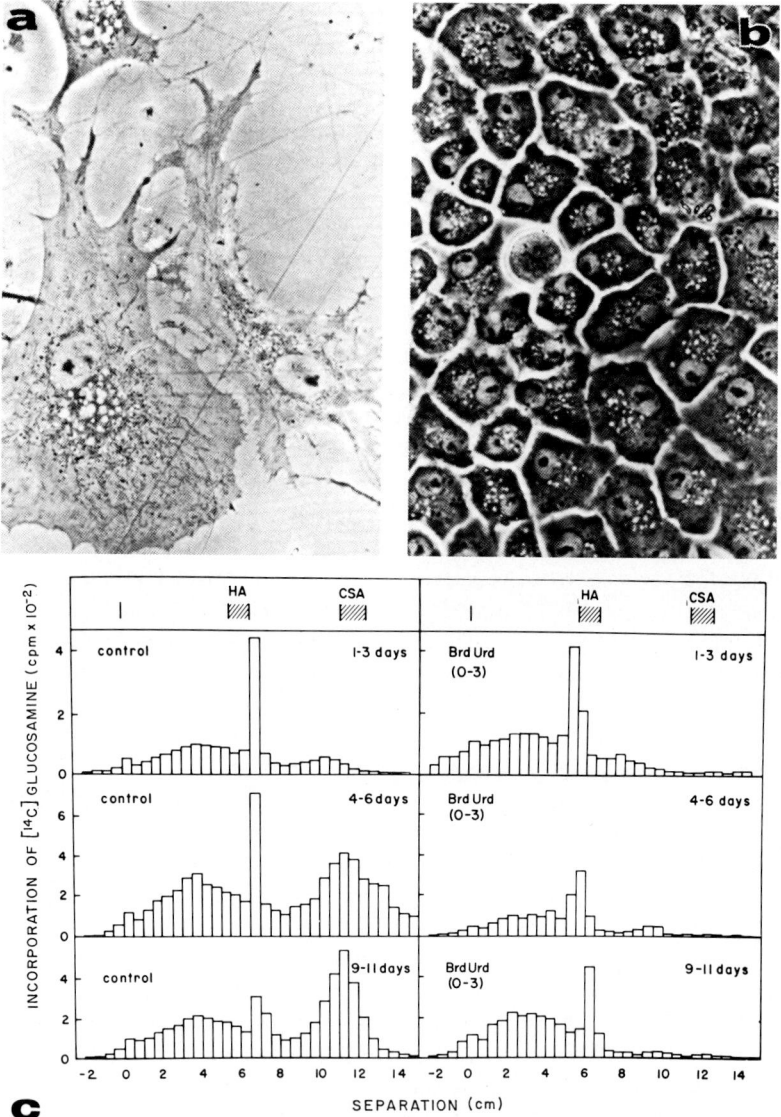

FIG. 10.   Effect of BrdUrd on cell differentiation. Living chondroblasts observed in a 3-week-old clonal culture, (a) Controls; (b) after 3 days of treatment with BrdUrd. (c) Separation by high-voltage electrophoresis of the glycosaminoglycan fraction obtained from notochord-somite cultures exposed to [³H]glucosamine on days 1–3, 4–6, or 9–11. Cultures were exposed to either BrdUrd (10 μg/ml) or thymidine (10 μg/ml) from days 0 to 3. Ha, hyaluronic acid; CSA, chondroitin sulfate. [(a,b) Courtesy of Drs. H. Holtzer and J. Abbott; (c) (Holtzer *et al.*, 1972).]

chondrogenesis, and erythrogenesis were discussed by Holtzer *et al.* (1972), who were among the first to show that the drug selectively suppresses the synthesis of luxury proteins. BrdUrd is an analog of thymidine and in order to be active, must be incorporated into DNA molecules during their replication. That the basic mode of action of BrdUrd is indeed a substitution in the DNA molecule is shown by the facts that it is active only in proliferating cells and that its inhibitory effects on cell differentiation are reversed by addition of thymidine to the medium.

Figure 10 shows the striking effects of BrdUrd on cartilage differentiation. The treated chondroblasts, in contrast to the controls, do not synthesize the chemical components of the extracellular matrix. Similarly, BrdUrd-treated melanoblasts do not form pigment granules unless thymidine is added to the medium. Suppression of myogenesis and erythrogenesis by the drug is accompanied by the absence, in the treated cells, of the corresponding marker proteins (muscle contractile proteins in the first case and hemoglobin in the second). In all of these experiments, BrdUrd was added at low concentrations that do not interfere with cell multiplication, but yet allow incorporation in the replicating DNA molecules. If BrdUrd is added to already differentiated cells (to late erythroblasts that are already synthesizing hemoglobin, for instance), it has no remarkable effects. Thus, in erythroblasts that are already expressing hemoglobin, further synthesis of this protein is BrdUrd resistant.

There are, as always, exceptions to general rules. In the present case, the most striking one is the induction of neurites in neuroblastoma cells by BrdUrd (Schubert and Jacob, 1970). Also, the results may vary according to the conditions used for the experiments. For instance, Ostertag *et al.* (1973) found that BrdUrd inhibits hemoglobin synthesis in DMSO-induced Friend erythroleukemia cells. In contrast, Adesnik and Smitkin (1978) reported that this same drug induces differentiation (thus, hemoglobin synthesis) in the same, but uninduced cells. To explain such paradoxical effects, Schubert and Jacob (1970) proposed that BrdUrd might exert secondary effects at the cell surface level, a possibility that should be seriously kept in mind. However, Wright and Aronoff (1983) have carefully studied hybrids between rat myoblasts (in which differentiation had been inhibited by BrdUrd) and differentiated chick myocytes; induction of rat myosin light chain was reduced five times compared to untreated controls. In these experiments, gene dosage and variations in BrdUrd incorporation into DNA were taken into account. The conclusion was that BrdUrd acts on DNA, not on the cell membrane.

The work just summarized has been done on cultured embryonic cells. Similar results have been obtained in organ cultures of small, still undifferentiated embryonic explants. For instance, Turkington *et al.* (1971) observed inhibition by BrdUrd of the *in vitro* differentiation of embryonic mammary glands. In embryonic limb buds, BrdUrd inhibits differentiation and cartilage proteoglycan synthesis without affecting collagen production (Levitt and Dorfman, 1972); it also inhibits differentiation and hemoglobin synthesis in *in vitro* cultured chick embryo blood islands (Miura and Wilt, 1971; Hagopian *et al.*, 1972), as well as in chicken embryos isolated at the primitive streak stage. If development is allowed to proceed for 10 hr more, BrdUrd no longer exerts negative effects on hemoglobin synthesis (Ingram *et al.*, 1974). According to Walther *et al.* (1974), the analog inhibits pancreas differentiation, provided that it is incorporated into DNA. Zimmerman *et al.* (1974) reported that BrdUrd inhibits melanin formation in somatic melanoblasts, but only partially in retina when pigmentation has already started. Moscona and Moscona (1979) showed that treatment with BrdUrd of retina from 5-day-old chicken embryos produces gross morphological abnormalities.

This leads us to the work done on whole embryos, which, as already mentioned, has led to rather disappointing results. At the time when the popularity of BrdUrd was at its peak (about 10 years ago), we studied its effects on a number of developmental systems (sea urchin, tunicate and amphibian eggs, and nucleate and anucleate fragments of *Acetabularia*). The results are summarized below (Tencer and Brachet, 1973). In sea urchin eggs, as previously found by Mazia and Gontcharoff (1964), development stops at the blastula stage; cleavage is delayed if treatment with BrdUrd begins before the 16-cell stage and is unaffected until the arrest at the blastula stage if it starts later. Cleavage arrest at the blastula stage can be explained by the occurrence of mitotic abnormalities in the treated eggs. In tunicate eggs, development stops at the gastrula stage in $10^{-3} M$ BrdUrd, but perfectly normal tadpoles (with eyes, brain, muscles, and notochord) are obtained in $10^{-4} M$ BrdUrd. In amphibian eggs, in which the drug was either added to whole eggs and explants of ectoderm or microinjected into fertilized eggs, the effects of the drug on embryonic differentiation were small and limited to an abnormal distribution of the muscle fibrils and lens fibers. In contrast, many cells underwent pycnotic degeneration; cleavage was quickly arrested if BrdUrd had been injected into fertilized eggs. Finally, in *Acetabularia,* where there is no chromosomal DNA synthesis so long as the vegetative nucleus is intact, BrdUrd had no effect on initiation of cap formation in both nucleate and anucleate fragments; later, the caps grew more slowly and often assumed a more irregular shape than the controls. These experiments led us to conclude that it was not possible to selectively suppress embryonic differentiation by treating whole eggs with BrdUrd and that the drug, after it was incorporated into DNA, exerted toxic side effects (chromosomal aberrations, pycnoses).

Similar conclusions have been drawn by Lee *et al.* (1974) and Garner (1974), who worked with early chick embryos and mouse eggs, respectively; only unspecific effects were obtained in these experiments. Bannigan and Langman (1979) injected BrdUrd into pregnant mouse females. They observed a slowdown of growth largely due to pycnotic degeneration of many nuclei, but no specific effect on embryonic differentiation was observed. More interesting are the results reported by Zagris and Eyal-Giladi (1982). They studied induction of the primitive streak in early chicken embryos and found that BrdUrd reduces the competence of the ectoblast without impairing the inducing power of the underlying hypoblast. A similar observation was made by Denis (1964), who studied the effects of actinomycin D on ectoblast competence and the inducing activity of the organizer in amphibians; it suggests that BrdUrd might interfere with transcription.

It is a striking fact, as we pointed out (Tencer and Brachet, 1973), that the BrdUrd-sensitive period coincide in sea urchin and amphibian eggs with stages in which mRNA synthesis is known to greatly increase. Although the molecular mechanisms underlying the remarkable effects of BrdUrd on *in vitro* cell differentiation are still poorly understood, it seems likely that incorporation of the drug into DNA affects regulatory sequences controlling the activity of the structural genes coding the differentiation-specific proteins. This hypothesis was proposed by Strom and Dorfman in 1976. BrdUrd would be incorporated into moderately repeated DNA sequences (perhaps amplified) that would control structural genes. Another possibility has been suggested by Bick and Devine (1977): the BrdUrd effects would result from a preferential binding of chromosomal proteins to the substituted DNA sequences; the result would be the cessation of their transcription.

Currently, BrdUrd has lost much of its popularity as a tool for suppressing differentiation because almost all possible differentiating systems have been treated with the drug. In addition, there are many new chemicals in use in the field. It has been found that phorbol esters, which promote cancer growth, also selectively affect cell differentiation.

3. The Newcomers: Phorbol Esters and Retinoids

The discovery by Berenblum (1954) that rubbing the ear of a mouse with croton oil (after local administration of a chemical carcinogen such as benzopyrene) speeds up the appearance of skin cancer is very important; later, we shall consider its impact on the field of carcinogenesis. The finding that croton oil acts, in this system, as a cocarcinogen (or promoter) has given rise to chemical studies that have lead to the identification of phorbol esters as the active principle of croton oil. Among these phorbol esters, the most potent and, therefore, the most widely used as tumor promoters are 12-0-tetradecanoylphorbol-13-acetate (TPA) and phorbol 12-myristate 13-acetate (PMA).

At very low concentrations (100 ng/ml or less), phorbol esters, which mimic the cell responses to EGF (Moon *et al.,* 1984), stimulate cell division and inhibit differentiation in cultures for a number of cell types (reviewed by Sivak, 1979). As in the case of BrdUrd, workers in H. Holtzer's laboratory have been particularly active in studying the effects of TPA on cell differentiation. Cohen *et al.*

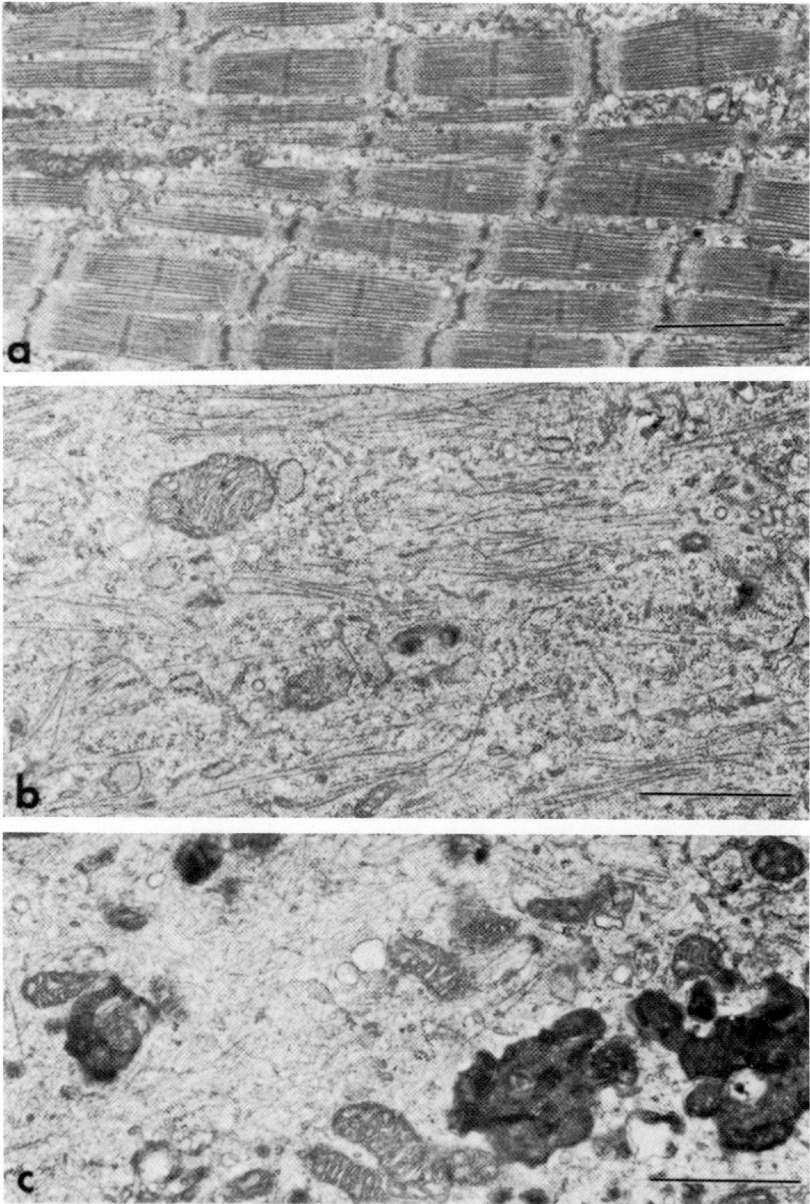

FIG. 11. Electron micrographs of muscle cultures maintained in the absence or presence of TPA (bar: 1 μm). (a) Section through an untreated myotube from a day 6 culture. (b) After 24 hr in TPA, most myofibrils are dispersed into their component thick and thin filaments. (c) After 3 days in TPA, the great majority of thick and thin filaments have disappeared; instead, the interior of the myotube consists of large numbers of intermediate filaments, mitochondria, and autophagosomes. [Croop *et al.*, 1982.]

(1977) showed that TPA suppresses myogenesis, and Pacifici and Holtzer (1977) found that it inhibits *in vitro* chondrogenesis. Further analysis of the effects of TPA on *in vitro*–grown chondroblasts showed that when terminal differentiation is arrested, the cells synthesize an abnormal sulfated proteoglycan (Pacifici and Holtzer, 1980). TPA is even able to induce a dedifferentiation in cultured myotubes (Cossu *et al.*, 1981) and to selectively and completely disrupt their myofibrils (Croop *et al.*, 1982) (Fig. 11). It also inhibits terminal differentiation in cultured epithelial cells (Sisskin and Barrett, 1981); keratin production does not take place, while DNA synthesis is stimulated.

Interesting effects, particularly for embryologists, have been obtained by treating explanted and cultured embryonic neural crests with TPA. According to Glimelius and Weston (1981), TPA at concentrations higher than $2.10^{-7} M$ irreversibly inhibits the differentiation of melanocytes without affecting the growth of undifferentiated cells; these authors noted that, in the presence of TPA, cell-to-cell adhesion is decreased. Similar results have been obtained by Sieber-Blum and Sieber (1981), who reported that TPA accelerates growth and inhibits differentiation in *in vitro*–cultured quail neural crest.

To date, the effects of TPA on whole embryos have been very little studied. We have observed (unpublished work in collaboration with Dr. Mahradjan) that fertilized sea urchin eggs reach the blastula stage in the presence of TPA; at this time, they lose cells and are reduced to very small swimming or immobile spherules. It might be interesting to study embryonic systems in which differentiation is more complex than that found in sea urchin larvae. A first step in that direction was made by Ellinger (1982). He treated frog eggs and early embryos with various concentrations of TPA and, as in sea urchin eggs, obtained a dissociation of the ectoderm cells. In addition, he observed an inhibition of muscle and adhesive organ differentiation. Unfortunately, there were no photographs of sections from the treated embryos, and it is thus impossible to exclude the possibility that cell death due to TPA toxicity was responsible for the inhibition of organ differentiation. What seems to be clear is that TPA breaks down cell–cell junctions in both sea urchin and frog embryos; as we shall see later, this has often been observed in tissues treated with phorbol esters.

Malignant cells, such as neuroblastoma and erythroleukemic Friend cells, have been treated with TPA. TPA inhibits differentiation in cultured mouse neuroblastoma cells (Ishii *et al.*, 1978). In murine erythroleukemia cells, phorbol esters block both spontaneous (Rovera *et al.*, 1977) and DMSO-induced (Yamasaki *et al.*, 1977, 1984; Fibach *et al.*, 1979) hemoglobin synthesis. Inhibition of hemoglobin synthesis can be reversed by addition of hemin, suggesting that TPA does not affect the early events induced by DMSO, but a late program characterized by globin mRNA accumulation (Mager and Bernstein, 1980).

However, there are exceptions to the general rule that TPA prevents cell differentiation. The most striking example, is of obvious interest for oncologists, namely, the induction of terminal differentiation in human promyelocytic leukemia cells by phorbol esters (Huberman and Callaham, 1979; Huberman *et al.*, 1981). In this case, TPA, like DMSO, acts as an inducer of cell differentiation. It has also been reported that, in epiderm, TPA induces differentiation in some basal cells (Yuspa *et al.*, 1982). TPA has also been reported to stimulate neurite outgrowth in sensory ganglions of chick embryos (Hsu *et al.*, 1984). In addition, TPA fails to affect growth and development in the nematod *Caenorhabditis elegans* (Yamasaki *et al.*, 1982). In a review paper, Holtzer *et al.* (1982) conclude that TPA has pleiotropic effects. It stops differentiation in chick myoblasts, chondroblasts, and melanoblasts, but it has no effect on fibroblasts, neurons, and cardiac cells. The authors point out that there is no such thing as a "normal undifferentiated cell," and we think that they are right. This is one of the reasons why substances that supposedly suppress or induce cell differentiation often have unexpected effects.

It has long been known that croton oil increases mitotic activity when it is rubbed on the ear skin of a mouse. A more recent and precise analysis by Dicker and Rozengurt (1980) on cultured fibroblasts has shown that TPA and the natural growth factors EGF and FGF (Chapter 5, Volume 1) have very similar effects. All induce several rounds of DNA replication even in the absence of serum addition to the medium. This suggests the possibility that the effects of TPA are mediated through EGF. However, work by Lee and Weinstein (1980) showed that, on the contrary, within a few minutes, TPA inhibits the binding of EGF to its receptors. They concluded that the phorbol esters change the microenvironment of the receptor and suggested that, in contrast to BrdUrd, they do not act on DNA, but on the cell membrane. That this conclusion is valid is supported by work showing that TPA affects cell-to-cell communications. For instance, in epithelial cells, addition of TPA leads to an opening of the tight junctions and to the loss of the microvilli (Ojakian, 1981), implying an effect on both the tight junctions and the membrane-associated cytoskeleton. The effects of TPA on metabolic cooperation between cells (see Chapter 3, Volume 1) are also under study. Warren *et al.* (1981) have reported that they may increase or decrease after TPA treatment according to the cell line used, but Guy *et al.* (1981) and Newbold and Amos (1981), using two different approaches, found a decrease in cell-to-cell communication. For instance, TPA prevents the transport of [³H]uridine nucleotides from a labeled to an unlabeled cell. D. R. Miller *et al.* (1982) studied the proliferation of epithelial cells grown on a "feeder layer" of x-irradiated fibroblasts. In this system, TPA arrested proliferation because it inhibited cell-to-cell communication and metabolic cooperation. The binding of phorbol esters to membrane receptors is receiving more and more attention and, according to Collins and Rozengurt (1982), the mitogenic activity of phorbol esters results from their binding to high-affinity sites. These sites (phorboid receptors) are saturable, specific, and subject to down regulation (i.e., their number decreases when the cells are cultured in the prolonged presence of phorbol esters); a factor (M_r 60,000) present in human serum inhibits the binding of the phorbol esters to the phorboid receptors (Horowitz et al., 1982). Binding of TPA to plasma membrane receptors exerts, as expected, various effects on the cytoskeleton. Bundling of the intermediary filaments, especially the cytokeratin network of epithelial cells (Fey and Penman, 1984), and redistribution of actin and vinculin in kidney epithelial cells (Schliwa et al., 1984) have been reported. These changes take place within 2 min in TPA-treated kidney epithelial cells. By 20–40 min, the stress fibers have disappeared and the cell adhesion properties of the cells have changed. The same changes occur in enucleated cells. They are not affected by the presence of cycloheximide, but they are inhibited by dinitrophenol and oligomycin. They are dependent upon energy production, but not on transcription; translation; fluctuations in the Ca^{2+} and cAMP levels; or changes in the organization of microtubules, intermediate filaments, and fibronectin (Schliwa et al., 1984).

Work done in several laboratories has clearly shown that the TPA receptor is protein kinase C (for details, see Weinstein, 1983, and Mitchell, 1983). Protein kinase C is a phospholipid- and Ca^{2+}- dependent enzyme that can phosphorylate a number of proteins (reviewed by Nishizuka, 1984). It is transiently activated by diacyl glycerol, a product of inositol phosphate turnover; if TPA is intercalated into the plasma membrane's double layer, it activates protein kinase C permanently. After binding, the TPA–protein kinase C complex moves from the cytosol to the cell membrane (Kraft and Anderson, 1983). Protein kinase C is probably the receptor of all the tumor promoters that will be discussed in the next section (R. C. Parker et al., 1984). One effect of the phorbol esters is to increase the turnover of the membrane phospholipids, in particular, inositol lipids. This results in the production of diacyl glycerol and inositol trisphosphate. The former activates protein kinase C, while the latter mobilizes Ca^{2+} from intracellular stores (mainly the endoplasmic reticulum). It is therefore not surprising that, as pointed out by Michell (1983), protein kinase C and Ca^{2+} are two synergistic cellular signals; calcium ions indeed increase TPA binding, which is inhibited by diacyl glycerol. There is thus competition for protein kinase C between phorbol esters and diacyl glycerol (Sharkey et al., 1984). Since protein kinase C is able to phosphorylate serine residues in a number of proteins, it is not surprising that the effects of TPA on target cells are very complex. For instance, TPA increases by four times the phosphorylation of the insulin and somatomedin C (insulin-like growth factor 1, IGF-1) receptors, suggesting that phosphorylation of these growth factor receptors is mediated by protein kinase C (Jacobs et al., 1983). According to Whiteley et al. (1984), phorbol esters inhibit the EGF stimulated Na^+/H^+ exchange in the malignant cell line A431; protein kinase C, which is activated by phorbol esters, apparently decreases the EGF stimulation of this exchange (which is believed to be important for cell division, as we have seen in

Chapter 5, Volume 1). In view of the complexity of the events that follow TPA binding to protein kinase C, it is no longer surprising that phorbol esters exert pleiotropic effects on cell differentiation.

It is obviously too early to draw specific conclusions. It is still uncertain whether the cell membrane is the only site where phorbol esters are acting, but the conclusion that it is an important one and probably the initial target seems unescapable. At first sight, it is surprising that BrdUrd and TPA exert almost identical inhibitory effects on terminal differentiation, despite the fact that they act on entirely different cellular compartments. The main target for the former is genomic DNA, although side effects on the cell membrane cannot be ruled out; on the other hand, TPA acts first on the cell membrane, and it is not yet clear, as we shall see when we discuss its effects on cell transformation, whether it induces chromosomal abnormalities. However, when leukemic cells are induced to differentiate into macrophages by TPA addition, protein changes are also induced, due perhaps either to the synthesis of new mRNAs or to a posttranscriptional control. It has been suggested that such changes in the pattern of protein synthesis are correlated with TPA-induced initial changes in cell attachment and spreading of the cells (Liebermann et al., 1981; Hoffman-Liebermann et al., 1981), but more work is needed to tell us whether BrdUrd and TPA are ultimately acting on a single key step that is absolutely required for cell differentiation.

In cancer cells, derivatives of vitamin A (retinoic acid and related retinoids) often antagonize the promoting effects of phorbol esters; they may even induce the differentiation of malignant cells (reviewed by Sporn and Roberts, 1983). The favorable effects on terminal differentiation are correlated with an almost predictable slowdown of cell proliferation in transformed fibroblasts (Lacroix et al., 1980). Particularly interesting are the favorable effects of retinoic acid on the differentiation of "nullipotent" embryonal carcinoma cells. Among the various cell lines isolated from these monstrous tumors, one of the most popular is the F9 cell line, which is composed of stem cells that actively proliferate, but have no potentiality for differentiation and are therefore called *nullipotent*. Solter et al. (1979) reported that retinoic acid induces the differentiation of embryo carcinoma cells in endoderm at concentrations at which it inhibits cell multiplication. Later, Hogan et al. (1981) transformed the nullipotent F9 teratocarcinoma cells into visceral endoderm after treatment with retinoic acid. This was ascertained by biochemical criteria (in particular, the synthesis of α-fetoprotein) in addition to morphological observations. The markers for parietal endoderm that are the major protein constituents of the basement membrane (laminin, type IV collagen) are missing in retinoic acid–treated F9 cells. Strickland et al. (1980) found that if F9 cells are treated with retinoic acid and then with agents that increase the intracellular cAMP content, they differentiate into parietal endoderm, synthesizing large amounts of type IV collagen and laminin. As pointed out by Strickland (1981) in a short review of the subject, the undifferentiated F9 stem cells provide excellent material for the study of embryogenesis, neoplasia (these cells are malignant), and the control of gene expression (synthesis of new specific proteins).

Induction of α-fetoprotein synthesis in embryocarcinoma cells by retinoic acid is accompanied by a genome-wide loss of DNA methylation. However, it is very doubtful that DNA demethylation is the actual cause of embryocarcinoma differentiation, since 5-azacytidine (the classic inhibitor of DNA methylation) induces neither α-fetoprotein synthesis nor differentiation (Young and Tilghman, 1984).

Retinoic acid also stimulates the *in vitro* differentiation of fetal hepatocytes. This is shown by an inhibition of α-fetoprotein synthesis and a concomitant increase in albumin synthesis (Chou and Ito, 1984).

However, it seems that retinoic does not always promote cell differentiation, since it has been reported that it inhibits the conversion of fibroblasts into adipocytes (Kuri-Harcuch, 1982). The results of further work on the effects of retinoids on other biological systems, including embryos, will be awaited with interest.

It is likely that the initial site of action of the retinoids, which are liposoluble compounds, is the cell membrane. It has, indeed, been shown that addition of retinoic acids to ghosts of red blood cells decreases by 50% the microviscosity of their membrane (Meeks *et al.*, 1981). Another significant finding is that retinoids, in sharp contrast to TPA, increase the number of EGF receptors on the cell membrane (Jetten and Jetten, 1979). Unfortunately, we do not know how the signal received at the cell membrane after treatment with either phorbol esters or retinoids is transmitted to the cell nucleus.

A likely possibility is that this signal is transmitted through a second messenger, for instance, a cyclic nucleotide or calcium ions. A review of the literature shows that there is a strong correlation between a high cellular content of cAMP and a slowdown of cell proliferation. In general, actively dividing cells have a low cAMP content. Increasing the cAMP content by adding dibutyryl-cAMP (which penetrates cells better than cAMP itself) or cAMP phosphodiesterase inhibitors (theophylline and other methylxanthines) slows down cell proliferation and, in some instances, induces or favors cell differentiation. However, it should be pointed out that the conclusion that a high cAMP content prevents cell proliferation and favors cell differentiation has many exceptions. It has also long been agreed that there is a kind of antagonism between cAMP and cGMP. A high cGMP content favors cell proliferation and prevents cell differentiation. Here again, exceptions to the rule are numerous (see the brief review by Goldberg, 1980).

Finally, release or sequestration of calcium ions might control any of the innumerable Ca^{2+}-dependent events, since these ions can affect the properties of the cell membrane, the organization of the cytoskeleton, and the assembly and disassembly of the mitotic spindle. As we saw in Chapter 2, release of membrane-bound Ca^{2+} triggers such important biological processes as amphibian oocyte maturation and sea urchin egg fertilization.

That the cyclic nucleotides and calcium ions are involved in cell differentiation is a very logical assumption, in view of their ubiquitousness and the precise regulation of their cellular content, and since most protein kinases are activated by either the cyclic nucleotides or calcium ions. As we have seen, protein phosphorylation is an exceedingly important and frequent mechanism for modifying the conformation of key proteins. It is almost impossible to believe that subtle changes in protein phosphorylation and dephosphorylation do not play a fundamental role in the induction of the morphological changes that characterize cell differentiation.

As an example of what has just been said, the addition of dibutyryl-cAMP induces the appearance of melanin granules in the cells of a "pale" melanoma (Johnson and Pastan, 1972). In contrast to this positive effect of cAMP on cell differentiation, it was reported, at about the same time, that cAMP and theophylline inhibit the fusion of myoblasts, thus arresting the myogenic differentiation program at an early stage (Wahrmann et al., 1973). Liu (1982) favors the idea that protein phosphorylation is involved in cell differentiation. He studied the differentiation of fibroblasts into adipocytes, which can be induced by various means (addition of insulin, glucocorticoids, methlyated xanthines) and found that, in all cases, an increase in cAMP-dependent protein kinase activity occurs. Of course, this is still an isolated observation. One would like to know whether an increase in protein phosphorylation during cell differentiation is a general phenomenon and which proteins were phosphorylated in Liu's (1982) experiments.

We have seen that the use of dibutyryl-cAMP has often been preferred to cAMP, because of its greater permeability in studies on in vitro cell differentiation. As a control, the possible effects of the butyrate moiety of the molecule were observed. It turned out that, in some systems at least, butyrate was able to strongly affect cell differentiation. Like DMSO, butyrate is a potent inducer of erythroid differentiation in Friend MEL cells. Reeves and Cserjesi (1979a) showed that this effect of butyrate on MEL cells is due, at least in part, to its well-known action on histone acetylases. Indeed, the authors found that the histones in butyrate-treated MEL cells were hyperacetylated, and concluded that in a large proportion of the cell population, this led to the synthesis of new mRNA and protein molecules. There is also a parallel between butyrate and DMSO in the case of myogenesis, since both inhibit the terminal differentiation of myoblasts (Blau and Epstein, 1979; Leibovitch et al., 1982). Whether this inhibition is related to histone acetylation is not known. Other sites of action for butyrate are, of course, conceivable, particularly incorporation in the membrane lipid bilayer. It is worth recalling here that, since the days of J. Loeb, butyrate has been one of the classic agents for inducing parthenogenesis in sea urchin eggs. It is curious that no one has studied how butyrate acts on sea urchin eggs using the methods now at our disposal. It is generally believed, without good reasons, that butyrate, in Loeb's method of parthenogenesis, mainly affects the membrane. This belief is probably correct, since, in a very different system (lymphocyte differentiation), it has been shown that cholesterol synthesis is essential for differentiation of a "naive" lymphocyte into a cytotoxic one; DNA synthesis is not required, showing that proliferation is not always a prerequisite for differentiation (Heiniger and Marshall, 1982).

Finally, a few attempts have been made to modify cell differentiation by the addition of "conditioned media" to differentiating cells. Conditioned media are culture media in which cells have been grown for a few hours or days and contain substances that have been released or secreted in the medium by the growing cells. Such media may exert positive or negative effects on growth and differentiation, and are very useful for the analysis of cell aggregation. This discussion will be limited to a few results obtained using that method for cartilage differentiation. Solursh and Meier (1973) found out that conditioned media from chondrocytes stimulate the differentiation of these cells into cartilage by specifically increasing the synthesis of chondroproteins and collagen. Possibly related to this finding is the observation of Kato *et al.* (1981) that fetal bovine cartilage possesses a somatomedin-like activity (thus, a growth factor) that increases not only DNA, but also proteoglycan synthesis. Work with conditioned media is easy to perform and is useful for a first analysis of a differentiating system. The isolation and identification of their active principle require additional skill.

F. GENE EXPRESSION AND CELL DIFFERENTIATION

It has often been said that the mystery of cell differentiation will be solved when we understand the regulation of gene activity in eukaryotes, as well as the control of the *lac* operon in *Escherichia coli*. A few years ago, it was generally assumed that there was a single key control mechanism operating at the genetic level during cell differentiation. Today, as was duly pointed out by Brown in his excellent review article (1981), we are faced with a host of possible controls of gene expression. This finding modifies the oversimplified view of the past but makes the problem of the control of gene expression in developing and differentiating embryos even more fascinating.

The program of protein synthesis changes significantly when a cell undergoes differentiation. In addition to the striking fact that a few marker proteins (Holtzer's luxury proteins) undergo disproportionate syntheses, minor quantitative or qualitative protein changes are often recorded during cell differentiation. At the time of differentiation, the cell undergoes a complete, usually irreversible, revolution. The differentiated cell, except in rare cases of dedifferentiation, will never again be what it was before its differentiation began. Its morphology, physiology, and entire molecular background greatly change, requiring complex, coordinate changes in the interactions between many structural and regulatory genes.

The many controls in gene expression during embryonic development were described at length in the preceding chapter; further discussion here would be redundant. Thus, our comments will be limited to the most important of these controls for cell differentiation.

The idea that the predominant synthesis of luxury proteins during cell differentiation results from gene amplification was stated by Holtzer et al. (1972); increasing the number of gene copies is an obvious and simple method for allowing the cell to produce larger amounts of the corresponding protein. However, this method is only rarely used by the differentiating cell. Besides the classic amplification of the ribosomal genes during Xenopus oogenesis, the only genes in which amplification is known to occur are those coding for the chorion proteins in Drosophila ovarian follicles (Spradling and Mahowald, 1980; reviewed by Chisholm, 1982, and Ish-Horowicz, 1982) and those coding for specific enzymes in drug-resistant cell lines (see Chapter 4, Volume 1). It would be farfetched to equate synthesis of rRNAs or chorion proteins during oogenesis to differentiation at much later stages of development. During oogenesis, the oocyte must suddenly cope, at a given stage, with the problem of synthesizing a very large amount of ribosomes or chorion proteins; gene amplification provides a solution to that problem. This same method allows cells to escape the lethal effects of a drug (methotrexate, cadmium, PALA) and to become drug resistant (reviewed by Schimke, 1984). These drugs induce selective gene amplification, causing changes in chromosome morphology. However, they do not induce a shift in cell differentiation comparable to the changes that take place during myogenesis, chondrogenesis, or erythropoiesis.

It is also unlikely that gene rearrangements, due to the random insertion of mobile genetic elements into the genome, are a general mechanism used by differentiating cells. Gannon et al. (1980) have shown that in the chicken, the sequence of the S region of the ovalbumin gene is exactly the same in cells that express the gene (oviduct) and in those that do not (red blood cells). Since similar results have been obtained for the silk fibroin, amylase (Schibler et al., 1982), albumin (Capetanaki et al., 1982), and globin genes, we can exclude gross DNA rearrangements as a general mechanism for tissue-specific gene expression. However, Farzaneh et al. (1982), Althaus et al. (1982), and Johnstone and Williams (1982), working on myoblasts, hepatocytes, and lymphocytes, respectively, have produced puzzling results that suggest that gene rearrangements might be involved in cell differentiation. They think that ADP-ribosyltransferase, an enzyme needed for DNA repair (as we saw in Chapter 5, Volume 1), is directly involved in cell differentiation (reviewed by Williams and Johnstone, 1983). Inhibition of this enzyme by 3-aminobenzamide or reduction of the NAD content of the cells by nicotinamide depletion stops fusion of myoblasts and synthesis of marker enzymes for myogenesis; the same inhibitor prevents the expression of two fetal functions in hepatocytes and the response of lymphocytes to mitogenic stimuli; it induces the differentiation of embryocercinoma cells (Okashi *et al.*, 1984). All of these experiments strongly suggest that DNA single-stranded breaking and rejoining are involved in gene expression during cytodifferentiation, and such processes could, of course, favor gene rearrangements, for instance, transpositions. It has indeed been found that DNA single-strand breaks accumulate during myoblast differentiation and that the activity of ADP ribosyltransferase increases during *Xenopus* development. Another possible function of this enzyme might be to modify histone H1 by ADP ribosylation. This could modify locally the higher-order structure of chromatin. Further work on DNA sequencing of genes involved in cell differentiation is obviously required before general conclusions can be drawn.

Posttranscriptional controls (maturation of pre-mRNA molecules) exist in all cells, whether differentiated or not. While translational controls are of fundamental importance during the very early steps of development, they lose their predominance by the late blastula–early gastrula stage. There is no further development (and no embryonic or *in vitro* cell differentiation) if gene transcription is inhibited with actinomycin D. However, a variable turnover rate of the individual cytoplasmic mRNAs might conceivably be an important factor in the expression of the differentiated phenotype.

This leaves us with transcriptional controls to explain the differential gene activation that supposedly explains cell differentiation. There is overwhelming evidence for the belief that this type of control is the major one in differentiating cells, stemming largely from the already mentioned fact that cells that differentiate and produce a specific protein have the same number of genes coding for this protein as other cells. For instance, the number of copies of the globin genes is the same in all cells; hematopoietic cells do not have increased numbers of genes. The large amount of work done in the laboratories of O'Malley and R. Palmiter on the selective stimulation of the ovalbumin gene in the oviduct of estrogen-treated chicken is particularly impressive [for reviews on the mode of action of steroid hormones on target cells, see Jensen and De Sombre (1973), O'Malley and Means (1974), and Baulieu, (1975)]. There is a single copy of the ovalbumin gene per haploid genome (Harris *et al.*, 1973), a gene that has been sequenced in the laboratories of P. Chambon and B. O. Malley. In the noninduced hen oviduct, there is less than one molecule of ovalbumin mRNA per cell; after estrogen administration, the rate of ovalbumin mRNA synthesis increases progressively culminating with 6×10^5 molecules/min/cell (Palmiter, 1975). Since the half-life of this mRNA is rather long (40–60 hr), and since each of its molecules can be translated 50,000 times, there is a tremendous accumulation of ovalbumin in the cells of the hormone-stimulated oviduct; ovalbumin, together with a few other hormone-dependent proteins (conalbumin, ovomucoid, lysozyme), makes up more than 70% of the total protein content of the cells (Harris *et al.*, 1975). This enormous increase is due entirely to the selective production and accumulation of the corresponding mRNAs in the estrogen-stimulated oviduct cells. According to Hynes *et al.* (1977), the hen oviduct contains about 13,000 different mRNA species, and the complexity of this mRNA population does not increase after hormonal stimulation. However, in the estrogen-treated chicken, the ovalbumin, ovomucoid, and lysozyme mRNAs represent 60% of the poly(A)$^+$ RNAs, in contrast to only 0.02% in the non-stimulated cells. Comparable results have been obtained for the vitellogenin genes, which are stimulated several thousand times in the liver of the rooster (Deeley *et al.*, 1977) or *Xenopus* after estrogen administration (Wahli *et al.*, 1981). According to Brock and Shapiro (1983), nuclei isolated from *Xenopus* liver synthesize less than one molecule of vitellogenin mRNA per gene per hour. After estrogen stimulation, overall RNA synthesis in liver nuclei is increased 20–60 times, while transcription of the vitellogenin gene is increased several thousands times. Erythropoietin increases the content of erythropoietic cells in globin mRNA 250-fold, reaching a level of 1800 molecules per cell (Ramirez *et al.*, 1975).

In the hen oviduct model, we begin to understand how the hormone-responsive genes are activated in the steroid-stimulated target cells. According to Bellard *et al.* (1982), there are no nucleosomes in a region that extends 2500 bp at the 5′-extremity of the ovalbumin gene in chicken oviduct, but nucleosomes are present in the same region in chicken red blood cell nuclei. This shows the existence of a tissue-specific alteration of chromatin. Results reported by Mulvihill *et al.* (1982) lend additional support to this view. They found that stimulation of the ovalbumin, conalbumin, ovomucoid, and the still mysterious *x* and *y* genes is correlated with a high-affinity binding of the steroid receptor to a region 250–300 nucleotides upstream of the mRNA's starting point.

Differentiating and hormone-stimulated cells are characterized by a selective and intensive stimulation of a small number of genes. This necessarily requires localized changes in chromatin structure. We have already mentioned the excellent review by Weisbrod (1982) on active chromatin. As we have seen, the main characteristics of active chromatin are DNase hypersensitivity of the coding and adjacent regions; the presence of specific non-histone proteins (in particular, those belonging to the HMG group), modification of the histone acetylation, phosphorylation, poly(AD) ribosylation, and methylation; binding of polyamines; and DNA modifications such as undermethylation or formation of left-handed (Z-form) DNA. This list demonstrates the complexity of the factors that control the transformation of chromatin in an active conformation; there is little doubt that other factors will be added to this list in the future.

It seems more and more likely that none of the aforementioned factors alone is sufficient to induce in chromatin the structural changes required for gene activation. While hypersensitivity to DNase I of active genes seems to be a general rule, it should be pointed out that the ovalbumin gene in hen oviduct nuclei remains sensitive to the enzyme even after the hormone has been withdrawn. At this time, the ovalbumin gene is no longer active, and ovalbumin production has dropped to its basal level (Palmiter *et al.*, 1977). According to Miller *et al.* (1978), the sensitivity to DNase I digestion of the mouse globin gene is the same in Friend erythroleukemic cells whether or not they have been induced to synthesize hemoglobin. In this system, DNase sensitivity of the adult β-globin genes is observed when the cells are induced by treatment with certain inducers, while no changes are found after treatment with others (Smith and Yu, 1984). As pointed out by Weisbrod (1982), the sensitivity to DNase I only reflects a potential for gene transcription rather than transcription itself.

Groudine and Weintraub (1982) have studied the propagation of DNase I hypersensitivity sites from one cell generation to the next. They induced the activity of globin genes in fibroblasts by injection with a virus or an NaCl shock and allowed the cells to divide 20 times. The sites remained DNase I sensitive, showing that they have the capacity to template their own structure independently of the initial inductive event. The DNase I sensitive sites are also sensitive to S1 nuclease (which cleaves single-stranded DNA), which might explain their propagation.

Of interest in terms of our present problem (cell differentiation) is a report by Carmon *et al.* (1982), who worked on myogenesis. They found that the genes encoding α-actin and myosin chain 2 are not sensitive to DNase I in myoblasts and that these two proteins are not expressed. They become sensitive to the enzyme when they become active during the transition toward differentiation.

Proteins associated with DNA undoubtedly play a role in the control of gene transcription. There is a striking correspondance between the regions of chromatin that are sensitive to DNase digestion and those that bind the non-histone proteins HMG-14 and -17 (Weisbrod and Weintraub, 1979). That non-histone proteins probably play a role in all types of differentiation is shown by the fact that they differ in cell lines of teratocarcinoma, which differentiate along different pathways (Blüthmann and Illmensee, 1981). Another chromosomal protein that is likely to draw attention in the future is histone H1°, which, as we saw in Chapter 5, Volume 1, accumulates when cells stop dividing. According to Gjerset *et al.* (1982), histone H1° appears when various kinds of cells (brain, retina, muscle, and liver cells) become fully differentiated; partial hepatectomy, which is followed by increased mitotic activity and decreased cell differentiation, is accompanied by a 66% drop in the H1° content of the liver cells. Similarly, Jackowski and Liew (1982) found that histone H1° accumulates in the nuclei when myocardial cells stop dividing and become completely differentiated.

That histone acetylation is accompanied by an increase in RNA synthesis has been known for many years (Pogo *et al.*, 1968). There is ample evidence that transcriptionally active chromatin is rich in acetylated histone. However, an increase in histone acetylation by treatment of living cells with butyrate does not increase the sensitivity of their chromatin to DNase I digestion (Simpson, 1978).

Other proteins could conceivably play a role in the control of gene activity. For instance, it has been reported that many proteins bind specifically to left-handed Z-DNA in *Drosophila* (Nordheim *et al.*, 1982).

An exciting field today for students of cell differentiation at the molecular level is that of DNA methylation, which has already been discussed more than once in this book. The evidence in favor of the idea that undermethylation of the cytidine residues present in C–G pairs has been presented and discussed by Razin and Riggs (1980) in a now classic review article. There is no doubt that, as a rule, expressed DNA sequences are undermethylated compared to total DNA (Naveh-Many and Cedar, 1981) and that, in many instances, DNA sequences become undermethylated during cell differentiation. This is the case for B-lymphocyte differentiation (Yagi and Koshland, 1981), for the crystalline genes in the lens (Jones *et al.*, 1981), for the embryonic trophoblast when it undergoes differentiation (Manes and Menzel, 1981), and for the α-fetoprotein gene in fetal liver and hepatoma (Andrews *et al.*, 1982). Interesting results have been obtained by Jähner *et al.* (1982), who injected viral DNA (in fact, a retroviral genome; see the next section) into the male pronuclei or into mouse morulae. This was followed by *de novo* methylation of the viral genome and blocking its expression. If the injection experiments are performed at later stages (in postimplantation embryos), there is no methylation and no restriction of viral expression. Thus, in early mouse embryos, *de novo* methylation activity seems to be one of the characteristics of gene regulation.

However, a word of caution is necessary. A close comparison of the methylation patterns of two genes (coding for human growth hormone and chorionic somatotropin) in tissues where they are and are not expressed shows that the correlation between undermethylation and gene expression is imperfect (Hjelle *et al.*, 1982). McKeon *et al.* (1982) took this one step further when they found that the expression levels of one of the collagen genes (the α_2I-collagen gene) is completely independent of its methylation pattern. Similarly, Cate *et al.* (1983) reported that the insulin I gene is less methylated than the insulin II gene; nevertheless, both are equally expressed. Stein *et al.* (1983) pointed out that the housekeeping genes required for nucleotide synthesis are completely methylated, except in 5′, in all tissues. It would thus be unwise to draw general conclusions about the correlation between DNA undermethylation and gene expression at our present stage of knowledge. The conclusion drawn by both Weisbrod (1982) and Hjelle *et al.* (1982) that, if undermethylation is necessary for gene expression, it is not sufficient, should remain essentially correct.

We have seen that BrdUrd, which is incorporated into replicating DNA molecules, is an excellent tool for the analysis of cell differentiation. Another abnormal nucleoside is becoming more popular among students of cell differentiation. This is 5-azacytidine, which inhibits DNA methylation by impeding DNA-methylase activity (reviewed by Razin and Riggs, 1980; see also Naveh-Many and Cedar, 1981; Jones, 1985; Christman, et al., 1985). We already know that 5-azacytidine (which is hardly mutagenic, according to Landolph and Jones, 1982) stimulates cell differentiation in various systems. For instance, 5-azacytidine activates the metallothionein-I gene in mouse lymphoid cells (Compere and Palmiter, 1981), and it induces the determination of preadipocytes and the following differentiation into adipocytes (Sager and Kovac, 1982). Other studies by Taylor and Jones (1982) and Jones et al. (1983) have yielded interesting results. 5-Azacytidine induces the differentiation not only of adipocytes, but also of myotubes and chondrocytes in cultured mouse embryo cells. In the case of adipocytes, a 5-min treatment during the S phase of the cell cycle is sufficient to induce phenotypic differentiation. Konieczny and Emerson (1984) observed the differentiation of a mouse embryonal cell line in chrondrocytes, adipocytes, and skeletal muscle, suggesting that this multiple differentiation might be due to hypomethylation of "determination" regulatory loci. In the neonatal rat, injection of 5-azacytidine stimulates the differentiation of the fetal hepatocytes, as shown by a decrease in α-fetoprotein synthesis (Cook and Chiu, 1983). Curiously, Charache et al. (1983) have reported that treatment of sickle cell anemia patients with 5-azacytidine increases the production of fetal hemoglobin (HbF), but not that of adult hemoglobin; the genome of the treated patients is undermethylated. However, 5-azacytidine exerts negative effects on the differentiation of cultured Drosophila embryonic cells, and it has been pointed out that the drug might exert teratogenic effects on embryos and fetuses (Bournias-Vardiabasis et al., 1983).

Perhaps the most interesting effect of 5-azacytidine is the previously mentioned reactivation of the X chromosome in mammals. Inactivation of one of the two X chromosomes (Lyons' phenomenon) has been the subject of brief reviews by Martin (1982) and by Blasi and Toniolo (1983). The two X chromosomes present in eggs (which will give rise to females) are still fully active in the inner cell mass of the mouse blastocysts (as well as in teratocarcinoma stem cells). Preferential inactivation of the paternal X takes place during later development at the time of endoderm formation: which is ascertained by the fact that the gene coding for the enzyme hypoxanthine guanine phosphoribosyltransferase (HGPRT) is located on the X chromosome; its activity can be assayed, and selection on (HAT) medium is possible. Mohandas et al. (1981) found that the X chromosome is reactivated when a mouse/human hybrid cell line is cultured in the presence of 5-azacytidine. This interesting finding has been confirmed by Jones et al. (1982), Marshall Graves (1982),and Hors-Cayla et al. (1983). Marshall Graves (1982) observed, in addition, that there is no X chromosome reactivation in diploid mouse fibroblasts (in which only one member of the X pair is active) when they are fused with teratocarcinoma stem cells (in which the two X chromosomes are active), thus negating the possibility that X activation and inactivation are mediated by cytoplasmic factors. Instead, it suggests that the factors responsible for X inactivation reside in chromatin, presumably in DNA itself. Experiments by Venolia *et al.* (1982) and Lester *et al.* (1982) strongly support this view. Using DNA-mediated gene transfer, these workers transferred the HGPRT gene into HGPRT⁻ recipient cells. These cells remain HGPRT⁻ if the donor DNA came from cells containing an inactive X chromosome, but they become HGPRT⁺ if the X chromosome has been reactivated by treatment with 5-azacytidine. As yet, there is no evidence that the reactivated DNA sequences were hypomethylated. In fact, D. A. Miller *et al.* (1982) were unable to detect differences in the 5-methylcytosine content of the two X chromosomes in monkeys; they stained metaphase chromosomes with a fluorescent antibody specific for 5-methylcytosine. It should be pointed out, however, that 5-azacytidine is not the only agent capable of reactivating the X chromosome; the same result has been found in cells treated with DMSO or sodium butyrate (Lester *et al.*, 1982). In fact, the three chemicals are definitely toxic, and cell death takes place at the concentrations used for the induction of cell differentiation. Bell and Jones (1982) and Sager and Kovacs (1982) have argued that cell toxicity is due to inhibition of methylation or alterations in the DNA methylation pattern. It should be added that X chromosome reactivation is not always obtained. Wolf and Migeon (1982) obtained negative results when they treated normal human fibroblasts with 5-azacytidine and concluded that differences in ubiquitous methylation cannot be the molecular basis for X chromosome reactivation.

It is easy to predict that, in the future, 5-azacytidine will be used more often for the study of cell differentiation. As with BrdUrd and phorbol esters, studies on the effects of this analog on egg development lag behind the work done on *in vitro*-cultured cells and somatic hybrids. This is unfortunate, since the idea that DNA methylation (and possibly other DNA modifications) plays an important role in embryonic differentiation was first suggested by Scarano (1969) on the basis of experiments done on developing sea urchin eggs. The effects of 5-azacytidine on sea urchin eggs are presently under study in Scarano's laboratory (personal communication). Preliminary results suggest that the analog might have side effects, such as decreases in the rates of DNA and RNA synthesis. This would not be surprising and should be kept in mind by those who plan to use 5-azacytidine to analyze the role played by DNA methylation in embryonic development and cell differentiation.

II. CELL TRANSFORMATION AND MALIGNANCY

A. General Background

"Nothing is more beautiful than an embryo developing in its harmonious way; nothing is uglier than a cancer growing in its malignant way. A cell has gone

mad, dividing without differentiating, the multiplication of its descendants ultimately killing the whole organism. Cancers are not under the control of organizers, gradients, and fields of differentiation, as are developing embryos. In some
respects, they resemble these ectoderm cells, which, having been acted upon by
abnormal evocators (killed organizers, implanted chemicals), may react by forming complex structures that look more like teratomas than well-differentiated
embryos."

These lines were written more than 25 years ago in "Biochemical Cytology,"
and they remain true today (Brachet, 1957). It was added that "the author has
never worked personally with cancer cells and he has no original ideas to present
here." This also remains true today. However, since cancer is a major problem
in cell biology, this section on cell transformation and malignancy will be much
longer than it was in the previous edition.

As everyone knows, cancer has become a major human problem for a number
of reasons: better diagnosis of the disease, increase in the average lifespan of
humans because of progress in the medical sciences, and an increase in chemical
and physical carcinogenic agents in our environment. The frequency of cancers
has apparently increased.

Cancer cells "go mad" and escape the regulatory mechanisms that control
proliferation and differentiation in normal cells because their DNA has been
injured. All carcinogenic agents (chemicals such as polycyclic hydrocarbons, γ,
x, or uv irradiation, DNA or RNA oncogenic viruses) modify DNA and are
therefore called *genotoxic*. They may bind to DNA (formation of DNA adducts
by carcinogenic chemicals), produce single-stranded or double-stranded breaks
in the DNA double helix (radiation), or integrate their own DNA into genomic
DNA sequences with, as a consequence, replication of viral DNA together with
host DNA (oncogenic viruses). If the oncogenic virus is a retrovirus, as is often
the case, its RNA genome is reverse transcribed by a viral reverse transcriptase
into complementary DNA; the resulting DNA provirus can then be integrated in
the host genome.

Two major pathways are open for a cell treated with, for instance, a chemical
carcinogen: death of the cell or replication, after repair (perfect or not), of the
damaged DNA. Replication of a cell in which DNA has been incorrectly repaired
after formation of a DNA adduct leads to a progeny of cells that are potentially
malignant. Expression of this malignant potentiality in humans will depend on
their "genetic predispositions" (Kopelovich, 1982; Knudson, 1985). This initiation step is only one in a series of events that lead to malignancy (Fig. 12). As
was shown by I. Berenblum as early as 1954, cancer is a multistep process
(reviewed by Ponder, 1980; Weinstein, 1981; Berenblum and Armuth, 1981;
Sabine, 1983). The initiated cell is still a "dormant, latent" cancer cell. The
next step is promotion by a cocarcinogen (e.g., the phorbol esters present in
croton oil) toward the neoplastic cell state; expression of this state and tumor

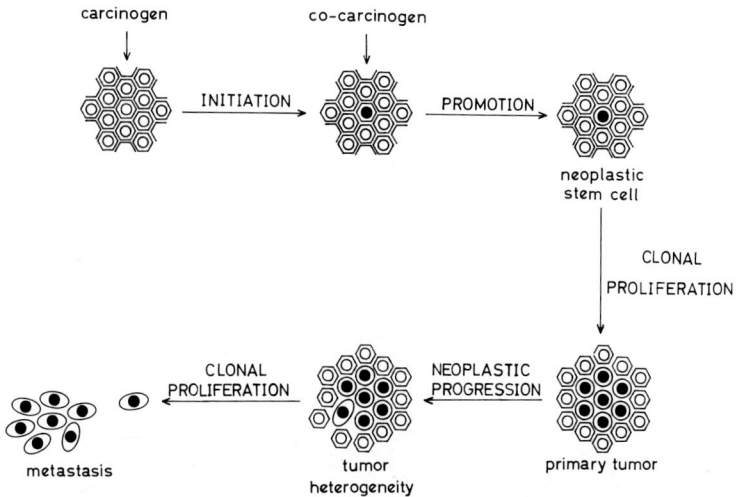

FIG. 12.   Schematic representation of the multistep process in cancer formation. [Original drawing by H. Alexandre.]

formation require clonal proliferation of the neoplastic stem cell. However, primary tumors seldom kill; death is more often due to metastases (reviewed by Fidler and Hart, 1982; Nicolson, 1982, 1984). It seems that each metastasis is also of clonal origin (Talmadge *et al.*, 1982) and is believed to be due to selective proliferation of a subpopulation of cells that preexists in heterogeneous tumors (Nicolson and Custead, 1982).

The potentiality for a cancer cell to metastasize seems to reside in altered properties of its cell surface (Poste and Nicolson, 1980; Roos, 1984). To produce a metastasis, a malignant cell must penetrate a vascular endothelium; this is made possible by the secretion or presence on the malignant cell membrane of various proteases (including plasminogen activator), collagenases, and glycosidases that destory the extracellular matrix of the endothelial cells (reviewed by Mullins and Rohrlich, 1983). For instance, Nakajima *et al.* (1983, 1984) found that invasive and metastatic sublines of a melanoma are more active than others in degrading heparan sulfate that is present in the extracellular matrix of vascular endothelial cells; this allows the destruction of the basal lamina that surrounds the walls of the pulmonary vessels. The fact that antibodies against plasminogen activator can inhibit the growth of metastases indicates that this protease plays an important role in basal lamina destruction (Ossowski and Reich, 1983). In a recent study, Carlsen *et al.* (1984) found a strong correlation between plasminogen activator activity and metastatic potential in various rat tumors. However, there is growing evidence that another important factor in metastasis is the enhanced capacity of the cell surface to adhere to substrates. Reich *et al.* (1984) found that metastatic cells adhere two times faster to the substratum than the others. In addition, metastatic cell lines bind selectively to the tissue they colonize. This could be shown by adding cryostat sections of host organs to cultures of malignant cells (Netland and Zetter, 1984). According to Steinemann *et al.* (1984), the invasive behavior of cultured mouse sarcoma cells is inhibited by antibodies against a 37,000-D membrane glycoprotein. The extracellular matrix glycoproteins probably play an important role in the increased adhesiveness of metastatic cells to basement membranes. Vollmers *et al.* (1984) found that monoclonal antibodies which inhibit the growth of lung metastases of melanoma cells block their attachment to laminin. Removal of fibronectin or addition of laminin increases the affinity of metastatic melanoma cells for basement membrane collagen; they also increase the metastatic potentialities of these cells (Terranova *et al.*, 1984). Finally, Grimstad *et al.* (1984) have reported that metastatic cells possess a laminin-like protein and α-D-galactopyranosyl end groups; both are important for attachment of the cells to their substratum. Increased production of hydrolytic enzymes and changes in the adhesive properties of the cell surface are thus two important properties of metastasis. They are particularly important because cancer patients are usually killed by metastases of a primary tumor, and it is therefore essential to find means to prevent their invasiveness.

Although the clonal origin of both primary tumors and metastases is generally accepted, it should be pointed out that changes in the clonal composition of the population often take place during the development of tumors and metastases (see the review on tumor heterogeneity by Heppner, 1984). Cancer cells usually display greater chromosomal lability than their normal counterparts, both *in vitro* and *in vivo*. Certain sublines, originating from variants of the original neoplastic cell, grow better than others in hosts in whom defense mechanisms (immunological and others) are at work (Howell, 1982; Poste *et al.*, 1982). For instance, in a case studied by Poste *et al.* (1982), 80% of the lung metastases of a malignant melanoma had unicellular origin; after 40–50 days, 90% of the metastatic lesions were populated by cells with heterogeneous metastatic phenotypes. Thus, variants with different metastatic properties had appeared during proliferation of the cells. In subclones of mouse fibroblasts which had undergone spontaneous transformation, heterogeneity appears rapidly (Rubin, 1984). The same conclusion has been drawn by Talmadge *et al.* (1984), who studied resistance to chemotherapeutic agents. It is generally believed that the fast appearance of heterogeneity and diversity in malignant cells is due to the well-known genetic instability of cancer cells (Nicolson, 1984; Heppner 1984), but several other possibilities have been discussed by Farber (1984).

To sum up, five stages can be distinguished in the development of human cancers (Sabine, 1983; Hicks, 1983): (1) *initiation* by genotoxic agents does not alter the phenotype of the affected cell; (2) *promotion* (which, as we shall see, occurs in two successive steps) produces a new phenotype; (3) *primary growth* of the modified cells leads to the formation of a single, usually benign tumor by clonal expansion; (4) a still poorly understood mechanism may then lead to malignant growth and formation of metastases; (5) finally, growth of *metastases,* if it cannot be controlled, kills the organism. Of course, this subdivision of events into five steps is as artificial as that of development into a series of stages; in both cases, we are dealing with continuous processes.

Since carcinogenesis is a multistep process, it is not easy to identify a single, truly initial step. However, there is little doubt that damage to chromosomal DNA molecules, which leads to chromosomal rearrangements, is an early—if not the first—step in the cascade of events that lead to cancer growth and invasion by metastases. Two questions, which have been discussed by Cairns (1981) and by Radman *et al.* (1982), immediately arise: are carcinogens always mutagens, and are chromosomal rearrangements responsible for cancerization? These questions cannot be answered by a single "yes" or "no" because the answers are not necessarily the same for all cancers. It seems clear that the vast majority of carcinogens are mutagenic in the classic Ames *Salmonella* test, which detects base substitution on frameshift mutations. This widely used test for the screening of mutagens detects the appearance of mutant colonies in cultures of *Salmonella typhimurium.* Liver microsomes are added to the cultures, since many polycyclic aromatic hydrocarbons are not carcinogenic unless they have been modified by the oxidizing enzymes present in liver microsomal membranes (see Chapter 3, Volume 1). According to McCann and Ames (1976), 90% of the carcinogenic substances are mutagenic in the Ames test; according to Radman *et al.* (1982), however, out of 61 carcinogens studied in detail, only 45 are Ames test positive (for a more detailed review of the relationships between carcinogenesis and mutagenesis, see Trosko and Cheng, 1978). It should be added that many investigators point out the obvious fact that humans are not bacteria, and have expressed doubts about the validity of the conclusions drawn from the quick Ames test. A more difficult, more time-consuming, but more reliable way to test whether or not cells treated *in vitro* with a presumptive carcinogen are malignant is to inject them into immunosuppressed animals and to see whether they produce a tumor after a specific amount of time (days or weeks). Such experiments are usually done with nude mice, which, having no thymus, are unable to produce T lymphocytes.

There is no doubt that chromosomal rearrangements, visible under the microscope, are a frequent occurrence in cancers. The first case of a chromosomal aberration specifically associated with a given form of cancer was called the Philadelphia chromosome; it was discovered by Nowell and Hungerford (1960), in patients with chronic myeloid leukemia. This aberration is a terminal deletion of chromosome 22 often associated with a translocation to chromosome 9 (Fig. 13). Since that time, many chromosomal rearrangements have been found in cancer-bearing human patients (reviewed by Van Den Berghe, 1980; Mitelman

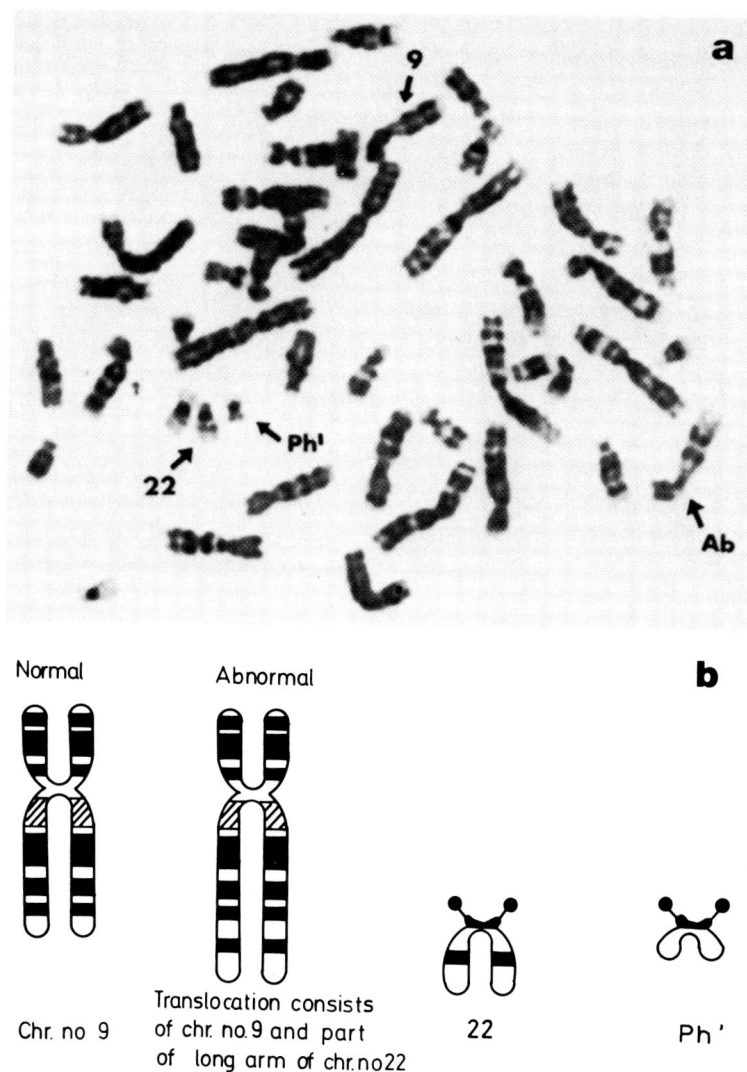

Fig. 13. Philadelphia (Ph[1]) chromosome in a patient with chronic granulocytic leukemia. (a) Translocation between chromosomes 22 and 9; Ab, aberrant chromosome. (b) The participating chromosomes and the translocation products are shown diagrammatically. [From Harnden, D. G. *In* "Genetics of Human Cancer" (J. J. Mulvihill, R. W. Miller, and J. G. Fraumeni, eds.), p. 87. Copyright 1977 by Raven Press, New York.]

and Levan, 1981). The more frequent cytogenetic abnormalities found in cancer patients are translocations, deletions, or the gain or loss of a whole chromosome, which leads to aneuploidy. Translocations are extremely frequent in human and mouse leukemias and lymphomas; their importance has been emphasized by

Klein (1981, 1983) and by Forman and Rowley (1982). In sarcomas, the loss of one chromosome band is frequent; trisomy has been found in some human cancers (Yunis, 1983). The presence of "double minutes" and of homogeneously staining regions in chromosomes, indicative of gene amplification, has been reported by Alitalo et al. (1983) and Schwab et al. (1983a,b) in various tumors (Alitalo, 1985). It should be added that a few "cancer-prone syndromes" in humans are also known (Bloom's syndrome, ataxia telangiectasia, Fanconi's anemia); they are characterized by an increased incidence of chromosomal rearrangements. The differential diagnosis of these three syndromes is based on chromosomal rearrangements that are typical for each of them. Patients of another human cancer-prone syndrome, xeroderma pigmentosum, do not display a high frequency of microscopically visible chromosomal aberrations, but their cells are defective in repair DNA synthesis after uv irradiation. As a result, they frequently display a high incidence of skin cancers.

Even if we are unable to detect chromosomal aberrations in all cancer patients, using the now available chromosome banding techniques, we cannot exclude more subtle aberrations of chromatin organization at the molecular level as one of the steps involved in carcinogenesis. This was suggested by the intriguing experiments of G. M. Cooper et al. (1981). They found that DNA from nonmalignant cells could induce (but with difficulty, and only if the DNA had been fragmented by sonication) malignant transformation in other nonmalignant cells by transfection; DNA extracted from cells transformed with fragmented DNA transformed other cells without sonication. This indicates that some kind of chromosomal rearrangement must have taken place in the transformed cells. These experiments also show that apparently normal cells contain "transforming genes." This suggests that, as was proposed by Bloch-Shtacher and Sachs (1976), malignancy is controlled by a balance between genes that regulate the expression or suppression of the malignant phenotype. We shall return to this important point when we discuss hybrids and cybrids between normal and malignant cells, and the induction of cancers by viruses. At this time, we can conclude that it is very likely that chromosomal instability leading to chromosomal rearrangements may not be sufficient to induce a tumor, as was shown by a classic experiment by McKinnell et al. (1969). They first induced triploidy (as a chromosomal marker) in a frog suppressing polar body emission in the fertilized egg using hydrostatic pressure. A frog tumor virus (Lucke's virus) was then injected into the triploid frog, resulting in the formation of a triploid kidney tumor. One of the nuclei isolated from the tumor was then injected into an enucleated frog egg, resulting in the formation of a normal triploid tadpole (Fig. 14), *not* a teratoma. Thus, the nucleus of a malignant cell is capable of giving rise, after transplantation into an egg, to a normal embryo.

Similar results have been obtained by Di Berardino et al. (1983), who injected into enucleated frog eggs triploid nuclei taken from epithelial cell cultures of frog

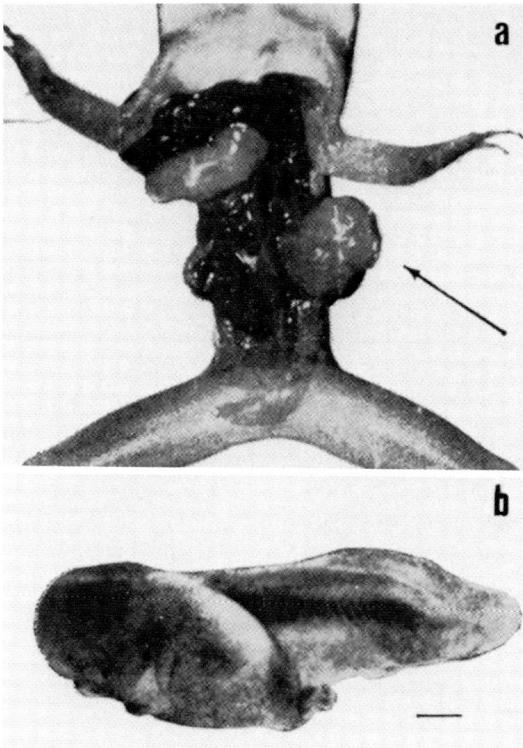

FIG. 14.   Renal tumor (arrow) in a recently metamorphosed triploid embryo. (b) Tadpole with a well-formed head, body, and tail after nuclear transplantation of a renal tumor nucleus. bar: 1mm. [McKinnell *et al.*, 1969, *Science* **165**, 394–395. Copyright 1969 by the AAAS.]

carcinomas induced by Lucké's frog herpes virus. Normal larvae without tumors were always obtained. This shows that the pronephric adenocarcinoma nuclei retained their genetic pluripotency and reversibility after being transferred into enucleated, unfertilized frog eggs. That DNA abnormalities do not necessarily lead to catastrophe was shown in the preceding chapter. Fragments of lethal hybrids are revitalized by transplantation into a normal host, and malignant teratocarcinoma cells may participate in the development of a mouse embryo. It has been reported that cells from Lewis lung carcinoma of the mouse lose their tumorigenicity if they have been cultured on chicken chorioallantoid membranes (Belin and Ossowski, 1983).

Transgenic mice have already brought a few interesting contributions to the cancer problem. Brinster *et al.* (1984) injected plasmids containing the early region of the SV40 oncogenic virus (coding for the T antigen) and a fused metallothionein/thymidine kinase gene into one of the pronuclei of fertilized mouse eggs and obtained transgenic mice with tumors of the choroid plexus. In these tumors, the viral T antigen was abundant and the SV40 genes were often amplified. Stewart *et al.* (1984a) injected a fused gene made of the promoter of the mouse mammary tumor virus and the *myc* oncogen (discussed later in this chapter) and obtained transgenic mice with mammary adenocarcinomas. Thus, the fused gene has been expressed in a tissue-specific manner in the breasts of the transgenic mice.

The second step in carcinogenesis (promotion) has been mainly studied in mouse skin. A single dose of a chemical carcinogen is not sufficient to develop a skin cancer, but if the skin has been rubbed with a promoting agent (the most active is the phorbol ester derivative TPA, which was mentioned in connection with cell differentiation), numerous skin cancers arise. Promoters (which are not carcinogens), in contrast to chemical carcinogens, do not act primarily on DNA, but rather on the cell surface, where, among other things, they interrupt cell-to-cell communication. The general (but, as we shall see, not universal) responses to TPA administration are as follows: activation of proteolytic activity (induction of plasminogen activator), stimulation of ornithine decarboxylase (ODC) and a concomitant increase in polyamine content, increase in prostaglandin synthesis, decrease in cAMP and increase in cGMP, stimulation of growth, and inhibition of terminal differentiation (reviewed by Berenblum and Armuth, 1981; Abrahm and Rovera, 1980; Mufson and Weinstein, 1981). It is a very general rule that promotion by phorbol esters is inhibited by vitamin A derivatives, the retinoids (reviewed by Lotan, 1980). According to a paper by Weeks *et al.* (1982), while initiation of skin cancer is produced by a single dose of a carcinogen, promotion occurs in two steps. The first is induced by TPA; the second is promoted by mezerein (a substance chemically unrelated to phorbol esters). This distinction is based on the fact that an inhibitor of polyamine synthesis, α-difluoromethyl ornithine, inhibits the growth of mezerein-induced skin papillomas, but not the induction of basal carcinomas by TPA.

Models of two-step promotion have been presented by Slaga *et al.* (1982), Marks *et al.* (1982), and Hicks (1983). Promotion I is induced by type I promoters, which, like TPA, react with cell membrane protein kinase C; they modify the cell phenotype and induce primitive stem cells. This first step can be arrested by the protease inhibitor TPCK. Promotion II (by type II promoters, mezerein, for instance) induces the selective proliferation of preneoplastic cells; this leads to the formation of a benign tumor. Type II promoters, like EGF and other growth factors, increase the activity of the polyamine-synthesizing enzymes [ODC, S-adenosylmethionide decarboxylase]; their effects are counterbalanced by retinoic acid. This model is valid for skin tumors, but it is probable that the same mechanisms operate in other tumors as well.

Other promoters (teleocidin, aplysiatoxin, debromoaplysiatoxin, etc.) have recently been discovered. All of them block the EGF-stimulated, tyrosine-specific phosphorylation of the EGF–membrane receptor and activate protein kinase C

(Friedman *et al.*, 1984; Fujiki *et al.*, 1984). In a recent brief review of the subject, Weinstein *et al.* (1984b) conclude that the target of the promoters is the cell membrane; in contrast, the target of the initiating carcinogens is DNA.

In discussing embryonic cell differentiation, we pointed out that differentiation can be studied either *in vitro* (in cultures of embryonic cells) or *in vivo* (in the intact embryos), and that the results obtained with these two approaches do not always coincide. The same is true in the field of cancer. For reasons of greater technical simplicity, a considerable amount of work is done on *in vitro* cell transformation. Treatment of cultured, apparently normal cells with oncogenic agents strongly modifies their phenotype (Fig. 15). The transformed phenotype has the following characteristics, which distinguish it sharply from the normal one: (1) the transformed cells display "immortality," i.e., they are capable of indefinite multiplication; (2) in contrast to normal cells, they are not subject to contact inhibition of growth and locomotion; (3) they are capable of "anchorage-independent" growth, i.e., they build colonies (foci) of transformed cells when cultivated in soft agar (Fig. 16); of these parameters, anchorage-independent growth is considered the most specific characteristic of malignant cell transformation. All of these changes reflect modifications in the cell surface of cells treated *in vitro* by carcinogens. Transformed cells have much less fibronectin than their normal counterparts; both kinds of cells differ in their ability to bind lectins and to agglutinate after lectin addition. In addition, transformed cells have a decreased requirement for serum factors and often express alkaline phosphatase on their cell surface. An enormous amount of work has been done, and is still being done, on chemically, physically, or virally transformed cells. However, there is growing, even overwhelming, evidence that *in vitro* transformation and *in vivo* tumorigenicity are not synonymous (Stiles *et al.*, 1975; Gee and Harris, 1979; Klein, 1979; Klinger, 1980). Thus, only if injection of cells in *nude* or x-irradiated mice is followed by tumor formation can one speak safely of tumorigenicity. This distinction between cell transformation and tumorigenicity is, of course, particularly important for those who work on human cancers with the hope of finding a cure for this disease.

Incidentally, the cause of most human cancers remains unknown. For Cairns (1981), environmental factors (pollution of air, water, and food with carcinogenic chemicals) play the most important role. Ames (1983) also concludes that environmental factors are responsible for 80% of human cancers; the most important among them is food, where carcinogenetic and mutagenic agents are abundant. They act by producing free radicals, and prophylaxy by absorption of antioxidants is recommended by Ames (1983). While the role played by cigarette smoking in lung cancer seems to be statistically well established, it is unlikely that the author's pipe (which he has been smoking for half a century without induction of any cancer so far) smoke is a serious danger for visitors in his office. Natural hormones are probably responsible for prostate cancer, which occurs sooner or later in almost all human males, provided that they live long enough.

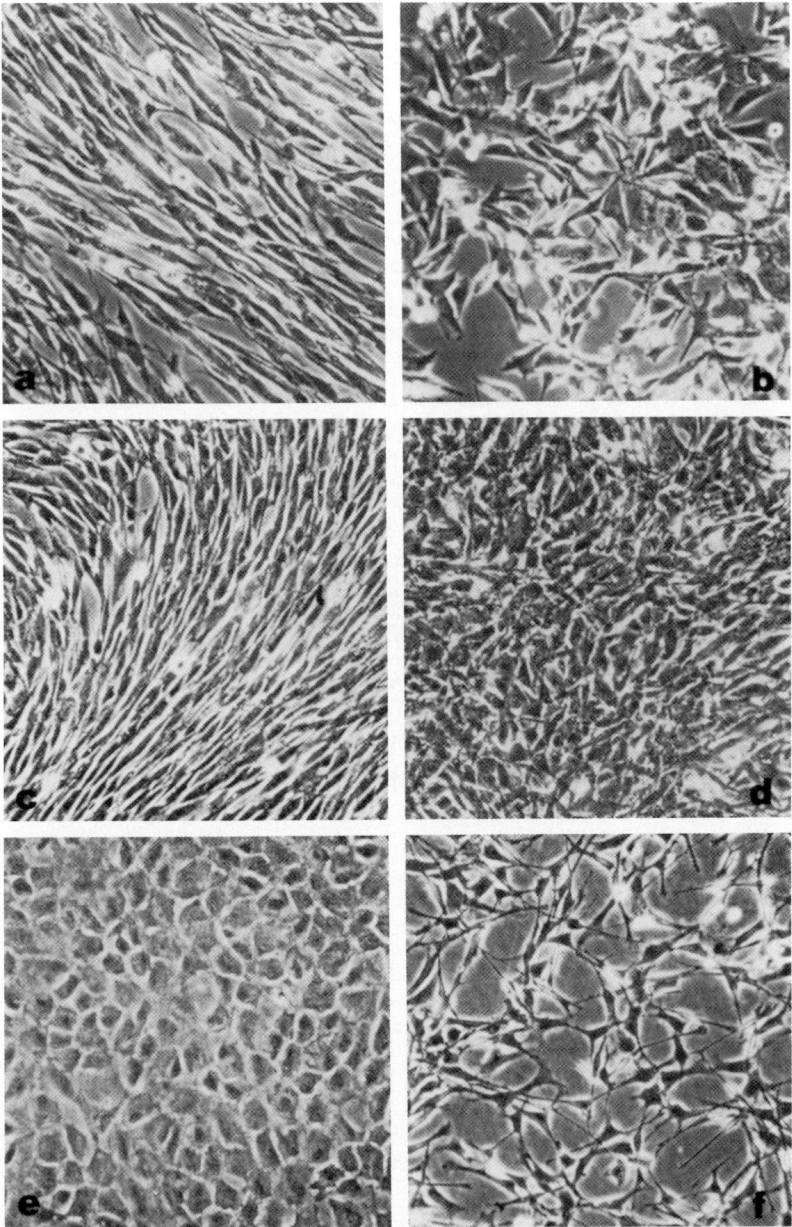

FIG. 15.   Growth patterns and morphological changes in transformed established cell lines. (a) BHK cells nearly confluent. (b) BHK cells transformed by polyoma virus. (c) RECL$_3$ cell culture. (d) same as (c), but transformed by polyoma virus. (e) 3T3 (Swiss), confluent cell culture. (f) Same as (e), but transformed by polyoma virus, low density. [Benjamin, 1974.]

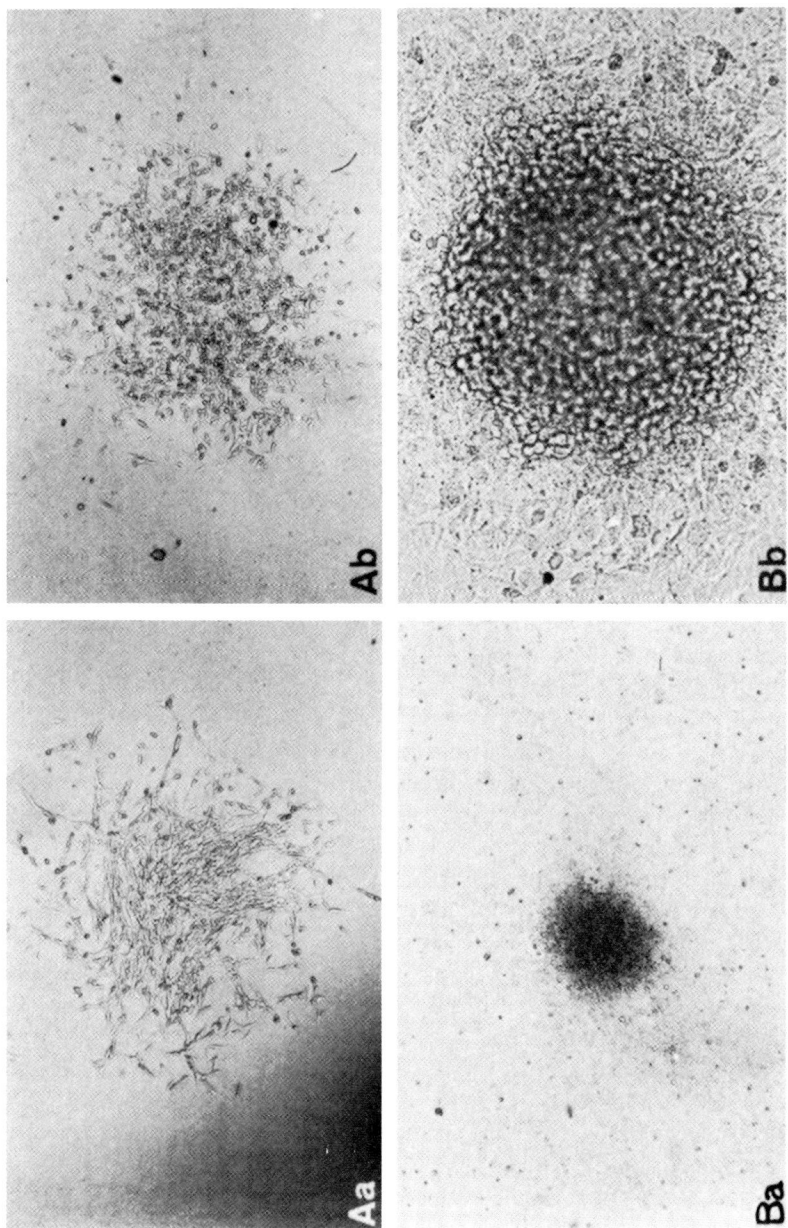

FIG. 16. Transformation assays. (A) colony morphology: (a) colony of normal BHK cells; (b) colony of polyomavirus–transformed BHK cells. (B) Focus assay. (a,b) Rous sarcoma virus–induced focus of chick embryo fibroblasts. [Benjamin, 1974.]

For example, diethylstilbestrol is a carcinogen, although it is not a mutagen. Ultraviolet radiation from the sun certainly induces skin tumors, but this will not prevent the modern habit of sunbathing. Even oxygen in the air is very suspicious, since formation of free radicals (especially the oxygen-activated species) plays an important role in both radiation damage and chemical carcinogenesis (Greenstock, 1981). Should we try to develop, like yeast cells, anaerobic fermentation processes that would allow us to avoid cancer by living in an anaerobic environment? This fantastic perspective is no more attractive than living in subterranean bunkers in order to be protected against atomic or uv radiation.

For A. B. Sabin (1981), human cancers are caused by a loss of specific regulatory genes, a hypothesis that belongs logically in a book that deals with molecular cytology and not with human cancer etiology. Sabin's (1981) hypothesis is based on the analysis of the numerous and often conflicting data that have emerged from intensive studies on somatic hybridization between normal and malignant cells. These data will be discussed later.

Before we leave generalities for more specific topics, a last point should be raised. Is there always opposition between differentiation and malignant proliferation? Are cancer cells fully undifferentiated? The general answer to both questions tends to be "yes." However, many pathologists, who have seen thousands of cancer biopsy specimens under a microscope, have serious reservations. As shown in Fig. 17, different cancers may have different morphologies; otherwise, it would not be possible for pathologists to distinguish one form of cancer from another. Cancer cells may retain some of the characteristics of their previous differentiated phenotype; others are almost undifferentiated and look like early embryonic cells. It therefore seems safer to speak of partial dedifferentiation in the case of cancer cells. In malignant tumors, cell division is very active and many mitotic figures (often abnormal) can be seen on stained sections. However, it would be incorrect to think that the division time is shorter for a cancer cell than for a normal one. In certain forms of leukemia, the cell cycle is longer for leukemic cells than for normal ones, but leukemic cells go on dividing continuously, while normal blast cells differentiate into mature white blood cells.

We have just said that cancer cells may look like embryonic cells under the microscope; this is also true at the biochemical and immunological levels. For instance, fetal liver synthesizes and secretes α-fetoprotein, while the adult liver synthesizes serum albumin instead, secreting it into the bloodstream; liver cancer (hepatoma) genes, like fetal liver, express the α-fetoprotein gene. As we have already seen, the albumin and α-fetoprotein genes are different but belong to the same gene family, and it is likely that they derive from the duplication of an ancestral gene. When the albumin gene becomes active during ontogeny, the α-fetoprotein gene is turned off. It is a striking fact that this gene is reactivated and reexpressed when the liver cells (hepatocytes) are transformed into hepatoma

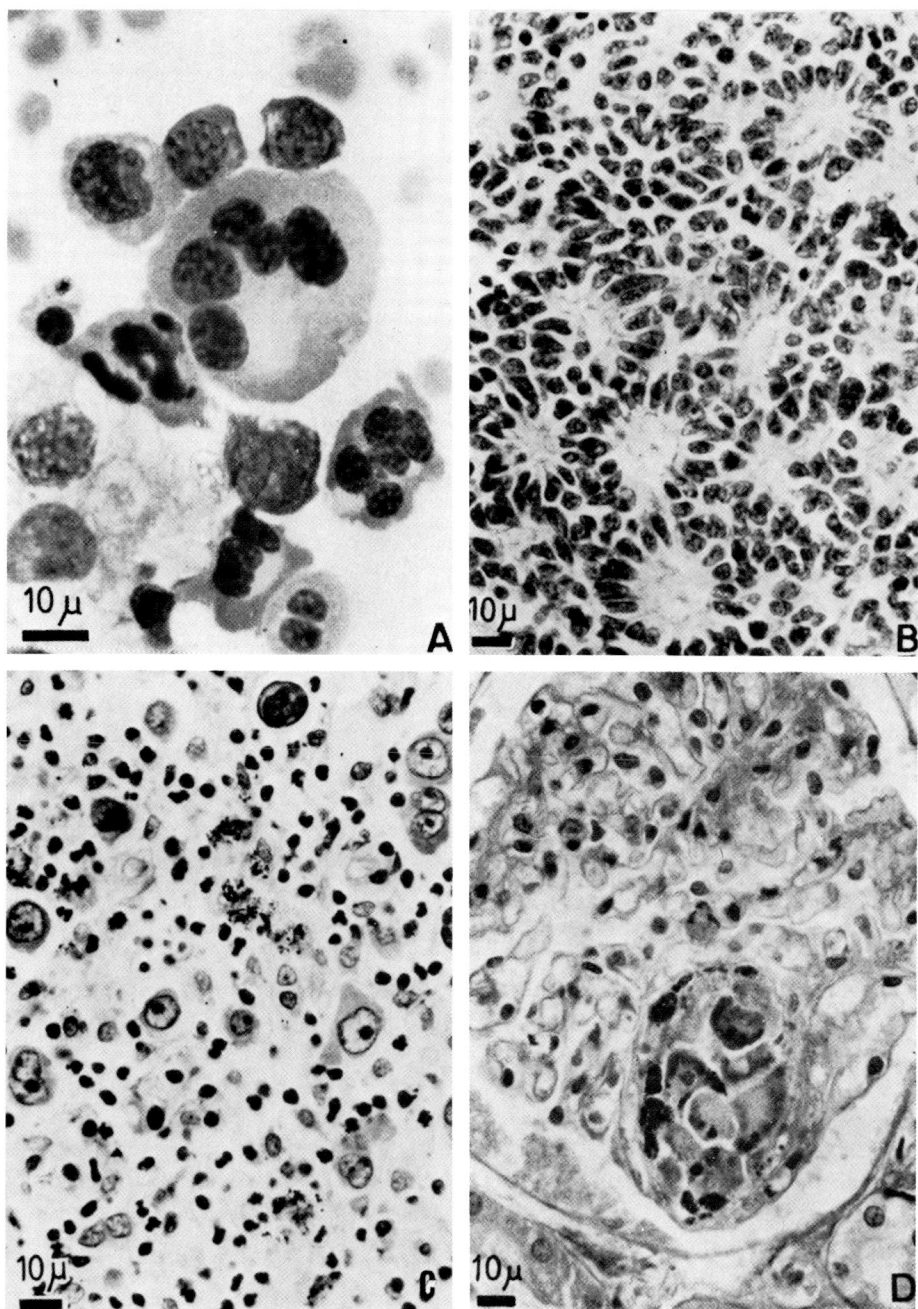

Fig. 17. Different phenotypes of tumor cells. (A) Neoplasic erythroblastosis. (B) Reti-noblastoma. (C) Lymphogranulomatosis. (D) Melanoma (metastatic cells). [Dustin, 1966.]

cells after chemical induction of a liver cancer. Similar observations have been made by immunologists who demonstrated the existence, in the cell membrane, of oncofetal or carcinoembryonic antigens that are shared by mammalian embryos and tumors (in particular, colon carcinomas) but cannot be detected in normal adult cells (Gold and Freedman, 1965). For instance, Gooding *et al.* (1976) reported the presence of a cell surface antigen in mouse eggs, morulae, and inner cell masses at the blastocyst stage; the same antigen is present in many tumors, but not in the trophoblast and in adult tissues. However, another antigen present in tumors (teratomas and hepatomas) cannot be detected in morulae, while a third antigen present in teratomas is not detectable in embryos. Clearly, some fetal genes, but not all, are reexpressed in cancer cells (reviewed by Ibsen and Fishman, 1979; Rogers, 1983). For this process, therefore, the term *retrodifferentiation* has been proposed. Sachs (1980) states that malignant cells no longer require the physiological inducers of differentiation, and this uncouples growth and differentiation leading him to suggest a model explaining the reappearance of fetal proteins in tumors. Finally, Chan (1981) has proposed that embryos possess growth genes, which would be reactivated in tumors where fetal genes are reexpressed. Thus, there are many theories to explain the reappearance of fetal proteins in tumor cells. Unfortunately, we still know very little about the molecular mechanisms that suppress embryonic genes during development (except for the well-studied case of the hemoglobin switches) and reactivate them in some tumors. It is likely that these mechanisms are the same as those discussed at the end of the preceding section, when we dealt with active chromatin.

It has often been suggested that if a cancer cell could be induced to differentiate, it would lose its malignancy. This idea is attractive and has been developed by Sachs (1982). We think that induction of differentiation in neoplastic cells might be a cure for those cancers in which the chromosomes are essentially normal, but that it would probably be of only limited value where there are strong chromosomal imbalances (aneuploidy, infection by oncogenic viruses) or major deficiencies in the DNA repair mechanisms. It is well established that there is a high cancer risk for patients in whom DNA repair is deficient (those with xeroderma pigmentosum, ataxia telangiectasia, or Bloom's syndrome). Fortunately, these hereditary diseases are infrequent, but clearly demonstrate the existence of a close link between cancer and imperfect DNA repair. It seems wishful to hope that patients suffering from such diseases or from marked aneuploidy would be permanently cured if tumor cells could be induced to differentiate. However, it should be recalled that, as we have seen in Chapter 2, fragments of lethal amphibian gastrulae or even lethal mouse embryos can be revitalized by being grafted onto a normal host. The problem of cancer and cell differentiation is very important and interesting for our understanding of carcinogenesis at the cellular and molecular levels. For this reason, we shall begin with the studies done on malignant cells that can be induced to differentiate (neuroblastoma, teratocarcinoma, and certain leukemic cells). This discussion will be brief because these cells have already been mentioned in the preceding chapter.

B. Induction of Differentiation in Cancer Cells

1. Neuroblastoma Cells

These cells of neuronal origin provide a good model for the study of neuronal terminal differentiation. Neuroblastoma cells divide actively in culture and are devoid of neurites (axons and dendrites). Treatments with many agents that slow down cell division may lead to neurite formation (Fig. 18A). Good inducers of neuronal differentiation are x-rays, DMSO, prostaglandins, dibutyryl-cAMP, methylxanthines, and 1-methylcyclohexane carboxylic acid (reviewed by Prasad, 1975). Electrophysiological studies have shown that the axons induced by such treatments display classic properties (Minna et al., 1972). It has been shown that arrest of cell division (probably due to an increase in cAMP) by the inducers of differentiation leads to a decrease in the synthesis of actin and tubulin. Since the axons contain many MTs, it is probable that their outgrowth takes place at the expense of a preformed pool of tubulin dimers. On the other hand, treatment of neuroblastoma cultured cells with the inducers increases the synthesis of four proteins, one of which is the intermediary filament protein vimentin (Portier et al., 1982). Treatment with DMSO is followed by a decrease in ODC activity, which was high during proliferation (Portier et al., 1982; Chen et al., 1982). It thus seems clear that in neuroblastoma cells, treatment with the inducers results in profound changes in the expression of the genetic program. However, as already mentioned, enucleation results in differentiation of neuroblastoma cells, suggesting that the nucleus contains or produces inhibitors of differentiation (Liebermann and Sachs, 1978).

Comparable results have been obtained in our laboratory with glial tumor cells (Heilporn et al., 1973); x radiation slowed down mitotic activity, and short neurites were formed (Fig. 18B). Despite this morphological differentiation, the cells remained malignant and formed tumors after injection into mice.

There are close similarities between neuroblastomas and retinoblastomas (reviewed by Robertson, 1984). Both tumors are of embryonic origin and result from profound genetic alterations (deletions and gene amplification). The cause of retinoblastomas seems to be the inactivation or deletion of the two alleles of a recessive gene that has a suppressor or regulatory function (Murphree and Benedict, 1984). The function of this gene in normal cells would be to turn off the embryonic phenotype, and thus the capacity for cell division; in its absence, the embryonic genes would remain active, and proliferation would go on continuously. In both neuroblastomas and retinoblastomas, there is marked amplification (25–700 times) of a particular gene called N-myc, which displays partial homology with the cellular oncogene c-myc, to be discussed in Section II,F

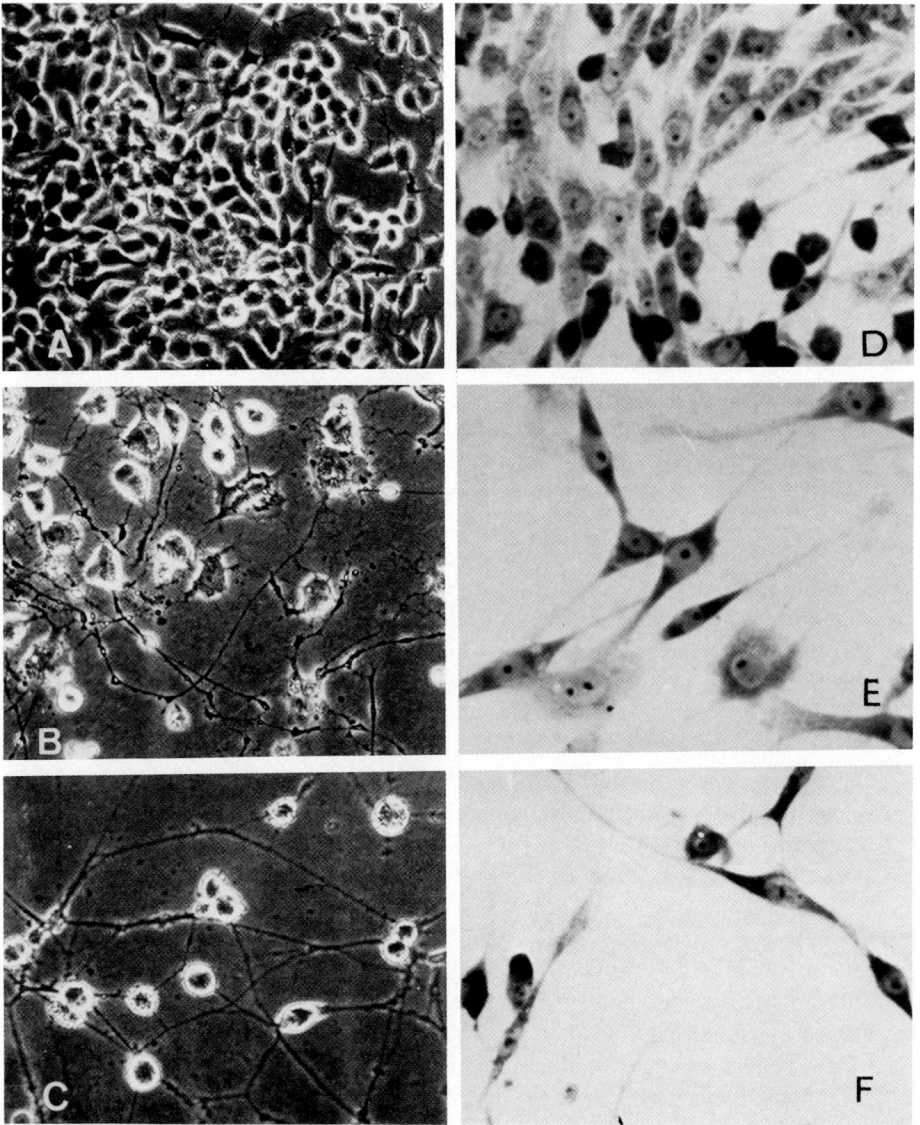

FIG. 18.   (A–C) Phase-contrast micrographs of mouse neuroblastoma cells in culture. (A) Control culture; cells grow in clumps, and some of them have short cytoplasmic processes. (B) Prostaglandin (PGE$_1$)-treated culture shows the formation of long neurites after a 4-day treatment. (C) PGE$_1$-treated culture 14 days after treatment; the remaining cells maintain their differentiated phenotype. (D–F) Light micrographs of glial tumor cells (Unna staining). (D) Control culture with typical bipolar cells. (E,F) Aspect of the cells 5 days after irradiation with 300 and 600 rads, respectively. Note the presence of long cytoplasmic processes and the size difference between the irradiated and control cells. [(A–C) Prasad, 1975; (D–F) Heilporn et al., 1973.]

(Kanda *et al.*, 1983; Kohl *et al.*, 1983; Lee *et al.*, 1984). The genetic instability of the neuroblastoma and retinoblastoma cells is further shown by the presence of double minute chromosomes in the cells and by the fact that, after amplification, the *N-myc* copies are translocated to several chromosomes (Schwab *et al.*, 1984). The function of *N-myc* might be to control the proliferation of neuroectodermal cells. Both neuroblastomas and glioblastomas have the same *neu* oncogene, coding a p185 protein which displays similarities with the EGF receptor (Schechter *et al.*, 1986).

2. Murine Erythroleukemia (MEL) Cells

These malignant cells derive from proerythroblast cells transformed by a complex of two viruses called *Friend virus*. Their main interest for cell biologists resides in the fact that they synthesize hemoglobin and several other proteins characteristic of red blood cells after treatment with chemical inducers (reviewed by Marks and Rifkind, 1978; Reuben *et al.*, 1980). The most popular inducer of erythroid differentiation in MEL cells is DMSO (Friend *et al.*, 1971), although many other substances (hexamethylene bisacetamide, butyrate, purine derivatives, ouabain, etc.) are equally active. A common characteristic of all inducers is that they slow down the G_1 phase of the cell cycle (Marks and Rifkind, 1978) in the transformed proerythroblasts which can no longer respond to erythropoietin stimulation of erythropoiesis. This arrest in G_1 might be due to a decrease in a nuclear protein ($M_r = 53,000$) called *p53* or *cellular tumor antigen* (Shen *et al.*, 1983). This protein is attracting attention because it apparently plays a role in cell proliferation and malignant transformation. It accumulates in both induced and spontaneous tumors, where it is found as a phosphoprotein. Traces of non-phosphorylated p53 protein can be detected in the thymus of healthy animals. This protein is present in 13-day-old mouse embryos, and its mRNA decreases 20 times when undifferentiated teratocarcinoma cells are induced to differentiate (Reich *et al.*, 1983); its expression can be induced by the phorbol tumor promoter PMA (Rotter, 1983). Curiously, p53 is located in the nucleus of malignant cells and in the cytoplasmic cytoskeleton of normal cells (Rotter *et al.*, 1983). This interesting protein might well play an important role in the control of cell proliferation and differentiation in normal and malignant cells. We shall return to it when we discuss *in vitro* malignant transformation.

Figure 19 shows uninduced and induced MEL cells after staining with the diaminobenzidine peroxidase reaction for hemoglobin. While the reaction is negative in noninduced cells, a majority of the DMSO-treated cells give a strong peroxidase reaction. In DMSO-treated cells, both the cytoplasm and the nucleus shrink prior to the synthesis of hemoglobin and membrane proteins characteristic of red blood cells.

The effects of the inducers on MEL cells are complex. Hemoglobin synthesis results from expression of previously silent globin genes producing globin mRNA. Synthesis is inhibited, after DMSO induction, by BrdUrd (Scher *et al.*,

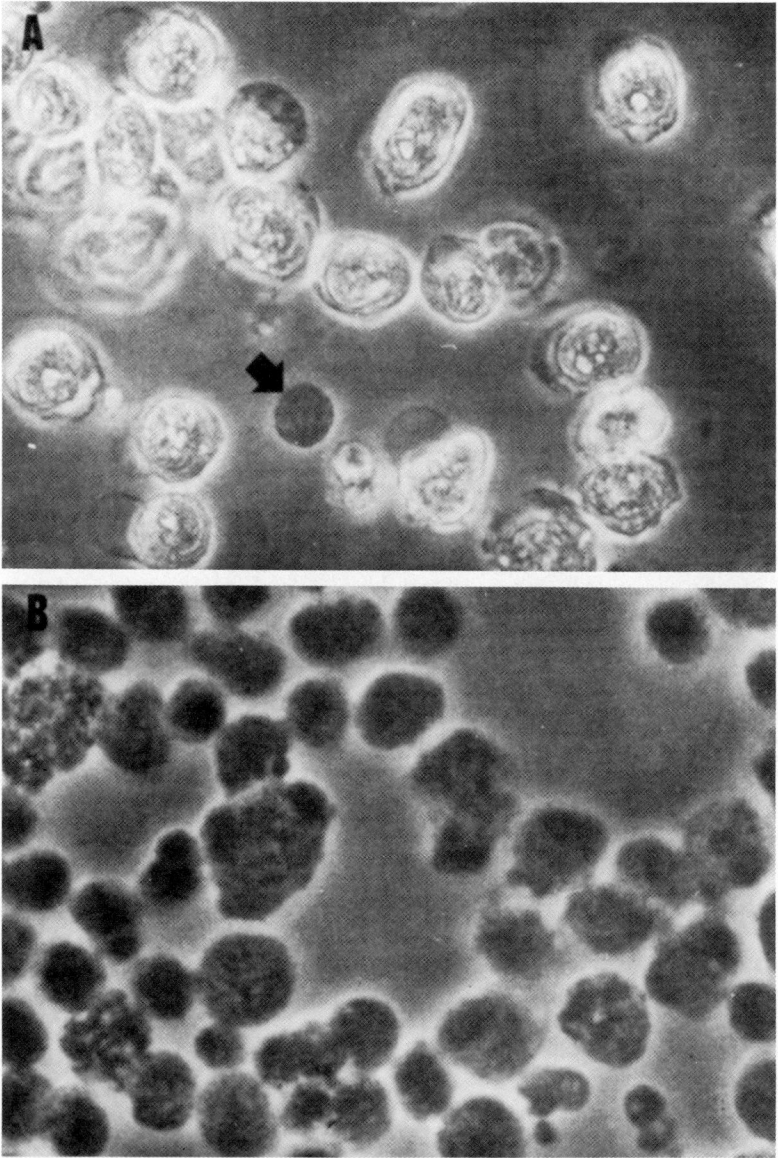

Fɪɢ. 19.   Murine erythroleukemic cells (MELC); benzidine staining for hemoglobin detection. (A) Uninduced MELC; arrow, a single positive cell. (B) Induced MELC (DMSO treatment for 5 days). All of the cells are positive. [Reeves and Cserjesi, 1979b.]

1973). Paradoxically, this thymidine analog is capable of acting as an inducer, as we have already seen (Adesnik and Smitkin, 1978). The target for the inducers in chromatin is not limited to the globin gene loci. In addition, the amount of spectrin increases, as does the activity of cholinesterase and carbonic anhydrase when MEL cells are treated with DMSO. The effects of DMSO are blocked not only by BrdUrd but also by the phorbol derivatives TPA and PMA (Yamasaki *et al.*, 1984).

It is also quite possible that the inducers do not act directly on the genome and that their primary site of action is the cell surface. This is indicated by the fact that DMSO, like ouabain, quickly inhibits the sodium pump localized in the plasma membrane, allowing a Ca^{2+} influx, mediated by a Na^+/Ca^{2+} exchange (R. L. Smith *et al.*, 1982). It has also been shown that treatment with DMSO increases the number of the adrenergic receptors associated with membrane adenylate cyclase (Schmitt *et al.*, 1980) and that this increase is prevented by agents that inhibit the Ca^{2+} flux. A likely consequence of increases in the Ca^{2+} content and in adenylate cyclase β-adrenergic receptors in the induced cells is an increase in their cAMP content. Indeed, Gazitt *et al.* (1978) have found that the chemicals that are known to increase the cellular cAMP content can induce hemoglobin synthesis in MEL cells. This increase in the cAMP content of DMSO-treated MEL cells is expected to slow down cell proliferation and to favor cell differentiation. The relationships between the two processes (proliferation and differentiation) have been analyzed by Tsiftsoglou and Sartorelli (1981). They found that cellular proliferation is required for the initiation of differentiation (or rather, the commitment to differentiate) in accordance with what we saw when we discussed the probable importance of critical (quantal) mitotic cycles for embryonic differentiation. Once the cells have been programmed to differentiate, cell proliferation loses much of its importance and is no longer required for the final accumulation of hemoglobin in the induced cells.

Also in favor of an important role for the cell membrane are the results of Bosman *et al.* (1982), who found that tunicamycin inhibits the differentiation of induced MEL cells. This suggests that glycosylation of proteins is a necessary step in this process, but the precise reasons for this requirement for protein glycosylation are not yet clear.

Since differentiation of induced MEL cells ultimately results from increased transcription of a limited number of genes, it is clear that the inducers, probably after a complex series of events, must modify the conformation of specific sites in chromatin. The existing data are not very numerous and are somewhat contradictory. Christman *et al.* (1977, 1980) have reported that DNA is undermethylated in induced MEL cells and that there is a close correlation in these cells between differentiation and undermethylation. According to a more recent paper by Sheffery *et al.* (1982), however, there are no alterations in the pattern of DNA methylation during differentiation of MEL cells; nevertheless, discrete, DNase-hypersensitive sites appear near both the α- and β-globin genes during differentiation. It has also been reported that treatment of MEL cells with DMSO results in the appearance, within 12 hr, of a DNase-hypersensitive site at the 5' end of the β-major globin gene (Balcarek and McMorris, 1983).

MEL cells were mentioned in the preceding section, when we discussed the use of cell hybridization techniques for the analysis of cell differentiation. It seems that fusion of MEL cells with any type of cell (fibroblasts, hepatoma cells, myeloma cells) suppresses transcription of the globin genes and hemoglobin synthesis in DMSO-induced heterokaryons (Deisseroth *et al.*, 1976; Conscience *et al.*, 1977, Ar-Rushdi *et al.*, 1982). That this repression is due to cytoplasmic factors was shown by the experiments of Gopalakrishnan *et al.* (1977). Fusion of a cytoplast from a normal cell with an induced MEL cell extinguished hemoglobin synthesis for many cell generations. Curiously, in cybrids between MEL cells and cytoplasts of hepatoma cells, cytoplasmic factors present in the hepatoma cytoplasts activated the phenylalanine hydroxylase gene in the MEL cell nucleus (Gopalakrishnan and Anderson, 1979). These experiments emphasize once more the complexity of nucleocytoplasmic interactions.

We cannot leave the subject of leukemic cells without saying a few words about the attempts made, mainly in the laboratory of L. Sachs, to induce differentiation in leukemic cells other than MEL cells. In this case, differentiation does not lead to hemoglobin synthesis, but to the formation of granulocytes and macrophages. In the following discussion, we shall follow mainly the review article of Sachs (1982). As already mentioned, chromosomal abnormalities (trisomy, monosomy, deletions, translocations) are frequent in patients suffering from acute leukemias, and these chromosome changes are specific for each type. However, it seems well established that in sarcomas, leukemias, and teratocarcinomas, the malignant cells have not lost the genes that regulate normal growth and differentiation, which is shown by the fact that the malignant phenotype can be reversed to a nonmalignant one by chromosome segregation. In the mouse, chromosomes 6 and 12 bear genes that control both malignancy and differentiation in leukemic cells (Azumi and Sachs, 1977). In certain clones of myeloid leukemia cells, reversion of malignancy has also been obtained by another mechanism: treatment with protein factors that induce the formation of colonies of granulocytes and macrophages (Fig. 20). These protein factors, which are present in culture supernatants, have been given different names: *colony-stimulating factor (CSF)*, *macrophage and granulocyte inducer (MGI)*, and *colony-stimulating activity (CSA)* (reviewed by Metcalf, 1985). The MGI factor studied in Sachs's laboratory is required by normal myeloblasts for cell viability, proliferation, and differentiation to mature macrophages or granulocytes. Leukemic cells, in contrast to normal cells, can multiply in the absence of MGI (just as MEL cells can grow in the absence of erythropoietin); thus, in leukemic cells, proliferation control is uncoupled from the processes that initiate differentiation; this is presumably due to the aforementioned chromosome changes. Sachs (1980,

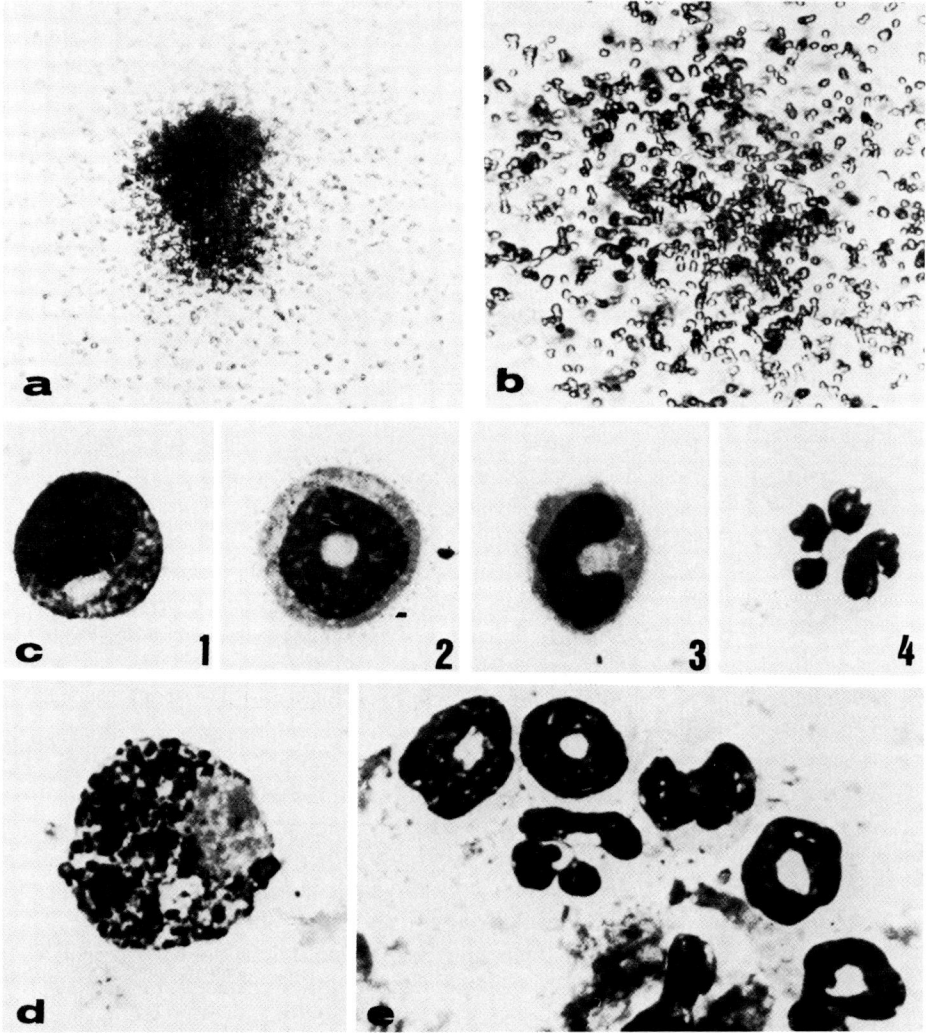

FIG. 20. (a,b) Colonies from normal hematopoietic cells induced by MGI: (a) granulocyte colony; (b) macrophage colony. (c–e) Differentiation of myeloid leukemic cells to mature granulocytes and macrophages: (c) leukemic cell (c_1)—stages of differentiation to mature granulocytes (c_2–c_4); (d) mature macrophage; (e) group of granulocytes in different stages of differentiation. [Sachs, 1982.]

1982) has proposed that myeloid leukemia results from changes that produce certain constitutive (instead of induced) pathways of gene expression, resulting in the uncoupling of the normal MGI requirement for growth and differentiation. There are several molecular forms of MGI (MGI-1 and MGI-2), which have different effects on proliferation and differentiation of normal and myeloid leukemic cells. The experiments of Sachs (1980, 1982) and his colleagues have shown that the MGI proteins can induce differentiation of certain myeloid leukemic cell clones to mature macrophages and granulocytes (Fig. 20) and that this differentiation is associated with changes in cell shape.

In a more recent paper, Lotem and Sachs (1983) reported that normal myeloid precursor cells depend on MGI-1 for survival and proliferation; MGI-1 induces the production of MGI-2, which is required for differentiation. In certain leukemic clones, MGI-1 is not required for proliferation; their differentiation can be induced by MGI-2. At that time, MGI-1 becomes necessary for survival and multiplication. In these clones, MGI-1 does not induce the formation of MGI-2; this explains why the cells remain undifferentiated. This analysis shows the complexity of the granulocyte–macrophage differentiation program and why it fails to occur in leukemic cells.

As in the case of MEL cells, the effects of MGI can be mimicked by a wide variety of chemically unrelated compounds, including some of the steroid hormones, actinomycin D, DMSO, the phorbol ester TPA, substances used for cancer chemotherapy (inhibitors of DNA synthesis) and x-rays. A long list of inducers and noninducers of differentiation is given by Sachs in his 1982 review of the subject. One of the many substances mentioned in this list is the phorbol ester cancer promoter TPA. According to Liebermann et al. (1981), TPA induces a rapid attachment of human myeloid leukemic cells to the substrate, followed by spreading and by a switchover of the granulocyte to the macrophage program. While cells in suspension can express either the macrophage or the granulocyte program, attachment and the following changes in cell shape result in a restriction of the developmental program to macrophages. TPA induces protein changes in these cells, regulating gene expression at the level of both mRNA transcription and mRNA translation (Hoffman-Liebermann et al., 1981). According to R. A. Cooper et al. (1982), the induction by TPA of macrophage-like differentiation in leukemic cells takes place while the phorbol ester is still retained on the cell surface, and it is suggested that differentiation takes place through a receptor-mediated transmembrane process. However, it would be a mistake to believe that induction of differentiation results solely from changes at the cell surface level, since early changes ultimately trigger modifications in gene expression and thus in chromatin conformation. This is shown, among other things, by the fact that treatment with the inhibitor of DNA methylation, 5-azacytidine, has allowed Boyd and Schrader (1982) to obtain macrophage-like cell lines from lymphoma cells.

For an embryologist, there is a striking parallel between the induction of differentiation in malignant cells and the induction of parthenogenesis and neural tube formation in sea urchin eggs and amphibian gastrulae, respectively (see Chapter 2). In both cases, many unrelated agents act as inducers. Their first site of action is the cell membrane, and by a cascade of still poorly understood events, specific genes become expressed.

This parallel between embryos and malignant cells is especially striking in the case of the teratocarcinomas. Although they were mentioned in the preceding chapter, they deserve a more complete treatment here.

3. Mouse Teratocarcinomas Cells[4]

Teratomas have been known for many years. They are a monstrous combination of all types of tissues, including teeth, hair, and fingers. Although teratomas are benign tumors, teratocarcinomas are malignant tumors because they contain, in addition to differentiated cells, stem cells [called *embryonal carcinoma (EC) cells*]; these cells are strikingly similar to those of early mouse embryos (Fig. 21). Teratomas may arise spontaneously in the ovary or the testis and can be induced when an early mouse embryo is grafted at an ectopic site (testis or kidney of an adult mouse). In the ovary, teratomas can arise if accidental, spontaneous parthenogenesis takes place in an egg. Although initial development is normal, the parthenogenetic embryo develops in an anarchic way, leading to the formation of a tumor (teratoma), which remains benign unless it contains undifferentiated stem cells. It should be mentioned that, according to Mintz *et al.* (1978), totipotent teratocarcinoma cells may arise from somatic and not from germinal cells.

Many cell lines have been isolated from teratocarcinomas. Some of these lines are nullipotent in that they are unable to differentiate under normal culture conditions. Other cell lines differentiate in either a single direction or are multipotent and give rise to a whole array of different tissues. Differentiation of a nullipotent malignant embryonal carcinoma cell line can now be obtained when the culture conditions are modified; culture in a serum-free medium leads to differentiation of certain malignant stem cells into neurons (Darmon *et al.,* 1982a). An EC cell line (1003 cell line) does not differentiate when it is cultured in a serum-containing medium and is thus taken as a nullipotent cell line, although it expresses surface antigens characteristic of early mouse embryos. If the same cell line is cultured in a serum-free medium, it progressively differentiates in nerve cells, forming first neuroepithelial cells, which possess vimentin intermediary-sized filaments, and, later, typical neurofilament proteins, which appear in neurites (Fig. 22). Simultaneously, vimentin and an embryonic surface antigen (called ECMA7) disappear (Darmon *et al.,* 1982b). Another teratocarcinoma

[4]Reviewed by Jacob (1978), Martin (1981), and Strickland (1981).

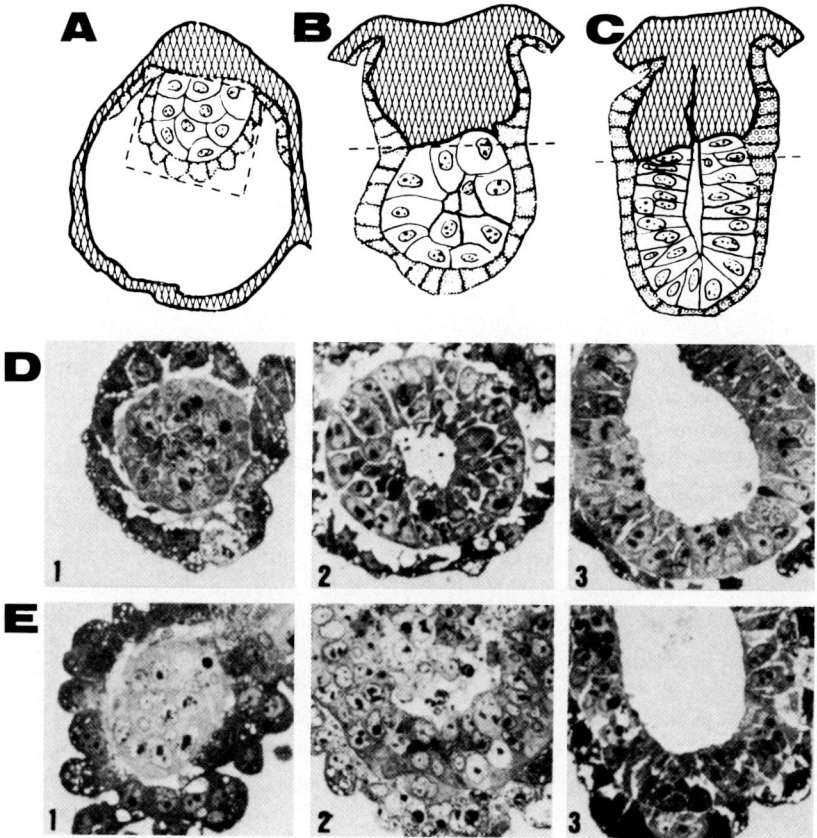

FIG. 21. Similarity between embryoid body development (derived from teratocarcinoma cells) and the early postimplantation development of the mouse embryo. (A–C) Normal embryos. (A) Blastocyst at the time of implantation, in which the embryo inner cell mass (dotted lines) is surrounded by the trophectoderm (crosshatched area); (B) After 5 days of development, the fetal portion of the embryo (below the dotted line) develops a ''proamniotic cavity.'' (C) After 6 days of development, the proamniotic cavity has enlarged and the embryonic ectoderm (clear area) is now organized into a columnar epithelium. (D) Inner cell mass isolated from a blastocyst prior to implantation (1) and cultured *in vitro* (2,3). (E) Embryonal carcinoma cell lines, when cultured as aggregates, differentiate to form a layer of endoderm and an inner cell mass (1). Subsequently, these two-layered ''embryoid bodies'' undergo internal changes (2,3) that parallel the development of the fetal portion of the embryo. [Martin, 1980, *Science* **209**, 768–776. Copyright 1980 by the AAAS.]

cell line is capable of forming bone (Nicolas *et al.*, 1980). Clearly, the concept of potentialities lacks precision in teratocarcinomas as well as in embryos. In an amphibian gastrula, the presumptive ectoderm gives rise to epidermis. If this region of the gastrula is explanted and submitted to a brief acid or alkaline shock, it can differentiate into neural tissue and, ultimately, into neurons. Here a com-

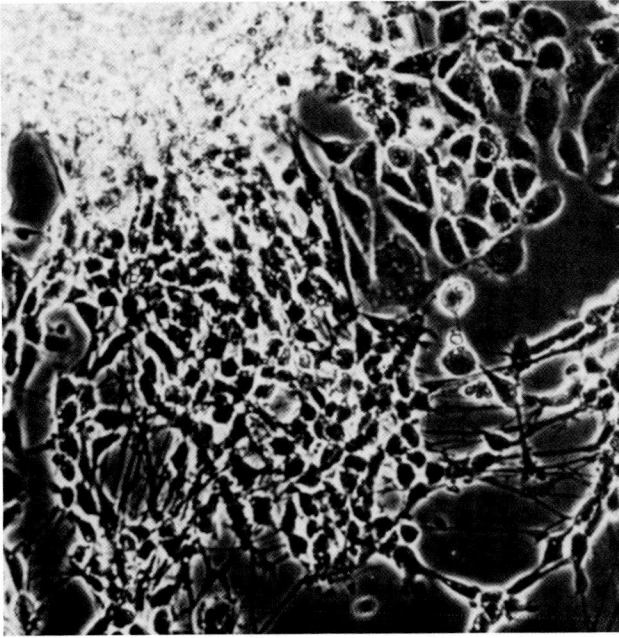

FIG. 22.   Phase-contrast photomicrograph of neuronal cells formed by attached embryoid bodies. [From Martin and Evans, 1975. Copyright 1975 by M. I. T.]

parable result can be obtained by culture of nullipotent cells in the absence of serum. It is impossible to know with certainty what the total potentialities of an embryonic explant or a teratocarcinoma cell line are because one can never exclude the possibility that new potentialities might awake under new experimental conditions.

Much work has been done, especially in F. Jacob's laboratory on the F9 cell line, in which the differentiation potentialities are apparently limited to the formation of endoderm-like cells. A great step forward was made when Artzt *et al.* (1973) discovered that these EC cells possess a surface antigen that they share in common with normal mouse morulae and blastocysts. This F9 antigen is not present in 9-day embryos and, in the adult, is found only in the male germ line, including spermatozoa (Fig. 23). Interestingly, this F9 antigen is found on morulae and spermatozoa of all mammals tested, but not in those of amphibians and birds (Gachelin *et al.,* 1977; Holden *et al.,* 1977). It has been suggested that the F9 antigen is encoded by the *T/t* locus (Buc-Caron *et al.,* 1974). This complex locus, as we have seen in Chapter 2, is important for early mouse embryogenesis, since it mediates several steps in early development. However, the evidence for the view that the F9 surface antigen is specified by genes located in the *T/t* complex is not yet compelling.

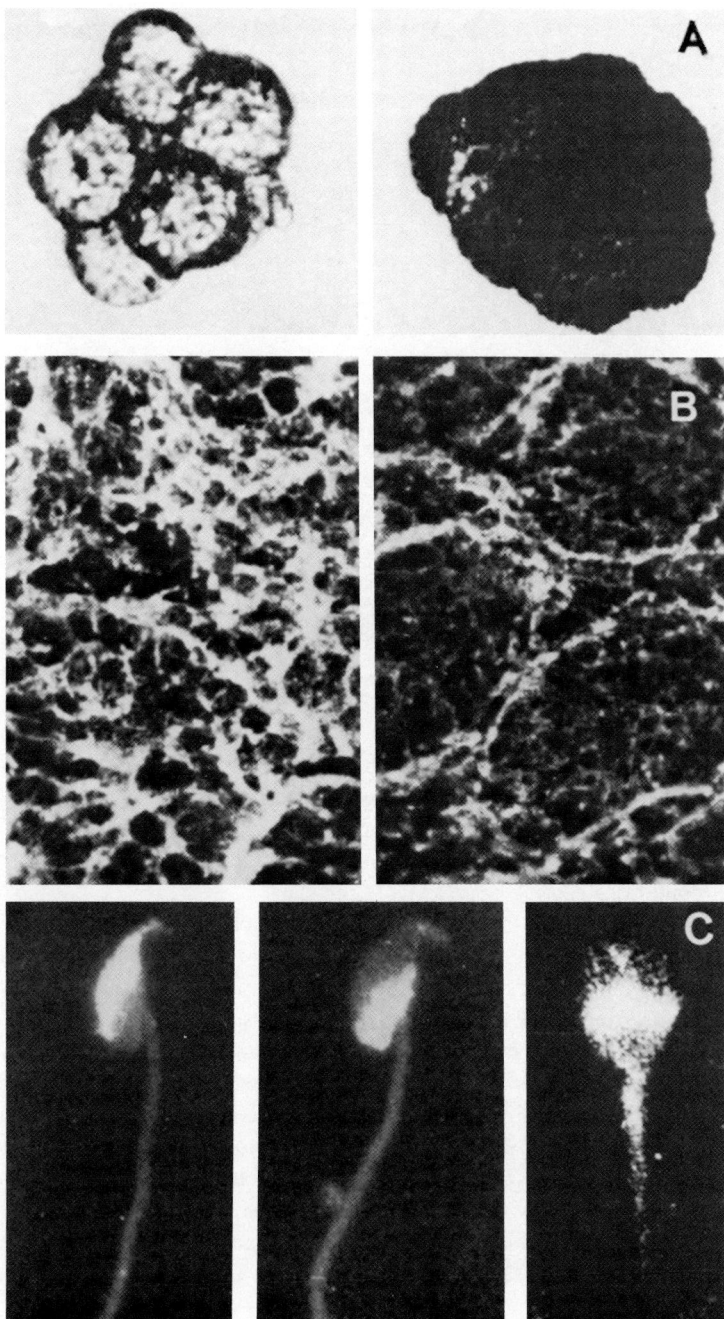

Although antisera against the F9 antigen had no effect on mouse development, positive results have been obtained with monovalent anti-F9 Fab fragments. Cleavage proceeded normally, but compaction of the morulae did not take place and no blastocysts could be obtained in the presence of such anti-F9 Fab fragments (Kemler et al., 1977). These experiments further showed that anti-F9 Fab fragments are not toxic, but that they loosen cell-to-cell interactions; they prevent compaction in morulae because this process requires the formation of gap and tight junctions.

That the chemical composition of the cell surface displays similarities in EC cells and early mouse embryos is further shown by the work of Muramatsu et al. (1978). Both synthesize a class of large fucoglycopeptides that are no longer synthesized by differentiated cells. These glycopeptides are a constituent of the cell surface molecules recognized by the antiserum against the F9 antigen (Muramatsu et al., 1979). It has also been shown that the plasma membrane is more fluid in undifferentiated embryonal carcinoma cell lines than in differentiated cell lines (Coulon-Morelec and Buc-Caron, 1981). In general, these differentiated cell lines are nonmalignant (see the review by Pierce, 1967, where the stem cell hypothesis of tumor formation is presented).

We shall not describe in detail the very interesting experiments showing that retinoic acid induces the differentiation of F9 cells into visceral endoderm cells (Strickland and Mahdavi, 1978) and that further treatment with agents that increase the cAMP level leads to differentiation into parietal endoderm (Kuff and Fewell, 1980; Strickland et al., 1980). Treatment with retinoic acid alone leads to the synthesis of α-fetoprotein. This marker of visceral endoderm is not synthesized when the same EC cell line (F9) is treated with both retinoic acid and cAMP; the basement membrane proteins are expressed instead. The conclusions drawn by Strickland (1981) are as follows: ''retinoic acid converts the F9 stem cells, which are equivalent to the inner-cell-mass cells, to a primitive endoderm cell. This cell can then, depending on external influences, form visceral (after aggregation) or parietal (after cAMP) endoderm.'' Perhaps more interesting for oncologists is the fact that when EC cells are injected into mice, they form a malignant carcinoma. If they have been treated with retinoic acid, they produce only benign teratomas that display multiple differentiations (Speers, 1982;

FIG. 23.   The teratocarcinoma F9 cell line possesses a surface F9 antigen, which it shares with normal mouse morulae and blastocysts, and with the male germ line in adults. F9 antigen is present in later embryos or in other adult tissues. (A) Left, morula treated with preimmunization serum control; right, morula treated with antiserum against F9 (peroxidase test). (B) Left, immunolabeling of frozen sections of a testis of a 6-day-old mouse by the anti-F9 serum; right, fluorescent pattern observed with anti-F9 serum absorbed on F9 cells. (C) Labeling of spermatozoa by antiprimitive carcinoma cell antibody. Left, acrosomal label of mouse spermatozoon; middle, postacrosomal label of mouse spermatozoon; right, labeling of human sperm cells. [(A) Artzt et al., 1973; (B) Gachelin et al., 1976; (C) Fellous et al., 1974.]

Speers and Altmann, 1984). At the cellular level, retinoic acid induces a G_1 phase and slows down the S and G_2 phases of the cell cycle in undifferentiated EC stem cells (Rosenstraus *et al.,* 1982).

Until recently, it was believed that F9 teratocarcinoma cells can differentiate only in endoderm after treatment with retinoic acid and dibutyryl-cAMP; however, Liesi *et al.* (1983) have found that an additional treatment with the nerve growth factor (NGF) may induce adrenergic neural differentiation. Addition of retinoic acid to the P19 teratocarcinoma cell line is also followed by differentiation in neurons; simultaneously, there is a decrease in an EC-specific surface antigen (Jones-Villcneuve *et al.,* 1983). The remarkable plasticity of teratocarcinoma cells is further shown by the fact that 5-azacytidine induces the differentiation of mesenchymal into cytokeratin-containing epithelial cells (Darmon *et al.,* 1984). The differentiated EC cells are, as expected different from the undifferentiated stem cells at the molecular level: Stacey and Evans (1984) found that a 3 kb poly $(A)^+$ RNA is present only in undifferentiated EC cells and in testis; it cannot be detected in visceral or parietal endoderm. Whether this RNA is involved in the synthesis of the F_9 antigen is not known; but it is already clear that genes which are expressed in undifferentiated EC cells are no longer transcribed in their differentiated progeny. On the contrary, the α-fetoprotein gene is expressed in differentiated endoderm, not in undifferentiated EC cells; as expected, this gene is nuclease-insensitive in the undifferentiated EC cells and sensitive to the enzyme in the endodermic cells deriving from them (Latchman *et al.,* 1984). Colberg-Poley *et al.,* (1985) have shown that EC stem cells possess a homeo box (see Chapter 2) which is not transcribed unless the cells are induced to differentiate.

We have seen that the main control of gene expression during normal differentiation takes place at the level of transcription. This is probably also the case when undifferentiated F9 cells of embryonal carcinoma are induced to differentiate in parietal endoderm by treatment with retinoic acid. According to Tabor and Oshima (1982), such treatment leads to the appearance of two new mRNA species encoding two cytoskeletal proteins (Endo A and Endo B) characteristic of extraembryonal endoderm. Oren *et al.* (1982) have also found that when F9 cells are induced to differentiate by treatment with retinoic acid plus dibutyryl-cAMP, they lose their p53-specific tumor antigen (a protein of M_r 53,000) due to a progressive decrease in p53 mRNA production. Malignant F9 cells contain 20 times more p53 mRNA than their differentiated counterparts (Reich *et al.,* 1983). In the treated F9 cells, there is a parallel synthesis of the basement membrane proteins collagen IV and laminin, and the corresponding mRNAs (Wang and Gudas, 1983).

Another interesting result has been reported by Wittig *et al.* (1983). They found that undifferentiated stem cells of teratocarcinomas do not express their heat shock genes after a heat shock; under the same conditions, teratocarcinoma cells that have differentiated in culture express those genes. This observation strengthens the analogy between carcinoma stem cells and early mouse embryos. The latter do not express the heat shock genes before the morula-blastocyst stages. According to Morange *et al.* (1984), EC cells and early mouse embryos synthesize heat shock proteins hsp 89 and 91 in the absence of stress; the expression of these proteins is not increased after a heat shock. Inducibility is restored after induction of differentiation by retinoic acid. In embryos, there is no synthesis of heat shock proteins at the eight-cell stage, but hsp 68 is synthesized in heated blastocysts. We have seen that a comparable situation exists in *Xenopus* oocytes and embryos.

Finally, Razin *et al.* (1984) found that teratocarcinoma cells undergo a genome-wide DNA demethylation (30%) when they differentiate; the same low level of DNA methylation is observed in mouse placenta and yolk sac. In contrast, there is *de novo* methylation in early mouse embryos.

Several authors have tried to analyze the malignancy of EC cells by somatic cell hybridization. For instance, Miller and Ruddle (1977) fused together teratocarcinoma and MEL cells. They obtained tumors of the teratocarcinoma type after injection into mice and found that the hybrid cells contained only traces of hemoglobin mRNA. Fusion between teratocarcinoma cells and thymus cells did not result in a loss of malignancy. The cell hybrids induced teratocarcinomas in mice (Rousset *et al.*, 1980). More recently Rousset *et al.* (1983) made somatic hybrids between nullipotent F9 teratocarcinoma cells and thymocytes; after injection in mice, they obtained multidifferentiated tumors. Hybridization with a normal thymocyte had thus allowed the expression of a pluripotency that was latent in the F9 cells.

Opposite results were obtained by Howe and Oshima (1982), who fused F9 cells with parietal endoderm and observed the extinction of tumorigenicity for 100 cell generations. Finally, Linder (1980) made cybrids between embryonal carcinoma cells and cytoplasts from myoblasts. Within 3 days these cybrids had regained their embryocarcinoma properties, showing that cytoplasmic factors from myoblasts are unable to suppress tumorigenicity permanently. We shall see that many other experiments of hybridization between malignant and normal cells (or cytoplasts) have led to conflicting results. This should not surprise those who believe that there are no such things as a normal, typical, totally undifferentiated cell and a ''tumor'' cell (Holtzer *et al.*, 1982).

We have already mentioned the exciting experiments done by Brinster (1974), who was the first to show that a teratocarcinoma cell injected into a mouse blastocyst can participate in the normal development resulting in a chimeric mouse. Using genetic markers, Mintz and Illmensee (1975) and Illmensee and Mintz (1976) clearly demonstrated that teratocarcinoma cells injected into blastocysts were totipotent and underwent normal differentiation; the mosaics (chimeras) obtained in this way did not form teratomas. One of these chimeras was remarkable in that there was a predominance of teratocarcinoma-derived cells in all tissues, including sperm. It thus appears as if the teratocarcinoma cells had lost their malignancy after injection into the blastocyst. Since these cells later developed normally, they were obviously subjected to embryonic regulation.

This conclusion is not valid when a larger number of teratocarcinoma cells are injected into blastocysts. Tumors were found in newly born mice when about 20 teratocarcinoma cells had been injected into a blastocyst (Papaioannou et al., 1975). The conclusion drawn by Papaioannou et al. (1978) is that teratocarcinoma cells injected into a blastocyst generally remain malignant, but that they can, in some cases, participate in normal development.

The question of suppression of malignancy by transfer into a very young mouse embryo has been examined in greater detail in B. Pierce's laboratory. It was found (Pierce et al., 1979) that no tumors developed when one to three embryocarcinoma cells were injected into a blastocyst, but that tumors arose when the number of injected embryocarcinoma cells was increased from three to five. More recently, Pierce et al. (1982) have shown that regulation of embryocarcinoma cells by the host blastocyst varies according to the cell lines. There is no control of malignancy when a single sarcoma or leukemia cell is implanted into a blastocyst; neuroblastoma cells are partially controlled by the host. It can be concluded that the environment plays an important role in the phenotypic expression of malignant cells. Wells (1982) has shown that the rate of cell proliferation decreases in EC cells after implantation into a blastocyst, and that only cells in the G_1 phase of the cell cycle respond quickly to the blastocyst for their reprogramming. There is no doubt that different results are obtained with different EC cell lines. For instance, Cronmiller and Mintz (1978) have shown that the only totipotent EC cells, after injection into a blastocyst, are those that possess a normal karyotype, suggesting that a normal embryonic environment is unable to correct malignancy when the chromosomes display visible abnormalities. Of interest is the fact that hybrids between two malignant cells (teratocarcinoma × hepatoma cell hybrids) can participate in normal development after implantation into a normal mouse blastocyst (Illmensee and Croce, 1979). These experiments have been repeated and extended by Duboule et al. (1982). Hybrids between mouse teratocarcinoma cells and rat hepatoma cells produce tumors after injection into *nude* mice; if one injects these hybrid cells into mouse blastocysts, a few of the fetuses (4 out of 61) show participation of the hybrid cells to development, although the rat chromosomes have been retained. This is again reminiscent of the "revitalization" of amphibian lethal hybrids after grafting onto normal hosts. Another interesting finding is that injection of normal embryonic (inner cell mass) cells induces teratocarcinomas in mice if they have been treated with a conditioned medium of teratocarcinoma cells (Martin, 1981).

The preceding chapter showed that teratocarcinoma cells can be incorporated in the germ line and that if they possess a foreign gene introduced by DNA-mediated gene transfer, this gene can be transmitted to the progeny (Stewart and

Mintz, 1981). However, according to Rossant and McBurney (1982), it is only with the cell line used by Stewart and Mintz (METT-1) that EC cells contribute to the germ line as well as to somatic tissues. With the cell line used by Rossant and McBurney (P19, which also has a normal karyotype), there is no regulation after injection of a single cell into a normal blastocyst. Chimeras are obtained where the injected cell has contributed to normal tissues, but has also produced tumors.

On the whole, the many facts elucidated by recent research on teratocarcinomas fit well with Pierce's (1967) developmental theory of cancer. Differentiation of stem cells within a tumor would deplete it of malignant cells and transform it into a benign tumor. Teratocarcinomas are—and will long remain—fascinating material for both embryologists and oncologists.

C. *In Vitro* Malignant Transformation and Reverse Transformation

We have already mentioned the main characteristics of the transformed phenotype observed after treatment with all kinds of carcinogenic agents (chemical, viral, physical) of "normal" cells. Most of the experiments have been done on fibroblasts cultured under classic conditions. In principle, these cells are diploid and possess a normal karyotype (ascertained with banding techniques). However, since cultured cells are necessarily exposed to mutagenic events, these conditions are not always fulfilled. Even in the most popular cell line currently used for the study of malignant cell transformation (the NIH 3T3 cell line), the appearance of heteroploidy and potential tumorigenicity in some cell lines cannot be completely excluded; heteroploidy enables the cells to overcome senescence and to grow indefinitely. This shows again how difficult it is to work, especially under *in vitro* culture conditions, with absolutely normal cells. In addition, nonfibroblast cells may behave differently from fibroblasts. For instance, normal bovine granulosa cells grow in anchorage-independent clones like transformed fibroblasts (Bertoncello *et al.*, 1982).

Despite these reservations, malignant transformation can readily be observed in fibroblast cultures using the three previously mentioned criteria: immortality (capacity for indefinite proliferation); lack of contact inhibition of growth and locomotion; and the capacity to grow in soft agar (anchorage-independent growth). The last of these criteria is widely used to ascertain whether or not the cells have been transformed. However, anchorage-independent growth does not necessarily mean that the cells are malignant. For instance, Mitrani (1984) has recently found that cultured cells from very early normal chick embryos (taken before the primitive streak stage) make colonies in soft agar.

Transformed fibroblasts tend to round up and to assume a more epithelial-like morphology (shown in Fig. 24). In 1971, Hsie and Puck showed that addition of dibutyryl-cAMP (or other agents that increase the cAMP content of the cells) changes the phenotypic appearance of transformed cells back to a fibroblast-like

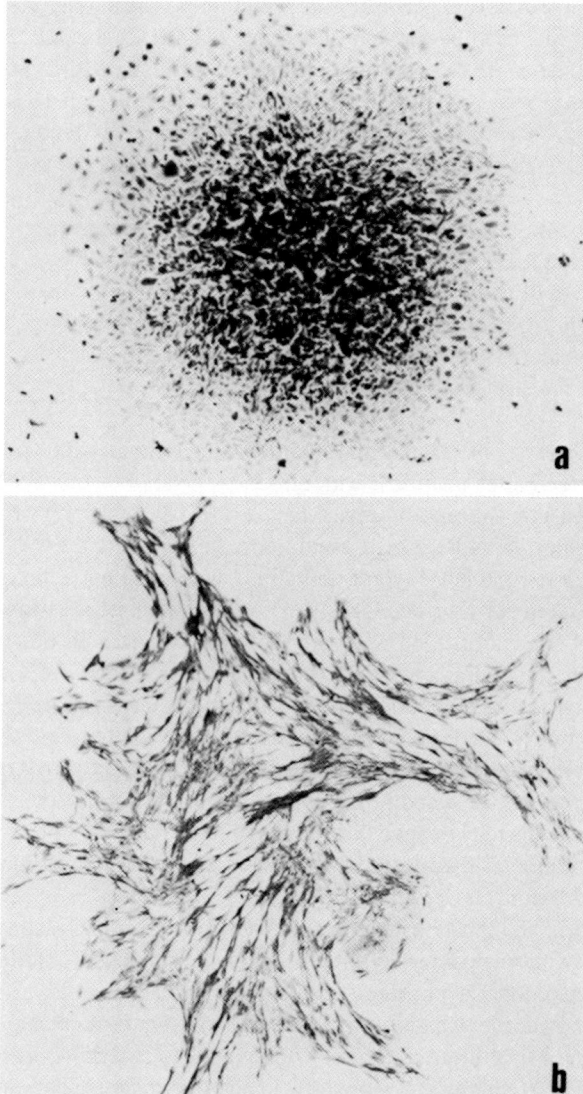

FIG. 24.   (a) Colony grown from a single Chinese hamster ovary (CHO) cell on standard medium for 6 days. (b) CHO colony grown in the presence of dibutyryl-cAMP; reversal of the transformed phenotype to a fibroblast-like morphology. [Hsie and Puck, 1971.]

morphology (Fig. 24). This cAMP-induced reversal of the transformed phenotype has been called *reverse transformation*. That changes in the cAMP and cGMP content are involved in transformation and reverse transformation is supported by the biochemical studies of Moens *et al.* (1975). They found that cAMP decreases and cGMP increases in transformed cells, in agreement with what was said in Chapter 5, Volume 1, about the role played by the two cyclic nucleotides in the control of cell proliferation.

It is generally believed that the transformed phenotype results from an initial genetic change that ultimately affects the cell surface and the underlying cytoskeleton. That a genetic change is the *primum movens* of cell transformation seems likely, since the transformed phenotype is induced by the innumerable carcinogenic agents that are known to damage DNA. It has been shown that the kinetics of transformation (acquisition of the capacity to form transformed foci) and of maturation (acquisition of ouabain resistance) by a carcinogen are the same (Backer *et al.*, 1982a,b). It seems certain that acquisition of immortality, which in general results from chromosomal imbalance (heteroploidy, aneuploidy), is a prerequisite for malignant transformation. In variants obtained after treatment with chemical carcinogens, Newbold *et al.* (1982) observed the appearance of immortality before anchorage independence, which is obviously linked to cell surface changes (Nicolson, 1976; Roos, 1984). As pointed out by Pastan and Willingham (1978), the transformed phenotype essentially results from a decrease in the adhesion of the cells to their substratum. This decrease is due, among other things, to a reduction in the production and secretion of extracellular matrix proteins (fibronectin, collagens) by transformed fibroblasts. Indeed, fibronectin was first called a "large, external, transformation-sensitive" protein to show that its production strongly decreased in transformed fibroblasts. According to Carter (1982), four major proteins (fibronectin and three glycoproteins called GP250, GP270, and GP410) are missing in transformed cells. The loss of or decrease in extracellular matrix proteins is due to a slowdown of transcription of the corresponding gene and is thus controlled at the transcriptional level. There is a very strong decrease (which may amount to 90% of the initial value) in the content in the mRNAs encoding fibronectin and the pro-α-1 and pro-α-2 collagens during transformation. This decrease is coordinated, suggesting that the genes coding for these three mRNAs are under the same general control (Fagan *et al.*, 1981; Sobel *et al.*, 1981).

It does not seem that transforming agents induce drastic changes in the pattern of protein synthesis, and it has even been suggested by Huberman *et al.* (1976) that transformation by chemical carcinogens might be the result of a mutation affecting a single gene. Baker *et al.* (1980) found that the differences among glycoproteins are quantitative rather than qualitative when one compares normal and transformed cells. According to Leavitt and Moyzis (1978), only 7 of 700 or more proteins change during transformation. In a more recent analysis of the pattern of protein synthesis during transformation, Bravo and Celis (1980) arrived at an even lower estimate. Out of 650 proteins detectable in fibroblasts, only 3 were specific to transformed cells. According to Crawford et al. (1981), all tumors (human and murine) are characterized by the presence of the already mentioned p53 nuclear phosphoprotein (M_r = 53,000), which is undetectable in normal cells (unless they are induced to proliferate). A 55,000-dalton protein (apparently the same as or related to p53) was discovered by Mora et al. (1980) in virus-transformed cells. Interestingly, this protein is present in 12- to 14-day mouse embryos and disappears at more advanced stages of fetal life. It is also found in the embryocarcinoma F9 cell line (Linzer and Levine, 1979). Injection of a monoclonal antibody against the transformation-related p53 protein into the nuclei of mouse fibroblasts inhibits the induction of DNA synthesis by serum addition (Mercer et al., 1982). According to Reich and Levine (1984), cultured transformed cells always contain a high level of the p53 cellular tumor antigen. In normal cells, synthesis of p53 mRNA and of p53 itself occurs in the late G_1 phase of the cell cycle; an increase in p53 might be necessary for the progression of the cells from the growth-arrested state to the actively dividing state.

Another protein is abundant in the nuclei of transformed and other rapidly dividing cells. It is called the *proliferating cell nuclear antigen (PCNA)* and can be identified because it reacts with antibodies present in the serum of lupus erythematosus patients. Mathews et al. (1984) found that PCNA is identical to cyclin, a nuclear protein that is also correlated with the proliferation state of the cells (Celis and Bravo, 1984); it accumulates in the nucleoli in G_1–S (Matthews et al., 1984). According to Celis et al. (1984), cyclin is a nuclear, transformation-sensitive, non-histone protein (M_r = 36,000); it increases very strongly in transformed cells. It was called *cyclin* because its level reaches a peak during the S phase of the cell cycle in all proliferating cells.

Another protein may be involved in cell transformation. According to Croy and Pardee (1983), transformation depends on the stabilization of a 68,000-dalton protein; this protein would be responsible for the regulation of cell growth. It has been found that its stability is increased in transformed cells.

Finally, we should mention the p21 protein, which is the product of a gene closely related to the so-called *ras* oncogene of certain retroviruses (see Section II,F). In contrast to p53, p21 is localized in the cell membrane and is not found in the nucleus; it has the ability to bind guanine nucleotides (GDP, GTP). Microinjection of a monoclonal antibody against p21 into a variety of cell types inhibits their proliferation; it blocks the passage from G_0 to G_1, but has no effect on mitosis itself (Mercer et al., 1984; Mulcahy et al., 1985).

It should be stressed that none of the transformation-related proteins so far identified is absolutely specific to the transformed state. They are present, in much lower concentrations, in all proliferating cells. There is evidence, to be reviewed later, that p21-like proteins are necessary for cell division in yeasts and

Drosophila eggs. Transformed cells thus differ from normal cells by having a higher content of the transformation-related proteins, suggesting that the transformed state results from overproduction of these proteins rather than from the presence of a highly specific transforming protein. Thus, one of the causes of the transformed state seems to be a deregulation of the synthesis of several proteins involved in normal cell proliferation.

The fact that there are no major changes in the pattern of protein synthesis during malignant transformation fits well with the existence of reverse transformation. The possibility of reversing the transformed pbenotype by increasing the cellular cAMP content indicates that phenotypic transformation is not an irreversible and highly specific process. In fact, Peehl and Stanbridge (1981) have found that it is possible to obtain anchorage-independent growth of normal fibroblasts. These cells grow in suspension if the medium is rich in serum or if hydrocortisone is added to it. Conversely, density-dependent inhibition of DNA synthesis can be restored to transformed cells if the content of magnesium ions is decreased (Rubin, 1982).

There has been a great deal of discussion about possible changes in the organization of the cytoskeleton during cell transformation, and the question is not yet resolved. De Mey *et al.* (1978) showed the difficulties of the immunocytochemical approach in comparing normal and transformed cells. As we have seen, transformed cells are thicker than their normal well-spread counterparts, which may easily lead to artifacts and errors in interpretation. There seems to be no good evidence for profound changes in the organization of the microtubule (MT) network in transformed cells (De Mey *et al.*, 1978). However, Rumsby and Puck (1982) have reported that colchicine inhibits the induction of ODC by phorbol esters (see Section II,D) in normal but not in transformed cells. This suggests that the control of ODC synthesis by the MTs is deficient in transformed cells. In addition, Meek (1982) has shown that reverse transformation by addition of dibutyryl-cAMP (in which spherical cells stretch and flatten) is accompanied by rearrangements of the cytoskeleton.

There is increasing evidence of changes in the actin MF network in transformed cells. For instance, Leonardi *et al.* (1982) found that, in virus-infected kidney cells, the stress fibers disappear with a concomitant reduction in the tropomyosin content. Hynes (1982) studied the phosphorylation of vimentin, vinculin, and filamin in normal and transformed cells and found that the phosphorylation of tyrosine residues increases 10 times in vinculin after the induction of transformation. This 130,000-dalton protein is, as we have seen in Chapter 3, Volume I, localized in the "attachment plates" (focal contacts between the cell surface and the substratum). It is believed that vinculin mediates the interactions between the MF bundles and the plasma membrane. A transmembrane complex, involving actin, α-actinin, and vinculin on the cytoplasmic side and fibronectin, collagen, and proteoglycans on the outside, breaks down when the configuration

of vinculin (and perhaps of other proteins) is changed as a result of phosphorylation. This agrees with the finding of Leonardi *et al.* (1982) that stress fibers
disappear when, in transformed cells, the adhesions between the cells and their
substratum are broken. Gerke and Weber (1984) recently showed that viral
transformation (by Rous sarcoma virus) affects the actin–spectrin network present in the cortical layer of the infected cells. This is due to the phosphorylation
of a cellular protein (p36) by a tyrosine-specific protein kinase produced by the
virus and to the subsequent binding of pp36 to Ca^{2+} and actin. There is no doubt
that the transformed phenotype results largely from changes in the actin MF
network.

We shall return to the question of the activation of tyrosine-specific protein
kinases when we discuss the oncogenic retroviruses. Suffice to say, for the
present, that normal cells possess one or two tyrosine-specific protein kinases
that are responsible for the phosphorylation of a limited number of proteins
(Martinez *et al.*, 1982) and that several oncogenic viruses posses genes that
encode similar enzymes. Viral infection, therefore, results in a large increase in
the phosphorylation of tyrosine residues in proteins. However, it would be a
mistake to believe that high tyrosine-specific protein kinase activity is a specific
marker of viral transformation. Swarup *et al.* (1983) have found that the spleen
of normal rats has a tyrosine-specific protein kinase activity as high as that of
fibroblasts transformed by infection with Rous sarcoma virus. Differences in the
identity of the protein substrates for tyrosine phosphorylation in normal and
transformed cells might be more important than the actual level of tyrosine-
specific protein kinase activity.

An indirect argument in favor of alterations of the actin cytoskeleton in transformed cells comes from studies on the differential effects of the cytochalasins
on normal and transformed cells. Cytochalasin B disrupts the contractile ring of
actin MFs, which plays a major role in furrowing at the end of cell division. In
normal cells, the result of cytochalasin B treatment is the formation of binucleate
cells, but in malignant cells, cytochalasin B induces multinucleation (Fig. 25). It
has even been proposed that this characteristic could be used as a test for
screening cancer cells (Somers and Murphey, 1980, 1982). These experiments,
of course, suggest that the distribution of the cortical actin MFs in transformed
cells is abnormal. Comparable observations have been made by Maness and
Walsh (1982). Dehydrocytochalasin B disrupts the actin cytoskeleton in normal,
but not in transformed, cells. In addition, it inhibits the stimulation of DNA
synthesis induced by addition of serum or growth factors (EGF, insulin) in
normal cells only.

A breakthrough in the field of malignant cell transformation was the discovery
by Todaro and his colleagues (1980) of transforming growth factors (TGFs)
(reviewed by Todaro *et al.*, 1981; Roberts *et al.*, 1983; Maciag, 1983; Racker,
1983; Lawrence, 1985). These factors, which are present in the conditioned

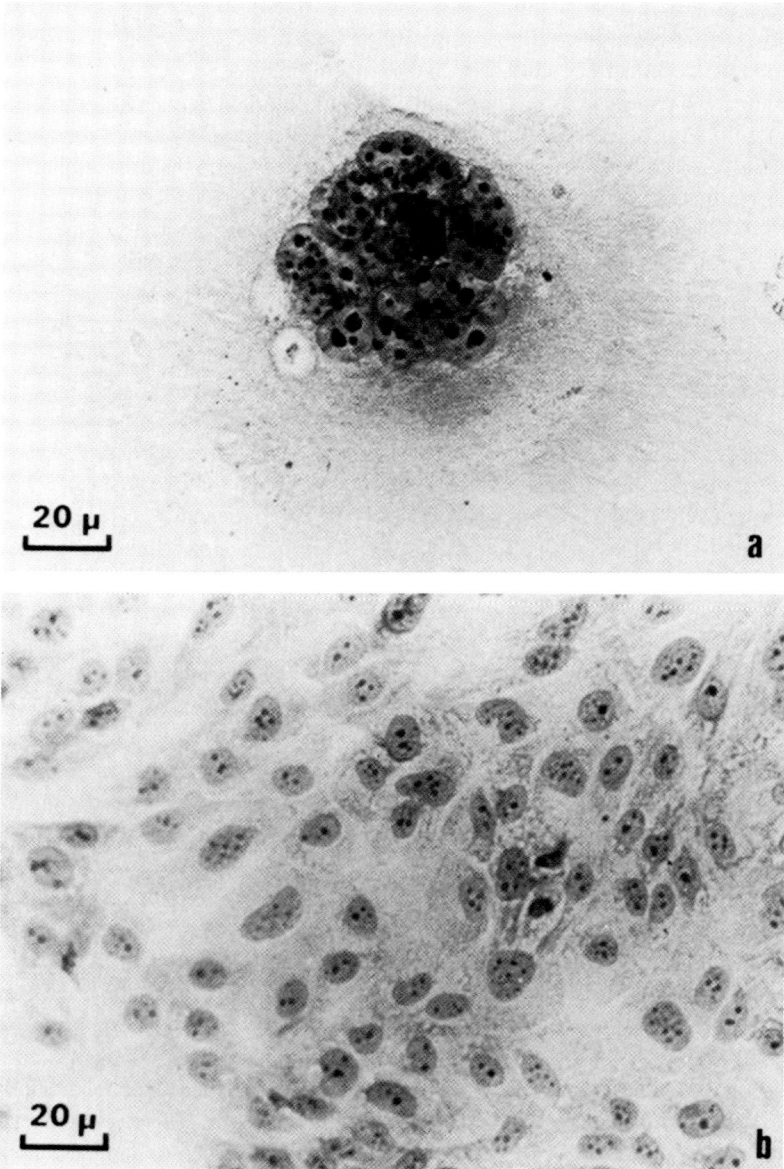

FIG. 25.   (a) Multinucleate WI-38 VA13 (malignant cells) after culture in 2 μg/ml of cytochalasin B medium for 13 days. (b) Untreated WI-38 VA 13 cells. [(B) Wright and Hayflick, 1972.]

media of cancer cells, induce the transformed phenotype in normal cells, stimulating their multiplication, changing their morphology, and allowing anchorage-independent growth. The TGFs are produced by cancer cells (murine sarcoma, F9 teratocarcinoma cells, etc.) and are released in the medium. However, numerous normal tissues possess and secrete TGFs. Their conditioned media induces growth in soft agar (anchorage-independent growth) in recipient cells, even in the absence of serum and growth factors (Roberts *et al.*, 1981; Kaplan *et al.*, 1982). TGFs are present in mammalian embryos (Twardzik *et al.*, 1982), and a purified TGF has been isolated from placenta by Frolik *et al.* (1983); it induces anchorage-independent growth and requires the presence of EGF for activity. Clearly, TGF activity is not specific for tumors. The fact that normal fibroblasts from several animal species release TGFs in an inactive form and that these factors are activated when the conditioned media are acidified (Lawrence *et al.*, 1984) still complicates the situation.

The relationships between TGFs and normal growth factors are still unclear; attempts to elucidate them are summarized in the reviews of Roberts *et al.* (1983), Maciag (1983), and Massaqué (1985). They concluded that TGFs should be separated into three distinct families: (1) TGF-α generally have a molecular weight of around 7400 [5.616 only for one of them, which, according to Marquardt *et al.* (1984), is closely related to EGF]. They require EGF for activity, bind to the EGF receptor, and increase tyrosine phosphorylation of the receptor; there is competition between TGF-α and EGF for EGF receptor sites (Marquardt *et al.*, 1983, 1984). (2) TGF-β ($M_r = 24,000$) do not compete with EGF for its receptors but nevertheless require EGF for full activity. According to Assoian *et al.* (1984), TGF-β controls the receptor levels for EGF in fibroblasts. It enhances EGF binding by increasing the number of receptors by a protein synthesis–requiring mechanism. (3) Finally, TGF-γ has nothing to do with EGF. One should perhaps place in this class a factor recently discovered by Nistér *et al.* (1984). Glioma and osteosarcoma cells release a PDGF-like factor, which binds to the PDGF receptor. Activation of this receptor might play a role in the growth of tumors of mesenchymal or glial origin. TGF-γ probably binds to specific receptors, distinct from the EGF receptor, in order to induce the transformed phenotype (Colburn and Gindhardt, 1981; Massagué *et al.*, 1982).

The complexity of the problem has been discussed by Heldin and Westermark (1984) in a recent review. They point out that transformed cells are characterized by growth factor independence and autonomous growth. This could be due to deregulation at the level of one of the controlling elements along the normal mitogenic pathway (the presence of growth factors in the medium, availability of membrane receptors, an intracellular signal inducing DNA synthesis in the nucleus). There are obvious similarities and synergisms between some of the TGFs and normal growth factors such as EGF and PDGF; the interactions between them are not yet clear, but it seems that the TGFs somehow potentiate the mechanisms that stimulate normal growth. For this reason, and because of their capacity to induce the transformed phenotype, TGFs deserve further study.

Possibly related to the TGF problem is the finding that transformed cells secrete a polypeptide capable of inducing rapid plasminogen activator synthesis in normal cells (Davies *et al.*, 1983). Since plasminogen activator is also induced by EGF, carcinogens, and tumor promoters, this factor might play an important role in tumor progression and metastasis.

D. ANALYSIS OF MALIGNANCY BY SOMATIC HYBRIDIZATION

This subject is mired in utter confusion and has been a highly controversial topic. H. Harris claims that it is possible to suppress malignancy by cell fusion. His opponents, led by H. Koprowski and C. M. Croce, disagree with this conclusion. What is clear is that opposite results may be obtained with different cell strains, and this should no longer be surprising. Before we discuss the factual evidence, a few important reviews by Harris (1971), Croce (1980), Klinger (1982), and Stanbridge *et al.* (1982a) should be mentioned; an important paper by A. B. Sabin (1981) discusses authoritatively the conflicting results obtained in several laboratories.

Harris *et al.* (1969) and Ephrussi *et al.* (1969) were the first to show that in murine somatic hybrids between a malignant and a normal cell, malignancy behaves like a recessive character. However, suppression of malignancy by cell fusion is seldom complete. It is usually necessary to inject a much larger number of hybrid cells compared to the initial malignant cells into mice in order to obtain growing tumors. In 1976, Stanbridge extended these studies to human malignant cells by fusing a human cancer cell with a diploid human fibroblast. He obtained the complete suppression of malignancy, although there was no suppression when two malignant cells were fused together. These results have been confirmed by Klinger *et al.* (1978) and Klinger (1980). Similarly, Marshall *et al.* (1982a,b) found that hybrids between anchorage-dependent and anchorage-independent hamster cells are anchorage-dependent. The transformed phenotype is thus suppressed in the hybrids, apparently linked to the loss of one copy of chromosome 1. Pereira-Smith and Smith (1983) fused immortal cell lines with normal fibroblasts. The hybrids had only limited division potential, showing that immortality is recessive in such hybrids. Marshall and Sager (1981) observed suppression of the transformed phenotype when transformed cells were fused together. On the other hand, Croce *et al.* (1975) did not observe suppression of malignancy in hybrids between human cells transformed with a DNA Virus (SV40) and normal mouse cells. A more recent study by Kucherlapati and Shin (1979) showed that there were differences between man × mouse and mouse × mouse hybrids. Malignancy was not suppressed in the former, in contrast to the latter. Avilès *et al.* (1980) found that some hybrids between normal fibroblasts and cancer cells were malignant, while others were not. More recently, Benham *et al.* (1983) reported that hybrids between various human cell lines and mouse teratocarcinoma cells retained the embryocarcinoma phenotype; these hybrids differentiated like embryocarcinoma cells after treatment with retinoic acid.

It now seems clear that the results vary with the cell strains used for fusion, with the carcinogen used (chemical or viral), and with chromosome loss during culture. Use of different cell lines is probably the main reason why conflicting results were obtained by Stanbridge (1976) and Kucherlapati and Shin (1979) in the hybids between human malignant and normal cells. A more recent study by Der and Stanbridge (1981) showed that fusion between HeLa cells (human cancer cells) and human fibroblasts suppressed tumorigenicity, but rare tumorigenic segregants have been found in the progeny of these hybrids. These hybrids are characterized by the presence of a 75,000-dalton surface glycoprotein characteristic of HeLa cells. The main difference between tumorigenic and non-tumorigenic hybrids is the loss, in the first case, of transmembrane interactions between MFs and fibronectin (Der *et al.,* 1981). This is exactly the conclusion we reached when discussing the transformed phenotype. It is likely, that, in the malignant hybrids, tyrosine phosphorylation of vinculin breaks down the links that normally exist between the actin cytoskeleton and the constituents of the extracellular matrix. That fusion between HeLa cells and human fibroblasts suppresses the extracellular matrix, except in rare tumorigenic fibroblasts, has been found by Howell (1982) in hybrids between virally transformed human cells and mouse fibroblasts. Suppression of malignancy thus seems to be the rule, but there may be a few exceptions; it is not an all-or-nothing phenomenon.

There is little doubt that the main reason why conflicting results are found in the literature is the well-known loss of chromosomes that frequently takes place during the multiplication of cultured somatic cell hybrids. This explanation seems to be universally accepted by all workers in the field. Certain chromosomes would bear genes responsible for tumorigenicity, while the presence of other chromosomes would suppress it. The imbalance between cancer genes and suppressors of malignancy genes, due to chromosome loss during hybrid cell multiplication, provides an easy and satisfactory explanation for the fact that opposite results have sometimes been obtained in different laboratories. In his review, Croce (1980) points out that human chromosome 7 is responsible for tumorigenicity, but that other chromosomes may suppress it. In their review, Stanbridge *et al.* (1982a) suggest that tumorigenicity would be controlled by a gene family; chromosomal instability in certain cell lines explains the disagreements found in the literature. Loss of chromosomes may lead to reexpression of tumorigenesis. The authors point out that in order to obtain reliable results, one should work with cell lines in which the chromosomes are stable and test malignancy by injecting the cells into *nude* (athymic) mice. According to Evans *et al.* (1982), in hybrids between mouse malignant and normal diploid cells, chromosome 4 of the diploid cells plays an important role in the suppression of malignancy. Since tumorigenicity reappears after the loss of this chromosome, it is suggested that chromosome 4 of the tumor plays a role in its malignancy. For Sabin (1981), human cancers of different somatic origins would be caused by a loss of different specific regulatory genes present in normal somatic cells. He correctly points out that the challenge is to determine, in molecular terms, what those missing genes are, how they function, and whether it may be possible to restore to the cancer cells what they have lost. He concludes by suggesting a number of experiments designed to test his hypothesis. Klein and Klein (1985) also emphasize the necessity of further studies on the suppressor genes (anti-oncogenes) present in hybrids between normal and malignant cells.

We do not know how many genes are involved in malignancy. We shall return to this question when we discuss the virus-induced cancers. Schäfer et al. (1981), pointed out that more than two mouse genes must be expressed in order to extinguish tumorigenicity in hybrids between tumorigenic hamster cells and normal mouse fibroblasts. In a more recent paper, Schäfer et al. (1983) concluded that suppression of tumorigenicity in hybrids between tumorigenic hamster cells and mouse fibroblasts depends on at least three different mouse chromosomes.

Of more direct interest for oncologists may be a paper by Stanbridge et al. (1982b) that shows the complexity of the problem. Classic clinical markers of malignancy are the presence of the carcinoembryonic antigen (Gold and Freedman, 1965), placental alkaline phosphatase, and human chorionic gonadotropin. In hybrids between HeLa cells and human diploid fibroblasts, tumorigenicity is lost, but rare malignant segregants appear in subcultures. Comparison between the two types of hybrids has shown that expression of the α subunit of human chorionic gonadotropin is specifically correlated with tumorigenicity; there is no such correlation between placental alkaline phosphatase and malignancy. This finding supports the multistep progression theory of cancer; whether or not it is due to somatic mutations or to epigenetic changes (or both) remains an open question.

The difficulty of analyzing malignancy (or suppression of malignancy) in somatic cell hybrids in molecular terms was made clear in a paper by Bravo et al. (1982). They worked on hybrids between malignant hamster cells and normal mouse fibroblasts, and found that these hybrids were not tumorigenic. Bravo et al. (1982) analyzed, with a high-resolution method, the pattern of protein synthesis in these hamster × mouse hybrids and found unexpectedly that they had lost the expression of about one-third of the mouse proteins, despite the fact that they had retained all of the mouse chromosomes. Thus, the genome of the cancer cell had suppressed the expression of a very large number of mouse proteins in the hamster–mouse cell hybrids, but there was no proof that these proteins are involved in the suppression of tumorigenicity. This suggests that a multiplicity of controls, more subtle than the mere presence or absence of one or several chromosomes, are operational in hybrids between normal and malignant cells. These controls might operate at the level of the gene itself and at each of the steps (transcription and postranscriptional events) needed for its expression. In fact, experiments with cybrids that will now be discussed indicate that the cytoplasm may also play a role in the expression of malignancy.

The results obtained from experiments in which a cytoplast from a normal cell was fused with a malignant cell (or vice versa) are, unfortunately, as conflicting as those found for hybrids between two whole cells. Howell and Sager (1978) reported that the cytoplasm of malignant cells in cybrids does not transmit tumorigenicity. On the other hand, the cytoplasm of normal cells suppresses malignancy in cancer cells. Since suppression of tumorigenicity could be transmitted during 20–30 cell generations, it was hypothesized that suppression might be due to cytoplasmic organelles endowed with genetic continuity—presumably the mitochondria. However, at the same time, Ziegler (1978) obtained quite different results. He found that in cybrids the cytoplast is responsible for chloramphenicol resistance (a marker of the mitochondrial genome), but not for the loss of malignancy. A reinvestigation of the problem by Shay and Clark (1980) led to the conclusion that in cybrids the nucleus always ultimately controls dominance or recessivity of tumorigenicity. Similar conclusions were drawn by Halaban *et al.* (1980), who compared hybrids and cybrids between melanoma cells and normal or malignant fibroblasts. Expression and extinction of both tumorigenicity and melanogenesis depend on the nucleus. In a more recent study, Hayashi *et al.* (1984) found no suppression of tumorigenesis when glioma cells were fused to cytoplasts from a variety of normal cells, and they concluded that mitochondrial DNA plays no role in tumorigenesis. We have already mentioned that cybrids between embryonal carcinoma cells and cytoplasts of myeloblasts regain properties of the embryocarcinoma within 3 days of culture (Linder *et al.,* 1981). However, more recent work on cybrids tends to give more importance to the cytoplasm than it once received. For instance, Giguère and Morais (1981) found that cybrids between tumorigenic cells and cytoplasts from normal cells generally give rise to nontumorigenic clones. Similar experiments by Shay *et al.* (1981) have led to the conclusion that nontumorigenic cytoplasm suppresses the tumorigenic phenotype in an inherited way; however, this is not always the case, as shown by the earlier experiments of Shay and Clark (1980). Finally, Sekiguchi *et al.* (1982) studied cybrids between karyoplasts (or whole cells) from an amelanotic mouse melanoma and cytoplasts from rat myoblasts. They found a strong overproduction of melanin in 30% of the clones that could be reversed by treatment with the tumor promoter TPA. They suggest that the first step in cancer would be mutation in an oncogene. The mutated oncogene would be inactivated by cytoplasmic factors, and TPA would suppress this cytoplasmic control and promote carcinogenesis. At present, this is little more than an ingenious hypothesis, but it might lead to new experiments and deserves mention here.

Somatic hybridization experiments seldom lead to simple, clear-cut results, due to the biological complexity of the system. For this reason, universal conclusions cannot be drawn from them. Nevertheless, they have provided us with information about the nucleocytoplasmic interactions that lead to cell differentiation and cell malignancy, and about the co-existence in the cells of oncogenes and anti-oncogenes. We must, therefore, be particularly thankful to the pioneers in this difficult but exciting field, B. Ephrussi and H. Harris.

E. ANALYSIS OF CARCINOGENESIS BY ADDITION OF CHEMICALS

As we have already seen, the initial event in malignant transformation is an inherited genetic change in a single cell. Since most chemical and physical carcinogens are also mutagens, it is generally assumed that this primary lesion in the DNA molecules is a mutation. However, we already know that only a small number of chromosomes are involved in the induction and suppression of malignancy, and it is therefore likely that the number of cancer genes (oncogenes) is small. It thus seems unlikely that any random damage to DNA sequences will necessarily lead to tumor production. If this were the case, since we are exposed from birth to innumerable carcinogenetic agents (reviewed by Weinstein, 1981), we should all die in infancy from cancer. Instead, cancer is a disease of old age and is probably due to a combination of factors: accumulation of mutations that finally affect the oncogenes; decreased efficiency and correctness of DNA repair mechanisms; and decreased efficiency of the host's defense mechanisms (production of antibodies, destruction of cancer cells by macrophages, etc). We should not forget that, according to the multistep theory of cancer, chemical carcinogens only induce initiation in the affected cell (Fig. 12); progression toward the neoplastic cell state requires the action of promoters. Finally, this single neoplastic cell will not produce a cancer unless it can proliferate and form a clone of cancer cells, which will eventually metastasize.

Although it is generally accepted that the initiating event is a gene (probably an oncogene) mutation, it is clear that other genetic mechanisms can be envisaged now that DNA is no longer the static, stable molecule it was believed to be only a few years ago. We shall see that insertion of viral promoters into nuclear DNA is an important factor in carcinogenesis (particularly in mice). Cairns (1981) has suggested that genetic transposition might play an important role in the genesis of human cancers. Spontaneous regression of tumors would result from excision of inserted genetic elements. We shall return to these important and exciting questions at the end of this section.

That DNA repair is an essential defense mechanism against cancer initiation is shown by the fact that people who are affected by genetic diseases in which DNA repair is deficient are cancer prone. Such diseases (xeroderma pigmentosum, ataxia telangiectasia, Fanconi's anemia, Bloom's syndrome, etc.) are eagerly studied by geneticists, oncologists, and radiobiologists; all of these diseases are rare, due to selection pressure. Patients with these diseases have an increased

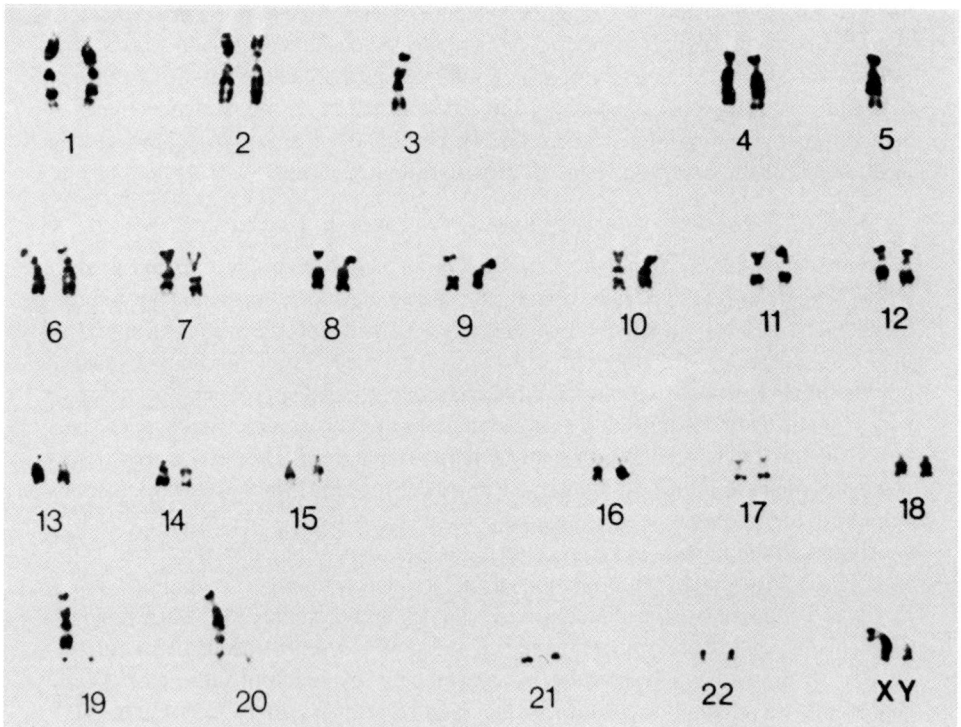

FIG. 26.   Karyotype of a G-banded cell from a patient with ataxia telangiectasia. Two spontaneously occurring dicentric chromosomes apparently resulted from telomeric binding. The chromosome pairs involved are 3/19 and 5/20. [Taylor *et al.*, 1981.]

number of chromosomal abnormalities, ranging from sister chromatid exchanges to more or less severe aneuploidy; chromosome rearrangements are frequent (Fig. 26). In the case of Bloom's syndrome, Emeritt and Cerutti (1981) have shown that the fibroblasts of patients release into the medium "clastogenic factors" that can induce chromosomal aberrations. Their effects can be corrected by the addition of superoxide dismutase, an enzyme that helps to rid human cells of oxygen-free radicals using the following mechanism:

$$O_2^\mp + O_2^\mp \xrightarrow[\text{superoxide dismutase}]{} H_2O_2 \qquad\qquad 2H_2O_2 \xrightarrow[\text{catalase}]{} 2H_2O + O_2\uparrow$$

Active oxygen (which is also responsible for lipid peroxidation in cell membranes) would thus be responsible for chromosome damage. Ames *et al.* (1981) found that substances that act as antioxidants (urate, reduced glutathione, ascorbic acid, etc.) protect red blood cells against lipid peroxidation by free radicals; these authors have suggested urate consumption for the prevention of cancer and aging. We are not sure that this would be beneficial and pleasant for patients affected with gout. However, as pointed out by Ames (1983), changing our dietary habits may be beneficial for cancer protection.

We have already discussed at length the effects of phorbol esters (particularly by TPA) and vitamin A derivatives (retinoids) on cell differentiation. We have seen that TPA inhibits *in vitro* cell differentiation (myogenesis, chondrogenesis, etc.) as efficiently as BrdUrd, and that retinoids induce differentiation of the F9 teratocarcinoma cell line. However, current interest in these chemicals arises from studies on cancer cells.

The phorbol esters are the active principles of croton oil, which, as was shown by Berenblum (1954) more than 30 years ago, are strong cancer promoters. TPA affects the cell in the following ways: it induces a wave of cell divisions when it is rubbed on the skin or added to cultured cells; in fibroblasts, it affects cell membrane permeability (it increases deoxyglucose uptake), decreases the adhesion of the cells to their substratum, modifies the cell surface glycoproteins, induces a loss of cell surface fibronectin, allows anchorage-independent growth, and, as shown repeatedly, induces ODC activity and prostaglandin synthesis. Stimulation of ODC activity results in an increase in the polyamine content in fibroblasts (O'Brien and Diamond, 1977), as well as in skin (Verma *et al.*, 1977). The increase in ODC activity after treatment of cells with TPA may be very great [up to 39-fold in bladder cancer cells, according to Izumi *et al.* (1981) and even 200 times in the skin (Verma *et al.*, 1979)]. An increase in polyamine synthesis and subsequent accumulation of putrescine and spermidine in the cells is, as we have seen in Chapter 5, Volume 1, likely to favor DNA synthesis and cell proliferation. However, another factor that might stimulate cell division is the induction, by phorbol esters possessing a cancer-promoting activity, of plasminogen activator production by fibroblasts. This serine protease is produced by most, if not all, cancer cells. We have seen that limited proteolysis of the cell surface often stimulates growth in cultured cells (Chapter 5, Volume 1); it might be an important factor in cancer growth and metastasis. In addition, induction by phorbol esters of plasminogen activator in fibroblasts is followed by changes in the morphology of the colonies; these modifications are suppressed by addition of serine protease inhibitors (Quigley, 1979).

Another response of many cells to phorbol ester promoters is a release of arachidonic acid and prostaglandins. This TPA-induced synthesis of prostaglandins is inhibited by indomethacin and retinoic acid (Mufson *et al.*, 1979; Fürstenberger and Marks, 1980). Since indomethacin also inhibits cell proliferation and ODC stimulation, it is now often believed that many of the effects exerted by TPA and other cancer promoters are mediated by prostaglandin synthesis and concomitant changes in membrane phospholipid metabolism (Mufson *et al.*, 1979; Wertz and Mueller, 1980). However, this is not always the case. For example, Crutchley *et al.* (1980) found that indomethacin does not affect the increase in plasminogen activator activity that is induced by TPA in HeLa cells.

The relationships between phorbol esters receptors, EGF receptor, and protein kinase C have been discussed when we dealt with cell differentiation. We may add that PMA (phorbol myristate acetate), as well as TPA, phosphorylates the EGF receptor (Davis and Czech, 1984) and that TPA and a diacylglycerol analog (1-oleyl 2-acetylglycerol, OADG) exert the same effects on protein kinase C in living cells: the activity of the cytosolic enzyme disappears and the enzyme moves to the cell membrane. Both TPA and OADG inhibit EGF-binding and tyrosine phosphorylation of the EGF receptor and increase its phosphorylation of serine and threonine residues. In short, TPA can substitute to diacylglycerol for protein kinase C activation (McCaffrey et al., 1984). Since this enzyme can phosphorylate many different proteins, the pleiotropic effects of the phorbol esters on living cells are not surprising. We have seen in Chapter 5 (Volume 1) that Moolenaar et al. (1984) also found that both TPA and OADG activate the Na^+/H^+ exchanger and raise the intracellular pH by a protein kinase C-dependent mechanism.

Another effect of TPA is the interruption of cell-to-cell communications. As was shown by Enomoto et al. (1981), treatment with TPA quickly interrupts electrical coupling between cultured cells. The electrophysiological data are supported by EM observations demonstrating that TPA decreases the number of gap junctions (Fig. 27). This explains the decrease in metabolic cooperation between adjacent cells (Yotti et al., 1979; Murray and Fitzgerald, 1979; Yancey et al., 1982; Fitzgerald and Murray, 1982). This clearly shows that the primary site of action of phorbol ester promoters is at the level of the cell surface, probably the plasma membrane itself. However, effects on the chromosomes (presumably of a secondary nature) have been reported. Kinsella and Radman (1978) have found that TPA induces spontaneous sister chromatid exchanges in cultured cells, and this effect is suppressed by protease inhibitors (Little et al., 1979). Fusenig and Dzarlieva (1982) observed chromosomal aberrations in cells cultured in the presence of TPA, while Hayashi et al. (1983) reported that this phorbol ester induces the amplification of the metallothionein-I genes. Emerit and Cerutti (1982) found that another phorbol ester (PMA) is "clastogenic," i.e., it induces chromosome aberrations. This detrimental effect is suppressed by addition of indomethacin, which prevents promotion by TPA (Borek and Troll, 1983), or by treatment with superoxide dismutase, indicating that prostaglandins and free radicals are involved in the induction of clastogenicity by PMA.

However, there are many discrepancies in the literature about this point. It seems reasonable, therefore, to conclude that if TPA induces chromosomal aberrations in certain cells and under certain conditions, this is not a general rule. These alterations of chromosome morphology are probably a consequence of the various effects produced by phorbol esters on the cell surface, and it is now widely accepted that chromosomal DNA is not the primary target (in contrast to the chemical carcinogens) of TPA and related promoting agents. This conclusion

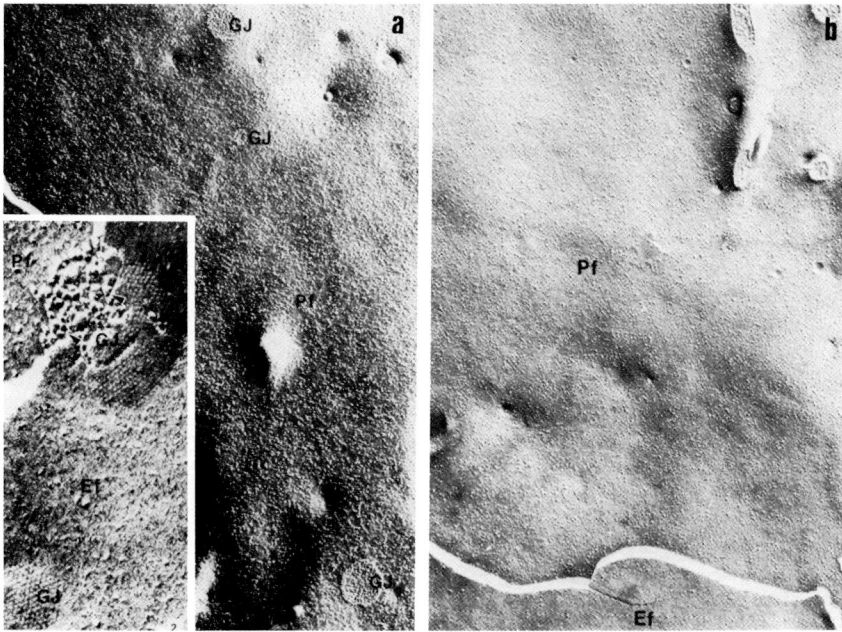

FIG. 27.   (a) Electron micrograph of the "plasmic" fracture face (Pf) of the plasma membrane of a Chinese hamster V-79 cell grown under standard conditions (control). Gap junctions (Gj) occurred frequently in these cultures. Inset: a higher-magnification view showing two gap junctions and external (Ef) and plasmic (Pf) fracture faces of the membranes of two adjacent cells. (b) An interface between two V-79 cells treated with TPA (phorbol ester). Gap junctions were absent from most of the membrane areas—where the cells appeared, nonetheless, to be closely apposed. [Yancey *et al.*, 1982.]

is reinforced by a study on the effects of TPA on enucleated cells. TPA induces the release of fibronectin from cytoplasts, an effect that is counteracted by retinoids; on the other hand, stimulation of ODC synthesis by TPA requires the presence of the nucleus, probably because this synthesis results from transcription of the ODC gene. The fact remains that one aspect, at least of promotion (fibronectin release), does not require the cell nucleus (Bolmer and Wolf, 1982). All of these findings fit with the fact that TPA, unlike the genotoxic initiators, is not a mutagen; this is perhaps the reason why phorbol esters do not transform benign papillomas into malignant carcinomas (Hennings *et al.*, 1983).

Although phorbol esters are best known as tumor promoters, it is important to remember that they exert pleiotropic effects on leukemic cells. According to the cell line used, they may increase proliferation and increase or decrease cell differentiation (Lotem and Sachs, 1979). Many reports show that, in certain cell lines at least, TPA induces normal differentiation in leukemic cells (Fig. 28). For instance, transformation of leukemic promyelocytes into adult macrophages was

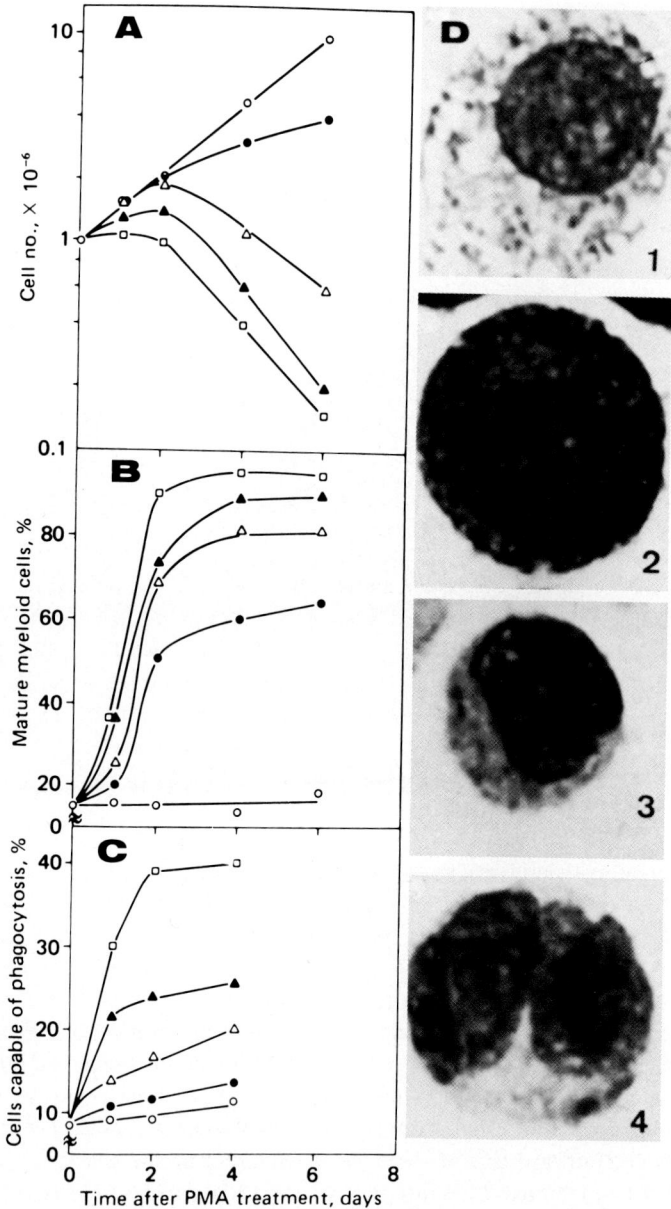

FIG. 28.   Cell growth (A), differentiation (B), and phagocytosis (C) in HL-60 cells at different times after treatment with various concentrations of the tumor promoter PMA. The mature myeloid cells were composed mainly of myelocytes and metamyelocytes. (D) Morphological myeloid differentiation in HL-60 cells: (1) myeloblast, (2) promyelocyte, (3) myelocyte, (4) metamyelocyte. [Huberman and Callaham, 1979.]

almost simultaneously observed by Lotem and Sachs (1979), Huberman and Callaham (1979), and Rovera et al. (1979a,b). The last group found that such a transformation is not inhibited by addition of retinoic acid, protease inhibitors, and indomethacin. As one can see, TPA-induced differentiation of leukemic cells does not seem to depend on the production of plasminogen activator or prostaglandins. According to Rovera et al. (1980), differentiation of leukemia promyelocytes in macrophages does not require even a single round of DNA synthesis. Differentiation of lymphocytic leukemic cells has also been obtained by TPA treatment (Nakao et al., 1982). Differentiation and inhibition of growth have been observed in human keratinocytes injected with an oncogenic virus (SV40) (Mufson et al., 1982).

The mechanism of action of phorbol esters on living cells is receiving increasing attention. Rosoff et al. (1984) studied the induction by TPA of the differentiation of a pre-B-lymphocyte cell line in immunoglobulin-producing lymphocytes. It activates a Na^+/H^+ system and raises the intracellular pH. In another system (PMA-induced proliferation of chicken heart mesenchymal cells), stimulation results from diacylglycerol-mediated activation of protein kinase C or mobilization of Ca^{2+} from intracellular stores. These are, as we have seen in Chapter 5, Volume 1, the biochemical mechanisms which induce cell proliferation in many, if not all, cells.

In complete contrast to the effects of the tumor promoters are those of the retinoids (reviewed by Lotan, 1980). They induce the differentiation of malignant embryonal and carcinoma cells, slowing down the growth of cancer cells and binding to chromatin. However, the fact that they counteract the release of fibronectin in TPA-treated cytoplasts shows that they also affect the cytoplasm (Bolmer and Wolf, 1982). In whole cells, retinoic acid counteracts the increase in ODC activity, the decrease in cell adhesion, and the alteration of cell surface glycoproteins, which are induced by treatment with TPA (Nagle and Blumberg, 1980). Of particular interest for oncologists is the observation by Dickens et al. (1979) that retinoids inhibit the malignant transformation of mammary glands; the content in the retinoid-binding proteins increases during cancerization of these glands (Mehta et al., 1980). Retinoic acid also inhibits the growth of several cell lines of osteosarcomas and chondrosarcomas, suppressing the expression of their transformed phenotype (Thein and Lotan, 1982). Retinoids prevent the mitogenic effects of lectins and phorbol esters on lymphocytes (Skinnider and Giesbrecht, 1979) and inhibit prostaglandin synthesis in TPA-treated cells (Mufson et al., 1979). In general, they slow down cell growth and proliferation; but their effect, in this respect, probably varies from one cell strain to another (Lotan and Nicolson, 1979). It is believed that retinoids act mainly on the G_1 phase of the cell cycle (Haddox et al., 1979). However, according to Dicker and Rozengurt (1979), retinoids potentiate the stimulatory effects of EGF, FGF, insulin, and even TPA in certain cell lines. This seems to be an isolated finding, since retinoic acid is thought to inhibit testosterone induced prostate hyperplasia (Chopra and Wilkoff, 1977) and to decrease the proliferation of melanoma cells; melanin simultaneously accumulates, and the number of cell processes increases (Lotan and Lotan, 1981).

As far as is known, retinoids are not widely used in cancer therapy, since it is well known that they are powerful teratogenic agents. This is probably due to the fact that they decrease the adhesion of embryonic cells (taken from the neural crests) to the substratum and reduce cell motility; the main effect of retinoids is to prevent the cells from migrating into the extracellular matrix in embryos (Thorogood *et al.*, 1982). It is regrettable that although so much work is being done throughout the world to test chemicals for mutagenic and potential carcinogenic activities, there has been only sporadic screening of chemicals for teratogenic activities. If this had been done, the thalidomide tragedy would have been avoided.

We will only briefly mention cancer therapy, since little can be added to what is already common knowledge. Most important, of course, is the early diagnosis of the disease by well-trained, well-equipped physicians. Besides surgery and radiation therapy, the great hope has been—and still remains—chemotherapy. Attempts to find a "wonder drug" by screening all possible chemicals for anticancer activity have thus far been unsuccessful. The few active drugs that are in current use were derived from studies on the inhibition of cell division in various kinds of cells by a number of drugs. These drugs interfere with either spindle formation (colchicine and *Vinca* alkaloids) or DNA replication. Chemicals that strongly bind to DNA (daunomycin, adriamycin, etc.) are also currently used for cancer therapy, although progress in this area has been slow. Unfortunately, the now available antimitotic drugs do not discriminate between malignant and normal dividing cells. Unpleasant side effects are frequent, although they are well accepted by patients who hope for a remission, if not a cure. All the antimitotic agents, including x-rays, are mutagenetic and carcinogenic. However, physicans are faced with the urgent problem of stopping a growing cancer as soon as possible, even at the cost of possibly inducing another 10 or 20 years hence.

Attempts are being made to reduce the toxicity of anticancer drugs by binding them to proteins or nucleic acids. Such complexes undergo endocytosis and are broken down in lysosomes, allowing a slower release of the anticancer drug (Trouet, 1978). It may perhaps be possible to devise similar complexes of anticancer drugs that would be targeted specifically toward a given organ (mammary glands, for instance), allowing direct action of the drug on the target organ and thereby reducing the side effects. Attempts are being made along this line by A. Trouet and his group, but it is not yet known how successful they will be. Obviously, a great deal of fundamental research in cell biology is still needed before we can find those drugs that will be specifically targeted for cancer cells.

As long as we do not fully understand how a malignant cell escapes the controls that govern cell proliferation and differentiation in normal cells, we will remain in the dark. Much progress has already been made in this area, and there is no reason to give up the hope that such agents will be found.

There is currently a great deal of interest (both scientific and commercial) in the interferons. In 1982, a complete issue of *Biochimica et Biophysica Acta* (**695,** 1) and a symposium published in *Philosophical Transactions of the Royal Society of London* (**299,** No. 1094) were devoted to these interesting polypeptides, possessing both antiviral and anticancer properties. There are three different types of interferons (IFN): α-IFN is induced in virus-infected leukocytes; β-IFN is made by fibroblasts treated with a double-stranded RNA [poly(rI:rC) is the classic inducer]; γ-IFN is produced by lymphocytes stimulated with mitogens or alloantigens. Recent work has shown that α- and β-IFNs can be induced in a variety of cells, but not in young mouse embryos or teratocarcinoma stem cells. Since IFNs slow the rate of tumor cell multiplication (Lin *et al.,* 1982), they are of major interest to both molecular biologists and virologists. The IFN are encoded by at least 13 nonallelic genes forming a gene family; there are no introns in the α- and β-IFN genes. At least six pseudogenes are related to the IFN genes, but they do not direct the synthesis of IFN polypeptides. IFN genes linked to prokaryotic promoters have been introduced into bacteria forcing them to synthesize and secrete the corresponding IFNs. Synthesis and secretion of IFN also follow injection of IFN mRNA into *Xenopus* oocytes. Secretion of IFN is related to the fact that these polypeptides possess a signal sequence (21 or 23 amino acids) in addition to a chain made up of 166 amino acids. In cells, IFN induces the synthesis of two enzymes: $2'$-$5'$ oligoadenylate synthetase and a protein kinase. $2'$-$5'$A ppp(A$2'$p)$_n5'$A activates a latent nuclease, inhibits DNA synthesis, and modifies the cell surface. The protein kinase phosphorylates initiation factor eIF$_2$ of protein synthesis. The investment in industrial production of the human IFNs has been huge, although it is doubtful that IFN is the wonder drug we are hoping for. Thus far, regression of some kinds of tumors (but not all) has been observed in only a limited number of patients, and the side effects (allergic reaction) have been unpleasant and even dangerous. A suspicion that IFN treatment might lead to serious cardiovascular troubles has recently led to the temporary halt of further trials on human patients in France. It is hoped that IFN will be used for more important medical applications than stopping a coryza by instillation in the nostrils. At present, it is probable that interferon will be a useful addition to the already existing arsenal of anticancer weapons, but that it will not be the definitive weapon. Due to the present interest in IFN as a potential antiviral and anticancer drug, we know more about the IFN gene family than about many others. Research on interferon, whatever its final outcome, will remain a cornerstone in the history of molecular biology, marking a shift from fundamental research to biotechnology.

Immunological therapy of tumors is another question that has been amply discussed. Several years ago, a well-known Belgian oncologist said, "We shall never cure cancer, but we shall vaccinate against it." This was at the time when many leukemia viruses were discovered in mice and when it was widely believed that most human cancers must also be of viral origin. Despite the existence of oncofetal antigens and the many changes that take place in the cell surface during malignant transformation, treatment of cancers by immunological methods has so far been disappointing. However, investigations in the area of immunology have made such great progress that this approach to cancer therapy remains promising. Every day we understand better the interactions of the various kinds of lymphocytes in the complex regulatory immunological network. If antibodies against specific membrane constituents of cancer cells quickly lose their capacity to kill these cells, this might be due to the appearance of suppressor lymphocytes. The problem that remains is the means to eliminate them in a specific way; only an immunological approach will allow us to reach this goal. Monoclonal antibodies produced by hybridomas will be essential tools in this immunological approach. Encouraging results have been reported by Vollmers and Birschmeier (1983) and Vollmus *et al.* (1984). Certain monoclonal antibodies against melanoma cells reduce the number of lung metastases. These antibodies prevent the adhesion of cancer cells, but not of normal cells, to the substratum; they cross-react with undifferentiated teratocarcinoma cells. The antigen common to melanoma and teratocarcinoma is still unknown; all we know is that it is not the p53 antigen, which is 1000 times more abundant in cancer cells than in normal cells. It has also been reported that two monoclonal antibodies are effective against solid mammary tumors growing in *nude* mice (Capone *et al.*, 1983). Monoclonal antibodies will perhaps been used in the future to transport drugs specifically to target cells. At present, the further experiments should be done on cultured malignant and control cells. An interesting model for such an approach is that of Boon and Kellermann (1977), who discovered that, if cancer cells (*tum*$^+$) are treated with mutagenic agents, nontumorigenic (*tum*$^-$) variants can be isolated. Since *tum*$^+$ and *tum*$^-$ cells, which originate from the same initial *tum*$^+$ cells, have different antigenic cell surface determinants, the system is amenable to a refined analysis by immunological methods.

We have mentioned the tumor viruses more than once, and discussion of these viruses is appropriate at this point. Viruses are not cells, but proliferate inside cells as obligatory parasites. They therefore belong marginally to molecular cytology.

F. Viral and Cellular Oncogenes[5]

It has been know since the work of Peyton Rous (1911) that the development of tumors in birds and mammals is often associated with infection by viruses.

[5]Reviewed by Cooper and Lane (1984) and Varmus (1984).

Since there is no compelling evidence to show that carcinogenic agents such as x-rays or aromatic polycyclic hydrocarbons induce cancers by activating latent cancer viruses, such viruses should be considered as only one of many other agents that can provoke malignant transformation and uncontrolled cell proliferation. This, in addition to the now well-accepted fact that cancer is a multistep process that may result from a variety of causes, is one of the reasons why the struggle against cancer is so difficult and, for many laymen, disappointing, despite the large amount of money and research efforts spent to solve this problem. In the 1950s, there was hope that all cancers might be due to viruses, making possible an immunological approach (vaccination) and the use of anti-viral drugs such as interferon in the struggle against cancer. Thus far, however, only a few human cancers have been shown to result from cancer virus infection. In addition, these cancers are found only in restricted areas of the world; an example is Burkitt's lymphoma (Uganda, New Guinea), which is believed to be due the Epstein-Barr virus. This virus, which belongs to the herpes virus family, is also implicated in other infections: mononucleosis, a disease whose course is long and unpleasant, but not lethal; and nasopharynx sarcoma, a disease found primarily in Southeast Asia. Herpes viruses have also been implicated in certain forms of cancers found in the genital tract of women. Adult T-cell leukemia, observed in Japan, and cutaneous T-cell lymphoma, found in the United States and Japan, is caused by a human T-cell leukemic virus (HTLV), which is similar to the bovine leukemia virus (Popovic *et al.*, 1982; Seiki *et al.*, 1982). One of the HTLV viruses is the causative agent of acquired immune deficiency syndrome (AIDS).

A lengthy discussion of the biology of tumor viruses is beyond the scope of this book. The interested reader will find a wealth of information in the book by Weiss *et al.* (1982) and in reviews by Kelly and Condamine (1982), Bishop (1983a,b), Duesberg (1983), and Varmus (1984).

Broadly speaking, there are two different classes of oncogenic viruses: those whose genetic material is a DNA molecule and those (called *retroviruses*) whose genetic material is a viral RNA molecule. Little will be said here about the DNA viruses, although two of them, polyomavirus and simian virus 40 (SV40), have been widely used in the past to induce malignant transformation in cultured cells. Of interest is the appearance on the cell surface of specific antigens, called the *T* antigens in polyomavirus-infected cells. According to Segawa and Ito (1982), one of the virus-induced T-antigens is associated with a tyrosine-specific protein kinase. The genomes of both SV40 virus and polyomavirus are relatively small and were completely sequenced several years ago; they cannot encode more than 5–10 proteins. Polyomavirus has been recently reviewed by Cuzin (1984) and Magnusson (1985). The three early T antigens (large, middle, and small) are all required for malignant transformation of normal fibroblasts; the middle T antigen is the membrane-bound tyrosine protein kinase; the large T antigen is necessary to maintain the transformed state and to immortalize the transformed cells.

Transformation by polyomavirus is thus a multigenic, multistep process. In contrast, the oncogenic viruses of the herpes family have a much larger genome, possessing, among other things, a gene encoding a viral thymidine kinase. Herpes simplex virus (HSV) has been used advantageously in DNA-mediated transfection experiments.

A considerable amount of work is now being done on the oncogenic RNA viruses, which are responsible for leukemias and sarcomas in many mammals, particularly mice. These viruses belong to the retroviruses family. They possess, as hypothesized by Temin in 1964 and shown by Baltimore in 1970, the enzyme reverse transcriptase, which can copy the RNA viral genome into DNA. This DNA copy is inserted into the host's genome, where it is called a *provirus*. Such DNA proviruses are integrated in all cells, often remaining latent and innocuous during the entire life of the organism. These endogenous retroviruses are transmitted vertically, that is, from parents to offspring, like normal genes. This is in contrast to contagious exogenous viruses of the influenza and measles types. Exogenous retroviruses also exist; their genetic information has nothing in common with that of the host, and their DNA sequences are found only in the infected cells, where they are transmitted horizontally.

The oncogene theory (Huebner and Todaro, 1969) states that a specific viral gene (the oncogene) is responsible for malignant transformation of the cells. We shall see that there is overwhelming evidence for the correctness of this theory; cancer would result from derepression of an oncogene. Temin (1981) has suggested that retroviruses originated from cellular movable elements comparable to those found in *Drosophila* (*copia*, etc.; see Chapter 4, Volume 1). This exciting hypothesis is receiving wide acceptance.

We will now briefly discuss the molecular organization of the retroviral particles (virions about 100 nm in diameter). We shall take, as an example, the classic Rous sarcoma virus (RSV), which is responsible for the development of a chicken sarcoma. Each viral particle possesses a genome composed of two identical RNA molecules (about 9500 nucleotides each). A tRNA^{Tryp} molecule is attached at one of the RNA fibers, which serves as a starting point for reverse transcription. Several thousand molecules of a protein called p12 ($M_r = 12,000$) and a few molecules of reverse transcriptase are associated with the viral RNA. This RNP complex is surrounded by several thousand molecules of two proteins called p27 and p25; the whole constitutes the viral nucleoid (Fig. 29), which can be isolated from whole virions by treatment with detergents. The nucleoid is surrounded by a first layer, composed of a 19,000-dalton phosphoprotein (pp19); a second layer is a lipid membrane, with the classical bilayer organization, from which about 1000 protuberances emerge. These protuberances are made up of two glycoproteins (gp 85 and gp 35) linked together by —SS— bridges. As one can see, the molecular anatomy of a large retrovirus like RSV is not as complicated as that of a cell. However, a high degree of complexity in the molecular organization of the virion clearly emerges from this brief description. Electron

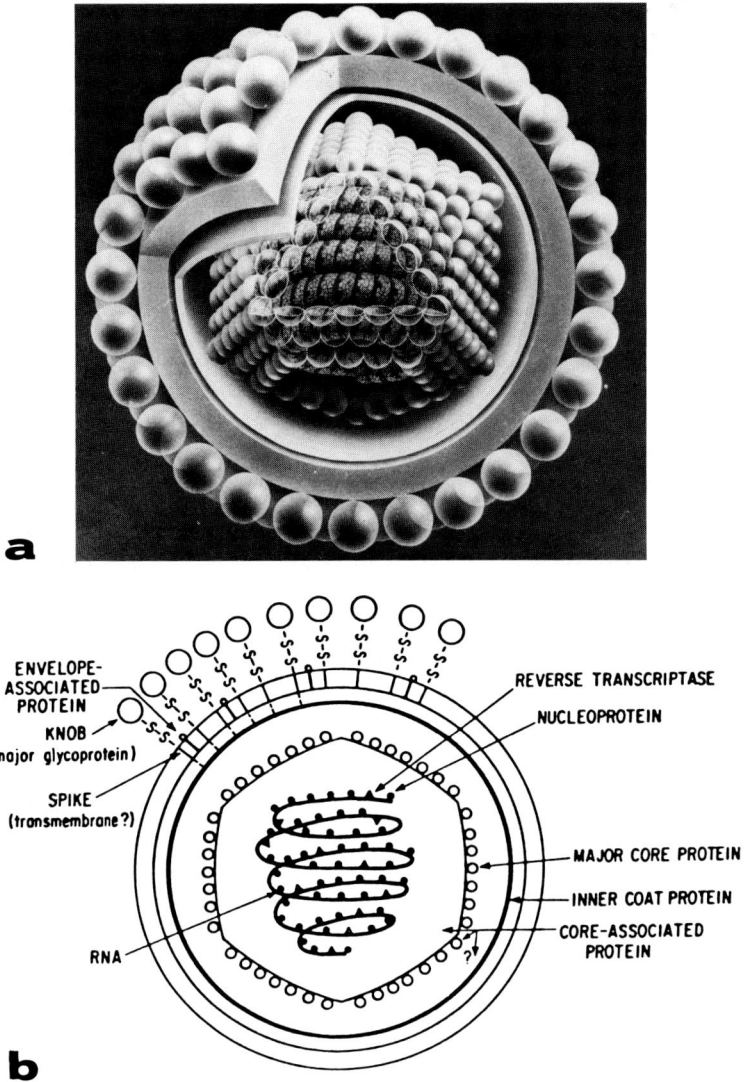

FIG. 29. (a) Tridimensional model of a type C oncornavirus Mu-LV particle (murine Friend leukemia virus). (b) Diagrammatic model of a type C oncornavirus particle. [(a) Courtesy of Dr. W. Schäffer; (b) Montelaro and Bolognesi, 1978.]

microscopic studies of cancer cells have shown that retroviruses can be classified as A, B, C, or D particles on the basis of morphological criteria (Fig. 30).

The RNA of RSV possesses four genes, called *gag, pol, env,* and *src.* The *gag* gene carries the information needed for the synthesis of the internal structural proteins of the virion responsible for group antigen specificity; the *pol* gene

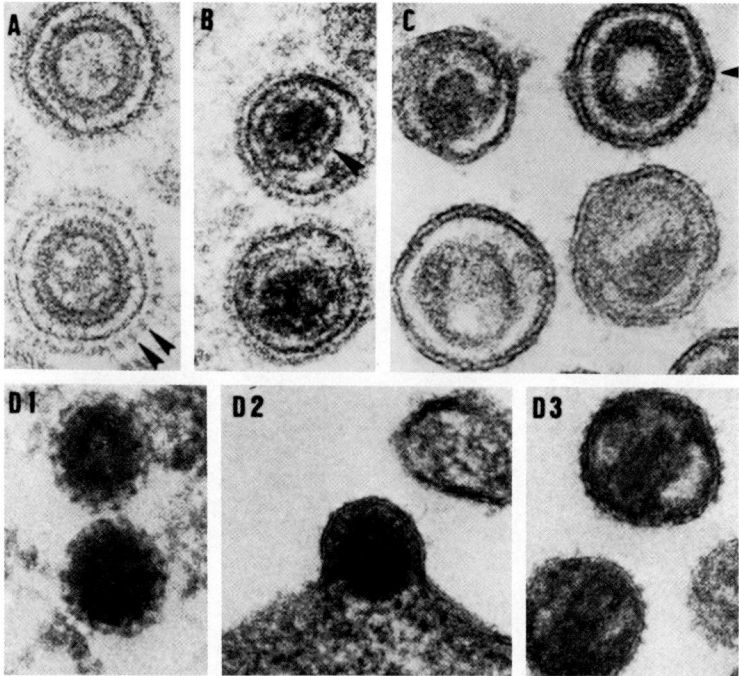

FIG. 30.   Electron micrographs of oncornavirus particles. (A) Type A mammary tumor virus (MTV) particles; arrowheads: surface spikes. (B) Type B MTV particles; there is a condensed, excentric core surrounded by an intermediate layer (arrowhead). (C) Type C murine Friend leukemia virus (MuLV) particles; condensed core without an intermediate layer. (D) Thin sections of type D MPHV (primate retroviruses) grown in cell cultures. (1) Immature intracytoplasmic precursor form of type D virus (type A particle); (2) immature type D virus budding from a cytoplasmic membrane; (3) mature type D virus. [(a–c) De Harven, 1974; (d) Fine and Schochetman, 1978.]

encodes reverse transcriptase, and *env* gene directs the synthesis of the surface glycoproteins. Most important is the *src* gene, since this is the oncogene responsible for malignant cell transformation. Viruses deficient in the *gag, pol,* or *env* genes cannot reproduce unless a "helper" virus supplies the missing proteins. All tumor viruses do not possess an oncogene comparable to the *src* gene of RSV; nevertheless, they may induce malignant transformation. This has led to the hypothesis that normal cells contain oncogenes, which are not expressed under ordinary conditions. The incomplete virus, after integration into the host's genome, would activate the silent cellular oncogene (proto-oncogene) by inserting at its proximity a strong promoter of its transcription. We shall consider the present status of this fascinating problem later.

We should first say a few words about the life cycle of a retrovirus when it comes in contact with a permissive target cell. The viral particle binds to the cell membrane and penetrates by endocytosis; due to the reverse transcriptase activity of the virion, the viral genomic RNA gives rise to a double-stranded linear DNA molecule (proviral DNA). This essential step of retroviral infection takes place in the host cell's cytoplasm. The proviral DNA then penetrates the host cell's nucleus and changes into a circular molecule that is finally integrated into the host's DNA. Reverse transcription of the viral DNA gives rise to two new DNA sequences about 600 bases long. They are repeated, in the same direction, at both ends of the proviral DNA and are called *long terminal repeats* (*LTR*) (reviewed by Temin, 1981; Groner and Hynes, 1982). In addition, integration of the provirus into nuclear DNA generates the appearance of two short DNA sequences (direct repeats) at the integration sites.

There is obviously a striking similarity in nucleotide structure between the sites of insertion of a provirus and of a transposable element (*copia,* for instance) in the *Drosophila* genome. The same similarity in nucleotide sequences exists in the dispersed repetitive sequences belonging to the human *Alu* family. These unexpected findings are currently leading to a great deal of speculation. Retroviral oncogenes might be normal genes that have escaped from the genome and have been transcribed into cRNA by the nuclear RNA polymerases (Temin, 1974). Mobile genetic elements of the *copia* type and *Alu* sequences might have originated from RNAs that, after reverse transcription, would be (like the retroviruses) integrated into the genome. Kugimiya *et al.* (1983) have indeed found that the LTR of two *copia*-like mobile genetic elements in *Drosophila* are very similar to those of an avian leukosis-sarcoma virus. It appears that a progenitor *Drosophila* had been infected by a retrovirus from which the avian leukosis-sarcoma virus would have derived. We shall probably never know for certain whether or not these are evolutionary changes, but it is probably no accident that proviral DNA, mobile genetic elements, and at least some of the *Alu* sequences are all inserted into very similar structures in genomic DNA. To this list one should probably add some of the pseudogenes that are particularly frequent in the small RNAs families. As we have seen, certain pseudogenes are "processed" genes that have no introns, but possess a poly(A) tail. The hypothesis that they originated from mRNAs that have been reverse transcribed and inserted into chromosomal DNA is an exciting one. We do not know for certain whether insertion of fresh genetic material into chromosomal DNA takes place at random or at specific sites. Mobile genetic elements—as implied by their name—seem to be capable of insertion at any site of the *Drosophila* genome. This might not be true for the retroviruses. Nusse and Varmus (1982) reported that the provirus of the mouse mammary tumor virus integrates in the same region of the host's genome in many tumors if this region lies close to a cellular oncogene; insertion of a provirus (even if it is defective and does not contain a viral oncogene) might, as pointed out by the authors, promote the expression of the cellular oncogene.

There is overwhelming evidence for the view that normal mammalian genomes contain "cellular oncogenes" (*c-onc* genes), also called *proto-oncogenes* (reviewed by Weinberg, 1983). This is shown by the fact that nucleotide sequences closely related to viral oncogenes (*v-onc* genes) are present in the DNA of apparently normal cells. The aforementioned *src* gene of RSV is only one of the representatives of the *v-onc* family; its cellular homolog is called *sarc* or c-*src*. Some of the evidence will be presented briefly.

Dalla Favera *et al.* (1981) found that the human genome contains a single constant *c-sis* locus that is closely related to the *v-sis* oncogene of a simian sarcoma virus. They suggested that, as had already been proposed by Duesberg (1980), malignant transformation, both *in vivo* and *in vitro,* is due to *onc* genes that originated from the normal genome by recombination between a parent provirus and host cellular DNA. The same authors think that neoplasic transformation results from the expression of cellular or viral *onc* genes. At the same time, Wong-Staal *et al.* (1981) discovered, in the human genome, three *c-onc* genes related to the *onc* DNA sequences of three different retroviruses (which induce sarcomas in monkeys, rats, and mice, respectively). These cellular oncogenes are also present in the chick and monkey genomes, showing that they have a widespread distribution among mammals. Another interesting finding (Lane *et al.,* 1981) was that the same transforming oncogene is activated in six different carcinomas induced in mice by either viruses or chemical carcinogens; this oncogene was also present in a human mammary carcinoma. Groudine *et al.* (1981) discovered that chicken cells possess two genes homologous to the *onc* genes of endogenous retroviruses. Only one of them (*ev-3*) is active, judging from DNase sensitivity, while the other (*ev-1*) becomes active when the cells are treated with 5-azacytidine. Such treatment induces DNA undermethylation and confers DNase sensitivity on the *ev-1* gene. Curiously, sequences homologous to the oncogenes of RNA tumor viruses affecting mammals are present in the *Drosophila* genome (Shilo and Weinberg, 1981a,b; M. A. Simon *et al.,* 1983; Hoffmann *et al.,* 1983). Recently, Lev *et al.* (1984) found that in *Drosophila,* the cellular homologs of the viral oncogenes *abl* and *src* are transcribed from genes called *Dash* and *Dsrc.* Interestingly, the transcripts of *Dash* are found in the maternal mRNAs of *Drosophila* eggs; they remain detectable during 4 days in embryos. The transcripts of *Dsrc* are also present in eggs and embryos; only small amounts are found in larvae and adults (Wadsworth *et al.,* 1985). This suggests that the *Dash* and *Dsrc* genes play an important role in the early stages of *Drosophila* embryogenesis. In addition, *Drosophila* possesses three genes homologous to one of the members of the *ras* family of oncogenes (Ha-*ras*). They code for a 21,000-dalton protein (p21) which displays 75% homology with the p21 protein expressed by the Ha-*ras* gene in vertebrates (Neuman-Silberberg *et al.,* 1984). In yeast the products of the two *ras* genes are precipitated by monoclonal antibodies directed against the p21 mammalian protein; the yeast *ras* products, like mammalian p21, bind GTP (Temeles *et al.,* 1984). Yeasts lacking functional *ras* genes (now called RAS[sc]-1 and -2 are not viable; they survive if they receive a mammalian *ras* gene (De Feo-Jones *et al.*, 1985).

Thus DNA sequences partially homologous to viral oncogenes have been found even in yeast cells. Yeasts possess genes coding for proteins that, like p21 in transformed cells, are localized in the plasma membrane and bind guanine nucleotides (Gallwitz *et al.*, 1983; DeFeo-Jones *et al.*, 1983; Papageorge *et al.*, 1984; Powers *et al.*, 1984). Dhar *et al.* (1984) recently gave the complete base sequences of two yeast genes related to the *ras* genes of murine sarcoma viruses. They encode proteins of 41,000 and 42,000 daltons in which the N-termini have been conserved and the C-termini have diverged during evolution when they are compared to the p21 protein of mammals. It seems that these genes play a role in the control of cell division in yeasts. Lörincz and Reed (1984) found that the product of the *start* gene, which is known to control cell division in yeasts, displays homologies with the products of the viral oncogenes that encode protein kinases. Strong homology between yeast cell cycle genes (CDC4 and DCD 36) and sequences of an avian erythroblastosis virus has also been reported by Peterson *et al.* (1984). This suggests that the cellular oncogenes (or at least some of them) have been highly conserved during evolution and that they must play an important—but still unknown—role in cell proliferation and differentiation. In agreement with the general trend of thought, Shilo and Weinberg (1981a,b) have suggested that association of a proto-oncogene (thus, the cellular counterpart of a transforming viral oncogene) with the genome of a retrovirus enhances the expression of the proto-oncogene, leading to malignant transformation. According to Chang *et al.* (1982a,b), the human genome contains a *c-ras* gene family that is homologous to the transforming *v-ras* genes of two murine sarcoma viruses.

However, there is a striking difference between the viral and cellular oncogenes at the molecular level. While the former have no introns, the cellular oncogenes possess the familiar *intron-exon* organization, which implies the processing of an mRNA precursor (Dalla Favera *et al.*, 1981; Wong-Staal *et al.*, 1981; Vennström and Bishop, 1982; Vennström *et al.*, 1982). For instance, the cellular *c-src* gene has 12 introns, in contrast to none in *v-src* (Takeya and Hanafusa, 1983). The cellular *c-fos* gene, which does not induce transformation, has three introns, in contrast to the intronless viral *v-fos* gene of the murine osteosarcoma virus; in addition, these two genes differ by a deletion (Van Straaten *et al.*, 1983).

We have just seen that, in agreement with many observations discussed in this book, undermethylation is one of the requirements for the expression of a cellular oncogene (Groudine *et al.*, 1981). Not unexpectedly, the methylation state plays a role in the infectiousness of the retroviruses themselves. For instance, an inactive Moloney murine leukemia virus becomes infectious if it has been demethylated by molecular cloning in *E. coli* or by other means (Harbers *et al.*, 1981; Stuhlmann *et al.*, 1981). However, experiments in which specific viral

DNA sequences have been methylated *in vitro* prior to infection have shown that the effects of DNA methylation on viral gene expression strictly depend on the genes that have been methylated (Waechter and Baserga, 1982). It seems to be a general rule that *in vitro* methylation inhibits the transforming activity of viral DNA in transfection experiments. This is true particularly for Moloney sarcoma virus; the inhibitory effect of *in vitro* methylation can be reversed if the transfected cells are treated with 5-azacytidine (McGeady *et al.,* 1983). D. Simon *et al.* (1983) found that retroviral genomes methylated *in vitro* with mammalian (but not bacterial) DNA methylases are noninfectious; treatment of the recipient cells with 5-azacytidine makes them infectious again. Interesting results for embryologists have been reported by Jaenisch *et al.* (1982). They found that Moloney leukemia virus is not infectious for mouse embryos before their implantation, while it is infectious for implanted embryos and adult mice; the resistance of the early embryos to viral infection is due to the fact that, in contrast to older embryos and adults, they contain a very powerful DNA-methylating activity. The embryonic DNA methylases methylate the virus, which loses its infectivity for adult mice. We are thus led to the conclusion, on the basis of ample evidence reviewed by Cooper (1982), that proto-oncogenes very similar to the transforming viral oncogenes occur widely in the genomes of apparently healthy organisms, including humans.

One question of interest for molecular cytologists is whether carcinogenesis is connected with oncogene amplification. Pall (1981) proposed that initiation of cancer might result from the tandem replication of cellular oncogenes; amplification might also derive from an unequal sister chromatid exchange in which three oncogene copies would be present on one of the chromatids, while the other would possess a single oncogene copy. Transformation would occur when a sufficient amount of the gene product has accumulated in the cell. Similarly, Lavi (1981) has suggested that chemical carcinogens, which perturb the DNA molecules, might activate replicons and produce, as for the dehydrofolate reductase genes in methotrexate-resistant cells (see Chapter 4, Volume 1), an amplification of transforming genes. The question has been tackled experimentally by Collins and Groudine (1982) and by Dalla Favera *et al.* (1982a) in human leukemia cell lines. Both groups found that DNA sequences (*c-myc*) homologous to the avian oncogene *v-myc* are amplified in certain cell lines, but not in others. They concluded that malignant transformation results from the enhanced expression of normal cellular oncogenes by either gene amplification or increased transcriptional activity. Still another possibility is the stabilization of the transcription products of a single gene copy. More recently, McCoy *et al.* (1983) found that cell lines from human colon or lung carcinoma contain oncogenes that have similarities to the oncogenes of two murine sarcoma viruses; in several of these tumor cell lines, the cellular oncogenes are amplified.

Schwab *et al.* (1983a) and George *et al.* (1984) have reported a 30- to 60-fold amplification of a cellular oncogene in adrenal tumors; the amplified oncogenes are localized in double minute chromosomes and a homogeneously staining chromosomal region. Similar observations have been made for human lung cancer cell lines by Little et al. (1983) and for colon carcinoma by Schwab et al. (1983b), who pointed out that double minutes and homogeneously staining regions are frequent in tumors. Amplification often takes place when a cellular oncogene is translocated to another chromosome, a phenomenon to which we shall soon return (Alitalo et al., 1983; Collins and Groudine, 1983). In human breast carcinoma cell lines, where the c-myc oncogene is not translocated, amplification seldom occurs (Kozbor and Croce, 1984). The c-myb oncogene, which is specifically expressed and regulated in hematopoietic cells, was amplified 5–10 times in a case of human acute myelogenous leukemia (Pelicci et al., 1984). The same oncogene was amplified (10 times) in two cell lines of colon carcinomas, but it was not expressed in other colon carcinomas (Alitalo et al., 1984). Strong amplification (up to 140 times) of the N-myc oncogene has been observed in neuroblastomas; amplification is limited to the tumor and increases with its growth (Brodeur et al., 1984; Schwab et al., 1984a). Oncogene amplification is certainly more frequent in tumor cells than gene amplification in differentiating embryonic cells. This is probably related to the well-known genetic instability of malignant cells. However, amplification is not the sole mechanism involved in malignant transformation. We shall soon see that other mechanisms, in particular gene activation by mutation or translocation, are of primordial importance for malignant transformation. It has been recently found by Taya et al. (1984) that in a human lung carcinoma case, the c-K-ras-2 and c-myc cellular oncogenes were both amplified 8–10 times, but there was also a point mutation in the c-K-ras-2 oncogene. Cooperation of two activated cellular oncogenes had been required for carcinogenesis in this patient. Although more information will be welcome, it seems unlikely that gene amplification is the cause of malignant transformation (Alitalo, 1985). Gene amplification is probably more frequent in cancer cells than in differentiating embryonic cells; this is not surprising, in view of the well-known genetic instability of malignant cells. We shall see that mechanisms other than amplification, in particular gene activation by mutation or translocation, are of primary importance for malignant transformation.

Another important question has not yet received a satisfactory answer. Are the cellular oncogenes transcribed and translated in the cells and, if so, what are the final gene products? Westin et al. (1982) found that, in human hematopoietic cells, the cellular homologs of several viral onc genes are transcribed into poly(A)$^+$ RNAs. It seems that transcription of the cellular oncogenes takes place in normal cells, but only at a very low level. Kirschmeier et al. (1982) found that cells transformed by RSV possess as many as 10,000–20,000 RNA copies of the src gene per cell; in contrast, only 5–10 RNA copies per cell of the homologous sarc cellular oncogene could be detected in normal cells. In the chicken genome, the c-myc gene (homolog of the v-myc oncogene of avian myelocytomatosis virus) is transcribed. At least five different transcripts are found in the nucleus, but only one in the cytoplasm (Vennström et al. (1982). Wang and Baltimore (1983) have shown that two RNAs homologous to the transforming gene of Abelson murine leukemia virus (v-abl) are present in all normal tissues of the mouse. These c-abl RNAs are polyadenylated and are localized in the cytoplasm, where they are presumably translated. They are more abundant in fibroblasts than in liver.

The product of the src gene of the Harvey murine sarcoma virus has been identified as the already mentioned p21 protein. Scolnick et al. (1981) found that this protein is present, but in very small amounts, in many kinds of normal cells. However, p21 is much more abundant in a hematopoietic, actively dividing cell line than in all of the other cells tested. Since the p21 protein present in normal cells is encoded by the cellular sarc and ras genes (Santos et al., 1983; Feig et al., 1984), it is possible that this cellular homolog of the retroviral src and ras genes plays a role in the control of normal cell proliferation. Further evidence for a role of oncogenes in normal cells has been presented by Mercer et al. (1982) and by Müller et al., (1982). As already mentioned, Mercer et al. (1982) found that a transformation-related protein (p53) is abundant in transformed and proliferating normal cells; in lymphocytes, this protein cannot be detected unless the cells are induced proliferate by addition of a mitogen. Müller et al. (1982) have shown that cellular c-onc genes are expressed during perinatal and postnatal development of the mouse in a stage- and tissue-specific manner. More recently, the same authors (Müller et al., 1983a,b) have reported that cellular oncogenes homologous to two murine leukemia viral transforming genes are expressed at a high level in the placenta, yolk sac, and amnion of both human and murine embryos; their transcripts are 100-fold more abundant in human term fetal membranes than in all other tissues. Similar results have been obtained by Adamson et al. (1983). It seems likely that the products of these oncogenes play a role in the protection and nourishment of mammalian fetuses or in the growth and differentiation of the extra-embryonal membranes. A recent study by Pfeiffer-Ohlsson et al. (1984) suggests a correlation between transcription of the c-myc proto-oncogene and cell proliferation. The transcripts appear in the young cytotrophoblast where mitoses are frequent.

It would be interesting to know whether these genes are involved in early stages of embryogenesis and whether similar genes operate during nonmammalian ontogenesis.

Since oncogenes are present in normal mammalian genomes, it should be possible to establish their localization on given chromosomes by somatic cell genetics analysis or by in situ hybridization. A few reports have been published in this area. According to Phillips et al. (1982), 3% of the Y chromosome is composed of retrovirus-related sequences in mice; utilization of a retroviral probe allowed them to distinguish between DNA from males and females. No other gene on the Y chromosome was previously known, with the exception of the gene that specifies the male transplantation antigen HY, which may play a role in sex determination. Prakash *et al.* (1982) found that human chromosome 8 bears the *c-mos* gene, which is homologous with the *v-mos* transforming gene of Moloney murine sarcoma virus. According to Dalla Favera *et al.* (1982b), the human analog *c-sis* of the transforming gene *v-sis* of the simian sarcoma virus is located on chromosome 22. On the other hand, Heisterkamp *et al.* (1982) have reported that the cellular homologs of two different viral oncogenes are localized on two different human chromosomes (numbers 15 and 9). There is thus no clustering on a single chromosome of these DNA sequences. It is obviously too early to attempt to correlate these scattered findings with the little we know from studies on somatic cell hybridization about the localization of "cancer genes." Progress in this important field is expected, but there is no doubt that the results on oncogene localization in specific chromosomes will reinforce the cancer gene concept.

Are the homologs of the retroviral transforming genes present in human cancers? The expected answer should be "yes," since we have seen that such genes can be detected in apparently normal, nontransformed cells. Indeed, Der *et al.* (1982) have found that cell lines derived from human bladder and lung carcinomas contain transforming genes that are homologous to the transforming *ras* genes of the retroviruses that induce two murine sarcomas. Similar results have been obtained by Parada *et al.* (1982) and Chang *et al.* (1982a,b), and have been critically discussed by Rigby (1982). This human oncogene induces the transformation of fibroblasts, in which it produces the p21 ras protein. It was concluded that high levels of a gene product encoded by a normal human oncogene can induce tumorigenic transformation. Similarly, the transforming gene of bladder human carcinoma is closely related to the BALB murine sarcoma virus–transforming *v-bas* gene. This transforming gene of human cancers is an activated form of the normal human homolog of the viral transforming gene (Santos *et al.*, 1982). Decisive proof that human cancers possess oncogenes has been presented by Slamon *et al.* (1984). They tested 20 different types of tumors from 54 patients for the presence of oncogenes and found that all of them had more than one transcriptionally active oncogene. In general, the transcriptional activity of the oncogenes was higher in malignant than in normal tissues from the same patients. The level of oncogene activation may be an important factor in tumor heterogeneity. Albino *et al.* (1984) studied the transcription of the *ras* oncogene in five cell lines originating from melanoma metastases present in the same individual and found that only one of them contained an activated *ras* gene.

The strongest evidence for the view that human cancers possess transforming genes comes from DNA-mediated transfection experiments, which clearly show that many malignant cell lines contain a tumor-inducing DNA (reviewed by Weinberg, 1981, 1982a,b; Newmark, 1982; Logan and Cairns, 1982). It was first shown by Shilo and Weinberg (1981a,b) that the transfer of DNA isolated from chemically transformed cells into normal cells resulted in the formation of foci of transformed cells. Since the same gene was transferred in all cases, the results indicate that the number of transforming genes induced by treatment with chemical carcinogens must have been small. In his brief reviews on the subject, Weinberg (1982a,b, 1983) concludes that different tumors have the same activated oncogene, but that there is some tissue specificity; there would be fewer than 20 cellular oncogenes (proto-oncogenes possessing the *c-onc* sequence), belonging to two different families (the *ras* family and the *src-myc* family). Shih *et al.* (1981) also showed that DNA extracted from methylcholanthrene (a potent chemical carcinogen)-transformed cells induces foci of transformed cells after transfection by the classic DNA-calcium phosphate coprecipitation method. Similar results were obtained with DNA isolated from human, mouse, and rat "spontaneous" tumors, and no species or tissue specificity was detected in these early experiments. At about the same time, Hynes *et al.* (1981), using the same methodology, transfected an endogenous proviral gene of a mouse mammary tumor. As already mentioned, Murray *et al.* (1981) showed, by transfection experiments, that three human malignant cell lines contained three different transforming genes. The existence of a tumor-inducing DNA was also clearly shown by the transfection experiments of Smith *et al.* (1982). Using human bladder sarcoma DNA, they transferred tumorigenicity (not only the induction of the transformed phenotype) to hamster fibroblasts; similar results were obtained by Pulciani *et al.* (1982). A further step forward was taken by Goldfarb *et al.* (1982), who worked with DNA isolated from a human bladder carcinoma cell line. The transforming gene had less than 5 kbp and was homologous to a 1000-bp poly(A)$^+$ RNA that was present in both bladder carcinoma and HeLa cells. Sequences homologous to the bladder carcinoma–transforming gene were present in the DNA of human placenta. A conclusion drawn from all this work by Weinberg (1982a) in his review was that the oncogenes present in human tumor cells may be tumor specific. They are different in transformed fibroblasts and in tumors of the colon or the mammary glands. He also speculated (Shih and Weinberg, 1982) that these oncogenes derive from sequences present in the cellular DNA of normal cells. However, the transforming genes present in human carcinoma cell lines are not necessarily homologous to the oncogenes of the retroviruses. Marshall *et al.* (1982a) found no homology between the transforming gene of human carcinoma cell lines and the *onc* genes of eight retroviruses. More recently, Balmain and Pragnell (1983) induced skin carcinomas in mice by the classic sequential treatment with initiators and promoters, isolated their DNA, and found that it was capable of transforming NIH 3T3 cells in transfection experiments; transformation was due to the transfer of an activated homolog of the viral Harvey-*ras* oncogene.

However, investigators began to wonder why transfection experiments that were so successful with NIH 3T3 fibroblasts gave negative results when other strains of fibroblasts (even the closely related 3T3 strain) were used. It was pointed out that NIH 3T3 fibroblasts are unique in being immortal; this led to the suspicion that they have already undergone a first step in malignant transformation. Indeed, Jariwalla et al. (1983) found that malignant transformation by a cloned DNA fragment of herpes simplex virus (HSV) takes place in two distinct stages: immortalization and tumorigenicity. Newbold and Overell (1983) reported that the cellular ras oncogene (homologous to the Harvey sarcoma virus oncogene), which does not transform normal fibroblasts, transforms cells that have been immortalized by treatment with a chemical carcinogen. Land et al. (1983a) reported similar findings. Transfection of embryonal fibroblasts by the human ras oncogene does not transform fibroblasts unless they have been previously immortalized. These transformed fibroblasts become tumorigenic if one introduces, together with the ras gene, a second oncogene such as v-myc or c-myc. These important findings reconcile the oncogene concept with the classic miltistep process (anaplasia, metaplasia, and neoplasia of pathologists) theory of cancer. Both groups of workers concluded that the NIH 3T3 immortal cells have probably undergone all the steps of malignant transformation except the last one.

As we have seen, the idea that cancer may result from translocation of a cellular oncogene from one chromosome to another (Klein, 1981; Klein and Klein, 1984, 1985) or, at the molecular level, from genetic transpositions (Cains, 1981) is currently widespread. Translocations could increase the expression of a cellular oncogene under the influence of neighboring regulatory sequences, such as a strong promoter; this could conceivably lead to cancer (Klein, 1981; Klein and Klein, 1984, 1985). Recent work has shown that specific chromosomal translocations are involved in the genesis of many B lymphocyte–derived tumors (plasmacytomas, various forms of leukemia, Burkitt's lymphoma) in both mice and humans. The breakpoint in the recipient chromosome is the immunoglobulin (Ig) locus; if an oncogene is inserted in this transcriptionally very active locus, it will be activated. In both mouse plasmacytomas and human Burkitt's lymphoma, the oncogene c-myc is transposed in the immediate neighborhood of the Ig genes (reviewed by Klein, 1983; Perry, 1983). A chromosomal translocation also occurs in chronic myeloid leukemia, but in this case the oncogene is c-abl instead of c-myc (reviewed by Rabbitts, 1983).

In lymphomas, translocation of a region of chromosome 8 to chromosome 14 is a frequent event that leads to a rearrangement of the c-myc gene (homologous to the avian myelocytomatosis viral oncogene) in the region of chromosome 14 where the Ig heavy chains genes are located; this increases the expression of c-myc, as shown by an increase in c-myc mRNA and c-myc protein production after translocation (Dalla Favera et al., 1982c, 1983; Giallongo et al., 1983; Erikson et al., 1983; Shen-Ong et al., 1982). In mice, c-myc is located on chromosome 15 and is translocated to the Ig heavy chain locus of chromosome 12 in plasmacytomas and other B-cell tumors (Crews *et al.*, 1982); this translocation, as well as the 8:14 translocation in human Burkitt's lymphoma, alters the expression of the *c-myc* gene (Adams *et al.*, 1983). The *c-myc* gene is strongly expressed in Burkitt's lymphoma if it has been translocated; in contrast, it is not expressed in normal B cells (Nishikura *et al.*, 1983). According to Dean *et al.*, 1983), translocation of the *c-myc* oncogene activates, in a reciprocal fashion, the expression of the Ig heavy chain genes. In chronic myelocytic leukemia, the human *c-abl* oncogene is translocated from a region of chromosome 9 to the Philadelphia chromosome 22; however, in this case, the position of the breakpoint is variable (Heisterkamp *et al.*, 1983). This 9:22 translocation has been confirmed by *in situ* hybridization. However, in certain particularly malignant forms of chronic myelocytic leukemias, there is no translocation of the *c-abl* gene; these leukemias must therefore have another origin (Bartram *et al.*, 1983).

The mechanisms that control the activation of the translocated *c-myc* genes in B-cell tumors are presently the subject of extensive work. It has been suggested that tissue-specific enhancer elements are associated with the human *Ig* heavy chain locus (Hayday *et al.*, 1984; reviewed by Perry, 1984). Such elements, which are also present in the LTR repeats of the oncogenic viruses, are short DNA sequences that stimulate transcription over wide stretches of the genome. It is likely that this stimulation is due to the selective binding to the enhancer sequences of transcription-stimulating factors (proteins that recognize the enhancer sequences). The fact that translocated *c-myc* oncogenes from Burkitt's lymphoma are transcribed in plasma cells, but not in lymphoblastoid cells, agrees with the concept that the translocated oncogenes are under the control of enhancer-like elements acting over long distances on the genome (Croce *et al.*, 1984). Since enhancers display tissue specificity, they might play a role in normal cell differentiation. It seems unlikely that a (8:14) translocation in B-cell lymphomas blocks irreversibly the capacity to differentiate. Benjamin *et al.* (1984) have recently reported that treatment of B-cell lymphoma cell lines by the tumor promoter PMA induces their differentiation into plasmacytoid cells.

In summary, there is now compelling evidence for the view that translocation of a *myc* or *abl* oncogene in the *Ig* heavy chain locus is a frequent event in lymphomas and leukemias, and that this results in oncogene activation. Oncogene translocation and activation might be the origin of these tumors; however, as we have just seen, certain leukemias apparently have a different origin. In addition, leukemias are only one form of cancer, and there is no evidence that translocation of oncogenes is responsible for the development of all other types of tumors. It has been concluded by Duesberg (1985) that there is no proof that activated cellular oncogenes are sufficient or even necessary for cancer production.

There must therefore be mechanisms of carcinogenesis other than chromosome translocation. A possibility was that a cellular oncogene undergoes mutation in a control region and acquires transforming properties. Much to the surprise of many workers in the field came two papers by Reddy *et al.* (1982) and Tabin *et al.* (1982). The work done in two of the leading laboratories in the field has proved that point mutations may play a decisive role in human bladder carcinoma. The two papers show that the oncogene of two different human bladder carcinoma cell lines derived from the alteration of a cellular proto-oncogene, and not from its activation. This conclusion was based on the following evidence. The flanking regulatory sequences are identical in the oncogene and its normal counterpart, but there is a difference in the DNA sequences of the two genes themselves; this change is a single point mutation from guanosine (GA) to thymidine (TA) (GC to TA transversion). This alteration of the gene does not affect its expression but does affect the structure of the oncogene-encoded p21 proteins, in which a glycine residue is replaced by a valine residue. This small change in the structure of p21 protein confers transforming properties. We do not know why this single amino acid substitution exerts a variety of effects on the cellular phenotype, but one may be reminded of sickle cell anemia, in which the shape of the red blood cells is profoundly changed, resulting from a replacement, due to a point mutation, of a single amino acid by another in hemoglobin. Tabin *et al.* (1982) have mentioned that a G–T transversion is one of the favorite mutations induced by bladder carcinogens. Reddy *et al.* (1982) pointed out that their discovery might have important consequences for early cancer diagnosis. Thus, scanning of human DNA for oncogene-like sequences and immunological detection of the modified protein are now possible.

The initial findings of Reddy *et al.* (1983) and Tabin *et al.* (1982) have been repeatedly confirmed and extended. Taparowsky *et al.* (1982) found that activation of the transforming gene of a bladder carcinoma is due to a single base mutation, which changes one amino acid in the encoded protein. The cellular homolog of *v-ras* (*c-ras*) and the viral *v-ras* oncogene thus differ by a single base. Capon *et al.* (1983) reported that the activated oncogene of a human bladder carcinoma cell line differs from the two alleles of the normal gene by a single mutation in the first exon; the proteins (p21) encoded by the normal and activated genes should differ by three amino acids. Similar results have been reported by Reddy (1983) and by Santos *et al.* (1983), who concluded that the transforming properties of the p21 protein present in this bladder carcinoma cell line result from the substitution of a single amino acid. Activation of the *N-ras* gene from a fibrosarcoma also results from a single base change (Brown *et al.*, 1984; Yuasa *et al.*, 1984). Another step forward was made when Yuasa *et al.* (1983) discovered that *two* point mutations may activate independently the same human cellular oncogene *c-bas/has* (*bas* and *has* are the oncogenes of the Balb and Harvey murine sarcoma viruses, respectively). Transformation activity of *c-bas/has* is due to a single point mutation in the second exon, while another mutation is responsible for the activation of two bladder carcinoma oncogenes. Taparowsky *et al.* (1983) also found that substitution of amino acids in two distinct regions of the *ras* gene product may activate its transforming potential.

These findings obviously fit well with the multistep theory of cancer. However, a word of caution is needed. Feinberg *et al.* (1983) have pointed out that mutations affecting the 12th amino acid of the p21 product encoded by the *c-ras* genes are rare events and that they play no role in most of the human epithelial cancers of the colon, bladder, and lung.

A striking difference between the p21 proteins encoded by the normal and activated *ras* genes has been recently discovered. Both gene products are located on the internal surface of the plama membrane and bind GTP, but only the normal protein has a GTPase activity and can thus hydrolyse bound GTP. A mutation, which selectively impairs the GTPase activity of p21 *ras*, activates its oncogenic potential. There are similarities between p21 *ras* and the guanine-binding regulatory protein (the G- or N-protein) of the ubiquitous enzyme adenylate cyclase (Mc Grath *et al.*, 1984; Sweet *et al.*, 1984; Gibbs *et al.*, 1984).

In conclusion, it can be said that about 24 cellular oncogenes have so far been discovered; it is unlikely that this number will increase greatly. These genes are heterogeneous structurally and functionally, but belong to two main families (*myc* and *ras*). There are two main types of activation: translocation in the neighborhood of sequences that "deregulate" the *c-myc* gene (Leder *et al.*, 1983) and alteration by mutational events of *c-ras* oncogenes. But as pointed out by Land *et al.* (1983b), there are at least five possible mechanisms for cellular oncogene activation: (1) over expression by acquisition of a new promoter; (2) amplification of a proto-oncogene; (3) action of viral enhancer sequences that increase the utilization of the transcriptional promoters to which they bind; (4) translocation of *c-myc* to the *Ig* locus; and (5) punctual mutations in the *ras* gene family. Activation of an oncogene is only one of the steps in a multistep process. Cooperation of two distinct genes, cellular and viral (for instance, the transforming gene of a DNA virus such as polyoma or SV40), is required for the transformation of a normal cell into a malignant one. Cells of chick lymphoma and Burkitt's human lymphoma have two different oncogenes: *Blym* and *myc; myc* does not transform normal cells unless *Blym* is also present (Leder *et al.*, 1983). Land *et al.* (1983b) further point out that there is a *ras–myc* synergism; the two genes achieve what neither *ras* nor *myc* alone can do. At the cellular level, *myc* and *ras* have different targets. The product of *myc* is located in the nucleus, that of *ras* on the inner surface of the cell membrane. The effect of the nucleus-targeted oncogene products (like the *myc* protein) would be to immortalize the cells; true malignant transformation would be due to the membrane-bound products of the *ras, src,* and other oncogenes. It would not be surprising to discover that each step of the multistep process leading to cancer is carried by a different activated oncogene; if so, cancer would result from the sequential activation of an array of oncogenes. According to Temin (1984), multiple genetic changes, spontaneous or induced by carcinogens, could explain multistep carcinogenesis. These changes would, as we have seen, be base pair changes, deletions, translocations, and amplification. A similar view is held by Caccia *et al.* (1984), while Weinstein *et al.* (1984b) propose that multistep carcinogenesis involves multiple genes and multiple mechanisms: Reaction of a carcinogen with DNA [a chemical carcinogen activates *in vitro* the c-Ha-*ras*-1 proto-oncogene, according to Marshall *et al.*, (1984)], pleiotropic effects of the cancer promoters due to stimulation of protein kinase C, activation of cellular oncogenes, and possibly changes in DNA methylation. This broad view should be acceptable to those who are reluctant to accept theories based solely on oncogene activation as an explanation for multistep carcinogenesis.

There is no doubt that somatic mutations play an important role in the activation of the cellular oncogenes of the *ras* family and that point mutations in these genes may lead to malignancy. The results is the production of a p21 protein (a membrane-bound protein that binds GTP) distinct from the normal one (Santos *et al.*, 1984; Feig *et al.*, 1984). It is important to note that chemical carcinogens, which are genotoxic agents, may activate oncogenes of the *ras* family (Parada and Weinberg, 1983; Sukumar *et al.*, 1983) by inducing single base mutations. Riggs and Jones (1983) have pointed out that chemical carcinogens inhibit DNA methylases and that aberrant cell division increases the chances of altered DNA methylation. Such changes in DNA methylation ''can masquerade as mutations.'' Olsson and Forchhammer (1984) found that 5-azacytidine, which inhibits DNA methylation, induces the metastatic phenotype in tumorigenic, but not metastatic, cell lines of Lewis lung carcinoma; simultaneously, a new antigen, probably related to metastatic potentialities, appears.

The work of Rabbitts *et al.* (1983, 1984) on the *c-myc* gene shows that one should be careful in opposing oncogene activation by translocation and mutation. They found that when the *c-myc* gene is translocated in Burkitt's lymphomas, no less than 25 base changes (leading to 16 codon alterations) may take place. Somatic mutations occur during and after translocation. These mutations are found in a noncoding exon (exon 1); in the Raji Burkitt's lymphoma cells, both the normal allele and the *c-myc* altered gene are transcribed. It was concluded (Rabbitts *et al.*, 1984) that activation of the *c-myc* gene in Burkitt's lymphoma results from the disruption of a normal transcriptional control mechanism in which the *c-myc* nuclear protein itself could be involved.

An important question for cell biologists still remains almost unanswered. How do the proteins encoded by the oncogenes function? It is now generally believed that the nuclear *myc*-like proteins induce the transcription of a set of genes necessary for proliferation; this would lead to immortalization, but not to tumorigenesis (Bernard *et al.*, 1983; reviewed by Weinberg, 1984). The membrane-bound *ras* protein would be responsible for the acquisition of the transformed phenotype. Craig and Bloch (1984) have reported, in agreement with this concept, that in human myeloblastic leukemia, the *c-myb* oncogene is expressed during the proliferation phase; it is no longer expressed when DNA synthesis stops and cell differentiation begins. Interestingly, the expression of *c-myc* is controlled by the cell cycle in normal, but not in transformed, cells; in contrast, the expression of *c-ras* is cell cycle controlled in both kinds of cells (Campisi *et al.*, 1984).

It is generally believed that the cellular *myc* and *ras* proto-oncogenes play some role in cell proliferation and/or cell differentiation. Evidence is accumulating in favor of a role for one of these genes (*myc*) in the control of cell proliferation. For instance, antibodies against the human *c-myc* protein precipitate a 68,000-dalton protein which has been strongly conserved during evolution; it is present in frogs as well as in mammals. This protein is strongly induced when resting T-lymphocytes are stimulated to divide by treatment with mitogens (Persson *et al.*, 1984). It is localized in the nucleus where it binds to double-stranded and single-stranded DNA (Persson and Leder, 1984). However the products of both *v-myc* and *c-myc* are absent from the nucleoli and metaphase chromosomes (Wingqvist *et al.*, 1984). During mitosis, they are diffuse in the cytoplasm; they reappear in chromatin at telophase. It was concluded that the *myc* proteins do not bind only to DNA and chromatin, but also to other nuclear structures such as the nuclear matrix (Eisenman *et al.*, 1985). It has been reported that *c-myc* mRNA is 30 times more abundant in plasmacytomas than in quiescent B-cells; but stimulation of cell division in the latter increases so much the expression of *c-myc* that they contain almost as many *c-myc* transcripts as the malignant cells. Goyette *et al.* (1984) reported that the transcripts of the *c-myc* and *c-ras* genes increase during rat liver regeneration. However, Stewart *et al.* (1984b) found a very low level of *c-myc* transcripts in spermatogenic cells. Cell proliferation is thus possible in the absence of *c-myc* transcription. Feramisco *et al.* (1984) used a more direct approach: they injected the *Ha-ras* p21 protein into cells from different cell lines and observed fast proliferation of the injected cells. Injection of p21 produced by the cellular *c-Ha-ras* gene had little or no effect on cell multiplication. Similar experiments have been done by Stacey and Kung (1984), who concluded that the p21 *ras* protein produced by the normal cellular genes has some transforming activity, but less than the mutated *Ha-ras* p21. We have seen that several cellular oncogenes are expressed in mouse embryos and human placenta, where there is a correlation between *c-myc* transcription and cell multiplication. There is thus a strong case for the involvment of the *c-myc* and possibly *c-ras* proto-oncogenes in cell proliferation. To date there is no evidence for a role of these genes in cell differentiation, but no work has been done so far on transcription of the cellular oncogenes during early embryonic development.

It would be premature to conclude that the problem of the origin of cancer is solved. As pointed out by Logan and Cairns (1982) and by Reddy *et al.* (1982), one should never forget that cancer is a multistep process. Whether the single base substitution found in the human bladder carcinoma oncogene occurs early or late in this process is still unknown. Activation of an oncogene, whether by activation of a cellular gene, by mutation, or by viral infection, is only one of the steps that lead to cancer. Logan and Cairns (1982) point out, as a further caveat, that the majority of the spontaneous tumors in humans and the cancers induced by x rays and chemical carcinogens do not yield oncogenic DNA upon transfection. Their skepticism is shared by Dunsberg (1985).

As one can see, the progress made in our understanding of cancer using molecular biological methodology is amazing. Yet Howard Temin (the discoverer, with D. Baltimore, of reverse transcriptases) wrote in 1983: "We still don't understand cancer, which remains the final result of a multistep process." Embryologists have no better understanding of another multistep (or multistage) process, embryonic development. There is no doubt that the remarkable progress achieved in cancer research will have an important impact on embryology. It will not be long before embryologists look for *myc* and *ras* cellular genes and their products in developing embryos. Such studies should increase our understanding of the role played by the *c-myc* and *c-ras* products in cell proliferation and differentiation.

The discovery of cellular and viral oncogenes may not be particularly encouraging for the layman who is hoping for a prompt and miraculous cure for cancer. If oncogene activation by translocation, mutation, or both is, as is likely, an essential step in tumorigenesis, only a therapy specifically affecting the structure or activity of these genes could be effective. Today no such therapy is available, but there is no reason to despair. Molecular biologists like Mulligan and Baltimore are already thinking about the possibilities of gene therapy with recombinant retroviruses (Kolata, 1984). A first attempt in that direction has given encouraging results: Williams *et al.* (1984) succeeded in transferring, with a retroviral vector, the bacterial gene of neomycin resistance into cultured pluripotent stem cells. Another approach would be the utilization of anti-oncogene (the suppressor genes of all hybrids) products. The study of these anti-oncogenes has so far been neglected (Klein and Klein, 1985).

G. Tyrosine-Specific Protein Kinases[6]

DNA sequencing is, of course, crucial for our understanding of the organization of genes implicated in such important biological processes as cell proliferation, cell differentiation, and malignant transformation. However, a perhaps even more important question for cell biologists concerns gene expression. How can the product of a single gene lead to profound morphological and biochemical transformations of normal cells?

Partial answers to this question have come, once more, from studies on the *src* gene of the Rous sarcoma virus. It has been found that the product of this gene is a 60,000-dalton protein called p60[src]. This protein possesses or is closely associated with protein kinase activity. The protein p60[src] is a substrate for protein kinases that phosphorylate one tyrosine and one serine residue, resulting in the

[6]Reviewed by Hunter (1984).

phosphoprotein pp60^{src}, which is the final product of the *src* gene responsible for neoplastic transformation in cells infected with RSV (Opperman *et al.*, 1981). pp60^{src} has a protein kinase activity that seemed unusual in that it is capable of phosphorylating tyrosine residues. Classic protein kinases, whether their activity is stimulated by cAMP or not, phosphorylate serine and threonine residues. Thus, pp60^{src} is a representative of a class of enzymes known as *tyrosine-specific protein kinases.* As shown by Parker *et al.* (1981), the cellular homolog of the transforming gene of RSV, *c-src,* is transcribed and translated into a pp60^{src} protein that seems to be slightly different from the viral pp60^{src} protein. Both proteins have a tyrosine-specific protein kinase activity and are phosphorylated on tyrosine and serine residues. It has been reported that both normal and malignant cells (carcinomas) possess two different forms of the natural pp60^{c-src} protein, one of which has an M_r of 60,000 and is mainly phosphorylated on tyrosine, while the other (M_r 59,000) contains serine phosphate residues (Shealy and Erikson, 1981). A sequence of 10 amino acids, containing the tyrosyl residue that is phosphorylated by both viral pp60^{v-src} and cellular pp60^{c-src} proteins, has been identified by Smart *et al.* (1981). It seems, however, that different sites are phosphorylated by pp60^{c-src} in living cells *in vitro.*

According to Levinson *et al.* (1981), the pp60^{src} transforming protein is localized in the cell membrane and, in particular, in the gap junctions of infected cells. It is therefore assumed that cell transformation originates from changes (tyrosine phosphorylation) induced in cell membrane proteins by tyrosine-specific protein kinases. However, it has been reported that a 1200 *g* sediment of malignant and transformed cells rich in mitochondria has a high tyrosine-specific protein kinase activity. This activity is also present, but at a very low level, in the 12,000 *g* sediment of normal embryonic cells of chicken and quails, as well as in diploid human fibroblasts (Montagnier *et al.*, 1982).

The pp60^{src} transforming protein induces tyrosine phosphorylation in more than 12 cellular proteins (Nakamura and Weber, 1982). However, according to Amini and Kaji (1983), the main targets of pp60^{v-src} in avian cells are a 36-kd (K) cell membrane phosphoprotein called p36, vinculin (130 kd), and a 50-kd protein. There is a correlation between phosphorylation of the 36K protein by the pp60^{src}-associated kinase activity and tumorigenicity in transformed cells and their revertants (Nawrocki *et al.*, 1984); however, phosphorylation of the 36K protein alone is not sufficient for the induction of the transformed phenotype. According to Nakamura and Weber (1982), phosphorylation of tne 36K protein does not induce loss of fibronectin, loss of density-dependent inhibition of growth, or increase in hexose transport; it correlates best with increased plasminogen activator activity. Like α-spectrin, p36 is located at the internal (cytoplasmic) face of the plasma membrane (Greenberg and Edelman, 1983; Lehto *et al.*, 1983). All this suggests that phosphorylation of the p36 membrane protein plays a role in the acquisition of the transformed phenotype, but that transformation also requires the phosphorylation of other target proteins.

Another main target protein for the pp60[src] protein kinase activity is vinculin, a fact that deserves considerable attention because, as we have seen in Chapter 3, Volume 1, this protein binds the actin filaments to the focal adhesion plaques (Sefton *et al.*, 1981). It was indeed found that the stress fibers disappear in transformed cells, presumably as a result of a change in the conformation of the vinculin molecules after tyrosine phosphorylation (Boschek *et al.*, 1981). Direct proof that pp60[src] acts on the cell membrane–associated cytoskeleton has been provided by Maness and Levy (1983). They injected purified pp60[src] into fibroblasts and observed a rapid disappearance of the actin stress fibers. In the injected cells, vinculin was phosphorylated, but only slightly; there was no phosphorylation of actin, myosin, and tropomyosin. However, Rohrschneider and Rosok (1983), who reinvestigated the localization of pp60[src] in infected cells, concluded that phosphorylation of vinculin is not sufficient to induce the dissolution of the stress fibers in adhesion plaques and that it is independent of fibronectin disappearance; however, vinculin phosphorylation is correlated with growth in soft agar. The protein pp60[src] is partly localized in the adhesion plaques, and its presence there is correlated with loss of fibronectin rather than with growth in soft agar. Beemon *et al.* (1982) found that many proteins are phosphorylated on tyrosine residues in noninfected embryonic fibroblasts, and that vinculin is not phosphorylated in cells transformed by viruses other than RSV. It is likely that tyrosine phosphorylation of several proteins, and not of vinculin alone, is responsible for the changes in cytoskeleton organization observed by Ball and Singer (1981). In RSV-transformed cells, pp 60[src] profoundly modifies the links between the MTs and the vimentin intermediate filaments (IFs). While MTs and IFs display the same pattern in normal cells, they fail to do so in transformed cells. There is now evidence, based on the use of anti-MT drugs such as nocodazole and taxol and of antibodies against tubulin, that the distribution of the IFs is determined by the MTs (De Brabander *et al.*, 1982; Green and Goldman, 1982; Blose *et al.*, 1982, reviewed by Schliwa *et al.*, 1982).

Although it is generally accepted that tyrosine phosphorylation of several proteins by pp60[src] plays an important role in malignant transformation by Rous sarcoma virus, a number of puzzling issues remain. For instance, it is clear that pp60[src] production is not specific to this virus. Reddy *et al.* (1983) found that the amino acid sequences of the various tyrosine-specific protein kinases produced by a number of different oncogenes (*src, abl, fes, fps, yes*) are very similar; this suggests that all of these transforming genes derive from different members of the same oncogene family. More surprising was a report by Schartl and Barnekow (1982) that the cellular *c-src* oncogene is probably ubiquitous. A pp60[c-src] activity is detectable, by immunological methods, in all the multicellular animals tested. According to Papkoff *et al.* (1983), the *v-mos* oncogene of Moloney murine sarcoma virus encodes a p37[mos] protein that is soluble and has no protein kinase activity, unlike other retroviral transforming proteins. Finally, Cross and Hanafusa (1983) submitted Rous sarcoma virus to local mutagenesis and obtained strains in which pp60^{src} was underphosphorylated on tyrosine or serine residues. After transfection of the mutated virus in cultured chicken cells, mutants deprived of the major sites of tyrosine or serine phosphorylation were isolated and found to be infectious. As one can see, the significance of tyrosine phosphorylation in malignant transformation remains nuclear.

Recent discoveries may shed new light on the subject. Sugimoto et al. (1984) found that the viral src product (pp60$^{v\text{-}src}$) phosphorylates phosphatidylinositol and diacyl-glycerol in addition to tyrosine residues. This was confirmed by Macara et al. (1984) for the transforming protein of another avian sarcoma virus (p68v$^{\text{-ros}}$). This protein has two kinase activities: tyrosine kinase and phosphatidylinositol kinase. The latter produces phosphatidylinositol-bis-phosphate; it can give rise to diacylglycerol, which activates the Ca^{2+}, phospholipid-dependent protein kinase C. We have seen that this kinase is activated by all tumor promoters, including, in addition to the phorbol esters, teleocidin and mezerein (Couturier et al., 1984; Miyake et al., 1984). All this strongly suggests that activation of protein kinase C is an important signal for the induction of mitotic activity. A recent paper by Mroczkowski et al. (1984) shows, for the first time, how this signal, which is received at the level of the cell membrane, might be transduced to the cell nucleus and induce DNA replication. The authors found that a DNA-nicking activity is associated with the tyrosine kinases; in particular, pp60^{src} interacts with supercoiled DNA. Translocation of membrane-associated proteins to the nucleus, where they may interact with DNA, is a possibility; interaction with different regions of the DNA molecules might explain the pleiotropic responses that accompany growth stimulation. It has recently been found that calmodulin and other Ca^{2+}-binding proteins inhibit protein kinase C. Calmodulin thus might play a central role in controlling the respective activities of Ca^{2+}-calmodulin protein kinases and protein kinase C. This could lead to the phosphorylation of one or another set of target proteins (Albert et al., 1984). No clear picture has yet emerged from these recent findings, which emphasize the great complexity of a problem of fundamental importance (the control of cell proliferation); future work will certainly lead to important advances in this field.

We have seen that many cells, if not all, possess the cellular homolog c-src of the viral v-src oncogene. What is its role in normal cells? The two genes exert different effects in transfection experiments. Only v-src has transforming activity, indicating that qualitative differences exist between pp60$^{v\text{-}src}$ and pp60$^{c\text{-}src}$ (R. C. Parker et al., 1984). Is the pp60$^{c\text{-}src}$ protein involved in cell proliferation or cell differentiation? We have no clear answer to that question. Barnekow and Bauer (1984) found no correlation between pp60$^{c\text{-}src}$ activity and cell proliferation. Sorge et al. (1984) made the interesting discovery that, in the neural retina, pp60$^{c\text{-}src}$ is the product of a developmentally regulated gene that seems to be more important for neuronal differentiation than for cell proliferation. Also of interest is the finding by Falcone et al. (1984) that infection of chick myoblasts by Rous sarcoma virus may induce either cell division and transformation or arrest of DNA replication and differentiation; the choice between these two alternatives probably depends on the protein kinase activity of pp60^{src}. It would be interesting to study the activity of the *c-src* gene in well-known embryological systems (development of *Drosophila,* sea urchin, amphibian, and mouse eggs). All we know about the early stages of development is that tyrosine phosphorylation activity increases two to four times a few minutes after sea urchin fertilization; the enzyme, as well as its substrates, is membrane bound. Several plasma membrane proteins are tyrosine phosphorylated in fertilized eggs (Ribot *et al.,* 1984). It is not known whether the sea urchin tyrosine kinase is the product of a *c-src*-like gene, but the findings speak for a role of tyrosine phosphorylation of plasma membrane proteins in the induction of DNA replication and cleavage.

We have already discussed the effects of the phorbol ester cancer promoters (TPA, in particular) on cells. It is worth mentioning that, according to a report by Pietropaolo *et al.* (1981), TPA treatment increases the protein kinase activity of a cellular *sarc* gene in normal avian fibroblasts three- to eightfold; however, this stimulation does not reach the high level of pp60^{src} kinase activity found in RSV-transformed cells.

Finally, one last question is whether tyrosine protein phosphorylation occurs by a kinase specific to malignant transformation. The answer is ''no,'' since it is now clear that many, if not all, growth factors induce tyrosine phosphorylation in cell membrane proteins and that this phosphorylation is mediated, as in RSV-transformed cells, by tyrosine-specific protein kinases.

Among the many growth factors discussed in Chapter 5, Volume 1, were EGF (reviewed by Carpenter and Cohen, 1979; Das, 1982), PDGF and insulin. Like pp60^{src}, EGF exerts pleiotropic effects on fibroblasts; an increase in the uptake of glucose, amino acids, nucleosides, and sodium ion fluxes is followed by a stimulation of RNA and DNA synthesis. The transforming factors of Todaro are EGF-like. An important analogy between EGF and pp60^{src} is that membrane proteins are phosphorylated on tyrosine residues when EGF binds to its membrane receptor, as first shown by Ushiro and Cohen (1980). This receptor is a 170,000-dalton glycoprotein and it is believed that this molecule possesses two distinct domains: one for EGF binding and the other for EGF stimulation of a tyrosine-specific protein kinase activity (Cohen *et al.,* 1982). There are even greater similarities between pp60^{src} and EGF, as was pointed out by Houslay (1981) and Das (1982). According to Erikson *et al.* (1981), EGF and the viral transforming gene product pp60^{src} phosphorylate the same proteins (via specific protein kinases). In addition, Harrison and Auersperg (1981) have reported that EGF favors the malignant transformation of ovarian granulosa cells by the Kirsten murine sarcoma virus. It appears that the product of the viral *src* gene and EGF have a synergistic effect, probably mediated by tyrosine-specific protein kinases. Tyrosine phosphorylation in human cancer cells is a very rapid process. A number of proteins, including the EGF membrane receptors, are phosphorylated within 1 min after EGF addition. This burst of protein phosphorylation is not induced by addition of insulin or the fibroblast growth factor (FGF) (Hunter and Cooper, 1981). In fibroblasts, the burst of protein phosphorylation induced by EGF addition seems to be somewhat slower, becoming detectable after 2 min and reaching a maximum within 5 min; this burst is due to tyrosine phosphorylation of several proteins (J. A. Cooper *et al.*, 1982).

Recent work has demonstrated that the transforming protein of the *erb-B* oncogene of avian erythroblastosis virus (AEV) is very similar to the human EGF receptor (Downward *et al.*, 1984); it appears that the virus once acquired a cellular gene coding for a truncated EGF receptor. In the *erb-B* protein, the EGF-binding domain of the receptor is missing, but both the transmembrane domain and the domain involved in stimulating cell proliferation are present. Merlino *et al.* (1984) think that the gene coding for the EGF receptor in normal cells is the cellular *c-erb B* gene; they point out that, in a human epidermoid carcinoma, there is an excess of EGF receptors and that the genes coding for them are amplified 30 times. Similar results have been reported by Ullrich *et al.* (1984). The amino acid sequence deduced for the EGF receptor from the corresponding cDNA sequence is very similar to that predicted for the product of the *v-Erb B* oncogene mRNA. In A431 carcinoma cells, the EGF receptor gene is amplified and probably rearranged; in these cells, it generates a truncated mRNA that codes for the extracellular EGF-binding domain. The A431 carcinoma cells overproduce a variety of RNAs that are homologous to the cDNA of the EGF receptor; it is probable that differential RNA processing generates this variety of RNAs (Xu *et al.*, 1984). Lin *et al.* (1984) succeeded in the expression cloning of the human gene for the EGF receptor and found a strong homology of the 3'-coding domain with the product of the viral *erb-B* gene. There is a more-than-100-fold amplification of the EGF receptor genes in the A431 malignant cell line characterized by an overproduction of EGF receptors.

The avian erythroblastosis virus possesses, in addition to *erb-B,* an *erb-A* oncogene. While *erb-B* displays homology with part of the EGF receptor and the *src* gene of Rous sarcoma virus, *erb-A* is a "potentiator" which blocks erythroid differentiation at an immature stage, allowing transformation by the product of the *erb-B* gene. The protein encoded by the *erb-A* gene bears similarities with carbonic anhydrase, a classical enzyme of red blood cells (Debuire et al., 1984).

The effects of the platelet-derived growth factor (PDGF) seem to be similar to those of EGF. This factor also stimulates a tyrosine-specific protein kinase, but the substrates are different for the two growth factors (Ek and Heldin, 1982). The kinetics of protein phosphorylation by PDGF and EGF are the same (J. A. Cooper *et al.*, 1982). PDGF had a pleasant surprise to molecular biologists working in the field of cancer (reviewed by Stiles, 1983). As shown by Doolittle *et al.* (1983), there is a remarkable similarity between PDGF and the transforming protein p28[sis], which is produced by the primate sarcoma virus oncogene *v-sis*. This demonstrates that oncogenes may be related to growth factor genes. Similar results have been reported by Waterfield *et al.* (1983), Deuel *et al.* (1983), and Robbins *et al.* (1983). In addition to the confirmation of the almost complete identity in amino acid sequence between PDGF and p28[sis], it was found that cells transformed by the simian sarcoma virus, in contrast to normal cells, possess PDGF activity; it was also reported that p28[sis] undergoes rapid processing steps that yield molecules very similar to active PDGF. The *c-sis* gene homolog of the viral *v-sis* oncogene encodes the B chain of PDGF, as was later shown by Josephs *et al.* (1984), Johnsson *et al.* (1984), and Chiu *et al.* (1984). All of these studies, which establish an important link between cell biology and molecular biology, strongly suggest that there is a subversion of normal growth regulation in cancer cells. However, it should be pointed out that according to Niman *et al.* (1984), many normal and transformed cells express PDGF-like molecules; some of them have high molecular weights. This suggests that PDGF production is not sufficient to induce neoplastic transformation.

Finally, Kasuga *et al.* (1982a) have found that another classic growth factor, insulin, increases the phosphorylation of tyrosine residues present in its own receptor, a finding made possible by the study of hepatoma cells. Since similar results have been obtained with adipocytes and human placenta cells (Petruzelli *et al.*, 1982), it is likely that we are dealing with a general phenomenon: the β-subunit of the receptor (reviewed by Czech, 1982) is phosphorylated after addition of insulin (Kasuga *et al.*, 1982b). The similarities that exist between the insulin receptors and those of the insulin-like growth factors (IGF-1 and IGF-2) have been pointed out by Czech (1982). It is likely that the IGFs and insulin have a similar mode of action in the induction of cell proliferation. A still unanswered question remains. Are the insulin and insulin-like growth factors protein kinases, or are protein kinases associated with the receptors? Further work will certainly provide an answer.

Another perhaps more important question, might be more difficult to answer. Is the phosphorylation of one key protein responsible for the stimulation of DNA synthesis and cell proliferation? Work in this area is only beginning. For instance, Nilsen-Hamilton *et al.* (1982) found that EGF, insulin, and insulin-like growth factors all stimulate the phosphorylation of ribosomal protein S6, but the authors concluded that phosphorylation of this protein is not the main regulatory event in the initiation of DNA synthesis. We discussed this S6 ribosomal protein when we reviewed cell division, maturation in oocytes, and fertilization. We may mention in passing that, although maturation is accompanied by a burst of protein phosphorylation, no tyrosine residues are phosphorylated during this process.

Finally, it has been found that EGF, PDGF, insulin, and the products of the *src* genes all increase tyrosine phosphorylation in a 170,000-dalton protein (Nishimura *et al.*, 1982). The authors conclude that this protein must play an important role in the initiation of DNA synthesis and cell proliferation. Further work on this protein (its distribution in various tissues, intracellular localization, etc.) is, of course, necessary before its real importance can be assessed.

In conclusion, there is little doubt that all growth factors, as well as pp60^{src}, induce the phosphorylation of tyrosine residues in a group of proteins. There is also little doubt that this is one of the events that trigger DNA replication and cell proliferation. However EGF, PDGF, insulin, and insulin-like factors, in contrast to pp60^{src} and similar products of *src* genes, do not induce cell transformation. We do not know why this occurs, but careful comparison between the effects, at the cellular and molecular levels, of the normal and transforming growth factors should lead to a better understanding of malignant growth. In this field, as in so many others, much remains to be done. This should be a comfort to young scientists who believe that there is little left to be accomplished.

III. CELL AGING AND DEATH

A. GENERAL BACKGROUND

Nothing was said about cell aging in "Biochemical Cytology" (Brachet, 1957). This was not due to the fact that the author was 25 years younger and perhaps less interested in senescence, but to the common belief that while cells age and die *in situ*, they are immortal when they are cultured *in vitro*. This belief came from the then classic tissue culture studies of Carrel (1912) and Ebeling (1913) on "the permanent life of tissues outside the organism." That cells (in this particular case, human diploid embryonic fibroblasts) age and die in culture was first shown by Hayflick and Moorhead in 1961. They found that, after about 50 cell generations, the culture stopped growing (even in a rich medium) and death ensued. These cells were thus capable of only a limited number of cell generations. This conclusion of Hayflick and Moorhead (1961) has been confirmed by innumerable studies dealing mainly with the same biological material (human diploid fibroblasts). Carrel and Ebeling probably believed that chick embryo fibroblasts were immortal because young cells were accidentally introduced into the cultures. Another possibility is that during years of culture, these chick fibroblasts underwent malignant transformation. As we saw in the preceding section, transformed cells—in contrast to their normal counterparts—are immortal and can give rise to "established" cell lines. As pointed out once more by Hayflick (1980), cell death is a normal phenomenon, and there are no immortal cell lines except for those of transformed cells.

Aging may be studied either *in vivo* or *in vitro*. *In vivo*, one can study the whole organism (this is the essence of gerontology) or cultures of cells from aged donors. The relationships between aging *in vivo* and *in vitro* (work on embryonic cells cultured *in vitro* for many generations) have been reviewed by Schneider and Smith (1981). It is clear that the two approaches should be correlated in order to understand aging. Goldstein *et al.* (1978) have pointed out that the life span of human fibroblasts often depends more upon the physiological and pathological status of the donor than on his age. If the donor suffers from diabetes, for instance, the life span of a culture of fibroblasts will be shortened. In experiments of this type (culture of cells from old subjects), it is important to have precise information about the medical record of the donor.

When one ages, it becomes increasingly clear that *all* cells become senescent, but not necessarily at the same rate. Small wounds heal slowly, suggesting a decline in mitotic activity; hairs are white because melanogenesis no longer takes place in hair follicles; wrinkles are probably due to abnormalities in collagen synthesis, which are also responsible for the so-called collagen diseases (arthrosis, rheumatism, etc.); gametogenesis stops and endocrine disturbances are frequent; immunity responses to bacterial or viral disease are sluggish, and production of antibodies against normal constituents of the body (autoantibodies) is far from infrequent; even the brain, despite the fact that the majority of our neurons live as long as we do (100 years or more, for some of us), does not work as rapidly as it did at earlier ages. That the rates of senescence are very different for different cell types, even under culture conditions, has been shown by La Rocca and Rafferty (1982), who compared the life spans of embryonic retinal pigment cells, fibroblasts, and chondroblasts.

There is good evidence for the view, first proposed by Hayflick (1974) and reemphasized by Röhme (1981), that there is a direct relationship between cellular life span and species longevity (Fig. 31). For instance, human fibroblasts become senescent and die, as we have seen, after about 50 population doublings; senescence and death occur after only 12 cell generations in mouse embryo fibroblasts cultured under the same conditions. The same difference between these two species is valid for the life span of red blood cells in a normal environment (blood serum). As was already pointed out, the age of the donor is important for studies on cellular aging. For example, Walford *et al.* (1981) found that human lymphocytes cultured *in vitro* were capable of 16.5 generations if the donor was less than 40 years old, but only of 6.5 generations if he was in his eighties. For this reason, most of the work on *in vitro* cellular aging (which will be the main topic of this section) has been done on cells (usually fibroblasts) taken from embryos.

Two main theories have been proposed to explain the arrest of cell growth, followed by cell death, which characterizes senescence. For stochastic theories, the main factor is the accumulation, due to random events such as somatic mutations, of abnormal molecules. Of these theories, the best known is Orgel's (1963, 1970, 1973) error catastrophe theory. According to this theory, aging results from a decrease in the accuracy of protein synthesis, leading to the production of abnormal, at first useless, and ultimately lethal proteins; successive errors made by the protein-synthesizing machinery finally lead to a catastrophe (death). Errors in both protein synthesis and mutations should have

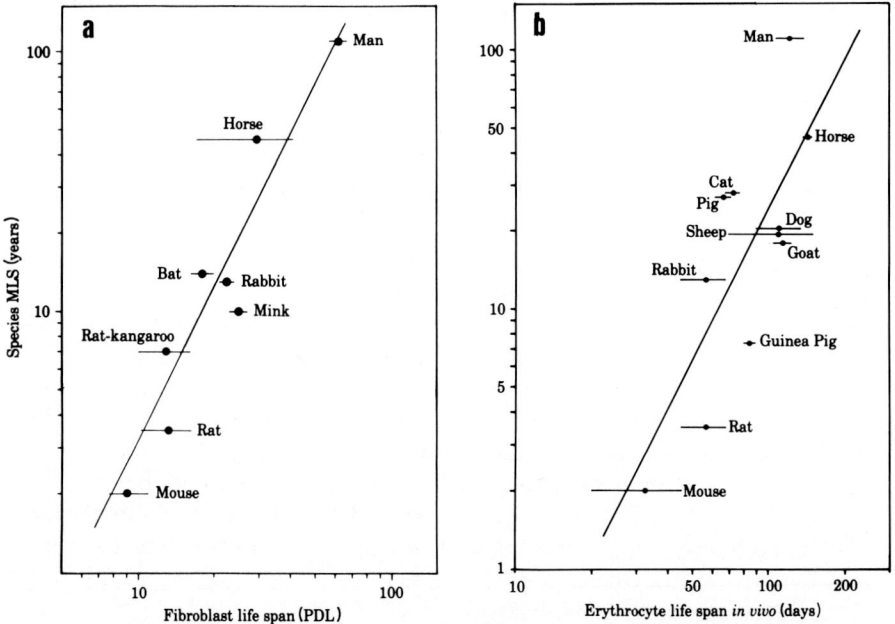

Fig. 31. (a) Linear correlation (log/log) between the maximal life span (MLS) of species and their fibroblast life spans *in vitro*. PDL, population doubling level. (b) Linear correlation (log/log) between MLS of species (in years) and their erythrocyte life span *in vivo* (in days). [Röhme, 1981.]

the same consequences (Holliday and Tarrant, 1972; Orgel, 1973). The opposite (genetically programmed) theories state that aging is the end of a genetic program starting at the moment of fertilization. Senescence is only the last stage of development, a view that would please most embryologists (certain genes would control the onset of senescence). This theory, of course, easily explains that the life span of different animals may vary, according to the species, from a few weeks to more than 100 years, since their DNA sequences differ. We should immediately point out that these two theories are not mutually exclusive.

The following discussion presents some of the facts that favor one or the other of these theories. In our opinion, none of these facts is sufficient to prove that one of the theories is right and the other wrong. In fact, both are probably right. Our life span is probably determined by the genome we received at the time of fertilization; on the other hand, accidents (mutations and others) occur during life that may strongly affect our life span. One of the main experts in the field, Hayflick (1979), concludes that aging results from a genetic instability superimposed on a genetic program; this conclusion seems to be reasonable.

A few words about the theories should be presented before we discuss the existing facts. Those who believe that the main factor in aging is genetic programming often think that there must be a clock or counting mechanism at work during successive generations. For Hayflick (1979, 1984), this clock must be localized in the cell nucleus. This conclusion is based on experiments (discussed later) wherein cytoplasts from young or old cells are fused with whole cells (or karyoplasts) of different ages. According to Shmookler-Reis *et al.* (1980), there is a strict proliferation limit for cells that must be determined by a counting process. We have seen that similar ideas have been expressed by embryologists studying the expression of the enzymes specific for the various germinal localizations in ascidian eggs, and that the concept of "quantal" mitoses is familiar to those who work on cell differentiation. It is not inconceivable that the final arrest of cell proliferation results from a quantal mitosis, and there are good reasons to believe, as we shall see, that acquisition of the senescent phenotype is a form of cell differentiation.

A commitment theory of aging has been proposed by Holliday *et al.* (1977), who have often stated that they prefer Orgel's theory to that of the genetic program. They suggest that potentially immortal cells give rise, after a given number of divisions (again, the clock idea), to cells that are committed to age, just as embryonic cells become committed to differentiate along one or another pathway. For Holliday *et al.* (1977), the death of cultured fibroblasts is not the end of a genetic program; we know of no overwhelming arguments to support this conclusion.

In a review of the various theories of aging, Walton (1982) proposed that aging might be due to a decrease in the replicative activity of stem cells that would be unable to offset cell losses. Finally, Gupta (1980) suggested that *in vitro* aging of human fibroblasts is controlled by one or a few genes and that their expression is modified by mutations. This view, which favors the theory of genetic programming, is based on experiments demonstrating that repeated mutagen treatment does not lead to a life-shortening effect in human fibroblasts. As we shall see, there is some evidence for the view that there are few differences in the patterns of protein synthesis when one compares young and senescent fibroblasts. If so, the situation prevailing in aging cultures of fibroblasts would be reminiscent of what we found when we examined the role played by oncogenes in malignant transformation.

Other theories of aging emphasize biochemical factors that might be responsible for senescence and death. Hart and Setlow (1974) have pointed out that there is a close correlation in a number of mammalian species between DNA excision-repair capacities and life spans. Many experiments have been done in order to test the idea that DNA repair is deficient in old cells, but their results are not conclusive. Another theory has been proposed by Harman (1981, 1984): the free radical theory of aging. This theory states that senescence results from damage to molecules due to free radical production. This is a version of the theory proposed by Orgel (1973), but is more specific in that it pinpoints free radicals as the agents of molecular damage. It is clear that we do not lack ideas and theories about aging *in vivo* and *in vitro*.

B. Characteristics of Cell Culture Aging[7]

There is general agreement on the facts that senescent cells are characterized by their large size and progressive loss of replication activity. As shown by Bowman *et al.* (1975), there is a strong correlation between these two parameters of cells aging *in vitro*. However, it should be pointed out that large size and decreased proliferation activity cannot be taken as specific markers of a senescent cell typical phenotype, since the same characteristics are found, for instance, in irradiated cells and in cells treated with aphidicolin. According to an analysis by Macieira-Coelho and Azzarone (1982), the percentage of rapidly and slowly dividing cells remains constant during the entire life span of the cultures, but that there is a progressive and then abrupt decline in the rate of entrance into DNA synthesis. Autoradiography after [^{3}H]thymidine incorporation has disclosed the existence of three phases in the life of the cultures. First, only 0–2% of the cells are unlabeled; later, there are 5–14% unlabeled cells. Finally, the percentage of unlabeled cells increases abruptly.

It seems certain that, in senescent cells, the length of the entire cell cycle is lengthened and that DNA replication (S phase) is not specifically affected. According to a study by Hasegawa *et al.* (1982, 1985), there are no marked changes in the rate of DNA chain elongation when one compares cultures of young and old human diploid fibroblasts; there is no increase in the minimum time required to reverse the S phase in senescent fibroblasts (Griffiths, 1984). From an earlier analysis of the cell cycle, Grove and Cristofalo (1977) concluded that the slowdown of growth in senescent cells is mainly due to a lengthening of the G_2 phase of the cell cycle. In cultured embryonic diploid mouse fibroblasts, 50% of the young cells are in the S phase; at the end of the culture, only 20% of the cells are undergoing DNA replication. Thus, 60% of the cells are in G_1 and only 20% are in G_2. This is in contrast to the results of Grove and Cristofalo (1977) on human diploid fibroblasts (Preumont *et al.*, 1983). The frequent decrease in [^{3}H]thymidine incorporation into DNA of old fibroblasts might be due to a decrease in EGF binding. Although processing and degradation of EGF remain normal during *in vitro* aging (Phillips *et al.*, 1983), the tyrosine kinase activity of the EGF receptors decreases in senescent fibroblasts; the autophosphorylating activity of the receptors is lost in old cells, which no longer respond to EGF (Carlin *et al.*, 1983). Puvion-Dutilleul *et al.* (1984) have recently confirmed that DNA and RNA synthesis remains essentially normal in senescent fibroblasts; however, they found that these cells contain low molecular weight DNA in addition to bulk DNA, and concluded that chromatin is more fragile in old than

[7]Reviewed by Reff and Schneider (1981), Macieira-Coelho (1983), and Smith and Lincoln (1984).

in young cells. Collins and Chu (1985) also think that the decrease in the rate of DNA synthesis in old cultures is due to alterations in the structure of the DNA template rather than to diminished DNA polymerase activity.

The proportion of giant cells increases during aging, both *in vivo* and *in vitro;* this is not necessarily due to polyploidy, although aneuploidy and polyploidy are not infrequent in late-passage cells (20% in the mouse embryonic fibroblasts studied by Van Gansen). In fact, Kaji and Matsuo (1981) found that polyploidy may reach a value as high as 128 C during aging of mouse fibroblasts, and they used this characteristic in an attempt to enrich old cells using the method of flow cytophotometry. The frequency of chromosomal abnormalities, leading to aneuploidy, also increases during *in vitro* aging (Sashela and Moorhead, 1963). According to Thompson and Holliday (1975), this could be a consequence rather than a cause of the aging processes; however, it is not always easy to distinguish unambiguously between cause and effect.

As we have seen, large size and slow growth of the cells are not specific markers of what one may call, with Van Gansen, the "terminal phenotype" of cells grown in culture during many generations. According to Vogel *et al.* (1981), in addition to a reduction in proliferation, late-passage fibroblasts show a decrease in the production of fibronectin, in gap junction formation, and in secretion; the synthesis of glycosaminoglycans, monitored by following the incorporation of radioactive sulfate into the cell surface components, remains normal but qualitative changes in the membrane glycoproteins have been observed by Blondal *et al.* (1985). Chandrasekar *et al.* (1983) have reported that in aged fibroblasts, fibronectin is altered in such a way that it does not bind efficiently to collagens I and II; this might explain why the adhesion of old fibroblasts to the substratum is decreased. Edick and Millis (1984) noticed that in early cultures of human fibroblasts, fibronectin is oriented parallel to the long axis of spread cells; in aged cultures, it forms networks of randomly oriented fibers. This might explain the well-known decreased motility of aged fibroblasts (Absher and Absher, 1976; Albrecht-Buehler, 1976).

Terminal mouse fibroblasts are being studied in detail by Van Gansen *et al.* (1977, 1979, 1984; Van Gansen, 1983), using electron microscopy and immunocytochemistry. As shown in Fig. 32, terminal cells are much larger than young fibroblasts, but they are also flatter, and their projected surface is about five times larger than that of early-passage cells. The number of filopodia increases progressively during *in vitro* aging and, in some cells, microvilli are very numerous. It is estimated that the cell surface, including filopodia and microvilli, has increased between 100 and 1000 times. At the end, blebs appear on the cell surface, suggesting that the cortical MF system is becoming disorganized; these blebs finally burst, and the cell dies. That the cortical MF system (cortical ring of MFs) is damaged in senescent cells is confirmed by the fact that toward the end of the period of culture, about 10% of the population is composed of binucleate cells. Immunocytochemistry shows that the cytoskeleton organization is pro-

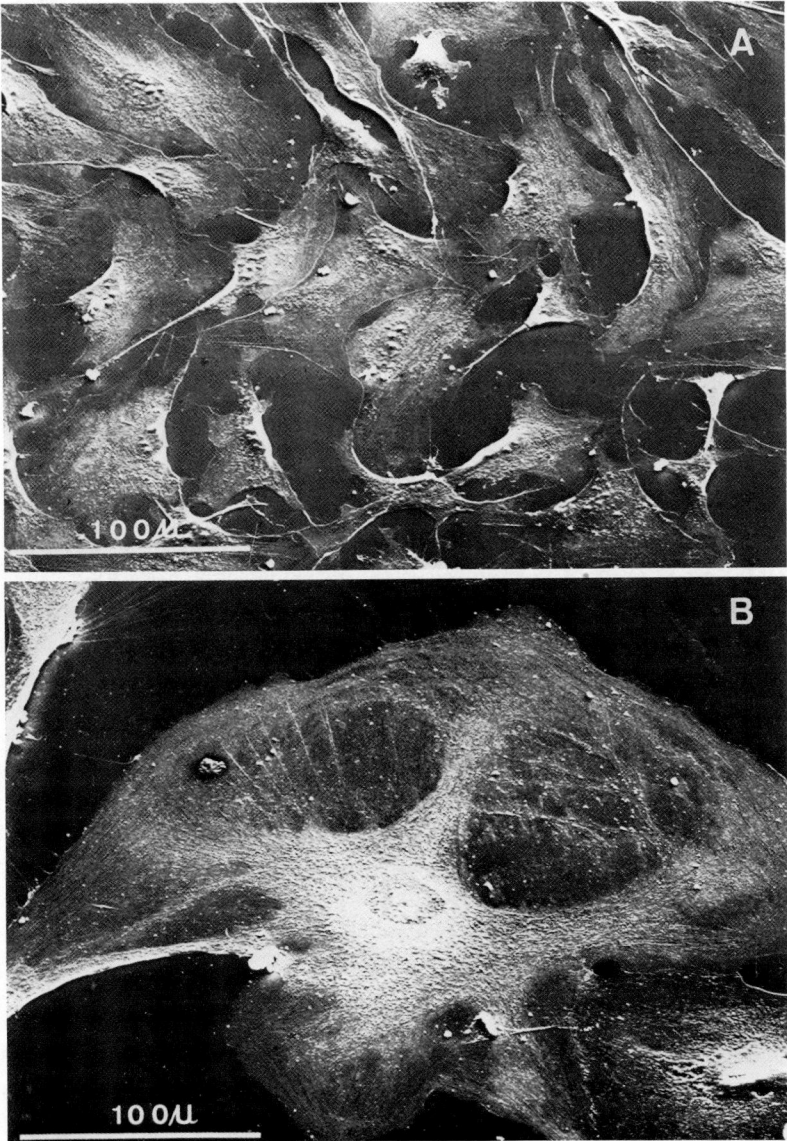

FIG. 32.   Scanning electron micrographs of murine fibroblasts derived from the same embryonic primary culture. (A) Early cells at the third population doubling level (PDL). (B) Late cell at the ninth PDL (subterminal). Same magnification in (A) and (B). [Van Gansen, 1983.]

foundly changed in *in vitro* aged mouse fibroblasts. Figure 33 shows that the cytoplasm of the terminal cells is filled with cables of actin filaments. The number of MTs also increases considerably, combined with a loss of the initial parallel orientation of the MTs, as shown by electron microscopy. Finally, the percentage of cells in which IFs can be detected by desmin-immunocytochemistry increases from 10 to 20% at the beginning of the culture to almost 100% toward its termination. The ribosomes remain very numerous during the entire life of *in vitro* cultured mouse fibroblasts, but the ER vesicles become much less numerous, and appear flattened in old cells. This suggests that their secretory activity is reduced, which is in agreement with the findings of Vogel *et al.* (1981). It is indeed possible that old fibroblasts no longer secrete collagens, since collagen fibers, which are frequent in young cultures, can no longer be seen under the electron microscope in old cultures. Wang and Gundersen (1984) have confirmed, for senescent human fibroblasts, Van Gansen's findings on mouse fibroblasts, observing increases in actin, microfilaments and intermediate filaments in large flat senescent fibroblasts. Wang (1985) reported that aged human fibroblasts contain large bundles of vimentin IF, while Kelley *et al.* (1985) found a decrease in the actin-binding protein filament. This might explain the decrease in locomotion and the cytoskeletal alterations found in senescent fibroblasts.

   A few isolated studies corroborate the findings made on mouse terminal fibroblasts. For instance, using chemical methods, Anderson (1978) found that the actin content increases in senescent fibroblasts. Naeim *et al.* (1981) reported that the lymphocytes of old people have an excess of MTs. Selkoe *et al.* (1982) discovered that, in old neurons, the neurofilaments (a specialized type of IF) are transformed into bundles of insoluble filaments due to a crosslinking of the neurofilaments by transglutaminase; this process is inhibited by cystamine, an inhibitor of this enzyme. It is, of course, too early to know whether the interesting findings made by Van Gansen and her colleagues can be generalized, but the fact that similar observations have been made on lymphocytes and neurons suggests that this may be so. However, a word of caution is needed here. The terminal phenotype of mouse fibroblasts in culture is never found in fibroblasts present in the skin of very old mice (personal communication of P. Van Gansen and N. Van Lerberghe). Fibroblasts that have aged *in situ* are surrounded by a thick layer of collagen fibers, in contrast to senescent cultured fibroblasts. It is possible that mouse skin fibroblasts that have been forced to divide repeatedly under artificial culture conditions age more quickly than fibroblasts that remained in the skin. This, as well as the presence or absence of surrounding collagen fibers, might explain why they express completely different phenotypes.

   In contrast to the marked changes observed in the cytoskeleton of old fibroblasts, no deep modifications can be seen in their mitochondria by high-voltage electron microscopy (except an accumulation of the senescence pigment lipofuscin) (Goldstein *et al.*, 1984). The number and size of the lysosomes increase

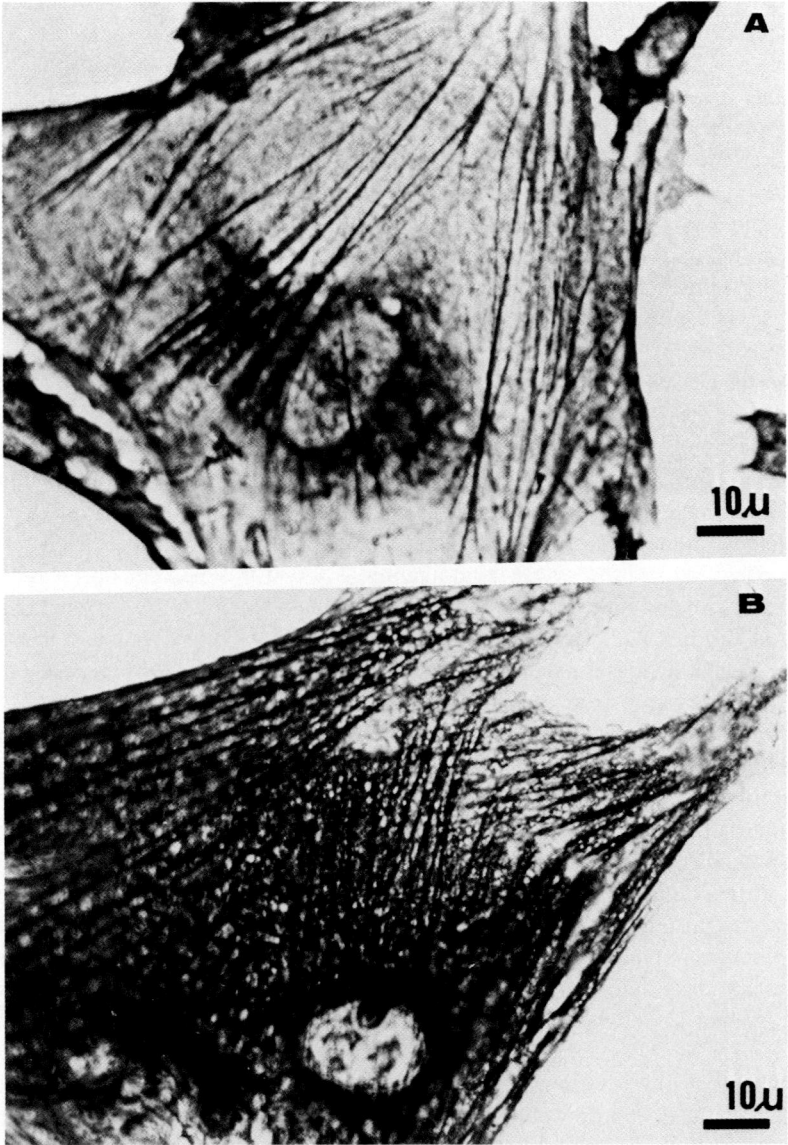

FIG. 33.   Actin MF distribution in mice fibroblasts from a primary culture (visualization by the PAP method after treatment with antiactin antibody). (A) Early fibroblast. (B) Late fibroblast. (A) and (B), same magnification. Cell size is a fluctuating parameter in all the population doubling levels of the culture; the mean values are increasing with age; the fibroblasts shown were selected for similarity in size [(A) is a large early cell and (B) a small late cell]. [Van Gansen, 1983.]

in the livers of aged rats (De Priester *et al.*, 1984). Large increases in the number of the lysosomes and in acid phosphatase activity have been found in many cells, including fibroblasts (Lipetz and Cristofalo, 1972; Brunk *et al.*, 1973; Van Gansen *et al.*, 1979). Lipofuscin accumulates in the lysosomes of old animals, particularly in their brain. It has been recently reported by Ivy *et al.* (1984) that accumulation of lipofuscin in brain lysosomes can be induced in young animals by administration of lysosomotropic agents such as leupeptine and chloroquine.

The changes that occur in the nucleus during *in vitro* aging are less impressive than those that take place in the cytoplasm. The diameter of the nucleus is enlarged, and its membrane becomes more irregular (Evans *et al.*, 1978). The number of nucleoli decreases, and one may observe, as in actinomycin D–treated cells, segregation between their granular and fibrillar components. There are good reasons to believe that the synthesis of the rRNAs is decreased in senescent cells. This conclusion can be confirmed by the fact that the number of nucleolar organizers that can be stained by the silver method (see Chapter 4, Volume 1) decreases in senescent cells. This has been shown in cells from donors who were 80–89 years old (Buys *et al.*, 1979; Denton *et al.*, 1981).

The organization of chromatin is also altered in old fibroblasts (Puvion-Dutilleul and Macieira-Coelho, 1982; Puvion-Dutilleul *et al.*, 1984), but the changes that have been observed seem to differ in human and mouse cultured fibroblasts. We have observed (Preumont *et al.*, 1978) that the chromatin of lymphocytes from very old human subjects binds less ^{3}H-labeled actinomycin D than that of lymphocytes from young ones; similar observations were made on senescent mouse fibroblasts by Preumont *et al.* (1983). As we have seen, ^{3}H-labeled actinomycin D binding to chromatin of fixed cells is a rough index of chromatin activity. The results of Preumont *et al.* suggest that a decrease in replication and transcription activities occurs during aging because the conformation of chromatin has changed. It has indeed been shown that the template activity of chromatin is decreased in senescent cells (Ryan and Cristofalo, 1975) and this would be the reason for the decreased DNA replication in old cells (Collins and Chu, 1985). Gaziev and Malakhova (1982) also concluded that the accessibility of chromatin to repair enzymes is decreased in hepatocytes from old animals. This suggests that the molecular structure of chromatin changes during aging. However, as far as is known, chromatin activity in nuclei from young and old cells has not yet been studied using the methods now available (susceptibility to DNase, methylation of DNA sequences, presence of H1° histone and HMG chromosomal proteins, etc). Such studies should be rewarding regardless of whether the results are positive or negative, provided that they give unambiguous answers. This has not been the case so far for DNA methylation. Working on different materials (human diploid fibroblasts and *Coenorhabditis*), Wilson and Jones (1983) and Klass *et al.* (1983) found, respectively, a decrease and an increase in DNA methylation during aging.

A final question may be asked. Is it possible to increase the life span of cultured cells by addition of hormones or growth factors? According to Rhein-wald and Green (1977), EGF increases the longevity of skin cells by slowing down keratin synthesis. More recently, Didinsky and Rheinwald (1981) found that hydrocortisone, EGF, and fibroblast growth factor (FGF) had no effect on *in vitro* senescence of fibroblasts. However, EGF increases by three times the life span of keratinocytes. There is a strong possibility that EGF will soon be readily available due to the miracles of genetic engineering. The question that remains is whether it will it be as efficient for wound healing as the magical ointments of medieval witches and fairies.

C. ANALYSIS OF AGING BY SOMATIC HYBRIDIZATION

Somatic hybridization is a very useful tool for the analysis of important biolog-ical processes. However, the results obtained using this method can seldom be broadly generalized. This is due to the fact that one cell is not another and that each cell has its own personality.

We are faced with the same difficulties when interpreting experiments of fusion of old and young cells (usually fibroblasts). In 1973, Littlefield concluded that, in hybrids between young and old cells, senescence is dominant; this suggests that old cells have undergone irreversible genetic changes (mutations or other DNA rearrangements). However, this conclusion is no longer universally accepted. Muggleton-Harris and Aroian (1982) conclude that the longevity of senescent fibroblasts is prolonged by fusion with young fibroblasts.

Closer analysis of cell hybrids shows that aging is a very complex phe-nomenon that affects both the nucleus and the cytoplasm. For instance, Rao (1976) found that only the cytoplasm of young cells can reactivate an inactive chick erythrocyte nucleus introduced by cell fusion. According to Yanishevsky and Stein (1980), fusion with an old cell does not arrest DNA synthesis in a young cell, provided that the latter was already in the S phase of the cell cycle; if the young cell was in G_1, entrance into the S phase would be prevented by fusion with an old, no longer dividing cell. These experiments suggest that the cyto-plasm of old cells contains factors that prevent the initiation of DNA replication in young cells. This conclusion emerges even more clearly from the experiments made on cybrids by Muggleton-Harris and Hayflick in 1976. They found that mitotic activity is quickly arrested when cytoplasts from old cells are fused with karyoplasts from young cells.

This early work, as well as experiments in which cytoplasts were fused with cells treated with mono-iodoacetate, led Hayflick (1975, 1984) and Wright and Hayflick (1975) to conclude that *in vitro* aging results from nuclear rather than cytoplasmic factors. This general conclusion is still valid, although the impor-tance of the cytoplasm has been made clearer in the more recent papers. For instance, Zorn *et al.* (1982) found that cultures formed by young cytoplasts and old karyoplasts are viable, but the reconstituted cells do not divide; in contrast, cultures formed by old cytoplasts and young karyoplasts undergo cell division. The conclusion was that cellular senescence is mainly controlled by the nucleus, but that the cytoplasm of an old cell can support the functions of a young nucleus. However, Drescher-Lincoln and Smith (1983) and Burmer et al. (1983) came to different conclusions. They found that fusion of cytoplasts from senescent cells with young fibroblasts delay entry in DNA synthesis in the young cells, and concluded that the cytoplasm of old cells contains factors that prevent the initiation of DNA synthesis. Taken together, these results suggest that the cytoplasm of old cells actually contains factors that prevent or slow down the onset of DNA replication, but that these factors have only a short life in reconstituted cells. Recently, Drescher-Lincoln and Smith (1984) treated senescent cells with cycloheximide and puromycin, then isolated cytoplasts from them. These cytoplasts are unable to inhibit DNA synthesis in senescent or young cybrids. The effects of the inhibitors are reversible by washing; the results suggest that the cytoplasmic inhibitor of DNA synthesis in senescent cells is a protein.

A paper by Nette et al. (1982) describes more complex experiments. They fused human fibroblasts (from a 82-year-old donor) that had aged in vitro during 21–28 cell population doublings with mouse L cells (an immortal cell line of fibroblasts). These experiments were performed not only on whole cells, but also on karyoplasts and cytoplasts from the two cell lines fused under all possible combinations. The authors found an increase in [^{3}H]thymidine incorporation in the old nuclei under all experimental conditions. The general conclusion of these experiments is that the cytoplasm contains factors required for the initiation of DNA synthesis, but that cytoplasts are less efficient than whole cells for reactivating DNA synthesis in senescent cells. Comparable experiments have been performed by Stein et al. (1982), who obtained intriguing results. They fused diploid senescent cells with transformed cells and found that the results differed depending on whether transformation had been obtained by infection with a DNA virus (polyoma, SV40) or treatment with an RNA virus or a chemical carcinogen. In the first case, DNA synthesis occurred in both nuclei of the heterokaryons; in the second, the nucleus of the transformed cell no longer synthesized DNA in the heterokaryons.

A possible explanation for such puzzling results may perhaps be found in experiments by Pendergrass et al. (1982). They discovered that reinitiation of DNA synthesis in a heterokaryon was linked to a threshold in DNA polymerase activity. There was no reinitiation of DNA synthesis in an old cell after fusion with a young cell if the latter had little DNA polymerase activity. Senescent cells decreased DNA polymerase activity in their young partner, with the result that DNA replication was not reinitiated in the heterokaryons. This explanation is satisfying at first sight, but whether it has general value remains to be seen. Studies by Fry et al. (1984) suggest that this might be the case. They found that liver regeneration was slowed in old animals due to a decrease in cell replication; there was a parallel decrease in DNA polymerase α-activity.

In this section, we have been dealing exclusively with mammalian cells. However, cell aging and cell death are universal phenomena except—in the Animal Kingdom—for didermic organisms such as *Hydra,* in which death is only accidental. Even in unicellular organisms, such as the ciliate *Paramecium,* aging and cell death occur after a number of fissions. According to Karino and Hiwatashi (1981), the life span in this organism is 750 fissions. In order to study aging in the nucleus and the cytoplasm, the Japanese workers transferred the micronucleus of aging paramecia into the cytoplasm of young recipient individuals. Micronuclei transferred after 650 fissions into young cytoplasm displayed normal functioning, but the functioning of micronuclei transferred from organisms that had undergone 700–750 fissions was altered, indicating that they had undergone some irreversible damage toward the end of the organism's life span.

Finally, we should recall what has been said in previous chapters in the discussions of *Acetabularia* and *Xenopus* eggs. We saw that transfer of an old *Acetabularia* nucleus into young cytoplasm was followed by its rejuvenation; in this case, the young cytoplasm clearly exerted a positive effect on the old nucleus. We also saw that an adult nucleus injected into an enucleate unfertilized *Xenopus* egg not only divided repeatedly, but even supported development until the tadpole stage; whether a nucleus taken from a very old *Xenopus* would do the same thing is not known. An interesting experiment could perhaps be done on fertilized mouse eggs, in which nuclear transfer experiments have been successful. It might be possible to inject into fertilized, enucleated mouse eggs nuclei isolated from young and "terminal" mouse fibroblasts. The outcome of such experiments might have an important bearing on the still unsolved problem of nuclear and cytoplasmic aging in cells grown under culture conditions.

D. BIOCHEMICAL STUDIES ON AGING

The literature in this field is enormous and too often full of contradictions. This discussion is limited to two theoretically important questions. Is there, as proposed by Orgel (1973), an accumulation of altered proteins, particularly enzymes, in senescent cells? How strong is the evidence for the view, first proposed by Hart and Setlow in 1974, that there is a correlation between efficiency of DNA repair mechanisms, senescence, and life span?

1. Accumulation of Abnormal Enzymes in Senescent Cells and Organisms[8]

A paper published in 1972 by Holliday and Tarrant stimulated considerable interest. It showed that there is an accumulation of abnormal enzymes in senes-

[8]Reviewed by Rothstein (1979) and Gershon (1979).

cent fibroblasts. Later, Zeelon *et al.* (1973) reported that part of aldolase becomes abnormal in old specimens of the nematode *Caenorhabditis elegans;* this finding was confirmed by Reznick and Gershon (1977), who found that aldolase activity in this nematode decreased by 50% during aging. These findings suggested that abnormal enzymes accumulate during aging both *in vitro* and *in vivo.* Holliday (1975) pointed out that they strongly support Orgel's (1963, 1970, 1973) theory of aging and that the evidence cannot be easily explained by the genetic program theory. How can one decide whether an enzyme is normal or not? The main criteria used by Holliday and his associates were a decrease in the overall enzymatic activity; an increase in the thermolability of the enzyme, which indicates a change in the structure of the protein; and the accumulation of an enzymatically inactive material that cross-reacts with antisera against the enzyme [cross-reacting material (CRM)]. Accumulation of CRM during aging shows that senescent cells possess large amounts of a protein that is immunologically related to an enzyme, but is devoid of catalytic activity. In the aforementioned case of *Caenorhabditis,* all three criteria have been met successfully for aldolase (Zeelon *et al.,* 1973; Reznick and Gershon, 1977).

We shall first discuss aldolase and see whether the findings made on a nematode can be generalized; we will then look at other enzymes. Unfortunately, the literature concerning aldolase is not free of contradictory reports. While Barrows and Kokkonen (1981) and Reznick *et al.* (1981) found that aldolase activity is reduced and that the enzyme is altered in the liver and other organs of old mice, Weber *et al.* (1976) found no change in aldolase activity in the livers of senescent mice. Steinhagen-Thiessen and Hilz (1976) observed in muscles of old animals a decrease in the activities of aldolase and creatine kinase, but there was no accumulation of the corresponding CRMs. Similarly, Burrows and Davison (1980) failed to find a decrease in the specific activities of aldolases A and B, and of superoxide dismutase in organs of old dogs and mice; in addition, there was no accumulation of the respective CRMs. On the other hand, Dovrat and Gershon (1983) observed a strong decrease of aldolase activity in the lenses of old animals; they concluded that there is an accumulation of defective enzyme molecules, probably partially denatured, in old lenses.

In view of these contradictory reports, it is impossible to conclude that accumulation of an abnormal aldolase is a specific and universal biochemical marker of senescence. A decrease in enzymatic activity might be due to changes in proteolytic activity in senescent cells and not necessarily to the synthesis of an abnormal protein. Unfortunately, the present data on proteolytic activity in young and old cells are still conflicting. While Shakespeare and Buchanan (1976) reported that senescent fibroblasts degrade their proteins faster (because they are abnormal) than young cells, Bosmann *et al.* (1976) found exactly the opposite. According to them, proteolytic activity almost vanished in senescent fibroblasts. The experiments of Dice (1982), who injected various enzymes into the cytoplasm of young and aged fibroblasts and observed a slowdown of their degradation in senescent cells, seem convincing. The accumulation of inactive cathepsin D molecules in the lysosomes of seven different organs in old rats also indicates a decrease of protein turnover in senescent organisms (Wiederanders and Oelke, 1984). Incidentally, a decrease in protein degradation, if it really exists, might be one of the reasons for the increase in the size of terminal fibroblasts; decreased proteolysis might also allow an accumulation of abnormal proteins, as pointed out by Tollefsbol et al. (1982). However, Kaftory et al. (1978) found that both the synthesis and degradation of proteins increased in senescent fibroblasts. The same conclusion has been drawn by Tollefsbol and Cohen (1985a) who studied lymphocytes from old donors.

Enzymes other than aldolase have also been studied, but less extensively. The results of these experiments generally support the predictions of Orgel's (1973) theory, i.e., enzymatic activity decreases in senescent cells. This has been shown by Gartler et al. (1981) to be the case in lymphocytes. They found that the activity of many enzymes decreased in lymphocytes from old people, and pointed out that there might be a correlation between this decrease and the well-known reduction in the immune defense response that also occurs. Glucose-6-phosphate dehydrogenase has been studied by Fulder and Holliday (1975), who made the interesting finding that, in human fibrolasts, the number of variant forms of this enzyme increases during aging; some of these variants might be altered forms of the enzyme, which is in agreement with Orgel's theory. More recently, Tollefsbol et al. (1982) studied triosephosphate dehydrogenase in cases of progeria and Werner's syndrome, genetic diseases in which the entire symptomatology of old age appears precociously. They found an increased thermolability of the enzyme in these patients, as well as in late-passage human fibroblasts in cell cultures. In the latter, it could even be shown that a labile constituent of the enzyme accumulated. Glycolysis decreases in lymphocytes from old people (Tollefsbol and Cohen, 1985b). As already mentioned, Tollefsbol et al. (1982) suggest that such an accumulation might be the result of a decrease in proteolytic activity, which would allow the accumulation of abnormal proteins. Although these proteins are perhaps also produced by young cells, they would be quickly degraded by intracellular (possibly lysosomal) proteases. We should also mention a paper by Dick and Wright (1982) dealing with ribonucleotide reductase. The activity of this enzyme, which is involved in DNA synthesis, decreases during in vitro aging of fibroblasts. This decrease is believed by the authors to be due to a mutation in the gene coding for the enzyme, since the resistance of the cells to hydroxyurea, a classic inhibitor of ribonucleotide reductase, increases with aging. However, since there is a reduction in the rate of cell proliferation in senescent fibroblasts, it is not altogether surprising that their resistance to hydroxyurea is increased. Finally, Sharma and Rothstein (1980) found that, during aging of the nematode C. elegans, enolase activity decreases, but this is due to a change in the conformation of the enzyme molecule and not to errors in base or amino acid sequences, as was postulated by Orgel. While Leray et al. (1983)

were unable to detect any modification of β-galactosidase during aging, Houben et al. (1984) observed an increase in the thermolability of glucose-6-phosphate dehydrogenase; however, in agreement with Rothstein (1977), they believe that the increased thermolability of this enzyme results from posttranslational changes, and not, as was postulated by Orgel in 1963, from errors in enzyme synthesis. As one can see, a decrease in enzymatic activities during aging, both *in vivo* and *in vitro*, seems to occur frequently. More work with highly purified, well-characterized enzyme preparations is clearly needed before it can be determined whether such a decrease is due to errors in transcription and translation or to other reasons not yet known.

Little is known about changes in enzyme activity during aging in the various cell organelles. Reiss and Sacktor (1983) have reported that senescent animals have an altered membrane maltase in their renal brush borders; the enzyme is antigenically different from that of young animals. On the other hand, 5'-nucleotidase, a classic cell membrane marker enzyme, increases 10 times in thymocytes of aged mice (Menahan and Kemp, 1982). In contrast, the activity of nucleoside triphosphatase (NTPase) decreases in the nuclear membranes of old animals; we have seen that this enzyme is probably involved in the transport of RNA molecules from the nucleus to the cytoplasm (Bernd et al., 1982). The activity of the liver microsomal enzymes also decreases during aging (Devasagayam et al., 1983). According to Hansford (1983), there are no marked changes in energy production by the mitochondria, although their membrane lipids probably undergo peroxidation; mitochondrial DNA remains unchanged during aging (White and Bunn, 1985). In glial cells of old animals, the lysosomes accumulate lipofuscins; their production is modulated by factors that increase or decrease lipid peroxidation (Thaw et al., 1984).

What are the quantitative or qualitative changes in overall protein synthesis during aging? According to Engelhardt et al. (1979), only two proteins out of about 500 polypeptides detectable by two-dimensional gel electrophoresis undergo marked changes (increases) during the aging of fibroblasts. The data, therefore, do not conform to Orgel's theory. However, according to Raes et al. (1981), out of 26 constituents of the plasma membrane, 5 underwent changes during aging. In contrast to the results of Menahan and Kemp (1982) on thymocytes, the activity of the classic plasma membrane marker enzymes, 5'-nucleotidase and alkaline phosphatase, decreased in senescent human diploid cells. In *Drosophila*, there is a decrease in the rate of overall protein synthesis during aging; it is preceded by a decrease in the synthesis of elongation factor EF-1 (Webster and Webster, 1983). During the *in vitro* aging of lens cells, a few changes in the protein pattern take place. In particular, vimentin disappears and protein glycosylation decreases; the changes undergone by the crystallines are not the same during *in vivo* and *in vitro* aging (Ramaekers et al., 1984). According to a recent report of Lincoln et al. (1984), senescent human fibroblasts that have ceased to divide contain two proteins that cannot be detected in young cells.

However, one of them appears when growth of young cells is arrested for a long period of time. The other protein might be specific for aging. But in *C. elegans,* no changes can be detected when the synthesis of 700 proteins is compared in young and old animals (Johnson and McCaffrey, 1985).

Some work has also been done on *in vitro* protein synthesis by extracts of young and old cells. Cook and Buetow (1981) found that, in the liver, the rate of *in vitro* protein synthesis decreased with the age of the donors; this decrease seems to be due to a multiplicity of factors, including senescent polysomes, tRNA synthetases, and tRNAs. Blaszejowski and Webster (1983) studied *in vitro* protein synthesis in extracts from several tissues of rats of different ages; they found a decrease during aging for all of them except cardiac muscle. A decrease in *in vitro* protein synthesis in liver and kidney has also been reported by Gabius *et al.* (1983), who concluded that it is due to decreased binding of the amino-acyl tRNAs to the ribosomes. Nakazawa *et al.* (1984) have confirmed that the activity of *in vitro* translation is decreased by 30–40% in the livers of aged rats but formation of the initiation complex decreases by 15–20% only. Thus, both *in vivo* and *in vitro* experiments show that protein synthesis is decreased in most tissues of old animals; however, the molecular mechanisms responsible for this reduction are not yet clear. One of them might be ribosome alteration, impeding their functioning (Egilmez and Rothstein, 1985).

However, the fidelity of *in vitro* translation, a crucial point in Orgel's theory, does not appear to be decreased. The number of errors made in protein synthesis does not increase during aging (Harley *et al.,* 1980), and the fidelity of poly(U) translation does not decrease when *in vitro* systems from young and old fibroblasts are compared (Wojtyk and Goldstein, 1980). More recently, Mori *et al.* (1983) found that the fidelity of *in vitro* translation of protamine mRNA does not decrease during aging; Mays-Hoopes *et al.* (1983b) reported that the tRNAs function normally in the livers of old rats and concluded that tRNAs are unlikely to produce protein errors. It would be premature to dismiss Orgel's theory on the basis of such limited evidence. However, the problem is an important one, and much more work on a variety of biological systems, on *in vivo* and *in vitro* protein synthesis, is obviously needed before definite conclusions can be drawn.

Errors might also occur at the level of DNA replication and transcription during senescence. According to Linn *et al.* (1976) and Murray and Holliday (1981), the frequency of errors made by DNA polymerases (especially G-T mispairing) increases in senescent fibroblasts. In addition, Murray (1981) reported that, in fibroblasts, the specific activity of DNA polymerase decreases in parallel with the slowdown of growth and with the increase in the frequency of *in vitro* errors. If these findings can be extended to other aging biological systems, a case for Orgel's theory could be made. However, Silber *et al.* (1985) failed to detect changes in the thermolability and fidelity of DNA polymerases α and β during aging.

The data found in the literature on RNA synthesis in young and senescent cells are, unfortunately, full of contradictions. According to Pochron *et al.* (1978) and Whatley and Hill (1979), RNA synthesis increases during aging, so that the large senescent fibroblasts contain excess RNA. Similar results have been obtained by Ono and Cutler (1978) for *in vivo* aging. They found an increase in the diversity of the mRNA population in aging livers and concluded that genes that are not expressed in young cells may be expressed in old ones. If this is the case, new proteins should be detectable in the livers of old animals using two-dimensional gel electrophoresis. In a more recent study by Chatterjee *et al.* (1981), it was reported that the mRNAs of two "senescence marker proteins" disappeared from the liver during the aging of male rats; a third mRNA was found in young and old rats, but not in the adults. Obviously, these data are still too fragmentary to allow conclusions to be drawn. On the other hand, the work of Evans *et al.* (1978) on RNA synthesis and RNA polymerase activity in mouse fibroblasts indicated a decrease in RNA synthesis. The RNA content was about the same in early and terminal mouse fibroblasts in culture, but the rate of RNA synthesis decreased by about 50% in the terminal populations, as shown by both biochemical methods and autoradiography after [^{3}H]uridine incorporation. In young fibroblasts, the three RNA polymerases are found in two different forms: "free and template bound; in senescent fibroblasts, free enzymes are almost nonexistent. In terminal cells, the synthesis of the rRNAs is much more strongly depressed (by 70%) than that of the mRNAs and tRNAs, in which the inhibition is only about 30%. In view of Orgel's theory, it was important to know whether the RNA polymerases, which are the key enzymes for transcription, are altered during aging. This point was studied by Evans (1976, 1977), who investigated the thermolability of the three RNA polymerases. He found that the overall thermolability of partially purified extracts containing all three enzymes was greater in old than in young fibroblasts. This was not due to an accumulation of altered enzyme molecules but resulted from the fact that the thermosensitivities of RNA polymerases I, II, and III were different, and that the proportions between these three enzymes changed during *in vitro* aging of mouse fibroblasts. Thus, the decrease in RNA synthesis, when these cells become senescent, is probably not due to an alteration in the molecules of the RNA polymerases, but rather to a decrease in the content or activity of RNA polymerase I (responsible for rRNA synthesis) in the nucleolar organizers.

We have tried to remain impartial in presenting the facts for and against Orgel's theory, and we think that it is still too early to give a verdict. It is likely that a decrease in the accuracy of protein synthesis occurs during aging, as was proposed by Orgel as early as 1963, but it seems unlikely that this is the unique and universal factor responsible for senescence and death. The essence of a theory is that it leads to new experiments, with the purpose of proving or disproving it; in this sense, Orgel's theory has been highly successful during the past 20 years and will probably remain so for many years to come. This conclusion is shared by Kirkwood *et al.* (1984), who critically reviewed the problem of translation errors during aging. They point out, among other things, that changes limited to one or a small number of essential enzymes (for instance, the membrane ATPases which control the ionic composition of the cell, or the nucleases) might be sufficient to provoke cell death. Alterations in the activity of a number of key enzymes should be studied before we can accept or dismiss Orgel's theory.

3. Life Span, Aging, and DNA Repair

We have already seen that, according to Hart and Setlow (1974) and Hart *et al.* (1979), there is a direct correlation between the life span of different animal species and the capacity of their cells to repair DNA after injury, for instance by uv irradiation. Can one extrapolate from species longevity to aging of individual cells? The fact that the life span of fibroblasts taken from patients suffering from progeria (Werner's syndrome, infantile progeria of Hutchinson-Gilford) is greatly reduced indicates an affirmative answer to this question; these patients suffer from dwarfism and precocious senility, and usually die at an early age. The disease results from semilethal gene mutations. It is obvious that all of our cells are exposed, during their entire life span, to many kinds of mutagenic agents. It is expected that the mutational load will increase during the life span of our cells, and it is obviously necessary that damage to DNA, induced by these mutagens, be continuously repaired.

We have already mentioned that chromosomal abnormalities, such as polyploidy, and aneuploidy, are more frequent in old than in young cells. At the molecular level, Beaupain *et al.* (1980) found that, during *in vitro* cell aging, the average molecular weight of the DNA molecules decreased and the number of breaks in these molecules increased. According to Shmookler-Reis and Goldstein (1980), many small deletions of tandem repeated DNA sequences occur during repeated cell divisions. These authors point out that this may have a bearing on the *in vitro* aging problem. More recently, Shmookler Reis *et al.* (1983) found that the "inter-*Alu*" sequence (i.e., the nucleotide sequence separating the *Alu*-repeated clusters) is progressively amplified in extrachromosomal DNA during *in vivo* and *in vitro* aging. It was also reported that the frequency of rearrangements in repetitive DNA sequences progressively decreases during life (Mays-Hoopes *et al.*, 1983a). We have seen that the present data concerning possible changes in DNA methylation are contradictory and that low molecular weight DNA accumulates in the nuclei of old fibroblasts (Puvion-Dutilleul *et al.*, 1984). Taken together, these papers suggest that DNA undergoes changes, presumably due to damage, during senescence. However, Thompson and Holliday (1978) have expressed doubts about the theory that aging results from somatic mutations and have presented a serious argument against this theory. In culture, the life spans of diploid and tetraploid fibroblasts are the same; the tetraploid cells would be expected to live longer if somatic mutations were the main cause of aging. On the other hand, Morley *et al.* (1982) found that the number of lymphocytes resistant to 6-thioguanine (they are probably mutants at the *HGPRT* locus) increased with age; there is thus an apparent relationship between mutagenesis and aging that might be due to a decrease, with age, in the fidelity of DNA replication or repair.

We have just seen that there is some evidence for the view that the frequency of errors made by DNA polymerase-α, which is involved in DNA replication and probably in DNA repair as well (see Chapter 5, Volume 1, increases with age (Linn *et al.*, 1976; Murray and Holliday, 1981; Murray, 1981). However, this is not the case for DNA polymerase-β, which certainly plays an important role in DNA repair. According to Fry *et al.* (1981), the fidelity of this enzyme does not decrease during aging.

Settling the question once and for all at first may appear to be an easy matter. By merely studying the repair capacities of young and old cells after damage (by uv irradiation, for instance) to the DNA we should be able to learn the answer. But, alas, there are too many contradictory results. A decrease in the DNA repair capacity of senescent cells has been found by Bowman *et al.* (1976) and Paffenholz (1978), who studied DNA repair after uv irradiation in mouse fibroblasts cultured *in vitro*. Similar results have been published by Meek *et al.* (1980) and Niedermüller (1982), who found in young and old animals a decrease with age in DNA repair after damage with chemical carcinogens, γ, or uv irradiation. Similarly, Gaziev and Malakhova (1982) reported that unscheduled (repair) DNA synthesis decreases in the hepatocytes of old animals; however, since they found no decrease in DNA polymerase-β and apurinic endonuclease activity, the authors suggest that the accessibility of the repair enzymes to the damaged DNA sites might be decreased in the chromatin of old cells. Finally, Treton and Courtois (1982) found, in agreement with the initial work of Hart and Setlow (1974), that there is a good correlation between DNA excision repair and life span in lens epithelial cells.

This evidence sounds very convincing, and it appears that a decrease in DNA repair capacities during aging is a very general, if not universal, phenomenon. However, Smith and Hanawalt (1976) found no difference in the DNA repair capacity after uv irradiation (Hanawalt is one of the leading experts in DNA repair mechanisms) in young and old fibroblasts. More recently, Henis *et al.* (1981) were unable to detect major differences in unscheduled DNA synthesis when they compared fibroblasts from young and old donors. Similarly, Turner *et al.* (1982) found no significant change in DNA repair in lymphocytes from young and old donors. Like Henis *et al.* (1981), Hall *et al.* (1982) were unable to find a difference in DNA repair when they compared fibroblasts from donors who were 3 days to 3 years old and donors aged 84 to 94 years. However, they noticed an increase in chromosome abnormalities during aging and suggested that some subtle repair mechanism might perhaps be affecting transcription.

Finally, Schneider *et al.* (1982), who pointed out that sister chromatid exchanges (SCE) are a good measure of DNA damage, were unable to find a difference in the SCE background in bone marrow cells from young and old rats. This entire issue is still very confused. While Plesko and Richardson (1984) found a decrease in unscheduled DNA synthesis in the hepatocytes of old rats, Hasegawa *et al.* (1984) reported an increase in aged human diploid fibroblasts. Cleaver (1984) compared senescence in cultured fibroblasts from normal individuals and patients of *xeroderma pigmentosum*, who are deficient in DNA repair, and found no differences. All one can conclude, in view of this conflicting evidence, is that a decrease in the DNA repair capacity is probably not a universal factor in cellular senescence.

We have previously mentioned the free radical theory of aging. It is related to aging in that free radicals are believed to play a very important role in DNA damage (particularly after damage by ionizing radiation). In addition, oxygen free radicals are powerful agents of membrane lipid peroxidation, leading to abnormalities in the structure and function of mitochondria, Golgi apparatus, ER, lysosomes, etc. Ames (1983) believes that the main causes of aging (and cancer) are damage to DNA and lipid peroxidation by oxygen radicals.

Superoxide dismutase is the main enzyme involved in riding the cell of these oxygen free radicals. Tolmasoff *et al.* (1980) found that there is a correlation between superoxide dismutase activity and the life span of various animal species (just as there is between life span and DNA excision repair). However, no correlation has been found between superoxide dismutase activity and the age of the subject.

Free radicals, as pointed out by Harman (1981, 1984), may be induced by respiration in mitochondria and the P-450 system of microsomes, in addition to ionizing radiations. Antioxidants should provide some protection against the free radicals produced by our own cells and those present in nature. They include tocopherols, carotene, cysteamine (a well-known protector against ionizing radiation), and the enzymes catalase, glutathione peroxidase, and superoxide dismutase. However, Allen *et al.* (1983) found that continuous inhibition of catalase has no visible effect on the longevity of flies.

It is likely that lipofuscins and similar pigments, which are good indicators of aging, result from lipid peroxidation by free radicals. Rattan *et al.* (1982) have reported that autofluorescence, probably due to lipofuscins, increases exponentially during serial passaging of human diploid fibroblasts. In contrast, a strong autofluorescence is observable during the first passages of fibroblasts in Werner's syndrome, which does not occur if these progeria fibroblasts are transformed and become an established cell line. In the near future, autofluorescence may become an elegant marker for separating young and senescent cells using a fluorescence-activated cell sorter.

It is likely that further work will emphasize the importance of free radicals in aging; it is also likely that they are not the only factor responsible for senescence both *in vitro* and *in vivo*. More likely, a multiplicity of factors, both genetic and epigenetic, are responsible for a process that affects all differentiated cells.

E. CELL DEATH[9]

People seldom die from senescence. They generally die from infections, cardiovascular diseases, cancer, accidents, etc., and, only too frequently, from wars and famines. The same is true for our cells, which, as we have seen, have very different life spans (in humans, neurons and red blood cells are at the two extremes). In culture, where artificial conditions prevail, death inevitably follows senescence. Cell death in cultures can be detected by simple, but perhaps not entirely reliable, tests. For example, one may stain with the dye trypan blue, which is excluded from living cells, or plate the cells on a solid substrate and count the colonies formed. In our bodies, cells in the skin and the intestinal tract continuously die; they are soon replaced by the multiplication of stem cells, followed by differentiation. Cell degeneration may be diffuse or may result from autophagy; it generally ends in pycnosis of the chromatin. Dead cells are eliminated by extrusion in the surrounding medium or by phagocytosis.

It is incorrect to believe that cell death is a late event in life. As pointed out by Saunders (1966), the "death clock" starts very early. Even during embryonic development, death of innumerable cells may occur at given stages of development in well-localized regions of the embryo; elimination of superfluous cells is an absolute necessity for the normal development of the limbs. In such cases, the term *programmed cell death* is often used.

A striking example of programmed cell death during embryogenesis is provided by the studies of Hedgecock *et al.* (1983) and Sulston *et al.* (1983) on *C. elegans*. In this nematode, 113 out of a total of 671 cells always die before hatching; they have precise locations in the larvae, and death occurs, during mitosis, at definite stages of development. Why these particular cells and not their neighbors die is not known.

There have been several reviews by Glücksmann on cell death during morphogenesis of the eye in amphibian embryos (1951, 1965). Glücksmann studied the eye (retina and lens) of 300 normal embryos of *Rana temporaria* from the neurula to the 10-day larva stage. He found numerous pycnoses in both the retina and the lens (Fig. 34) of apparently normal embryos. Glücksmann's (1965) careful analysis has disclosed that in the embryonic retina there are three successive waves of pycnotic degeneration. In the chick embryo retina, cell death occurs in three well-defined regions (Garcia-Porrero *et al.*, 1984). Glücksmann believes that cell death in the eye of the frog embryo results from a homeostatic mechanism. Intense mitotic activity produces too many cells. Those cells that are in excess must die. Why a given cell dies while its neighbor becomes a well-differentiated retina cell is not known, nor do we know anything about the

---

[9]Reviewed by Wyllie *et al.* (1980) and Beaulaton and Lockshin (1982).

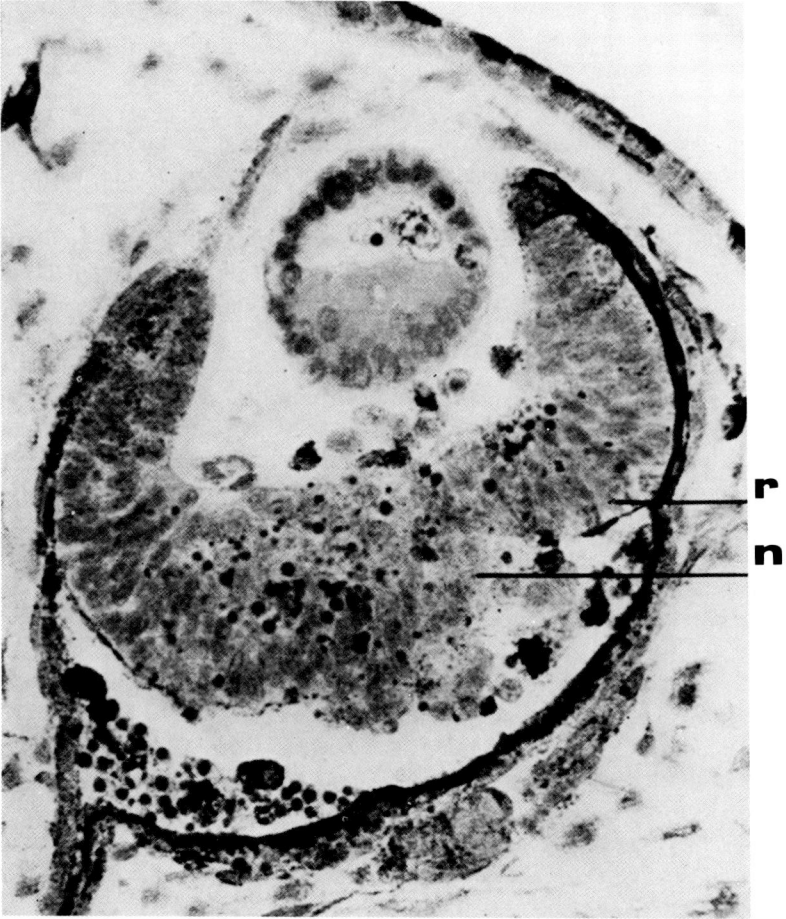

FIG. 34. Cell death in normal development. A necrotic center (n) in the retina extrudes into the
lumen of a normal differentiating eye in a rat embryo. Regeneration by a clearly demarcated
germinative zone (r) is in progress. The pigment layer also contains degenerating cells. [Glücks-
mann, 1965.]

stimuli that induce three successive waves of cell death in the embryonic retina.
Why pycnotic degeneration affects many cells in the nervous system of the
amphibian embryos or in the limb of a mouse embryo (Fig. 35), and not elsewhere, is also unknown. All we can say is that some embryonic cells are much
more fragile than others and, therefore, die easily during normal development.
This weakness must somehow be related to the position these cells occupy in the
embryo. What is certain, however, is that cells in mitosis are more fragile than
those in interphase.

   One of the most striking and famous cases of programmed cell death on a large

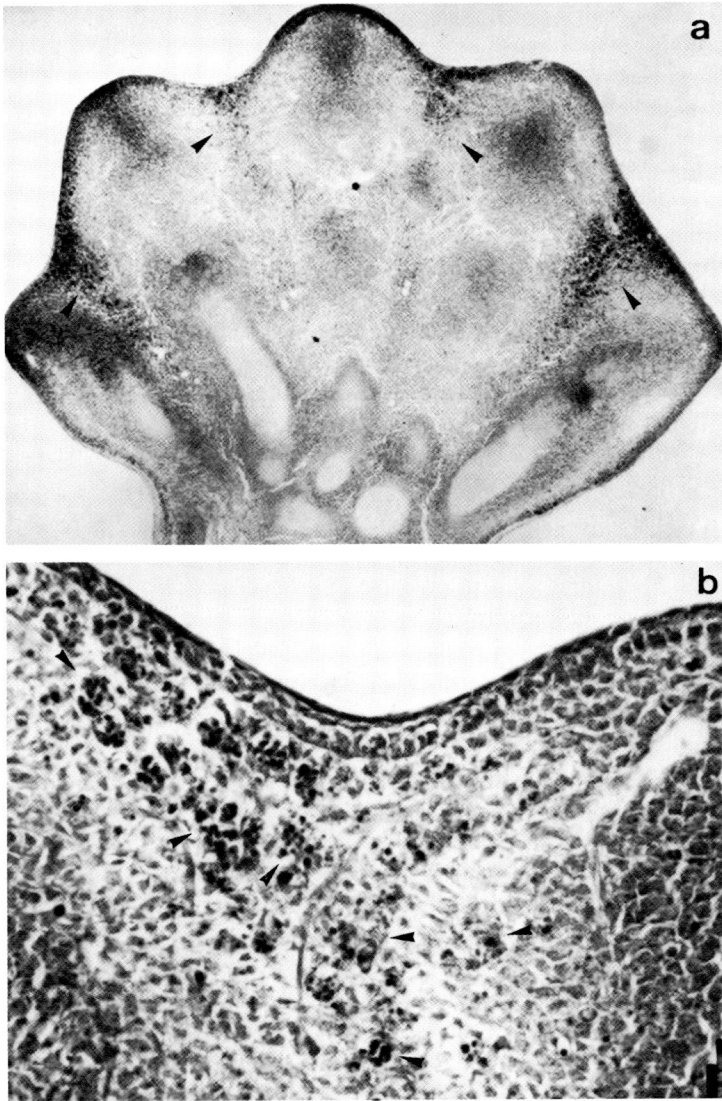

FIG. 35. Cell death during limb morphogenesis in a rat embryo. (a) Posterior limb of a 16-day rat embryo. The four interdigital necrotic areas (arrowheads) are visualized by their high acid phosphatase activity (Gomori staining). (b) Higher magnification of the first interdigital space of an anterior limb in a 13-day mouse embryo (Unna staining). Arrowheads: necrotic cells. [Courtesy of Dr. J. Milaire.]

scale occurs during the limb morphogenesis, when a row of cells die (Fig. 35). The purpose of this self-destruction of part of the limb bud is to ensure the formation of the digits; cell death is required for remolding the bud. The regression of interdigital material during limb morphogenesis has been mainly studied in chick embryos; regression of this material (and programmed cell death) do not take place in the duck limb bud, where it will form the webbing. Remarkable experiments by Zwilling (reviewed in 1964) have shown that if the ectoderm of a chick limb bud is replaced by ectoderm from a duck limb bud, there is no cell degeneration. In this case, species-specific factors obviously play a major role. The wing limb bud of the chick embryo has been studied by Saunders et al. (1962, reviewed by Saunders, 1966). In this case, programmed cell death occurs on the ventral side of the limb bud, resulting in the separation of the wing from the trunk. If the presumptive necrotic zone is grafted onto the dorsal side of the wing limb bud, there are no cell degenerations. If, however, the same zone is grafted onto the somites, it undergoes necrosis (Saunders et al., 1962). Survival or death of the cells clearly results, in this case, from their position in the embryo. Therefore, positional information clearly plays a very important role in the outcome of the grafting experiments. Unfortunately, we still know nothing about the molecular bases of positional information except that exchanges of substances from cell to cell along morphogenetic gradients are probably involved. However, a paper by Toné et al. (1983) has paved the way to further analysis of the limb bud system. They found that programmed cell death in the interdigital mesenchyme of the chicken leg bud was suppressed by treatment with BrdUrd; the treated embryos formed a web-like structure. Since development is normal if thymidine is added together with BrdUrd, it is likely that the suppression of programmed cell death in this system resulted from BrdUrd incorporation into DNA. It has long been known that a mutation called talpid reduces programmed cell death in the limb buds; this strongly suggests that there is a genetic origin for programmed cell death in chicken limb buds. This explanation is less likely for the motoneurons of the lumbar lateral motor column. Their natural cell death can be prevented by treatment with dibutyryl-cGMP (Weill and Greene, 1984) suggesting that it results from cytoplasmic biochemical changes.

In a few cases, extensive cell death results from hormonal stimulation. This is the case when insect larvae undergo metamorphosis under the influence of the hormone ecdysterone. Remolding is very extensive, and most of the larval cells (with the exception of the imaginal disks) die. This is also the case for the larval salivary glands, in which the polytene chromosomes undergo complete degeneration. Another interesting case of hormone-induced programmed cell death is the regression of the müllerian ducts in chick embryos. In male embryos, these ducts regress on the ninth day of development; regression is due to extensive cytolysis. Many years ago, when lysosomes were believed to be ''suicide bags,'' we tested the hypothesis that regression of the müllerian ducts and concomitant cell death might be due to a release of the lysosomal hydrolases. An increase in the activity of the lysosomal enzymes in the regressing ducts (Brachet *et al.*, 1958) indeed occurred. A closer analysis of this same problem by Scheib-Pfleger and Wattiaux (1962) confirmed our results and showed, in addition, that the increase in lysosomal enzyme activity during hormone-induced regression of the müllerian ducts was due to a release of the hydrolases that had accumulated inside the lysosomes. Cell fractionation studies showed that in regressing ducts a large proportion of the acid hydrolases could no longer be sedimented with intact lysosomes by centrifugation; instead, their activity was now present in the supernatant of extracts from regressing müllerian ducts. Unfortunately, there have been no recent reinvestigations of this problem, particularly since we now know that the lysosomes are digestive vacuoles. It seems likely that the release of the lysosomal enzymes in soluble form is the consequence rather than the cause of cell death.

A similar trend of thought was the basis of the work done by Weber (1963, 1967) and Tata (1971) on tail regression during metamorphosis in frogs. In this case, the stimulus that induced regression of the larval tissues and cell death was thyroid hormone. Interestingly, a tail sectioned from a tadpole and cultured *in vitro* in a simple medium regressed if thyroid hormone was added to this medium (Fig. 36). Weber (1963, 1967) found that, as in the müllerian ducts, hormone-induced regression of a tadpole's tail is accompanied by a sharp increase in the activity of two lysosomal marker enzymes: cathepsin and acid phosphatase (Fig. 37). Closer analysis showed that breakdown of the lysosomes and release of their acid hydrolases was not the only factor involved in this resorption. A complicating factor was the invasion of the regressing tail by macrophages, which have a very high lysosomal enzyme content. Their role is, of course, to destroy and digest the dying cells. Additional work by Weber (1967) and Tata (1971) showed that programmed cell death was due to much more complex factors than a mere release of lysosomal enzymes. As shown in Fig. 38, resorption of tails treated *in vitro* with thyroid hormone is completely inhibited by the addition of actinomycin D, puromycin, and cycloheximide to the medium. Obviously, RNA and protein syntheses are required for tail regression during metamorphosis. Unfortunately, we know nothing about the identity of the RNAs and proteins that are required for the induction of extensive cell death after hormonal stimulation during tail resorption.

In fact, we know surprisingly little about the molecular mechanisms of cell agony and death. Studies by Pollak and Fallon (1974, 1976) on the posterior necrotic zone of stage 24 chick embryo have shown the following sequence of events in dying cells: (1) arrest of DNA and RNA synthesis, (2) progressive decrease in and finally arrest of protein synthesis, and (3) cell death. This is essentially the sequence of events we found in anucleate cytoplasm from eggs or unicellular organisms. Protein synthesis will continue for some time because

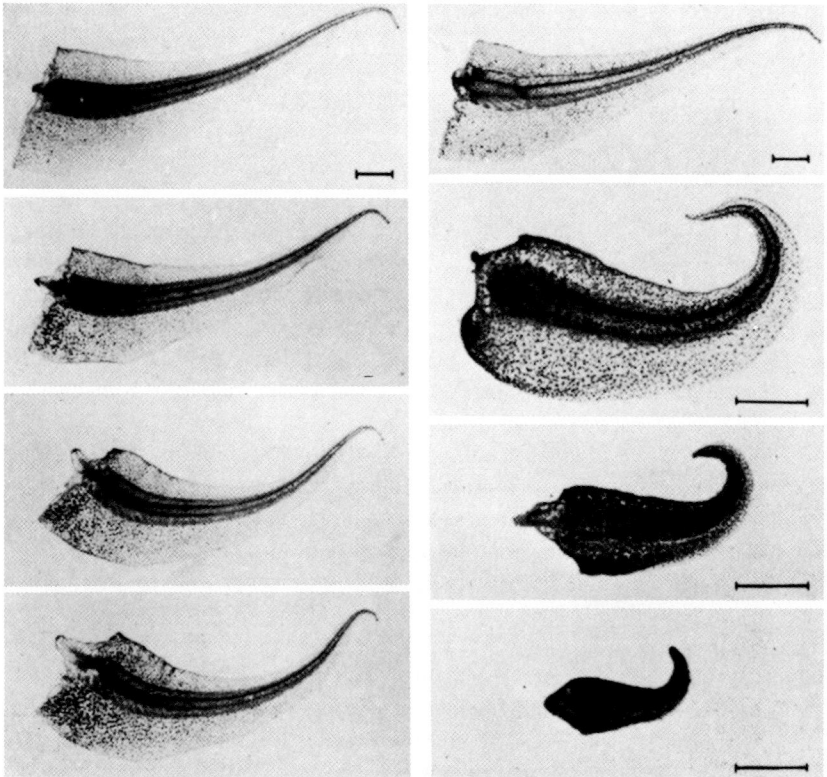

FIG. 36.   Involution of isolated tail tips of *Xenopus* larvae under the influence of thyroxin (from top to bottom). Left: control tips maintained in Holtfreter solution for 6, 8, 10, and 12 days after amputation. Right: tail tips of the same age as the controls, but treated for 3, 5, 7, and 9 days with thyroxine (1:5.10^6). Bar: 1 mm. [Weber, 1967.]

previously synthesized RNAs can still support it. Eventually, synthesis decreases, and it is probable that death of anucleate fragments occurs when some key proteins are no longer available in sufficient amounts.

Kane *et al.* (1980) have pointed out that cell death is not due to the rupture of lysosomal membranes and release of their hydrolases, but rather to an influx of calcium ions entering the cells from the surrounding medium; this would be the cause of traumatic cell death (see the review by Beaulaton and Lockshin, 1982). Since, as we have seen, calcium ions play so many important roles in cell structure and metabolism, it would not be surprising that an overflow of free Ca^{2+} in a cell would cause death. A large increase in free Ca^{2+} is likely to deeply affect important regulatory mechanisms such as cyclic nucleotide content

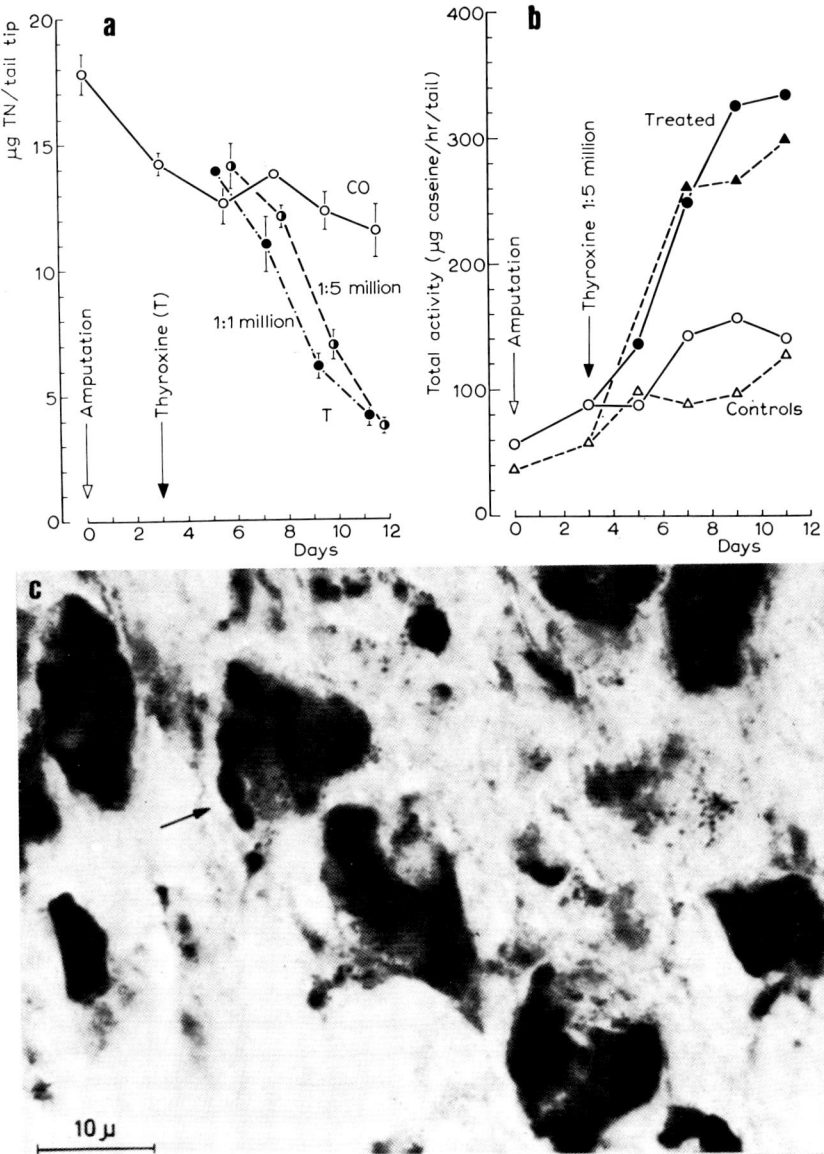

FIG. 37.   Tail atrophy in amphibian metamorphosis. Tissue involution and proteolytic activity in isolated tail tips after treatment with thyroxine. (a) loss of total nitrogen (TN); CO, control tips; T, tips treated with thyroxine 1:10⁶ and 1:5.10⁶, respectively. (b) Changes in the total activity of cathepsins in isolated tail tips. (c) Histochemical demonstration of acid hydrolases in tail rudiments of *Xenopus* larva undergoing spontaneous metamorphosis (in this case, organophosphorus-resistant esterases). The enzyme activity appears to be confined to phagosomes (arrow). [Weber, 1967.]

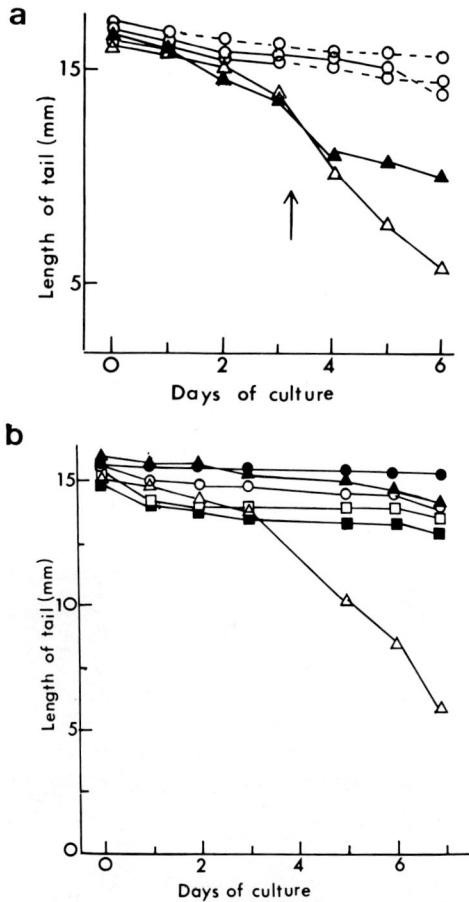

Fig. 38.   Tail atrophy in amphibian metamorphosis. (a) Irreversibility of the inhibition by actinomycin D of thyroxine-induced regression of the isolated tadpole tails and the effectiveness of actinomycin D when added (arrows) during the course of tissue resorption. Open circles: actinomycin (A) and thyroxine (T) were present in the medium on day 0 of the culture, which was then cultivated with T in the presence (solid lines) or absence (dotted lines) of A. Triangles: culture begun in the control medium, and T was added after 1 day; white triangles, no actinomycin; black triangles, A added after 3 days (vertical arrow). (b) Inhibition by puromycin and cycloheximide of thyroxine-induced regression of the isolated tadpole tail. The hormone and inhibitors were added 18 hr after culture was begun. White triangles: with T; black triangles: without T; white and black squares: cycloheximide with and without T; white and black circles: puromycin with and without T. [Tata, 1966].

and protein kinase activity. In this field, as in so many others, we do not lack hypotheses or ideas, but rather the facts. Cell death—like any death—is a dramatic event; this should not prevent molecular cytologists from studying it and providing us with more information about its molecular mechanisms.

REFERENCES

Abrahm, J., and Rovera, J. (1980). *Mol. Cell. Biochem.* **31**, 165.

Adams, J. M., Gerondakis, S., Webb, E., Corcoran, L. M., and Cory, S. (1983). *Proc. Natl. Acad. Sci. U.S.A.* **80**, 1982.

Absher, P. P. and Absher, R. G. (1976). *Exp. Cell Res.* **103**, 247.

Adamson, E. D., Müller, R., and Verma, I. (1983). *Cell Biol. Int. Rep.* **7**, 557.

Adesnik, M., and Smitkin, H. (1978). *J. Cell. Physiol.* **95**, 307.

Albert, K. A., Wu, W. C.-S., Nairn, A. C., and Greengard, P. (1984). *Proc. Natl. Acad. Sci. U.S.A.* **81**, 3622.

Albino, A. P., Le Stange, R., Oliff, A. I., Furth, M. E., and Old, L. J. (1984). *Nature (London)* **308**, 69.

Albrecht-Buehler, G. (1976). *J. Cell Biol.* **69**, 275.

Alitalo, K. (1985). *Trends Biochem. Sci.* **10**, 194.

Alitalo, K., Schwab, M., Lin, C. C., Varmus, H. E., and Bishop, J. M. (1983). *Proc. Natl. Acad. Sci. U.S.A.* **80**, 1707.

Alitalo, K., Winqvist, R., Lin, C. C., de la Chapelle, A., Schwab, M., and Bishop, J. M. (1984). *Proc. Natl. Acad. Sci. U.S.A.* **81**, 4543.

Allan, M., and Harrison, P. (1980). *Cell (Cambridge, Mass.)* **19**, 437.

Allen, R. G., Farmer, K. J., and Sohal, R. S. (1983). *Biochem. J.* **216**, 503.

Althaus, F. R., Lawrence, S. D., He, Y. Z., Sattler, G. L., Tsukada, Y., and Pitot, H. C. (1982). *Nature (London)* **300**, 366.

Ames, B. N. (1983). *Science* **221**, 1256.

Ames, B. N., Cathcart, R., Schniers, E., and Hochstein, P. (1981). *Proc. Natl. Acad. Sci. U.S.A.* **78**, 6858.

Amini, S., and Kaji, A. (1983). *Proc. Natl. Acad. Sci. U.S.A.* **80**, 960.

Anderson, P. J. (1978). *Biochem. J.* **169**, 169.

Andrews, G. K., Dziadek, M., and Tamaoki, T. (1982). *J. Biol. Chem.* **257**, 5148.

Ar-Rushdi, A., Tan, K. B., and Croce, C. M. (1982). *Somatic Cell Genet.* **8**, 151.

Assoian, R. K., Frolik, C. A., Roberts, A. B., Miller, D. M., and Sporn, M. B. (1984). *Cell (Cambridge, Mass.)* **36**, 35.

Artzt, K., Dubois, P., Bennett, D., Condamine, A., Babinet, C., and Jacob, F. (1973). *Proc. Natl. Acad. Sci. U.S.A.* **70**, 2988.

Avilès, D., Ritz, E., and Jami, J. (1980). *Somatic Cell Genet.* **6**, 171.

Azumi, J., and Sachs, L. (1977). *Proc. Natl. Acad. Sci. U.S.A.* **74**, 253.

Backer, J. M., Boersig, M. R., and Weinstein, I. B. (1982a). *Biochem. Biophys. Res. Commun.* **105**, 855.

Backer, J. M., Boerzig, M., and Weinstein, I. B. (1982b). *Nature (London)* **299**, 458.

Baker, S. R., Blithe, D. L., Buck, C. A., and Warren, L. (1980). *J. Biol. Chem.* **255**, 8719.

Balcarek, J. M., and McMorris, F. A. (1983). *J. Biol. Chem.* **258**, 10622.

Balk, S. D., Morisi, A., and Gunther, H. S. (1984). *Proc. Natl. Acad. Sci. U.S.A.* **81**, 6418.

Ball, E. H., and Singer, S. J. (1981). *Proc. Natl. Acad. Sci. U.S.A.* **78**, 6986.

Balmain, A., and Pragnell I. B. (1983). *Nature (London)* **303**, 72.

Baltimore, D. (1970). *Nature (London)* **226**, 1209

Bannigan, J., and Langman, J. (1979). *J. Embryol. Exp. Morphol.* **50**, 123.

Barnekow, A., and Bauer, H. (1984). *Biochim. Biophys. Acta* **782**, 94.

Barrows, C. H., Jr., and Kokkonen, G. C. (1981). *Age* **4**, 1.

Bar-Sagi, D., and Prives, J. (1983). *J. Cell Biol.* **97**, 1375.

Bartram, C. R., de Klein, A., Hagemeijer, A., Van Agthoven, T., Geurts van Kessel, A., Bootsma, D., Grosveld, G., Ferguson-Smith, M. A., Davies, T., Stone, M., Heisterkamp, N., Stephenson, J. R., and Groffen, J. (1983). *Nature (London)* **306**, 277.

Baulieu, E. E. (1975). *Mol. Cell. Biochem.* **7**, 157.

Beaulaton, J., and Lockshin, R. A. (1982). *Int. Rev. Cytol.* **79**, 215.

Beaupain, R., Icard, C., and Marcieira-Coelho, A. (1980). *Biochim. Biophys. Acta* **606**, 251.

Beemon, K., Ryden, T., and McNelly, E. A. (1982). *J. Virol.* **42**, 742.

Belin, D., and Ossowski, L. (1983). *Cancer Res.* **43**, 3263.

Bell, P. A., and Jones, C. N. (1982). *Biochem. Biophys. Res. Commun.* **104**, 1202.

Bellard, M., Dretzen, G., Bellard, F., Oudet, P., and Chambon, P. (1982). *EMBO J.* **1**, 223.

Benda, P., and Davidson, R. L. (1971). *J. Cell. Physiol.* **78**, 209.

Benham, J. F., Wiles, M. V., and Goodfellow, P. N. (1983). *Mol. Cell. Biol.* **3**, 2259.

Benjamin, D. (1974). *Methods Cell Biol.* **8**, 367.

Benjamin, D., Magrath, I. T., Triche, T. J., Schroff, R. W., Jensen, J. P., and Korsmeyer, S. J. (1984). *Proc. Natl. Acad. Sci. U.S.A.* **81**, 3547.

Bennett, D. C. (1983). *Cell (Cambridge, Mass.)* **34**, 445.

Benya, P. D., and Shaffer, J. D. (1982). *Cell (Cambridge, Mass.)* **30**, 215.

Berenblum, I. (1954). *Cancer Res.* **14**, 471.

Berenblum, I., and Armuth, V. (1981). *Biochim. Biophys. Acta* **651**, 51.

Bernard, O., Cory, S., Gerondakis, S., Webb, E., and Adams, J. M. (1983). *EMBO J.* **2**, 2375.

Bernd, A., Schröder, H. C., Zahn, R. K., and Müller, W. E. G. (1982). *Mech. Ageing Dev.* **20**, 331.

Berridge, M. J. (1984). *Biochem. J.* **220**, 345.

Bertolotti, R., Rutishauser, U., and Edelman, G. M. (1980). *Proc. Natl. Acad. Sci. U.S.A.* **77**, 4831.

Bertoncello, I., Bradley, T. R., Chamley, W. A., and Hodgson, G. S. (1982). *J. Cell. Physiol.* **113**, 224.

Bick, M. D., and Devine, E. A. (1977). *Nucleic Acids Res.* **4**, 3687.

Bishop, J. M. (1983a). *Cell* **32**, 1018.

Bishop, J. M. (1983b). *Annu. Rev. Biochem.* **52**, 301.

Blasi, F., and Toniolo, D. (1983). *Mol. Biol. Med.* **1**, 271.

Blau, H. M., and Epstein, C. J. (1979). *Cell (Cambridge, Mass.)* **17**, 95.

Blaszejowski, C. A., and Webster, G. C. (1983). *Mech. Ageing Dev.* **21**, 345.

Bloch-Shtacher, N., and Sachs, L. (1976). *J. Cell Physiol.* **87**, 89.

Blondal, J. A., Dick, J. E., and Wright, J. A. (1985). *Mech. Ageing Dev.* **30**, 273.

Blose, S. H., Meltzer, D. I., and Feramisco, J. R. (1982). *J. Cell Biol.* **95**, 229a.

Blüthmann, H., and Illmensee, K. (1981). *Wilhelm Roux' Arch. Dev. Biol.* **190**, 374.

Bolmer, S. D., and Wolf, G. (1982). *Proc. Natl. Acad. Sci. U.S.A.* **79**, 6541.

Boon, T., and Kellermann, O. (1977). *Proc. Natl. Acad. Sci. U.S.A.* **74**, 272.

Borek, C., and Troll, W. (1983). *Proc. Natl. Acad. Sci. U.S.A.* **80**, 1304.

Boschek, C. B., Jockusch, B. M., Friis, P. R., Back, R., Grundmann, E., and Bauer, H. (1981). *Cell (Cambridge, Mass.)* **24**, 175.

Bosman, G. J., Boer, P., and Steyn-Parvé, E. P. (1982). *Biochim. Biophys. Acta* **696**, 285.

Bosmann, H. B., Gutheil, R. L., Jr., and Case, K. R. (1976). *Nature (London)* **261**, 499.

Boucaut, J. C., Darribère, T., Boulek-Bache, H., and Thiéry, J. P. (1984). *Nature (London)* **307**, 364.

Bournias-Vardiabasis, N., Buzin, C. H., and Reilly, J. G. (1983). *Wilhelm Roux's Arch. Dev. Biol.* **192**, 299.

Bower, D. J., Errington, L. H., Cooper, D. N., Morris, S., and Clayton, R. M. (1983). *Nucleic Acids Res.* **11**, 2513.

Bowman, P. D., Meek, R. L., and Daniel, C. W. (1975). *Exp. Cell Res.* **93**, 183.

Bowman, P. D., Meek, R. L., and Daniel, C. W. (1976). *Mech. Ageing Dev.* **5**, 251.

Boyd, A. W., and Schrader, J. W. (1982). *Nature (London)* **297**, 691.

Brachet, J. (1950). "Chemical Embryology." Wiley (Interscience), New York.

Brachet, J. (1957). "Biochemical Cytology." Academic Press, New York.

Brachet, J., and Malpoix, P. (1971). *Adv. Morphog.* **9**, 263.

Brachet, J., Decroly-Briers, M., and Hoyez, J. (1958). *Bull. Soc. Chim. Biol.* **40**, 2039.

Brachet, J., Denis, H., and de Vitry, F. (1964). *Dev. Biol.* **9**, 398.

Bravo, R., and Celis, J. E. (1980). *Exp. Cell Res.* **127**, 249.

Bravo, R., Schafer, R., Willecke, K., MacDonald-Bravo, H., Freys, S. J., and Celis, J. E. (1982). *Proc. Natl. Acad. Sci. U.S.A.* **79**, 2281.

Brinster, R. L. (1974). *J. Exp. Med.* **140**, 1049.

Brinster, R. L., Chen, H. Y., Messing, A., Van Dyke, T., Levine, A. J., and Palmiter, R. D. (1984). *Cell* **37**, 367.

Brock, M. L., and Shapiro, D. J. (1983). *J. Biol. Chem.* **258**, 5449.

Brodeur, G. M., Seeger, R. C., Schwab, M., Varmus, H. E., and Bishop, J. M. (1984). *Science* **224**, 1121.

Brown, D. D. (1981). *Science* **211**, 667.

Brown, R., Marshall, C. J., Pennie, S. G., and Hall, A. (1984). *EMBO J.* **3**, 1321.

Brunk, C. F. (1979). *Differentiation* **14**, 95.

Brunk, U., Ericsson, J. L. E., Pontén, J., and Westermark, B. (1973). *Exp. Cell Res.* **79**, 1.

Buc-Caron, M. H., Gachelin, G., Hofnung, M., and Jacob, F. (1974). *Proc. Natl. Acad. Sci. U.S.A.* **71**, 1730.

Buckingham, M. E., Cohen, A., and Gros, F. (1976). *J. Mol. Biol.* **103**, 611.

Budzik, G. P., Powell, S. M., Kamagata, S., and Donahoe, P. K. (1983). *Cell (Cambridge, Mass.)* **34**, 307.

Burmer, G. C., Motulsky, H., Zeigler, C. J., and Norwood, T. H. (1983). *Exp. Cell Res.* **145**, 79.

Burrows, R. B., and Davison, P. F. (1980). *Mech. Ageing Dev.* **13**, 307.

Buys, C. H., Osinga, J., and Anders, G. J. (1979). *Mech. Ageing Dev.* **11**, 55.

Caccia, N. C., Mak, T. W., and Klein, G. (1984). *J. Cellul. Physiol. Suppl.* **3**, 199.

Cairns, J. (1981). *Nature (London)* **289**, 353.

Campisi, J., Gray, H. E., Pardee, A. B., Dean, M., and Sonenshein, G. E. (1984). *Cell (Cambridge, Mass.)* **36**, 241.

Capetanaki, Y. G., Flytzanis, C. M., and Alonso, A. (1982). *Mol. Cell. Biol.* **2**, 258.

Capon, D. J., Chen, E. Y., Levinson, A. D., Seeburg, P. H., and Goeddel, D. V. (1983). *Nature (London)* **302**, 33.

Capone, P. M., Papsidero, L. D., Croghan, G. A., and Chu, T. M. (1983). *Proc. Natl. Acad. Sci. U.S.A.* **80**, 7328.

Carlin, C. R., Phillips, P. D., Knowles, B. B., and Cristofalo, V. J. (1983). *Nature (London)* **306**, 647.

Carlsen, S. A., Ramshaw, I. A., and Warrington, R. C. (1984). *Cancer Res.* **44**, 3012.

Carmon, Y., Czosnek, H., Nudel, U., Shani, M., and Yaffe, D. (1982). *Nucleic Acids Res.* **10**, 3085.

Carpenter, G., and Cohen, S. (1979). *Annu. Rev. Biochem.* **48**, 193.

Carrel, A. (1912). *J. Exp. Med.* **15**, 576.

Carter, W. G. (1982). *J. Biol. Chem.* **257**, 13805.

Cate, R. L., Chick, W., and Gilbert, W. (1983). *J. Biol. Chem.* **258**, 6645.

Celis, J. E., and Bravo, R. (1984). *FEBS Lett.* **165**, 21.

Celis, J. E., Fey, S. J., Larsen, P. M., and Celis, A. (1984). *Proc. Natl. Acad. Sci. U.S.A.* **81**, 3128.

Chan, G. L. (1981). *Int. Rev. Cytol.* **70**, 101.

Chandrasekar, S., Sorrentino, J. A., and Millis, A. J. T. (1983). *Proc. Natl. Acad. Sci. U.S.A.* **80**, 4747.

Chang, E. H., Furth, M. E., Scolnick, E. M., and Lowy, D. R. (1982a). *Nature (London)* **297**, 479.

Chang, E. H., Gonda, M. A., Ellis, R. W., Scolnick, E. M., and Lowy, B. R. (1982b). *Proc. Natl. Acad. Sci. U.S.A.* **79,** 4848.

Charache, S., Dover, G., Smith, K., Talbot, C. C., Jr., Moyer, M., and Boyer, S. (1983). *Proc. Natl. Acad. Sci. U.S.A.* **80,** 4842.

Chatterjee, B., Nath, T. S., and Roy, A. K. (1981). *J. Biol. Chem.* **256,** 5939.

Chen, K. Y., Presepe, V., Parken, N., and Liu, A. Y. C. (1982). *J. Cell. Physiol.* **110,** 285.

Chisholm, R. (1982). *Trends Biochem. Sci.* **7,** 161.

Chiu, I. M., Reddy, E. P., Givol, D., Robbins, K. C., Tronick, S. R., and Aaronson, S. A. (1984). *Cell (Cambridge, Mass.)* **37,** 123.

Chopra, D. F., and Wilkoff, L. J. (1977). *Nature (London)* **265,** 339.

Chou, J. Y., and Ito, F. (1984). *Biochem. Biophys. Res. Commun.* **118,** 168.

Christman, J. K., Price, P., Pedrinan, L., and Acs, G. (1977). *Eur. J. Biochem.* **81,** 53.

Christman, J. K., Wiech, N., Schoenbrun, B., Schneiderman, N., and Acs, G. (1980). *J. Cell Biol.* **86,** 366.

Christman, J. K., Schneiderman, N., and Acs, G. (1985). *J. Biol. Chem.* **260,** 4059.

Cleaver, J. E. (1984). *Mech. Ageing Dev.* **27,** 189.

Coffino, P., Knowles, B., Nathenson, S., and Scharff, M. (1971). *Nature (London)* **231,** 87.

Cohen, R., Pacifici, M., Rubinstein, N., Biehl, J., and Holtzer, H. (1977). *Nature (London)* **266,** 538.

Cohen, S., Fava, R. A., and Sawyer, S. T. (1982). *Proc. Natl. Acad. Sci. U.S.A.* **79,** 6237.

Colberg-Poley, A. M., Voss, J. D., Chowdhury, K., and Gruss, P. (1985). *Nature (London)* **314,** 713.

Colburn, N. H., and Gindhardt, T. D. (1981). *Biochem. Biophys. Res. Commun.* **102,** 799.

Collins, J. M., and Chu, A. K. (1985). *J. Cell. Physiol.* **124,** 65.

Collins, M. K. L., and Rozengurt, E. (1982). *J. Cell. Physiol.* **112,** 42.

Collins, S. J., and Groudine, M. (1982). *Nature (London)* **298,** 679.

Collins, S. J., and Groudine, M. T. (1983). *Proc. Natl. Acad. Sci. U.S.A.* **80,** 4813.

Compere, S. J., and Palmiter, R. D. (1981) *Cell (Cambridge, Mass.)* **25,** 233.

Conscience, J. F., Miller, R. A., Henry, J., and Ruddle, F. H. (1977). *Exp. Cell Res.* **105,** 401.

Cook, J. R., and Buetow, D. E. (1981). *Mech. Ageing Dev.* **17,** 41.

Cook, J. R., and Chiu, J. F. (1983). *Biochem. Biophys. Res. Commun.* **116,** 939.

Coon, H. G., and Cahn, R. D. (1966). *Science* **153,** 1116.

Cooper, G. M. (1982). *Science* **218,** 801.

Cooper, G. M., and Lane, M. A. (1984). *Biochim. Biophys. Acta.* **738,** 9.

Cooper, G. M., Okenquist, S., and Silverman, L. (1981). *Nature (London)* **284,** 418.

Cooper, J., Nakamura, K. D., Hunter, T., and Weber, M. J. (1983). *J. Virol.* **46,** 15.

Cooper, J. A., Bowen-Pope, D. F., Raines, E., Ross, R., and Hunter, T. (1982). *Cell (Cambridge, Mass.)* **31,** 263.

Cooper, R. A., Braunwald, E. D., and Kuo, A. L. (1982). *Proc. Natl. Acad. Sci. U.S.A.* **79,** 2865.

Copenhaver, W. M. (1955). *In* "Analysis of Development" (B. H. Willier, P. A. Weiss, and V. Hamburger, eds.), p. 452. Saunders, Philadelphia, Pennsylvania.

Cossu, G., Pacifici, M., Adamo, S., and Molinaro, M. (1981). *Cell Biol. Int. Rep.* **5,** 337.

Coulon-Morelec, M. J., and Buc-Caron, M. H. (1981). *Dev. Biol.* **83,** 278.

Couturier, A., Bazgar, S., and Castagna, M. (1984). *Biochem. Biophys. Res. Commun.* **121,** 448.

Craig, R. W., and Bloch, A. (1984). *Cancer Res.* **44,** 442.

Crawford, L. V., Pim, D. C., Gurney, E. G., Goodfellow, P., and Taylor-Papadimitriou, J. (1981). *Proc. Natl. Acad. Sci. U.S.A.* **78,** 41.

Crews, S., Barth, R., Hood, L., Prehn, J., and Calame, K. (1982). *Science* **218,** 1319.

Croce, C. M. (1980). *Biochim. Biophys. Acta* **605,** 411.

Croce, C. M., Aden, D., and Koprowski, H. (1975). *Proc. Natl. Acad. Sci. U.S.A.* **72,** 1397.

Croce, C. M., Erikson, J., Ar-Rushdi, A., Aden, D., and Nishikura, K. (1984). *Proc. Natl. Acad. Sci. U.S.A.* **81,** 3170.

Cronmiller, C., and Mintz, B. (1978). *Dev. Biol.* **67,** 465.

Croop, J., Dubyak, G., Toyama, Y., Dlugosz, A., Scarpa, A., and Holtzer, H. (1982). *Dev. Biol.* **89,** 464.

Cross, F. R., and Hanafusa, H. (1983). *Cell (Cambridge, Mass.)* **34,** 597.

Croy, R. G., and Pardee, A. B. (1983). *Proc. Natl. Acad. Sci. U.S.A.* **80,** 4699.

Crutchley, D. J., Conaman, L. B., and Maynard, J. R. (1980). *Cancer Res.* **40,** 849.

Cuzin, F. (1984). *Biochim. Biophys. Acta* **781,** 193.

Czech, M. P. (1982). *Cell (Cambridge, Mass.)* **31,** 8.

Dalla-Favera, R., Gelmann, E. P., Gallo, R. C., and Wong-Staal, F. (1981). *Nature (London)* **292,** 31.

Dalla-Favera, R., Gallo, R. C., Giallongo, A., and Croce, C. M. (1982a). *Science* **218,** 686.

Dalla-Favera, R., Wong-Staal, F., and Gallo, R. C. (1982b). *Nature (London)* **299,** 61.

Dalla-Favera, R., Bregni, M., Erikson, J., Patterson, D., Gallo, R. C., and Croce, C. M. (1982c). *Proc. Natl. Acad. Sci. U.S.A.* **79,** 7824.

Dalla-Favera, R., Martinotti, S., Gallo, R. C., Erikson, J., and Croce, C. M. (1983). *Science* **219,** 963.

Darmon, M., Buc-Caron, M. H., Paulin, D., and Jacob, F. (1982a). *EMBO J.* **1,** 901.

Darmon, M., Stallcup, W. B., and Pittman, Q. J. (1982b). *Exp. Cell Res.* **138,** 73.

Darmon, M., Nicolas, J. F., and Lamblin, D. (1984). *EMBO J.* **3,** 961.

Das, M. (1982). *Int. Rev. Cytol.* **78,** 233.

Davidson, R. L. (1971). *Fed. Proc., Fed. Am. Soc. Exp. Biol.* **30,** 926.

Davidson, R. L., Ephrussi, B., and Yamamoto, K. (1966). *Proc. Natl. Acad. Sci. U.S.A.* **56,** 1437.

Davies, R. L., Rifkin, D. B., Tepper, R., Miller, A., and Kucherlapati, R. (1983). *Science* **221,** 171.

Davis, R. J., and Czech, M. P. (1984). *J. Biol. Chem.* **259,** 8545.

Davis, T. J., and Harris, H. (1975). *J. Cell Sci.* **18,** 207.

Dean, M., Kent, R. B., and Sonenshein, G. E. (1983). *Nature (London)* **305,** 443.

De Brabander, M., Geuens, G., Nuydens, R., and De Mey, J. (1982). *J. Cell Biol.* **95,** 226a.

Debuire, B., Henry, C., Benaissa, M., Biserte, G., Claverie, J. M., Saule, S., Martin, P., and Stehelin, D. (1984). *Science* **224,** 1456.

Deeley, R. G., Gordon, J. I., Burns, A. T., Mullinik, K. P., Binastein, M., and Goldberger, R. F. (1977). *J. Biol. Chem.* **252,** 8310.

DeFeo-Jones, D., Scolnick, E. M., Koller, R., and Dhar, D. (1983). *Nature (London)* **306,** 707.

DeFeo-Jones, D., Tatchell, K., Robinson, L. C., Sigal, I. S., Vass, W. C., Lowy, D. R., and Scolnick, E. M. (1985). *Science* **228,** 179.

DeHarven, E. (1974). *Adv. Virus Res.* **19,** 242, 338.

Deisseroth, A., Burk, R., Picciano, D., Minna, J., Anderson, W. F., and Nienhuis, A. (1975). *Proc. Natl. Acad. Sci. U.S.A.* **72,** 1102.

Deisseroth, A., Velez, R., Burk, R. D., Minna, J., Anderson, W. F., and Nienhuis, A. (1976). *Somatic Cell Genet.* **2,** 373.

De Mey, J., Joniau, M., De Brabander, M., Moens, W., and Geuens, J. (1978). *Proc. Natl. Acad. Sci. U.S.A.* **75,** 1339.

Denis, H. (1964). *Dev. Biol.* **9,** 435.

Denton, T. E., Liem, S. L., Cheng, K. M., and Barrett, J. V. (1981). *Mech. Ageing Dev.* **15,** 1.

De Priester, W., Van Manen, R., and Knook, D. L. (1984). *Mech. Ageing Dev.* **26,** 205.

Der, C. J., and Stanbridge, E. J. (1981). *Cell (Cambridge, Mass.)* **26,** 429.

Der, C. J., Ash, J. E., and Stanbridge, O. (1981). *J. Cell Sci.* **52,** 151.

Der, C. J., Krontiris, T. G., and Cooper, G. M. (1982). *Proc. Natl. Acad. Sci. U.S.A.* **79,** 3637.

Deuel, T. F., Huang, J. S., Huang, S. S., Stroobant, P., and Waterfield, M. D. (1983). *Science* **221,** 1348.

Devasagayam, T. P. A., Pushpendran, C. K., and Eapen, J. (1983). *Mech. Ageing Dev.* **21,** 365.

Dhar, R., Nieto, A., Koller, R., DeFeo-Jones, D., and Scolnick, E. M. (1984). *Nucleic Acids Res.* **12,** 3611.

Di Berardino, M. A., Mizell, M., Hoffner, N. J., and Friesendorf, D. G. (1983). *Differentiation* **23,** 213.

Dice, J. F. (1982). *J. Biol. Chem.* **257,** 14624.

Dick, J. E., and Wright, J. A. (1982). *Mech. Ageing Dev.* **20,** 103.

Dickens, M. S., Custer, R. P., and Sorof, S. (1979). *Proc. Natl. Acad. Sci. U.S.A.* **76,** 5891.

Dicker, P., and Rozengurt, E. (1979). *Biochem. Biophys. Res. Commun.* **91,** 1203.

Dicker, P., and Rozengurt, E. (1980). *Exp. Cell Res.* **130,** 474.

Dickson, J. G., Flanigan, T. P., and Walsh, F. S. (1983). *EMBO J.* **2,** 283.

Didinsky, J. B., and Rheinwald, J. G. (1981). *J. Cell. Physiol.* **109,** 171.

Dlugosz, A. A., Tapscott, S. J., and Holtzer, H. (1983). *Cancer Res.* **43,** 2780.

Doolittle, R. F., Hunkapiller, M. W., Hood, L. E., Aaronson, S. A., Robbins, K. C., Devare, S. G., and Antoniades, H. N. (1983). *Science* **221,** 275.

Dovrat, A., and Gershon, D. (1983). *Biochim. Biophys. Acta* **757,** 164.

Downward, J., Yarden, Y., Mayes, E., Scrace, G., Totty, N., Stockwell, P., Ullrich, A., Schlessinger, J., and Waterfield, M. D. (1984). *Nature (London)* **307,** 521.

Drescher-Lincoln, C. K., and Smith, J. R. (1983). *Exp. Cell Res.* **144,** 455.

Drescher-Lincoln, C. K., and Smith, J. R. (1984). *Exp. Cell Res.* **153,** 208.

Duboule, D., Croce, C. M., and Illmensee, K. (1982). *EMBO J.* **1,** 1595.

Duesberg, P. H. (1980). *Cold Spring Harbor Symp. Quant. Biol.* **44,** 13.

Duesberg, P. H. (1983). *Nature (London)* **304,** 219.

Duesberg, P. H. (1985). *Science* **228,** 669.

Duprat, A. M., Gualandris, L., and Rouge, P. (1982). *J. Embryol. Exp. Morphol.* **70,** 171.

Dustin, P. (1966). "Leçons d'Anatomie pathologique générale," pp. 498, 521, 535, 547. Presses Acad. Eur., Bruxelles.

Ebeling, A. H. (1913). *J. Exp. Med.* **17,** 273.

Edelman, G. M. (1983). *Science* **219,** 450.

Edelman, G. M., Gallin, W. J., Delouvée, A., Cunningham, B. A., and Thiéry, J. P. (1983). *Proc. Natl. Acad. Sci. U.S.A.* **80,** 4384.

Edelman, G. M. (1984). *Proc. Natl. Acad. Sci. U.S.A.* **81,** 1460.

Edick, G. F., and Millis, A. J. T. (1984). *Mech. Ageing Dev.* **27,** 249.

Egilmez, N. K., and Rothstein, M. (1985). *Biochem. Biophys. Acta* **840,** 335.

Eguchi, G., and Okada, T. S. (1973). *Proc. Natl. Acad. Sci. U.S.A.* **70,** 1495.

Eisenman, R. N., Tachibana, C. Y., Abrams, H. D., and Hann, S. R. (1985). *Mol. Cell Biol.* **5,** 114.

Ek, B., and Heldin, C. H. (1982). *J. Biol. Chem.* **257,** 10486.

Ellinger, M. S. (1982). *Cancer Res.* **42,** 2804.

Emerit, I., and Cerutti, P. (1981). *Proc. Natl. Acad. Sci. U.S.A.* **78,** 1868.

Emerit, I., and Cerutti, P. A. (1982). *Proc. Natl. Acad. Sci. U.S.A.* **79,** 7509.

Engelhardt, D. L., Lee, G. T.-Y., and Moley, J. F. (1979). *J. Cell Physiol.* **98,** 193.

Enomoto, T., Sasaki, Y., Shiba, Y., Kanno, Y., and Yamasaki, H. (1981). *Proc. Natl. Acad. Sci. U.S.A.* **58,** 5628.

Ephrussi, B. (1956). *In* "Enzymes: Units of Biological Structure and Function" (H. O. Gaebler, ed.), p. 29. Academic Press, New York.

Ephrussi, B. (1972). "Hybridization of Somatic Cells." Princeton Univ. Press, Princeton, New Jersey.

Ephrussi, B., and Weiss, M. C. (1965). *Proc. Natl. Acad. Sci. U.S.A.* **53,** 1040.

Ephrussi, B., Scaletta, L. J., Stenchever, M. A., and Yoshida, M. C. (1964). *In* "Cytogenetics in Cells in Culture" (H. Harris, ed.), p. 13. Academic Press, New York.

Ephrussi, B., Davidson, R. L., and Weiss, M. C. (1969). *Nature (London)* **224,** 1314.

Erikson, E., Shealy, D. J., and Erikson, R. L. (1981). *J. Biol. Chem.* **256**, 11381.

Erikson, J., Ar-Rushdi, A., Drwinga, H. L., Nowell, P. C., and Croce, C. M. (1983). *Proc. Natl. Acad. Sci. U.S.A.* **80**, 820.

Errington, L. H., Cooper, D. N., and Clayton, R. M. (1983). *Differentiation* **24**, 33.

Erwin, B. G., Ewton, D. Z., Florini, J. R., and Pegg, A. E. (1983). *Biochem. Biophys. Res. Commun.* **114**, 944.

Evans, C. H. (1976). *Differentiation* **5**, 101.

Evans, C. H. (1977). *Exp. Gerontol.* **12**, 169.

Evans, C. H., Van Gansen, P., and Rasson, I. (1978). *Biol. Cell.* **33**, 117.

Evans, E. P., Burtenshaw, M. D., Brown, B. B., Hennion, R., and Harris, H. (1982). *J. Cell Sci.* **56**, 113.

Fagan, J. B., Sobel, M. E., Yamada, K. M., de Crombrugghe, B., and Pastan, I. (1981). *J. Biol. Chem.* **256**, 520.

Falcone, G., Boettiger, D., Alemà, S., and Tató, F. (1984). *EMBO J.* **3**, 1327.

Fallat, M. E., Hutson, J. M., Budzik, G. P., and Donahoe, P. K. (1983). *Dev. Biol.* **100**, 358.

Farber, E. (1984). *J. Cellul. Physiol. Suppl.* **3**, 123.

Farley, J. R., and Baylink, D. J. (1982). *Biochemistry* **21**, 3502.

Farzaneh, F., Zalin, R., Brill, D., and Shall, S. (1982). *Nature (London)* **300**, 362.

Feig, L. A., Bast, R. C., Jr., Knapp, R. C., and Cooper, G. M. (1984). *Science* **223**, 698.

Feinberg, A. P., Vogelstein, B., Droller, M. J., Baylin, S. B., and Nelkin, B. D. (1983). *Science* **220**, 1175.

Fellous, M., et al. (1974). *Dev. Biol.* **41**, 333, 335.

Feramisco, J. R., Gross, M., Kamata, T., Rosenberg, M., and Sweet, R. W. (1984). *Cell* **38**, 109.

Fibach, E., Gambari, R., Shaw, P., Maniatis, G., Reuben, R., Sassa, S., Rifkin, R., and Marks, P. (1979). *Proc. Natl. Acad. Sci. U.S.A.* **76**, 1906.

Fey, E. G., and Penman, S. (1984). *Proc. Natl. Acad. Sci. U.S.A.* **81**, 4409.

Fidler, I. J., and Hart, I. R. (1982). *Science* **217**, 998.

Fine, D., and Schochetman, G. (1978). *Cancer Res.* **38**, 3139.

Finer, M. H., Gerstenfeld, L. C., Young, D., Doty, P., and Boedtker, H. (1985). *Mol. Cell Biol.* **5**, 1415.

Fisher, P. B., Bozzone, J. H., and Weinstein, I. B. (1979). *Cell (Cambridge, Mass.)* **18**, 695.

Fisher, P. B., Cogan, U., Horowitz, A. D., Schachter, D., and Weinstein, I. B. (1981). *Biochem. Biophys. Res. Commun.* **100**, 370.

Fitzgerald, D. J., and Murray, A. W. (1982). *Cell Biol. Int. Rep.* **6**, 235.

Forman, D., and Rowley, J. (1982). *Nature (London)* **300**, 403.

Fougère, C., and Weiss, M. C. (1978). *Cell (Cambridge, Mass.)* **15**, 843.

Fougère, C., Ruiz, F., and Ephrussi, B. (1972). *Proc. Natl. Acad. Sci. U.S.A.* **69**, 330.

Franke, W. W., and Keenan, T. W. (1979). *Differentiation* **13**, 81.

Friedman, B., Frackelton, A. R., Jr., Ross, A. H., Connors, J. M., Fujiki, H., Sugimura, T., and Rosner, M. R. (1984). *Proc. Natl. Acad. Sci. U.S.A.* **81**, 3034.

Friend, C., Sher, W., Holland, J., and Soto, T. (1971). *Proc. Natl. Acad. Sci. U.S.A.* **68**, 378.

Frolik, C. A., Dart, L. L., Meyers, C. A., Smith, D. M., and Sporn, M. B. (1983). *Proc. Natl. Acad. Sci. U.S.A.* **80**, 3676.

Fry, M., Loeb, L. A., and Martin, G. M. (1981). *J. Cell Physiol.* **106**, 435.

Fry, M., Silber, J., Loeb, L. A., and Martin, G. M. (1984). *J. Cell. Physiol.* **118**, 225.

Fujiki, H., Tanaka, Y., Miyake, R., Kikkawa, U., Nishizuka, Y., and Sugimura, T. (1984). *Biochem. Biophys. Res. Commun.* **120**, 339.

Fulder, S. J., and Holliday, R. (1975). *Cell (Cambridge, Mass.)* **6**, 67.

Fürstenberger, G., and Marks, F. (1980). *Biochem. Biophys. Res. Commun.* **92**, 749.

Fusenig, N. E., and Dzarlieva, R. T. (1982). *Carcinogenesis* **7**, 201.

Gabius, H. J., Engelhardt, R., Deerberg, F., and Cramer, F. (1983). *FEBS Lett.* **160**, 115.

Gachelin, G., Fellous, M., Guenet, J-L., and Jacob, F. (1976). *Dev. Biol.* **50**, 316.

Gachelin, G., Kemler, R., Kelly, F., and Jacob, F. (1977). *Dev. Biol.* **57**, 199.

Gallwitz, D., Donath, C., and Sander, C. (1983). *Nature (London)* **306**, 704.

Gannon, F., Jeltsch, J. M., and Perrin, F. (1980). *Nucleic Acids Res.* **8**, 4405.

Garcia-Porrero, J. A., Colvée, E., and Ojeda, J. L. (1984). *J. Embryol. Exp. Morphol.* **80**, 241.

Garner, W. (1974). *J. Embryol. Exp. Morphol.* **32**, 849.

Gartler, S. M., Hornung, S. K., and Motulsky, A. G. (1981). *Proc. Natl. Acad. Sci. U.S.A.* **78**, 1916.

Gaziev, A. I., and Malakhova, L. V. (1982). *Int. J. Radiat. Biol.* **42**, 435.

Gazitt, Y., Reuben, R. C., Deitch, A. D., Marks, P. A., and Rifkind, R. A. (1978). *Cancer Res.* **38**, 3779.

Gearhart, J. D., and Mintz, B. (1972). *Dev. Biol.* **29**, 27.

Gee, C. J., and Harris, H. (1979). *J. Cell Sci.* **36**, 223.

George, D. L., Scott, A. F., de Martinville, B., and Francke, U. (1984). *Nucleic Acids Res.* **12**, 2731.

Gerke, V., and Weber, K. (1984). *EMBO J.* **3**, 227.

Gershon, D. (1979). *Mech. Ageing Dev.* **9**, 189.

Giallongo, A., Appella, E., Ricciardi, R., Rovera, G., and Croce, C. M. (1983). *Science* **222**, 430

Gibbs, J. B., Sigal, I. S., Poe, M., and Scolnick, E. M. (1984). *Proc. Natl. Acad. Sci. U.S.A.* **81**, 5704.

Giguère, L., and Morais, R. (1981). *Somatic Cell Genet.* **7**, 457.

Gjerset, R., Gorka, C., Hasthorpe, S., Lawrence, J. J., and Eisen, H. (1982). *Proc. Natl. Acad. Sci. U.S.A.* **79**, 2333.

Glimelius, B., and Weston, J. A. (1981). *Dev. Biol.* **82**, 95.

Glücksmann, A. (1951). *Biol. Rev. Cambridge Philos. Soc.* **26**, 59.

Glücksmann, A. (1965). *Arch. Biol.* **76**, 419.

Gold, P., and Freedman, S. O. (1965). *J. Exp. Med.* **122**, 467.

Goldberg, N. D. (1980). *Adv. Cyclic Nucleotide Res.* **12**, 147.

Goldfarb, M., Shimizu, K., Perucho, M., and Wigler, M. (1982). *Nature (London)* **296**, 404.

Goldstein, S., Moerman, E. J., Soeldner, J. S., Gleason, R. E., and Barnett, D. M. (1978). *Science* **199**, 781.

Goldstein, S., Moerman, E. J. and Porter, K. (1984). *Exptl. Cell Res.* **154**, 101.

Gooding, L. R., Hsu, Y. C., and Edidin, M. (1976). *Dev. Biol.* **49**, 479.

Gopalakrishnan, T. V., and Anderson, W. F. (1979). *Proc. Natl. Acad. Sci. U.S.A.* **76**, 3932.

Gopalakrishnan, T. V., Thompson, E. B., and Anderson, W. F. (1977). *Proc. Natl. Acad. Sci. U.S.A.* **74**, 1642.

Goshima, K., Kaneko, H., Wakabayashi, S., Masuda, A., and Matsui, Y. (1984). *Exp. Cell Res.* **151**, 148.

Goyette, M., Petropoulos, C. J., Shank, P. R. and Fausto, N. (1984). *Mol. Cell. Biol.* **4**, 1493.

Green, K., and Goldman, R. D. (1982). *J. Cell Biol.* **95**, 235a.

Greenberg, M. E., and Edelman, G. M. (1983). *Cell (Cambridge, Mass.)* **33**, 767.

Greenstock, C. L. (1981). *Radiat. Res.* **86**, 196.

Griffiths, T. D. (1984). *Mech. Ageing Dev.* **24**, 273.

Grimstad, I. A., Varani, J., and McCoy Jr., J. P. (1984). *Exp. Cell Res.* **155**, 345.

Grobstein, C. (1953). *Nature (London)* **172**, 869.

Grobstein, C., and Dalton, A. J. (1957). *J. Exp. Zool.* **135**, 57.

Groner, B., and Hynes, N. E. (1982). *Trends Biochem. Sci.* **7**, 400.

Groudine, M., and Weintraub, H. (1982). *Cell (Cambridge, Mass.)* **30**, 131.

Groudine, M., Eisenman, R., and Weintraub, H. (1981). *Nature (London)* **292**, 311.

Grove, G. L., and Cristofalo, V. J. (1977). *J. Cell. Physiol.* **90**, 415.

Grumet, M., and Edelman, G. M. (1984). *J. Cell Biol.* **98**, 1746.

Grumet, M., Hoffman, S., and Edelman, G. M. (1984). *Proc. Natl. Acad. Sci. U.S.A.* **81**, 267.

Grunz, H. (1972). *Wilhelm Roux's Arch. Dev. Biol.* **169**, 41.

Grunz, H. (1983). *Wilhelm Roux's Arch. Dev. Biol.* **192**, 130.

Gupta, R. S. (1980). *J. Cell. Physiol.* **103**, 209.

Guy, G. R., Tapley, P. M., and Murray, A. W. (1981). *Carcinogenesis* **2**, 223.

Haddox, M. K., Scott, K. F., and Russell, D. H. (1979). *Cancer Res.* **39**, 4930.

Hadorn, E. (1968). *Scient. Am.* **219**, 110.

Hagopian, H. K., Lippke, J. A., and Ingram, V. M. (1972). *J. Cell Biol.* **54**, 98.

Halaban, R., Moellmann, G., Godawska, E., and Eisenstadt, J. M. (1980). *Exp. Cell Res.* **130**, 427.

Hall, J. D., Almy, R. E., and Scherer, K. L. (1982). *Exp. Cell Res.* **139**, 351.

Hansford, R. G. (1983). *Biochim. Biophys. Acta* **726**, 41.

Harbers, K., Schnicke, A., Stuhlmann, H., Jähner, D., and Jaenisch, R. (1981). *Proc. Natl. Acad. Sci. U.S.A.* **78**, 7609.

Harding, J. D., and Rutter, W. J. (1978). *J. Biol. Chem.* **253**, 8736.

Harley, C. B., Pollard, J. W., Chamberlain, J. W., Stanners, C. P., and Goldstein, S. (1980). *Proc. Natl. Acad. Sci. U.S.A.* **77**, 1885.

Harman, D. (1981). *Proc. Natl. Acad. Sci. U.S.A.* **78**, 7124.

Harman, D. (1984). *Age* **7**, 111.

Harnden, D. G. (1977). *In* "Genetics of Human Cancer" (J. J. Mulvihill, R. W. Miller, and J. F. Fraumeni, Jr., eds.), p. 87. Raven Press, New York.

Harris, H. (1971). *Proc. R. Soc. London, Ser. B* **179**, 1.

Harris, H., Miller, O. J., Klein, G., Worst, P., and Tachibana, T. (1969). *Nature (London)* **223**, 363.

Harris, S. E., Means, A. R., Mitchell, W. M., and O'Malley, B. W. (1973). *Proc. Natl. Acad. Sci. U.S.A.* **70**, 3776.

Harris, S. E., Rosen, J. M., Means, A. R., and O'Malley, B. W. (1975). *Biochemistry* **14**, 2072.

Harrison, J., and Auersperg, N. (1981). *Science* **213**, 218.

Hart, R. W., and Setlow, R. B. (1974). *Proc. Natl. Acad. Sci. U.S.A.* **71**, 2169.

Hart, R. W., D'Ambrosio, S. M., and Kwokei, J. N. G. (1979). *Mech. Ageing Dev.* **9**, 203.

Hasegawa, N., Hanaoka, F., Hori, T., and Yamada, M. A. (1982). *Exp. Cell Res.* **140**, 443.

Hasegawa, N., Hanaoka, F., and Yamada, M. A. (1984). *Mech. Ageing Dev.* **25**, 297.

Hasegawa, N. Hanaoka, F., and Yamada, M. (1985). *Exp. Cell Res.* **156**, 478.

Hayashi, K., Fujiki, H., and Sugimura, T. (1983). *Cancer Res.* **43**, 5433.

Hayashi, J. I., Tagashira, Y., Watanabe, T., and Yoshiba, M. C. (1984). *Cancer Res.* **44**, 3957.

Hayday, A. C., Gillies, S. D., Saito, H., Wood, C., Wiman, K., Hayward, W. S., and Tonegawa, S. (1984). *Nature (London)* **307**, 334.

Hayflick, L. (1974). *J. Am. Geriatr. Soc.* **22**, 1.

Hayflick, L. (1975). *Fed. Proc., Fed. Am. Soc. Exp. Biol.* **34**, 9.

Hayflick, L. (1979). *Fed. Proc., Fed. Am. Soc. Exp. Biol.* **38**, 1847.

Hayflick, L. (1980). *Mech. Ageing Dev.* **14**, 59.

Hayflick, L. (1984). *Mech. Ageing Dev.* **28**, 177.

Hayflick, L., and Moorhead, P. S. (1961). *Exp. Cell Res.* **25**, 585.

Hedgecock, E. M., Sulston, J. E., and Thomson, J. N. (1983). *Science* **220**, 1277.

Heilporn, V., Lievens, A., Limbosch, S., Zampetti-Bosserler, F., and Steinert, G. (1973). *Radiat. Res.* **54**, 252.

Heiniger, H. J., and Marshall, J. D. (1982). *Proc. Natl. Acad. Sci. U.S.A.* **79**, 3823.

Heisterkamp, N., Groffen, J., Stephenson, J. R., Spurr, N. K., Goodfellow, P. N., Solomon, E., Carritt, B., and Bodmer, W. F. (1982). *Nature (London)* **299**, 747.

Heisterkamp, N., Stephenson, J. R., Groffen, J., Hansen, P. F., de Klein, A., Bartram, C. R., and Grosveld, G. (1983). *Nature (London)* **306**, 239.

Heldin, C. H., and Westermark, B. (1984). *Cell (Cambridge, Mass.)* **37**, 9.

Henis, H. L., III, Braid, H. L., and Vincent, R. A., Jr. (1981). *Mech. Ageing Dev.* **16**, 355.

Hennings, H., Shores, R., Wenk, M. L., Spangler, E. F., Tarone, R., and Yuspa, S. H. (1983). *Nature (London)* **304**, 67.

Heppner, G. H. (1984). *Cancer Res.* **44**, 2259.

Hicks, R. M. (1983). *Carcinogenesis* **4**, 1209.

Hjelle, B. L., Phillips, J. A., III, and Seeburg, P. H. (1982). *Proc. Natl. Acad. Sci. U.S.A.* **10**, 3459.

Hoffman, S., Chuong, C. M., and Edelman, G. M. (1984). *Proc. Natl. Acad. Sci. U.S.A.* **81**, 6881.

Hoffmann, F. M., Fresco, L. D., Hoffman-Falk, H., and Shilo, B. Z. (1983). *Cell (Cambridge, Mass.)* **35**, 393.

Hoffman-Liebermann, B., Liebermann, D., and Sachs, L. (1981). *Int. J. Cancer* **28**, 615.

Hogan, B. L. M., Taylor, A., and Adamson, E. (1981). *Nature (London)* **291**, 235.

Holden, S., Bernard, O., Artzt, K., Whitmore, W. F., Jr., and Bennett, D. (1977). *Nature (London)* **270**, 518.

Holliday, R. (1975). *Fed. Proc., Fed. Am. Soc. Exp. Biol.* **34**, 51.

Holliday, R., and Pugh, J. E. (1975). *Science* **187**, 226.

Holliday, R., and Tarrant, G. M. (1972). *Nature (London)* **238**, 26.

Holliday, R., Huschtscha, L. I., Tarrant, G. M., and Kirkwood, T. B. L. (1977). *Science* **198**, 366.

Holtfreter, J. (1933). *Biol. Zbl.* **53**, 404.

Holtfreter, J. (1939). *Arch. Exp. Zellforsch. Besonders Gewebezuecht.* **23**, 169.

Holtfreter, J. (1945). *J. Exp. Zool.* **98**, 161.

Holtzer, H., Weintraub, H., Mayne, R., and Mochan, B. (1972). *Curr. Top. Dev. Biol.* **7**, 229.

Holtzer, H., Rubinstein, N., Fellini, S., Yeoh, G., Chi, J., Birnbaum, J., and Okayama, L. (1975). *Q. Rev. Biophys.* **8**, 523.

Holtzer, H., Pacifici, M., Payette, R., Croop, J., Dlugosz, A., and Toyama, Y. (1982). *Carcinogenesis* **7**, 347.

Horowitz, A. D., Greenebaum, E., Nicolaides, M., Woodward, K., and Weinstein, I. B. (1982). *Mol. Cell. Biol.* **2**, 545.

Hors-Cayla, M. C., Heuertz, S., and Frezal, J. (1983). *Somatic Cell Genet.* **9**, 645.

Houben, A., Raes, M., Houbion, A., and Remacle, J. (1984). *Mech. Ageing Dev.* **25**, 35.

Houslay, M. D. (1981). *Biosci. Rep.* **1**, 19.

Howe, W. E., and Oshima, R. G. (1982). *Mol. Cell. Biol.* **2**, 331.

Howell, A. N., and Sager, R. (1978). *Proc. Natl. Acad. Sci. U.S.A.* **75**, 2358.

Howell, N. (1982). *Cytogenet. Cell Genet.* **34**, 215.

Hsie, A. W., and Puck, T. T. (1971). *Proc. Natl. Acad. Sci. U.S.A.* **68**, 358.

Hsu, L., Natyzak, D., and Laskin, J. D. (1984). *Cancer Res.* **44**, 4607.

Huberman, E., and Callaham, M. F. (1979). *Proc. Natl. Acad. Sci. U.S.A.* **76**, 1293.

Huberman, E., Mager, R., and Sachs, L. (1976). *Nature (London)* **264**, 360

Huberman, E., Weeks, C., Herrmann, A., Callaham, M., and Slaga, R. (1981). *Proc. Natl. Acad. Sci. U.S.A.* **78**, 1062.

Huebner, R. J., and Todaro, G. J. (1969). *Proc. Natl. Acad. Sci. U.S.A.* **64**, 1087.

Hunter, T. (1984). *Sci. Am.* **251**(2), 60.

Hunter, T., and Cooper, J. A. (1981). *Cell (Cambridge, Mass.)* **24**, 741.

Hutson, J. M., Fallat, M. E., Kamagata, S., Donahoe, P. K., and Budzik, B. P. (1984). *Science* **223**, 586.

Hynes, N. E., Groner, B., Sippel, A. E., Nguyen-Huu, M. C., and Schütz, G. (1977). *Cell (Cambridge, Mass.)* **11**, 923.

Hynes, N. E., Kennedy, N., Rahmsdorf, V., and Groner, B. (1981). *Proc. Natl. Acad. Sci. U.S.A.* **78**, 2038.

Hynes, R. (1982). *Cell (Cambridge, Mass.)* **28**, 437.

Ibsen, K. H., and Fishman, W. H. (1979). *Biochim. Biophys. Acta* **560**, 243.

Illmensee, K., and Croce, C. M. (1979). *Proc. Natl. Acad. Sci. U.S.A.* **76**, 879.

Illmensee, K., and Mintz, B. (1976). *Proc. Natl. Acad. Sci. U.S.A.* **73**, 549.

Ingram, V. M., Chan, N. L., Hagopian, H. K., Lippke, J. A., and Wu, L. (1974). *Dev. Biol.* **36**, 411.

Ish-Horowicz, D. (1982). *Nature (London)* **296**, 806.

Ishii, D. N., Fibach, E., Yamasaki, H., and Weinstein, I. B. (1978). *Science* **200**, 556.

Ivy, G. O., Schottler, F., Wenzel, J., Baudry, M. and Lynch, G. (1984). *Nature* **226**, 985.

Izumi, L., Hirao, Y., Hopp, L., and Oyasu, R. (1981). *Cancer Res.* **41**, 405.

Jackowski, G., and Liew, C. C. (1982). *Cell Biol. Int. Rep.* **6**, 867.

Jacob, F. (1978). *Proc. R. Soc. London, Ser. B* **201**, 249.

Jacobs, S., Sahyoun, N. E., Saltiel, A. R., and Cuatrecasas, P. (1983). *Proc. Natl. Acad. Sci. U.S.A.* **80**, 6211.

Jacodzinski, L. L., Sargent, T. D., Yang, M., Glackin, C., and Bonner, J. (1981). *Proc. Natl. Acad. Sci. U.S.A.* **78**, 3521.

Jaenisch, R., Harbers, K., Jähner, D., Stewart, C., and Stuhlmann, H. (1982). *J. Cell Biochem.* **20**, 131.

Jähner, D., Stuhlmann, H., Stewart, C. L., Harbers, K., Löhler, J., Simon, I., and Jaenisch, R. (1982). *Nature (London)* **298**, 623.

Janeczek, J., Born, J., John, M., Scharschmidt, M., Tiedemann, H., and Tiedemann, H. (1984a). *Eur. J. Biochem.* **140**, 257.

Janeczek, J., John, M., Born, J., and Tiedemann, H. (1984b). *Roux's Arch. Dev. Biol.* **193**, 1.

Jariwalla, R. J., Aurelian, L., and Ts'o, P. O. P. (1983). *Proc. Natl. Acad. Sci. U.S.A.* **80**, 5902.

Jensen, E. V., and DeSombre, E. R. (1973). *Science* **182**, 126.

Jetten, A. M. (1980). *Nature (London)* **284**, 626.

Jetten, A. M., and Jetten, M. E. R. (1979). *Nature (London)* **278**, 180.

Jetten, A. M., Jetten, M. E. R., and Sherman, M. (1979). *Exp. Cell Res.* **124**, 381.

Johnson, T. E., and McCaffrey, G. (1985). *Mech. Ageing Dev.* **30**, 285.

Johnson, G. S., and Pastan, I. (1972). *Nature (London), New Biol.* **237**, 267.

Johnsson, A., Heldin, C. H., Wasteson, Å., Westermark, B., Deuel, T. F., Huang, J. S., Seeburg, P. H., Gray, A., Ullrich, A., Scrace, G., Stroobant, P., and Waterfield, M. D. (1984). *EMBO J.* **3**, 921.

Johnstone, A. P., and Williams, G. T. (1982). *Nature (London)* **300**, 368.

Jones, P. A. (1985). *Cell (Cambridge, Mass.)* **40**, 485.

Jones, P. A., Taylor, S. M., Mohandas, T., and Shapiro, L. J. (1982). *Proc. Natl. Acad. Sci. U.S.A.* **79**, 1215.

Jones, P. A., Taylor, S. M., and Wilson, V. (1983). *J. Exp. Zool.* **228**, 287.

Jones, R. E., De Feo, D., and Piatigorsky, J. (1981). *J. Biol. Chem.* **256**, 8172.

Jones-Villeneuve, E. M. V., Rudnicki, M. A., Harris, J. F., and McBurney, M. W. (1983). *Mol. Cell. Biol.* **3**, 2271.

Josephs, S. F., Guo, C., Ratner, L., and Wong-Staal, F. (1984). *Science* **223**, 487.

Kaftory, A., Hershko, A., and Frey, M. (1978). *J. Cell. Physiol.* **94**, 147.

Kahn, C. R., Betolotti, R., Ninio, M., and Weiss, M. C. (1981). *Nature (London)* **290**, 717.

Kaji, K., and Matsuo, M. (1981). *Exp. Cell Res.* **131**, 410.

Kanda, N., Schreck, R., Alt, F., Bruns, G., Baltimore, D., and Latt, S. (1983). *Proc. Natl. Acad. Sci. U.S.A.* **80**, 4069.

Kane, A. B., Stanton, R. P., Raymond, E. G., Dobson, M. E., Knafelc, M. E., and Farber, J. L. (1980). *J. Cell Biol.* **87**, 643.

Kaplan, P. L., Anderson, M., and Ozanne, B. (1982). *Proc. Natl. Acad. Sci. U.S.A.* **79**, 485.

Karino, S., and Hiwatashi, K. (1981). *Exp. Cell Res.* **136**, 407.

Kasuga, M., Zick, Y., Blithe, D. L., Crettaz, M., and Kahn, C. R. (1982a). *Nature (London)* **298**, 667.

Kasuga, M., Zick, Y., Blith, D. L., Karlsson, F. A., Häring, H. V., and Kahn, C. R. (1982b). *J. Biol. Chem.* **257**, 9891.

Kato, Y., Nomura, Y., Tsuji, M., Kinoshita, M., Ohmae, H., and Suzuki, F. (1981). *Proc. Natl. Acad. Sci. U.S.A.* **78**, 6831.

Keath, E. J., Kelekar, A., and Cole, M. D. (1984). *Cell* **37**, 521.

Kelley, R. O., Mann, P. L., Perdue, B. D., and Marek, L. F. (1985). *Mech. Ageing Dev.* **30**, 79.

Kelly, F., and Condamine, H. (1982). *Biochim. Biophys. Acta* **651**, 105.

Kemler, R., Babinet, C., Eisen, H., and Jacob, F. (1977). *Proc. Natl. Acad. Sci. U.S.A.* **74**, 4449.

Killary, A. M. and Fournier, R. E. K. (1984). *Cell* **38**, 523.

Kinsella, A. R., and Radman, M. (1978). *Proc. Natl. Acad. Sci. U.S.A.* **75**, 6149.

Kirkwood, T. B. L., Holliday, R., and Rosenberger, R. F. (1984). *Intern. Rev. Cytol.* **22**, 93.

Kirschmeier, P., Gattoni-Celli, S., Dina, D., and Weinstein, I. B. (1982). *Proc. Natl. Acad. Sci. U.S.A.* **79**, 2773.

Klass, M., Nguyen, P. N., and Dechavigny, A. (1983). *Mech. Ageing Dev.* **22**, 253.

Klein, G. (1979). *Proc. Natl. Acad. Sci. U.S.A.* **76**, 2442.

Klein, G. (1981). *Nature (London)* **294**, 313.

Klein, G. (1983). *Cell (Cambridge, Mass.)* **32**, 311.

Klein, G., and Klein, E. (1984). *Carcinogenesis* **5**, 429.

Klein, G., and Klein, E. (1985). *Nature (London)* **314**, 190.

Klinger, H. P. (1980). *Cytogenet. Cell Genet.* **27**, 254.

Klinger, H. P. (1982). *Cytogenet. Cell Genet.* **32**, 68.

Klinger, H. P., Baim, A. S., Eun, C. K., Shows, T. B., and Ruddle, F. H. (1978). *Cytogenet. Cell Genet.* **22**, 245.

Knudson, A. G., Jr. (1985). *Cancer Res.* **45**, 1437.

Kohl, N. E., Kanda, N., Schreck, R. R., Bruns, G., Latt, S. A., Gilbert, F., and Alt, F. W. (1983). *Cell (Cambridge, Mass.)* **35**, 359.

Köhler, G., and Milstein, C. (1975). *Nature (London)* **256**, 495.

Kolata, G. (1984). *Science* **223**, 1376.

Konieczny, S. F., and Emerson, Jr., C. P. (1984). *Cell (Cambridge, Mass.)* **38**, 791.

Kopelovich, L. (1982). *Int. Rev. Cytol.* **77**, 63.

Kozbor, D., and Croce, C. M. (1984). *Cancer Res.* **44**, 438.

Kraft, A. S., and Anderson, W. B. (1983). *Nature (London)* **301**, 621.

Krawisz, B. R., and Scott, R. E. (1982). *J. Cell Biol.* **94**, 394.

Kucherlapati, R., and Shin, S. (1979). *Cell (Cambridge, Mass.)* **16**, 639.

Kuff, E. L., and Fewell, J. W. (1980). *Dev. Biol.* **77**, 103.

Kugimiya, W., Ikenaga, H., and Saigo, K. (1983). *Proc. Natl. Acad. Sci. U.S.A.* **80**, 3193.

Kühn, A. (1971). "Lectures on Developmental Physiology," p. 235. Springer-Verlag, Berlin and New York.

Kuri-Harcuch, W. (1982). *Differentiation* **23**, 164.

Lacroix, A., Anderson, G. D., and Lippman, M. E. (1980). *Exp. Cell Res.* **130**, 339.

Land, H., Parada, L. F., and Weinberg, R. A. (1983a). *Nature (London)* **304**, 596.

Land, H., Parada, L. F., and Weinberg, R. A. (1983b). *Science* **222**, 771.

Landolph, J. R., and Jones, P. A. (1982). *Cancer Res.* **42**, 817.

Lane, M. A., Sainten, A., and Cooper, G. M. (1981). *Proc. Natl. Acad. Sci. U.S.A.* **78**, 5185.

La Rocca, P. J., and Rafferty, K. A., Jr. (1982). *J. Cell. Physiol.* **113**, 203.

Lash, J. W. (1969). *Ann. Embryol. Morphol.*, *Suppl.* **1**, 255.

Latchman, D. S., Brzeski, H., Lovell-Badge, R. and Evans, M. J. (1984). *Biochim. Biophys. Acta* **783**, 130.

Lavi, S. (1981). *Proc. Natl. Acad. Sci. U.S.A.* **78**, 6144.

Lawrence, D. A. (1985). *Biol. Cell* **53**, 98.

Lawrence, D. A., Pircher, R., Krycève-Martinerie, C., and Jullien, P. (1984). *J. Cell Physiol.* **121,** 184.

Leavitt, J., and Moyzis, R. (1978). *J. Biol. Chem.* **253,** 2497.

Leder, P., Battey, J., Lenoir, G., Moulding, C., Murphy, W., Potter, H., Stewart, T., and Taub, R. (1983). *Science* **222,** 765.

Lederberg, J. (1952). *Physiol. Rev.* **32,** 403.

Le Douarin, N. (1980). *Curr. Top. Dev. Biol.* **16,** 31.

Lee, H., Deshpande, A. K., and Kalmus, G. W. (1974). *J. Embryol. Exp. Morphol.* **32,** 835.

Lee, L. S., and Weinstein, I. B. (1980). *Carcinogenesis* **1,** 669.

Lee, W. H., Murphree, A. L., and Benedict, W. F. (1984). *Nature (London)* **309,** 458.

Lehto, V. P., Virtanen, I., Paasivuo, R., Ralston, R., and Alitalo, K. (1983). *EMBO J.* **2,** 1701.

Leibovitch, M. P., Leibovitch, S. A., Harel, J., and Kruh, J. (1982). *Differentiation* **22,** 106.

Leonardi, C. L., Warren, R. H., and Rubin, R. W. (1982). *Biochim. Biophys. Acta* **720,** 154.

Leray, G., Guenet, L., Le Treut, A., and Le Gall, J. Y. (1983). *Biol. Cell.* **47,** 235.

Lester, S. C., Korn, N. J., and DeMars, R. (1982). *Somatic Cell Genet.* **8,** 265.

Lev, Z., Leibovitz, N., Segev, O., and Shilo, B. Z. (1984). *Mol. Cell. Biol.* **4,** 982.

Levilliers, J., and Weiss, M. C. (1983). *Somatic Cell Genet.* **9,** 407.

Levi-Montalcini, R., and Angeletti, P. U. (1968). *Physiol. Rev.* **48,** 534.

Levine, S., Pictet, R., and Rutter, W. J. (1973). *Nature (London), New Biol.* **246,** 49.

Levinson, A. D., Courtneidge, S. A., and Bishop, J. M. (1981). *Proc. Natl. Acad. Sci. U.S.A.* **78,** 1624.

Levitt, D., and Dorfman, A. (1972). *Proc. Natl. Acad. Sci. U.S.A.* **69,** 1253.

Levitt, D., and Dorfman, A. (1974). *Curr. Top. Dev. Biol.* **8,** 103.

Liebermann, D., and Sachs, L. (1978). *Exp. Cell Res.* **113,** 383.

Liebermann, D., Hoffman-Liebermann, B., and Sachs, L. (1981). *Int. J. Cancer* **28,** 285.

Liesi, P., Rechardt, L., and Wartiovaara, J. (1983). *Nature (London)* **306,** 265.

Lin, C. R., Chen, W. S., Kruiger, W., Stolarsky, L. S., Weber, W., Evans, R. M., Verma, I. M. Gill, G. N., and Rosenfeld, M. G. (1984). *Science* **224,** 843.

Lin, S. L., Greene, J. J., Ts'O, P. O. P., and Carter, W. A. (1982). *Nature (London)* **297,** 417.

Lincoln II, D. W., Braunschweiger, K. I., Braunschweiger, W. R. and Smith, J. R. (1984). *Exp. Cell Res.* **154,** 136.

Linder, S. (1980). *Exp. Cell Res.* **130,** 159.

Linder, S., Zuckerman, S. H., and Ringertz, N. R. (1981). *Proc. Natl. Acad. Sci. U.S.A.* **78,** 6286.

Linn, S., Kairis, M., and Holliday, R. (1976). *Proc. Natl. Acad. Sci. U.S.A.* **73,** 2818.

Linzer, H. I., and Levine, A. J. (1979). *Cell (Cambridge, Mass.)* **17,** 43.

Lipetz, J., and Cristofalo, V. J. (1972). *J. Ultrastr. Res.* **39,** 43.

Little, C. D., Nau, M. M., Carney, D. N., Gazdar, A. F., and Minna, J. D. (1983). *Nature (London)* **306,** 194.

Little, J. B., Nagasawa, H., and Kennedy, A. R. (1979). *Radiat. Res.* **79,** 241.

Littlefield, J. W. (1973). *J. Cell. Physiol.* **82,** 129.

Liu, A. Y. C. (1982). *J. Biol. Chem.* **257,** 298.

Logan, J., and Cairns, J. (1982). *Nature (London)* **300,** 104.

Lörincz, A. T., and Reed, S. I. (1984). *Nature (London)* **307,** 183.

Lotan, R. (1980). *Biochim. Biophys. Acta* **605,** 33.

Lotan, R., and Lotan, D. (1981). *J. Cell. Physiol.* **108,** 179.

Lotan, R., and Nicolson, G. L. (1979). *Cancer Res.* **39,** 4767.

Lotem, J., and Sachs, L. (1979). *Proc. Natl. Acad. Sci. U.S.A.* **76,** 5158.

Lotem, J., and Sachs, L. (1983). *Int. J. Cancer* **32,** 127.

Lowe, M., Pacifici, A., and Holtzer, H. (1978). *Cancer Res.* **38,** 2350.

Macara, I. G., Marinetti, G. V., and Balduzzi, P. C. (1984). *Proc. Natl. Acad. Sci. U.S.A.* **81,** 2728.

MacCann, J., and Ames, B. N. (1976). *Proc. Natl. Acad. Sci. U.S.A.* **73,** 950.

McCaffrey, P. G., Friedman, B., and Rosner, M. R. (1984). *J. Biol. Chem.* **259**, 12502.

McCoy, M. S., Toole, J. J., Cunningham, J. M., Chang, E. H., Lowy, D. R., and Weinberg, R. A. (1983). *Nature (London)* **302**, 79.

McGeady, M. L., Jhappan, C., Ascione, R., and Vande Woude, G. F. (1983). *Mol. Cell. Biol.* **3**, 305.

McGrath, J. P., Capon, D. J., Goeddel, D. V., and Levinson, A. D. (1984). *Nature* **310**, 644.

Maciag, T. (1983). *Trends Biochem. Sci.* **8**, 265.

Macieira-Coelho, A. (1983). *Int. Rev. Cytol.* **83**, 183.

Macieira-Coelho, A., and Azzarone, B. (1982). *Exp. Cell Res.* **141**, 325.

McKeehan, M. S. (1951). *J. Exp. Zool.* **117**, 31.

McKeon, C., Ohkubo, H., Pastan, I., and de Crombrugghe, B. (1982). *Cell (Cambridge, Mass.)* **29**, 203.

McKinnell, R. G., Deggins, B. A., and Labat, D. D. (1969). *Science* **165**, 394.

Madsen, K., Friberg, U., Roos, P., Edén, S., and Isaksson, O. (1983). *Nature (London)* **304**, 545.

Mager, D., and Bernstein, A. (1980). *J. Cell. Physiol.* **105**, 519.

Magnusson, G. (1985). *Exp. Cell Res.* **157**, 1.

Manes, C., and Menzel, P. (1981). *Nature (London)* **293**, 589.

Maness, P. F., and Levy, B. T. (1983). *Mol. Cell. Biol.* **3**, 102.

Maness, P. F., and Walsh, R. C., Jr. (1982). *Cell (Cambridge, Mass.)* **30**, 253.

Marks, F., Berry, D. L., Bertsch, S., Fürstenberger, G., and Richter, H. (1982). *Carcinogenesis* **7**, 331.

Marks, P. A., and Rifkind, R. A. (1978). *Annu. Rev. Biochem.* **47**, 419.

Marquardt, H., Hunkapiller, M. W., Hood, L. E., Twardzik, D. R., DeLarco, J. E., Stephenson, J. R., and Todaro, G. J. (1983). *Proc. Natl. Acad. Sci. U.S.A.* **80**, 4684.

Marquardt, H., Hunkapiller, M. W., Hood, L. E., and Todaro, G. J. (1984). *Science* **223**, 1079.

Marshall, C. J., and Sager, R. (1981). *Somatic Cell Genet.* **7**, 713.

Marshall, C. J., Hall, A., and Weiss, R. A. (1982a). *Nature (London)* **299**, 171.

Marshall, C. J., Kitchin, R. M., and Sager, R. (1982b). *Somatic Cell Genet.* **8**, 709.

Marshall, C. J., Vousden, K. H., and Phillips, D. H. (1984). *Nature* **310**, 586.

Marshall Graves, J. A. (1982). *Exp. Cell Res.* **141**, 99.

Martin, G. R. (1980). *Science* **209**, 770.

Martin, G. R. (1981). *Proc. Natl. Acad. Sci. U.S.A.* **78**, 7634.

Martin, G. R. (1982). *Cell (Cambridge, Mass.)* **29**, 721.

Martin, G. R., and Evans, M. J. (1975). *Cell (Cambridge, Mass.)* **6**, 470.

Martinez, R., Nakamura, K. D., and Weber, M. J. (1982). *Mol. Cell. Biol.* **2**, 653.

Massagué, J. (1985). *Tibs* **10**, 237.

Massagué, J., Czech, M. P., Iwata, K., DeLarco, J. E., and Todaro, G. J. (1982). *Proc. Natl. Acad. Sci. U.S.A.* **79**, 6822.

Mathews, M. B., Bernstein, R. M., Franza, B. R., Jr., and Garrels, J. I. (1984). *Nature (London)* **309**, 374.

Mays-Hoopes, L. L., Brown, A., and Huang, R. C. C. (1983a). *Mol. Cell. Biol.* **3**, 1371.

Mays-Hoopes, L. L., Cleland, G., Bochantin, J., Kalunian, D., Miller, J., Wilson, W., Wong, M. K., Johnson, D., and Sharma, O. K. (1983b). *Mech. Ageing Dev.* **22**, 135.

Mazia, D., and Gontcharoff, M. (1964). *Exp. Cell Res.* **35**, 14.

Medford, R. M., Nguyen, H. T., and Nadal-Ginard, B. (1983). *J. Biol. Chem.* **258**, 11063.

Meek, R. L., Rebeiro, T., and Daniel, C. W. (1980). *Exp. Cell Res.* **129**, 265.

Meek, W. D. (1982). *Mol. Cell. Biol.* **2**, 863.

Meeks, R. G., Zaharevitz, D., and Chen, R. F. (1981). *Arch. Biochem. Biophys.* **207**, 141.

Mehta, R. G., Cerny, W. L., and Moon, R. C. (1980). *Cancer Res.* **40**, 47.

Melton, D. A. (1985). *Proc. Natl. Acad. Sci. U.S.A.* **82**, 144.

Menahan, L. A., and Kemp, R. G. (1982). *Mech. Ageing Dev.* **20**, 195.

Mercer, W. E., Nelson, D., Deleo, A. B., Old, L. J., and Baserga, R. (1982). *Proc. Natl. Acad. Sci. U.S.A.* **79,** 6309.

Mercer, W. E., Avignolo, C., and Baserga, R. (1984). *Mol. Cell. Biol.* **4,** 276.

Merlino, G. T., Xu Y.-H., Ishii, S., Clark, A. J. L., Semba, K., Toyoshima, K., Yamamoto, T., and Pastan, I. (1984). *Science* **224,** 417.

Metcalf, D. (1985). *Science* **229,** 16.

Mével-Ninio, M., and Weiss, M. C. (1981). *J. Cell Biol.* **90,** 339.

Michell, B. (1983). *Trends Biochem. Sci.* **8,** 263.

Miller, D. A., Okamoto, E., Erlanger, B. F., and Miller, O. J. (1982). *Cytogenet. Cell Genet.* **33,** 345.

Miller, D. M., Turner, P., Nienhuis, A. W., Axelrod, D. E., and Gopalakrishnan, T. V. (1978). *Cell (Cambridge, Mass.)* **14,** 511.

Miller, D. R., Hamby, K. M., and Slaga, T. J. (1982). *J. Cell. Physiol.* **112,** 76.

Miller, R. A., and Ruddle, F. H. (1977). *Dev. Biol.* **56,** 157.

Minna, J., Glazer, D., and Nirenberg, M. (1972). *Nature (London), New Biol.* **235,** 225.

Mintz, B. (1971). *In* "Methods in Mammalian Embryology" (J. L. Daniel, Jr., ed.), p. 191. Freeman, San Francisco, California.

Mintz, B., and Illmensee, K. (1975). *Proc. Natl. Acad. Sci. U.S.A.* **72,** 3585.

Mintz, B., Cronmiller, C., and Custer, R. P. (1978). *Proc. Natl. Acad. Sci. U.S.A.* **75,** 2834.

Mitelman, F., and Levan, G. (1981). *Hereditas* **95,** 79.

Mitrani, E. (1984). *Exp. Cell Res.* **152,** 148.

Miura, Y., and Wilt, F. H. (1971). *J. Cell Biol.* **48,** 523.

Miyake, R., Tanaka, Y., Tsuda, T., Kaibuchi, K., Kikkawa, U., and Nishizuka, Y. (1984). *Biochem. Biophys. Res. Commun.* **121,** 649.

Moens, W., Vokaer, A., and Kram, R. (1975). *Proc. Natl. Acad. Sci. U.S.A.* **72,** 1063.

Mohandas, T., Sparkes, R. S., and Shapiro, L. J. (1981). *Science* **211,** 393.

Montagnier, L., Chamaret, S., and Dauguet, C. (1982). *C.R. Hebd. Seances Acad. Sci.* **295,** 375.

Montelaro, R. C., and Bolognesi, D. P. (1978). *Adv. Cancer Res.* **28,** 79.

Moolenaar, W. H., Tertoolen, L. G. J., and De Laat, S. W. (1984). *Nature (London)* **312,** 371.

Moon, S. O., Palfrey, H. C., and King, A. C. (1984). *Proc. Natl. Acad. Sci. U.S.A.* **81,** 2298.

Mora, P. T., Chandrasedaran, K., and McFarland, V. W. (1980). *Nature (London)* **288,** 722.

Morange, M., Diu, A., Bensaude, O., and Babinet, C. (1984). *Mol. Cell. Biol.* **4,** 730.

Mori, N., Hiruta, K., Funatsu, Y., and Goto, S. (1983). *Mech. Ageing Dev.* **22,** 1.

Morley, A. A., Cox, S., and Holliday, J. (1982). *Mech. Ageing Dev.* **19,** 21.

Moscona, A. A. (1957). *Proc. Natl. Acad. Sci. U.S.A.* **83,** 184.

Moscona, A. A., and Degenstein, L. (1981). *Cell Differ.* **10,** 39.

Moscona, A. A., Brown, M., Degenstein, L., Fox, L., and Soh, B. M. (1983). *Proc. Natl. Acad. Sci. U.S.A.* **80,** 7239.

Moscona, M., and Moscona, A. A. (1979). *Differentiation* **13,** 165.

Moscona, M., Frenke, L., and Moscona, A. A. (1972). *Dev. Biol.* **28,** 229.

Mroczkowski, B., Mosig, G., and Cohen, S. (1984). *Nature (London)* **309,** 270.

Mufson, R. A., and Weinstein, I. B. (1981). *In* "Biochemistry of Cellular Regulation" (M. J. Clemens, ed.), Vol. 3, p. 179. CRC Press, Cleveland, Ohio.

Mufson, R. A., DeFeo, D., and Weinstein, I. B. (1979). *Mol. Pharmacol.* **16,** 569.

Mufson, R. A., Steinber, M. L., and Defendi, V. (1982). *Cancer Res.* **42,** 4600.

Muggleton-Harris, A. L., and Aroian, M. A. (1982). *Somatic. Cell Genet.* **8,** 41.

Muggleton-Harris, A. L., and Hayflick, L. (1976). *Exp. Cell Res.* **103,** 321.

Mulcahy, L. S., Smith, M. R.., and Stacey, D. W. (1985). *Nature (London)* **313,** 241.

Müller, R., Slamon, D. J., Tremblay, J. M., Cline, M. J., and Verma, I. M. (1982). *Nature (London)* **299,** 640.

Müller, R., Verma, I. M., and Adamson, E. D. (1983a). *EMBO J.* **2,** 679.

Müller, R., Tremblay, J. M., Adamson, E. D., and Verma, I. M. (1983b). *Nature (London)* **304**, 454.

Mullins, D. E., and Rohrlich, S. T. (1983). *Biochim. Biophys. Acta* **695**, 177.

Mulvihill, E. R., Le Pennec, J. P., and Chambon, P. (1982). *Cell (Cambridge, Mass.)* **26**, 621.

Muramatsu, T., Gachelin, G., Nicolas, J. F., Condamine, H., Jakob, H., and Jacob, F. (1978). *Proc. Natl. Acad. Sci. U.S.A.* **75**, 2315.

Muramatsu, T., Gachelin, G., Damonneville, M., Delarbre, C., and Jacob, F. (1979). *Cell (Cambridge, Mass.)* **18**, 183.

Murphree, A. L., and Benedict, W. F. (1984). *Science* **223**, 1028.

Murray, A. W., and Fitzgerald, D. J. (1979). *Biochem. Biophys. Res. Commun.* **91**, 395.

Murray, M., Shilo, B., Shih, C., Cowing, D., Hsu, H. W., and Weinberg, R. A. (1981). *Cell (Cambridge, Mass.)* **25**, 355.

Murray, V. (1981). *Mech. Ageing Dev.* **16**, 327.

Murray, V., and Holliday, R. (1981). *J. Mol. Biol.* **146**, 55.

Naeim, F., Bergmann, K., and Walford, R. L. (1981). *Age* **4**, 5.

Nagle, D. S., and Blumberg, B. M. (1980). *Cancer Res.* **40**, 1066.

Nakajima, M., Irimura, T., Di Ferrante, D., Di Ferrante N., and Nicolson, G. L. (1983). *Science* **220**, 611.

Nakajima, M., Irimura, T., Di Ferrante, N., and Nicolson, G. L. (1984). *J. Biol. Chem.* **259**, 2283.

Nakamura, K. D., and Weber, M. J. (1982). *Mol. Cell. Biol.* **2**, 147.

Nakao, Y., Matsuda, S., Fujica, T., Watanabe, S., Morikawa, S., Saida, T., and Ito, Y. (1982). *Cancer Res.* **42**, 3843.

Nakatsuji, N., and Johnson, K. E. (1984). *Nature (London)* **307**, 453.

Nakazawa, T., Mori, N., and Goto, S. (1984). *Mech. Ageing Dev.* **26**, 241.

Naveh-Many, T., and Cedar, H. (1981). *Proc. Natl. Acad. Sci. U.S.A.* **78**, 4246.

Nawrocki, J. F., Lau, A. F., and Faras, A. J. (1984). *Mol. Cell. Biol.* **4**, 212.

Nelson, W. J., and Lazarides, E. (1983). *Nature (London)* **304**, 364.

Netland, P. A. and Zetter, B. R. (1984). *Science* **224**, 1113.

Nette, E. C., Sit, H. L., and King, D. W. (1982). *Mech. Ageing Dev.* **18**, 75.

Neuman-Silberberg, F. S., Schejter, E., Hoffmann, F. M., and Shilo, B. Z. (1984). *Cell* **37**, 1027.

Newbold, R. F., and Amos, J. (1981). *Carcinogenesis* **2**, 243.

Newbold, R. F., and Overell, R. W. (1983). *Nature (London)* **304**, 648.

Newbold, R. F., Overell, R. W., and Connell, J. R. (1982). *Nature (London)* **299**, 633.

Newmark, P. (1982). *Nature (London)* **296**, 393.

Nicolas, J. F., Gaillard, J., Jakob, H., and Jacob, F. (1980). *Nature (London)* **286**, 716.

Nicolson, G. L. (1976). *Biochim. Biophys. Acta* **458**, 1.

Nicolson, G. L. (1982). *Biochim. Biophys. Acta* **695**, 113.

Nicolson, G. L. (1984). *Exp. Cell Res.* **150**, 3.

Nicolson, G. L., and Custead, S. E. (1982). *Science* **215**, 176.

Niedermüller, H. (1982). *Mech. Ageing Dev.* **19**, 259.

Nilsen-Hamilton, M., Hamilton, R. T., Allen, W. R., and Potter-Perigo, S. (1982). *Cell (Cambridge, Mass.)* **31**, 237.

Niman, H. L., Houghten, R. A., and Bowen-Pope, D. F. (1984). *Science* **226**, 701.

Nishikura, K., Ar-Rushdi, A., Erikson, J., Watt, R., Rovera, G., and Croce, C. M. (1983). *Proc. Natl. Acad. Sci. U.S.A.* **80**, 4822.

Nishimura, J., Huang, J. S., and Devel, T. F. (1982). *Proc. Natl. Acad. Sci. U.S.A.* **79**, 4303.

Nishizuka, Y. (1984). *Nature (London)* **308**, 693.

Nistér, M., Heldin, C.-H., Wasteson, Å., and Westermark, B. (1984). *Proc. Natl. Acad. Sci. U.S.A.* **81**, 926.

Nomura, K. (1982). *Differentiation* **22**, 179.

Nordheim, A., Tesser, P., Azorin, F., Kwon, Y. H., Möller, A., and Rich, A. (1982). *Proc. Natl. Acad. Sci. U.S.A.* **79,** 7729.

Nowell, P. C., and Hungerford, D. A. (1960). *Science* **132,** 1497.

Nusse, R., and Varmus, H. E. (1982). *Cell (Cambridge, Mass.)* **31,** 99.

O'Brien, T. G., and Diamond, L. (1977). *Cancer Res.* **37,** 3895.

Ojakian, G. K. (1981). *Cell (Cambridge, Mass.)* **22,** 95.

Okada, T. S. (1973). *Proc. Natl. Acad. Sci. U.S.A.* **70,** 1496.

Okada, T. S. (1980). *Curr. Top. Dev. Biol.* **16,** 349.

Okashi, Y., Veda, K., Hayaishi, O., Ikai, K., and Niwa, O. (1984). *Proc. Natl. Acad. Sci. U.S.A.* **81,** 713.

Okayama, M., Pacifici, M., and Holtzer, H. (1976). *Proc. Natl. Acad. Sci. U.S.A.* **73,** 3224.

Olsson, L., and Forchhammer, J. (1984). *Proc. Natl. Acad. Sci. U.S.A.* **81,** 3389.

O'Malley, B. W., and Means, A. R. (1974). *Science* **183,** 610.

Ono, T., and Cutler, R. G. (1978). *Proc. Natl. Acad. Sci. U.S.A.* **75,** 4431.

Oppermann, H., Levinson, W., and Bishop, J. M. (1981). *Proc. Natl. Acad. Sci. U.S.A.* **78,** 1067.

Oren, M., Reich, N. C., and Levine, A. J. (1982). *Mol. Cell. Biol.* **2,** 443.

Orgel, L. E. (1963). *Proc. Natl. Acad. Sci. U.S.A.* **49,** 517.

Orgel, L. E. (1970). *Proc. Natl. Acad. Sci. U.S.A.* **67,** 1476.

Orgel, L. E. (1973). *Nature (London)* **243,** 441.

Ossowski, L., and Reich, E. (1983). *Cell (Cambridge, Mass.)* **35,** 611.

Ostertag, W., Crozier, T., Kruge, N., Melderis, H., and Dube, S. (1973). *Nature (London), New Biol.* **243,** 203.

Pacifici, M., and Holtzer, H. (1977). *Am. J. Anat.* **150,** 207.

Pacifici, M., and Holtzer, H. (1980). *Cancer Res.* **40,** 2461.

Paffenholz, V. (1978). *Mech. Ageing Dev.* **7,** 131.

Pall, M. L. (1981). *Proc. Natl. Acad. Sci. U.S.A.* **78,** 2465.

Palmiter, R. D. (1975). *Cell (Cambridge, Mass.)* **4,** 189.

Palmiter, R. D., Mulvihill, E. R., McKnight, G. S., and Senear, A. W. (1977). *Cold Spring Harbor Symp. Quant. Biol.* **42,** 639.

Papageorge, A. G., Defeo-Jones, D., Robinson, P., Temeles, G., and Scolnick, E. M. (1984). *Mol. Cell. Biol.* **4,** 23.

Papaioannou, D. E., Gardner, R. L., McBurney, M. W., Babinet, C., and Evans, M. J. (1978). *J. Embryol. Exp. Morphol.* **44,** 93.

Papaioannou, V. E., McBurney, M. W., Gardner, R. L., and Evans, M. J. (1975). *Nature (London)* **258,** 70.

Papkoff, J., Nigg, E. A., and Hunter, T. (1983). *Cell (Cambridge, Mass.)* **33,** 161.

Parada, L. F., and Weinberg, R. A. (1983). *Mol. Cell. Biol.* **3,** 2298.

Parada, L. F., Tabin, C. J., Shih, C., and Weinberg, R. A. (1982). *Nature (London)* **297,** 476.

Parfett, C. L. J., Jamieson, J. C., and Wright, J. A. (1981). *Exp. Cell Res.* **136,** 1.

Parker, P. J., Stabel, S., and Waterfield, M. D. (1984). *EMBO J.* **3,** 953.

Parker, R. C., Varmus, H. E., and Bishop, J. M. (1981). *Proc. Natl. Acad. Sci. U.S.A.* **78,** 5842.

Parker, R. C., Varmus, H. E., and Bishop, J. M. (1984). *Cell (Cambridge, Mass.)* **37,** 131.

Pastan, I., and Willingham, M. (1978). *Nature (London)* **274,** 645.

Paterson, B. M., and Bishop, J. O. (1977). *Cell (Cambridge, Mass.)* **12,** 751.

Peehl, D. M., and Stanbridge, E. J. (1981). *Proc. Natl. Acad. Sci. U.S.A.* **78,** 3053.

Pelicci, P. G., Lanfrancone, L., Brathwaite, M. D., Wolman, S. R. and Dalla-Favera, R. (1984). *Science* **224,** 1117.

Pendergrass, W. R., Saulewicz, A. C., Burmer, G. C., Rabinovitch, P. S., Norwood, T. H., and Martin, G. M. (1982). *J. Cell. Physiol.* **113,** 141.

Pereira-Smith, O. M., and Smith, J. R. (1983). *Science* **221,** 964.

Perle, M. A., Leonard, C. M., and Newman, S. A. (1982). *Biochemistry* **21**, 2379.

Perry, R. P. (1983). *Cell (Cambridge, Mass.)* **33**, 647.

Perry, R. P. (1984). *Nature (London)* **310**, 14.

Persson, H., and Leder, P. (1984). *Science* **225**, 718.

Persson, H., Hennighausen, L., Taub, R., Degrado, W., and Leder, P. (1984). *Science* **225**, 687.

Peterson, T. A., Yochem, J., Byers, B., Nunn, M. F., Duesberg, P. H., Doolittle, R. F., and Reed, S. I. (1984). *Nature (London)* **309**, 556.

Petruzelli, L. M., Ganguly, S., Smith, C. J., Cobb, M. H., Rubin, C. S., and Rosen, O. M. (1982). *Proc. Natl. Acad. Sci. U.S.A.* **79**, 6792.

Pfeifer-Ohlsson, S., Goustin, A. S., Rydnert, J., Wahlström, T., Bjersing, L., Stehelin, D., and Ohlsson, R. (1984). *Cell* **38**, 585.

Phillips, P. D., Kuhnle, E., and Cristofalo, V. J. (1983). *J. Cell. Physiol.* **114**, 311.

Phillips, S. J., Birkenmeier, E. H., Callahan, R., and Eicher, E. M. (1982). *Nature (London)* **297**, 241.

Piatigorsky, J. (1981). *Differentiation* **19**, 134.

Piatigorsky, J. (1984). *Cell (Cambridge, Mass.)* **38**, 620.

Pierce, G. B. (1967). *Curr. Top. Dev. Biol.* **2**, 223.

Pierce, G. B., Pantazis, C. G., Caldwell, J. E., and Wells, R. S. (1982). *Cancer Res.* **42**, 1082.

Pierce, J. B., Lewis, S. H., Miller, G. J., Moritz, E., and Miller, P. (1979). *Proc. Natl. Acad. Sci. U.S.A.* **76**, 6649.

Pietropaolo, C., Laskin, J. D., and Weinstein, I. B. (1981). *Cancer Res.* **41**, 1565.

Plesko, M. M., and Richardson, A. (1984). *Biochem. Biophys. Res. Commun.* **118**, 730.

Pochron, S. F., O'Meara, A. R., and Kurtz, M. J. (1978). *Exp. Cell Res.* **116**, 63.

Pogo, B. G. T., Pogo, A. O., Allfrey, V. G., and Mirsky, A. E. (1968). *Proc. Natl. Acad. Sci. U.S.A.* **59**, 1337.

Pollak, R. D., and Fallon, J. F. (1974). *Exp. Cell Res.* **86**, 9.

Pollak, R. D., and Fallon, J. F. (1976). *Exp. Cell Res.* **100**, 15.

Ponder, B. A. J. (1980). *Biochim. Biophys. Acta* **605**, 369.

Popovic, M., Reitz, M. S., Jr., Sarngadharan, M. G., Robert-Guraff, M., Kabyanaraman, V. S., Nakao, Y., Miyoshi, I., Minowada, J., Yoshida, M., Ito, Y., and Gallo, R. C. (1982). *Nature (London)* **300**, 63.

Portier, M. M., Croizat, B., and Gros, F. (1982). *FEBS Lett.* **146**, 283.

Poste, G., and Nicolson, G. L. (1980). *Proc. Natl. Acad. Sci. U.S.A.* **77**, 399.

Poste, G., Tzeng, J., Doll, J., Greig, R., Rieman, A., and Zeidman, I. (1982). *Proc. Natl. Acad. Sci. U.S.A.* **79**, 6574.

Powers, S., Kataoka, T., Fasano, O., Goldfarb, M., Strathern, J., Broach, J., and Wigler, M. (1984). *Cell (Cambridge, Mass.)* **36**, 607.

Prakash, K., McBride, O. W., Swan, D. C., Devare, S. G., Tronick, S. R., and Aaronson, S. A. (1982). *Proc. Natl. Acad. Sci. U.S.A.* **79**, 5210.

Prasad, K. N. (1975). *Biol. Rev. Cambridge Philos. Soc.* **50**, 129.

Preumont, A. M., Van Gansen, P., and Brachet, J. (1978). *Mech. Ageing Dev.* **7**, 25.

Preumont, A. M., Capone, B., and Van Gansen, P. (1983). *Mech. Ageing Dev.* **22**, 167.

Pulciani, S., Santos, E., Lauver, A. V., Long, L. K., Robbins, K. C., and Barbacid, M. (1982). *Proc. Natl. Acad. Sci. U.S.A.* **79**, 2845.

Puvion-Dutilleul, F., and Macieira-Coelho, A. (1982). *Exp. Cell Res.* **138**, 423.

Puvion-Dutilleul, F., Puvion, E., Icard-Liepkalns, C., and Macieira-Coelho, A. (1984). *Exp. Cell Res.* **151**, 283.

Quigley, J. P. (1979). *Cell (Cambridge, Mass.)* **17**, 131.

Rabbitts, T. H. (1983). *Mol. Biol. Med.* **1**, 275.

Rabbitts, T. H., Hamlyn, P. H., and Baer, R. (1983). *Nature (London)* **306**, 760.

Rabbitts, T. H., Forster, A., Hamlyn, P., and Baer, R. (1984). *Nature (London)* **309**, 592.

Racker, E. (1983). *Biosci. Rep.* **3**, 507.

Radman, M., Jeggo, P., and Wagner, R. (1982). *Mutat. Res.* **98**, 249.

Raes, M., Houbion, A., and Remacle, J. (1981). *Biochim. Biophys. Acta* **642**, 313.

Ramaekers, F. C. S., Hukkelhoven, M. W. A. C., Groeneveld, A., and Bloemendal, H. (1984). *Biochim. Biophys. Acta* **799**, 221.

Ramirez, F., Gambino, R., Maniatis, G. M., Rifkind, R. A., Marks, P. A., and Bank, A. (1975). *J. Biol. Chem.* **250**, 6054.

Rao, M. V. N. (1976). *Exp. Cell Res.* **102**, 25.

Rattan, S. I. S., Keeler, K. D., Buchanan, J. H., and Holliday, R. (1982). *Biosci. Rep.* **2**, 561.

Razin, A., and Riggs, A. D. (1980). *Science* **210**, 604.

Razin, A., Webb, C., Szyf, M., Yisraeli, J., Rosenthal, A., Naveh-Many, T., Sciaky-Gallili, N., and Cedar, H. (1984). *Proc. Natl. Acad. Sci. U.S.A.* **81**, 2275.

Reddy, E. P. (1983). *Science* **220**, 1061.

Reddy, E. P., Reynolds, R. K., Santos, E., and Barbacid, M. (1982). *Nature (London)* **300**, 149.

Reddy, E. P., Smith, M. J., and Srinivasan, A. (1983). *Proc. Natl. Acad. Sci. U.S.A.* **80**, 3623.

Reeves, R., and Cserjesi, C. (1979a). *J. Biol. Chem.* **254**, 4283.

Reeves, R., and Cserjesi, C. (1979b). *Dev. Biol.* **69**, 584.

Reff, M., and Schneider, E. L. (1981). *Mol. Cell. Biochem.* **36**, 169.

Reich, N. C., and Levine, A. J. (1984). *Nature (London)* **308**, 199.

Reich, N. C., Oren, M., and Levine, A. J. (1983). *Mol. Cell. Biol.* **3**, 2143.

Reich, S., Rosin, H., Levy, M., Karkash, R., and Raz, A. (1984). *Exp. Cell Res.* **153**, 556.

Reiss, U., and Sacktor, B. (1983). *Proc. Natl. Acad. Sci. U.S.A.* **80**, 3255.

Reuben, R. C., Rifkind, R. A., and Marks, P. A. (1980). *Biochim. Biophys. Acta* **605**, 325.

Reynolds, F. H., Todaro, G. J., Fryling, C., and Stephenson, J. R. (1981). *Nature (London)* **292**, 259.

Reznick, A. Z., and Gershon, D. (1977). *Mech. Ageing Dev.* **6**, 345.

Reznick, A. Z., Lavie, L., Gershon, H. E., and Gerhson, D. (1981). *FEBS Lett.* **128**, 221.

Rheinwald, J. G., and Green, H. (1977). *Nature (London)* **265**, 421.

Ribot, H. D., Jr., Eisenman, E. A., and Kinsey, W. H. (1984). *J. Biol. Chem.* **259**, 5333.

Rigby, P. W. J. (1982). *Nature (London)* **297**, 451.

Riggs, A. D., and Jones, P. A. (1983). *Adv. Cancer Res.* **40**, 1.

Ringertz, N. R., Krondahl, U., and Coleman, J. R. (1978). *Exp. Cell Res.* **113**, 233.

Robbins, K. C., Antoniades, H. N., Devare, S. C., Hunkapiller, M. W., and Aaronson, S. A. (1983). *Nature (London)* **305**, 605.

Roberts, A. B., Anzano, M. A., Lamb, L. C., Smith, J. M., and Sporn, M. B. (1981). *Proc. Natl. Acad. Sci. U.S.A.* **78**, 5339.

Roberts, A. B., Frolik, C. A., Anzano, M. A., and Sporn, M. B. (1983). *Fed. Proc., Fed. Am. Soc. Exp. Biol.* **42**, 2621.

Robertson, M. (1984). *Nature (London)* **309**, 512.

Rogers, G. T. (1983). *Biochim. Biophys. Acta* **695**, 227.

Röhme, D. (1981). *Proc. Natl. Acad. Sci. U.S.A.* **78**, 5009.

Rohrschneider, L., and Rosok, M. J. (1983). *Mol. Cell. Biol.* **3**, 731.

Roos, E. (1984). *Biochem. Biophys. Acta* **783**, 263.

Rosenstraus, M. J., Sundell, C. L., and Liskay, R. M. (1982). *Dev. Biol.* **89**, 516.

Rosoff, P. M., Stein, L. F., and Cantley, L. C. (1984). *J. Biol. Chem.* **259**, 7056.

Rossant, J., and McBurney, M. W. (1982). *J. Embryol. Exp. Morphol.* **70**, 99.

Rothstein, M. (1977). *Mech. Ageing Dev.* **6**, 241.

Rothstein, M. (1979). *Mech. Ageing Dev.* **9**, 197.

Rotter, V. (1983). *Proc. Natl. Acad. Sci. U.S.A.* **80**, 2613.

Rotter, V., Abutbul, H., and Ben-Ze'ev, A. (1983). *EMBO J.* **2,** 1041.

Rous, P. (1911). *J. Exp. Med.* **13,** 397.

Rousset, J. P., Jami, J., Dubois, P., Avilès, D., and Ritz, E. (1980). *Somatic Cell Genet.* **6,** 419

Rousset, J. P., Bucchini, D., and Jami, J. (1983). *Dev. Biol.* **96,** 331.

Rovera, G., O'Brien, T. G., and Diamond, L. (1977). *Proc. Natl. Acad. Sci. U.S.A.* **74,** 2894.

Rovera, G., O'Brien, T. G., and Diamond, L. (1979a). *Science* **204,** 868.

Rovera, G., Santoli, D., and Damsky, G. (1979b). *Proc. Natl. Acad. Sci. U.S.A.* **76,** 2779.

Rovera, G., Olashaw, N., and Meo, P. (1980). *Nature (London)* **284,** 69.

Rubin, H. (1982). *Cancer Res.* **42,** 1761.

Rubin, H. (1984). *Proc. Natl. Acad. Sci. U.S.A.* **81,** 5121.

Rumsby, G., and Puck, T. T. (1982). *J. Cell. Physiol.* **111,** 133.

Rutishauser, U. (1984). *Nature* **310,** 549.

Rutter, W. J., Pictet, R. L., and Morris, P. W. (1973). *Annu. Rev. Biochem.* **42,** 601.

Ryan, J. M., and Cristofalo, V. (1975). *Exp. Cell Res.* **90,** 456.

Sabin, A. B. (1981). *Proc. Natl. Acad. Sci. U.S.A.* **78,** 7129.

Sabine, J. R. (1983). *Trends Biochem. Sci.* **8,** 234.

Sachs, L. (1980). *Proc. Natl. Acad. Sci. U.S.A.* **77,** 6152.

Sachs, L. (1982). *J. Cell. Physiol., Suppl.* **1,** 151.

Sager, R., and Kovac, P. (1982). *Proc. Natl. Acad. Sci. U.S.A.* **79,** 480.

Saidapet, C. R., Munro, H. N., Valgeirsdottir, K., and Sarkar, S. (1982). *Proc. Natl. Acad. Sci. U.S.A.* **79,** 3087.

Sampath, T. K., Nathanson, M. A., and Reddi, A. H. (1984). *Proc. Natl. Acad. Sci. U.S.A.* **81,** 3419.

Saneto, R., and Johnson, H. M. (1982). *Biochem. Biophys. Res. Commun.* **106,** 373.

Santos, E., Tronick, S. R., Aaronson, S. A., Pulciani, S., and Barbacid, M. (1982). *Nature (London)* **298,** 343.

Santos, E., Reddy, E. P., Pulciani, S., Feldmann, R. J., and Barbacid, M. (1983). *Proc. Natl. Acad. Sci. U.S.A.* **80,** 4679.

Santos, E., Martin-Zanca, O., Reddy, E. P., Pierotti, M. A., Della Porta, G., and Barbacid, M. (1984). *Science* **223,** 661.

Sashela, E., and Moorhead, P. S. (1963). *Proc. Natl. Acad. Sci. U.S.A.* **50,** 390.

Sato, T. (1930). *Roux's Arch.* **122,** 451.

Saunders, J. W. (1966). *Science* **154,** 604.

Saunders, J. W., Jr., Gasseling, M. T., and Saunders, L. C. (1962). *Dev. Biol.* **5,** 147.

Saxén, L., and Toivonen, S. (1962). "Primary Embryonic Induction." Academic Press, New York.

Saxén, L., Koskimies, P., Lahti, A., Miettinen, H., Rapola, J., and Wartiovaara, J. (1968). *Adv. Morphog.* **7,** 251.

Scarano, E. (1969). *Ann. Embryol. Morphog., Suppl.* **1,** 51.

Scarpelli, D. G., and Rao, M. S. (1981). *Proc. Natl. Acad. Sci. U.S.A.* **78,** 2577.

Schäfer, R., Doehmer, J., Drüge, P. M., Rademacher, I., and Willecke, K. (1981). *Cancer Res.* **41,** 1214.

Schäfer, R., Hoffmann, H., and Willecke, K. (1983). *Cancer Res.* **43,** 2240.

Schartl, M., and Barnekow, A. (1982). *Differentiation* **23,** 109.

Schechter, A. L., Stern, D. F., Vardyanathan, L., Decker, S. J., Drebin, J. A., Greene, M. I., and Weinberg, R. A. (1984). *Nature (London)* **312,** 513.

Scheib-Pfleger, D., and Wattiaux, R. (1962). *Dev. Biol.* **5,** 205.

Scher, W., Preisler, H. D., and Friend, C. (1973). *J. Cell Physiol.* **81,** 63.

Schibler, U., Pittet, A. C., Young, R. A., Hagenbüchle, O., Tosi, M., Gellman, S., and Wellauer, P. K. (1982). *J. Mol. Biol.* **155,** 247.

Schimke, R. T. (1984). *Cancer Res.* **44,** 1735.

Schliwa, M., Pryzwawsky, K. B., and Van Blerkom, J. (1982). *Philos. Trans. R. Soc. London, Ser. B* **299**, 199.

Schliwa, M., Nakamura, T., Porter, K. R., and Euteneuer, U. (1984). *J. Cell Biol.* **99**, 1045.

Schmitt, H., Guyaux, M., Pochet, R., and Kram, R. (1980). *Proc. Natl. Acad. Sci. U.S.A.* **77**, 4065.

Schneider, E. L., and Smith, J. R. (1981). *Int. Rev. Cytol.* **69**, 261.

Schneider, E. L., Bickings, C. K., and Sternberg, H. (1982). *Cytogenet. Cell Genet.* **33**, 249.

Schubert, D., and Jacob, F. (1970). *Proc. Natl. Acad. Sci. U.S.A.* **67**, 247.

Schwab, M., Alitalo, K., Varmus, H. E., Bishop, J. M., and George, D. (1983a). *Nature (London)* **303**, 497.

Schwab, M., Alitalo, K., Klempnauer, K. H., Varmus, H. E., Bishop, J. M., Gilbert, F., Brodeur, G., Goldstein, M., and Trent, J. (1983b). *Nature (London)* **305**, 245.

Schwab, M., Ellison, J., Busch, M., Rosenau, W., Varmus, H. E., and Bishop, J. M. (1984a). *Proc. Natl. Acad. Sci. U.S.A.* **81**, 4940.

Schwab, M., Varmus, H. E., Bishop, J. M., Grzeschik, K. H., Naylor, S. L., Sakaguchi, A. Y., Brodeur, G., and Trent, J. (1984b). *Nature (London)* **308**, 288.

Schwartz, R. J., and Rothblum, K. N. (1981). *Biochemistry* **20**, 4122.

Scolnick, E. M., Weeks, M. O., Shih, T. Y., Ruscetti, S. K., and Dexter, T. M. (1981). *Mol. Cell. Biol.* **1**, 66.

Scott, R. E., Florine, D. L., Wille, J. J., Jr., and Yun, K. (1982a). *Proc. Natl. Acad. Sci. U.S.A.* **79**, 845.

Scott, R. E., Hoerl, B. J., Wille, J. J., Jr., Florine, D. L., Krawisz, B. R., and Yun, K. (1982b). *J. Cell Biol.* **94**, 400.

Sefton, B. M., Hunter, T., Ball, E. H., and Singer, S. J. (1981). *Cell (Cambridge, Mass.)* **24**, 165.

Segawa, K., and Ito, Y. (1982). *Proc. Natl. Acad. Sci. U.S.A.* **79**, 6812.

Seiki, M., Hattori, S., and Yoshida, M. (1982). *Proc. Natl. Acad. Sci. U.S.A.* **79**, 6899.

Sekiguchi, T., Tosu, M., Yoshida, M. C., Oikawa, A., Ishihara, K., Fujiki, H., Tumuraya, M., and Kameya, T. (1982). *Somatic Cell Genet.* **8**, 605.

Selkoe, D. J., Abraham, C., and Ihara, Y. (1982). *Proc. Natl. Acad. Sci. U.S.A.* **79**, 6070.

Shakespeare, V., and Buchanan, J. H. (1976). *Exp. Cell Res.* **100**, 1.

Sharkey, N. A., Leach, K. L., and Blumberg, P. M. (1984). *Proc. Natl. Acad. Sci. U.S.A.* **81**, 607.

Sharma, H. K., and Rothstein, M. (1980). *Proc. Natl. Acad. Sci. U.S.A.* **77**, 5865.

Shay, J. W., and Clark, M. A. (1980). *Proc. Natl. Acad. Sci. U.S.A.* **77**, 381.

Shay, J. W., Lorkowski, G., and Clark, M. A. (1981). *J. Supramol. Struct. Cell Biochem.* **16**, 75.

Shealy, D. J., and Erikson, R. L. (1981). *Nature (London)* **293**, 666.

Sheffery, M., Rifkind, R. A., and Marks, P. A. (1982). *Proc. Natl. Acad. Sci. U.S.A.* **79**, 1180.

Shen, D. W., Real, F. X., Deleo, A. B., Old, L. J., Marks, P. A., and Rifkind, R. A. (1983). *Proc. Natl. Acad. Sci. U.S.A.* **80**, 5919.

Shen-Ong, G. L. C., Keath, E. J., Piccoli, S. P., and Cole, M. D. (1982). *Cell (Cambridge, Mass.)* **31**, 443.

Shih, C., and Weinberg, R. A. (1982). *Cell (Cambridge, Mass.)* **29**, 161.

Shih, C., Padhy, L. C., Murray, M., and Weinberg, R. A. (1981). *Nature (London)* **290**, 261.

Shilo, B. Z., and Weinberg, R. A. (1981a). *Proc. Natl. Acad. Sci. U.S.A.* **78**, 6789.

Shilo, B. Z., and Weinberg, R. A. (1981b). *Nature (London)* **289**, 607.

Shmookler Reis, R. J., and Goldstein, S. (1980). *Cell (Cambridge, Mass.)* **21**, 739.

Shmookler Reis, R. J., Goldstein, S., and Harley, C. D. (1980). *Mech. Ageing Dev.* **13**, 393.

Shmookler Reis, R. J., Lumpkin, C. K., Jr., McGill, J. R., Riabowol, K. T., and Goldstein, S. (1983). *Nature (London)* **301**, 394.

Sieber-Blum, M., and Sieber, F. (1981). *Differentiation* **20**, 117.

Silber, J. B., Fry, M., Martin, G. M., and Loeb, L. A. (1985). *Biochem. Biophys. Acta* **840**, 335.

Silvers, W. K., Gasser, D. L., and Eicher, E. M. (1982). *Cell (Cambridge, Mass.)* **28,** 439.
Simon, D., Stuhlmann, H., Jähner, D., Wagner, H., Werner, E., and Jaenisch, R. (1983). *Nature (London)* **304,** 275.
Simon, M. A., Kornberg, T. B., and Bishop, J. M. (1983). *Nature (London)* **302,** 837.
Simpson, R. T. (1978). *Cell (Cambridge, Mass.)* **13,** 691.
Sisskin, E. E., and Barrett, J. C. (1981). *Cancer Res.* **41,** 593.
Sivak, A. (1979). *Biochim. Biophys. Acta* **560,** 67.
Skinnider, L. F., and Giesbrecht, K. (1979). *Cancer Res.* **39,** 3332.
Slack, J. M. W. (1984a). *J. Embryol. Exp. Morphol.* **80,** 289.
Slack, J. M. W. (1984b). *J. Embryol. Exp. Morphol.* **80,** 321.
Slaga, T. J., Fischer, S. M., Weeks, C. E., Nelson, K., Mamrack, M., and Klein-Szanto, A. J. P. (1982). *Carcinogenesis* **7,** 19.
Slamon, D. J., de Kernion, J. B., Verma, I. M., and Cline, M. J. (1984). *Science* **224,** 256.
Smart, J. E., Oppermann, H., Czernilofsky, A. P., Purchio, A. F., Erikson, R. L., and Bishop, J. M. (1981). *Proc. Natl. Acad. Sci. U.S.A.* **78,** 6013.
Smith, B. L., Anisowicz, A., Chodosh, L. A., and Sager, S. (1982). *Proc. Natl. Acad. Sci. U.S.A.* **79,** 1964.
Smith, C. A., and Hanawalt, P. C. (1976). *Biochim. Biophys. Acta* **447,** 121.
Smith, R. C., and Knowland, J. (1984). *Dev. Biol.* **103,** 355.
Smith, J. R., and Lincoln, II, D. W. (1984). *Intern. Rev. Cytol.* **89,** 151.
Smith, R. D., and Yu, J. (1984). *J. Biol. Chem.* **259,** 4609.
Smith, R. L., Macara, I. G., Levenson, R., Hausman, D., and Cantley, L. (1982). *J. Biol. Chem.* **257,** 773.
Sobel, M. E., Yamamoto, T., de Crombrugghe, B., and Pastan, I. (1981). *Biochemistry* **20,** 2678.
Solter, D., Shevinsky, L., Knowles, B., and Strickland, S. (1979). *Dev. Biol.* **70,** 515.
Solursh, M., and Meier, S. (1973). *Dev. Biol.* **30,** 279.
Somers, K. D., and Murphey, M. M. (1980). *Cancer Res.* **40,** 4410.
Somers, K. D., and Murphey, M. M. (1982). *Cancer Res.* **42,** 2575.
Sorge, L. K., Levy, B. T., and Maness, P. F. (1984). *Cell (Cambridge, Mass.)* **36,** 249.
Sparkes, R. S., and Weiss, M. C. (1973). *Proc. Natl. Acad. Sci. U.S.A.* **70,** 377.
Speers, W. C. (1982). *Cancer Res.* **42,** 1843.
Speers, W. C., and Altmann, M. (1984). *Cancer Res.* **44,** 2129.
Spemann, H. (1938). ''Embryonic development and Induction''. Yale Univers. Press.
Spemann, H. (1968). ''Experimentelle Beiträge zu einer Theorie der Entwicklung,'' p. 51. Springer-Verlag, Berlin and New York (originally published, 1936).
Spemann, H., and Mangold, H. (1924). *Arch. Mikrosk. Anat. U. Entw. Mech.* **100,** 599.
Sperling, L., and Weiss, M. C. (1980). *Proc. Natl. Acad. Sci. U.S.A.* **77,** 3412.
Spiegelman, B. M., and Farmer, S. R. (1982). *Cell (Cambridge, Mass.)* **29,** 53.
Spiegelman, B. M., Frank, M., and Green, H. (1983). *J. Biol. Chem.* **258,** 10083.
Sporn, M. B., and Roberts, A. B. (1983). *Cancer Res.* **43,** 3034.
Spradling, A. C., and Mahowald, A. P. (1980). *Proc. Natl. Acad. Sci. U.S.A.* **77,** 1096.
Stacey, A. J., and Evans, M. J. (1984). *EMBO J.* **3,** 2279.
Stacey, D. W., and Kung, M. F. (1984). *Nature* **310,** 508.
Stanbridge, E. J. (1976). *Nature (London)* **260,** 17.
Stanbridge, E. J., Der, C. J., Doersen, C. J., Nishimi, R. Y., Peehl, D. M., Weissman, B. E., and Wilkinson, J. E. (1982a). *Science* **215,** 252.
Stanbridge, E. J., Rosen, S. W., and Sussman, H. H. (1982b). *Proc. Natl. Acad. Sci. U.S.A.* **79,** 6242.
Stanley, J. R., and Yuspa, S. H. (1983). *J. Cell Biol.* **96,** 1809.

Stein, G. H., Yanishevsky, R. M., Gordon, L., and Beeson, M. (1982). *Proc. Natl. Acad. Sci. U.S.A.* **70** 5287.

Stein, R., Sciaky-Gallili, N., Razin, A., and Cedar, H. (1983). *Proc. Natl. Acad. Sci. U.S.A.* **80,** 2422.

Steinberg, M. M., and Brownstein, B. L. (1982). *J. Cell. Physiol.* **113,** 359.

Steinemann, C., Fenner, M., Binz, H., and Parish, R. W. (1984). *Proc. Natl. Acad. Sci. U.S.A.* **81,** 3747.

Steinhagen-Thiessen, E., and Hilz, H. (1976). *Mech. Ageing Dev.* **5,** 447.

Stewart, T. A., and Mintz, B. (1981). *Proc. Natl. Acad. Sci. U.S.A.* **78,** 6314.

Stewart, T. A., Pattengale, P. K., and Leder, P. (1984a). *Cell* **38,** 627.

Stewart, T. A., Bellvé, A. R., and Leder, P. (1984b). *Science* **226,** 707.

Stiles, C. D. (1983). *Cell (Cambridge, Mass.)* **33,** 653.

Stiles, C. D., Desmond, W., Jr., Sato, G., and Saier, M. H. (1975). *Proc. Natl. Acad. Sci. U.S.A.* **72,** 4971.

Strickland, S. (1981). *Cell (Cambridge, Mass.)* **24,** 277.

Strickland, S., and Mahdavi, V. (1978). *Cell (Cambridge, Mass.)* **15,** 393.

Strickland, S., Smith, K. K., and Marotti, K. R. (1980). *Cell (Cambridge, Mass.)* **21,** 347.

Strom, C. M., and Dorfman, A. (1976). *Proc. Natl. Acad. Sci. U.S.A.* **73,** 1019.

Stuhlmann, H., Jähner, D., and Jaenisch, R. (1981). *Cell (Cambridge, Mass.)* **26,** 221.

Suda, T., Suda, J., and Ogawa, M. (1984). *Proc. Natl. Acad. Sci. U.S.A.* **81,** 2520.

Sugimoto, Y., Whitman, M., Cantley, L. C., and Erikson, R. L. (1984). *Proc. Natl. Acad. Sci. U.S.A.* **81,** 2117.

Sukumar, S., Notario, V., Martin-Zanca, D., and Barbacid, M. (1983). *Nature (London)* **306,** 658.

Sulston, J. E., Schierenberg, E., White, J. G., and Thomson, J. N. (1983). *Dev. Biol.* **100,** 64.

Swarup, G., Dasgupta, J. D., and Garbers, D. L. (1983). *J. Biol. Chem.* **258,** 10341.

Sweet, R. W., Yokoyama, S., Kamata, T., Feramisco, J. R., Rosenberg, M., and Gross, M. (1984). *Nature* **311,** 273.

Szpirer, J., Szpirer, C., and Wanson, J. C. (1980). *Proc. Natl. Acad. Sci. U.S.A.* **77,** 6616.

Tabin, C. J., Bradley, S. M., Bargmann, C. I., Weinberg, R. A., Papageorge, A. G., Scolnick, E. M., Dhar, R., Lowy, D. R., and Chang, E. H. (1982). *Nature (London)* **300,** 143.

Tabor, J. M., and Oshima, R. G. (1982). *J. Biol. Chem.* **257,** 8771.

Taketani, Y., and Oka, T. (1983). *Proc. Natl. Acad. Sci. U.S.A.* **80,** 1646.

Takeya, T., and Hanafusa, H. (1983). *Cell (Cambridge, Mass.)* **32,** 881.

Talmadge, J. E., Woman, S. R., and Fidler, I. J. (1982). *Science* **217,** 361.

Talmadge, J. E., Benedict, K., Madsen, J., and Fidler, I. J. (1984). *Cancer Res.* **44,** 3801.

Taparowsky, E., Suard, Y., Fasano, O., Shimizu, K., Goldfarb, M., and Wigler, R. (1982). *Nature (London)* **300,** 762.

Taparowsky, E., Shimizu, K., Goldfarb, M., and Wigler, M. (1983). *Cell (Cambridge, Mass.)* **34,** 581.

Tata, J. R. (1966). *Dev. Biol.* **13,** 87, 88.

Tata, J. R. (1971). *Curr. Top. Dev. Biol.* **6,** 79.

Taya, Y., Hosogai, K., Hirohashi, S., Shimosato, Y., Tsuchiya, R., Tsuchida, N., Fushimi, M., Sekiya, T., and Nishimura, S. (1984). *EMBO J.* **3,** 2943.

Taylor, S. M., and Jones, P. A. (1982). *J. Cell. Physiol.* **111,** 187.

Taylor, S. M., Oxford, J. M., and Metcalfe, J. A. (1981). *Int. J. Cancer* **27,** 317.

Temeles, G. L., Defeo-Jones, D., Tatchell, K., Ellinger, M. S., and Scolnick, E. M. (1984). *Molec. Cell Biol.* **4,** 2298.

Temin, H. M. (1964). *Natl. Cancer Inst. Monogr.* **17,** 557.

Temin, H. M. (1974). *Cancer Res.* **34,** 2835.
Temin, H. M. (1981). *Cell (Cambridge, Mass.)* **27,** 1.
Temin, H. M. (1983). *Nature (London)* **302,** 656.
Temin, H. M., and Mizutani, S. (1970). *Nature (London)* **226,** 1211.
Tencer, R., and Brachet, J. (1973). *Differentiation* **1,** 51.
Terranova, V. P., Williams, J. E., Liotta, L. A., and Martin, G. R. (1984). *Science* **226,** 982.
Thaw, H. H., Collins, V. P., and Brunk, U. T. (1984). *Mech. Ageing Dev.* **24,** 211.
Thein, R., and Lotan, R. (1982). *Cancer Res.* **42,** 4771.
Thompson, K. V. A., and Holliday, R. (1975). *Exp. Cell Res.* **96,** 1.
Thompson, K. V. A., and Holliday, R. (1978). *Exp. Cell Res.* **112,** 281.
Thorogood, P., Smith, L., Nicol, A., McGinty, R., and Garrod, D. (1982). *J. Cell Sci.* **57,** 331.
Tiedemann, H. (1968). *J. Cell. Physiol.* **72,** Suppl. 1, 129.
Till, J. E., and McCulloch, E. A. (1980). *Biochim. Biophys. Acta* **605,** 431.
Todaro, G. J., Fryling, C., and DeLarco, J. E. (1980). *Proc. Natl. Acad. Sci. U.S.A.* **77,** 5258.
Todaro, G. J., DeLarco, J. E., Fryling, C., Johnson, P. A., and Sporn, M. B. (1981). *J. Supramol. Struct. Cell Biochem.* **15,** 287.
Toivonen, S. (1950). *Rev. Suisse Zool.* **57,** Suppl. 1, 41.
Toivonen, S. (1979). *Differentiation* **15,** 177.
Toivonen, S., and Saxén, L. (1955). *Exp. Cell Res. Suppl.* **3,** 346.
Tollefsbol, T. O., Zaun, M. R., and Gracy, R. W. (1982). *Mech. Ageing Dev.* **20,** 93.
Tollefsbol, T. O., and Cohen, H. J. (1985a). *Mech. Ageing Dev.* **30,** 53.
Tollefsbol, T. O., and Cohen, H. J. (1985b). *J. Cell Physiol.* **123,** 417.
Tolmasoff, J. M., Ono, T., and Cutler, R. G. (1980). *Proc. Natl. Acad. Sci. U.S.A.* **77,** 2777.
Toné, S., Tanaka, S., and Kato, Y. (1983). *Dev., Growth Differ.* **25,** 381.
Tong, P. S. L., and Marcelo, C. L. (1983). *Exp. Cell Res.* **149,** 215.
Treton, J. A., and Courtois, Y. (1982). *Cell Biol. Int. Rep.* **6,** 253.
Trosko, J. E., and Cheng, C. C. (1978). *Photochem. Photobiol.* **28,** 157.
Trouet, A. (1978). *Eur. J. Cancer.* **14,** 105.
Tsiftsoglou, A. S., and Sartorelli, A. C. (1981). *Biochim. Biophys. Acta* **653,** 226.
Turkington, R. W., Majumder, G. C., and Riddle, M. (1971). *J. Biol. Chem.* **246,** 1814.
Turner, D. R., Griffith, V. C., and Morley, A. A. (1982). *Mech. Ageing Dev.* **19,** 325.
Twardzik, D. R., Ranchalis, J. E., and Todaro, G. J. (1982). *Cancer Res.* **42,** 590.
Ullrich, A., Coussens, L., Hayflick, J. S., Dull, T. J., Gray, A., Tam, A. W., Lee, J., Yarden, Y., Libermann, T. A., Schlessinger, J., Downward, J., Mayes, E. L., (1984). *Nature (London)* **309,** 418.
Urist, M. R., Kuo, Y. K., Brownell, A. G., Hohl, W. M., Buyske, J., Lietze, A., Tempst, P., Hunkapiller, M., and De Lange, R. J. (1984). *Proc. Natl. Acad. Sci. U.S.A.* **81,** 371.
Ushiro, H., and Cohen, S. (1980). *J. Biol. Chem.* **255,** 8363.
Van den Berghe, H. (1980). *Arch. Biol.* **91,** 473.
Van Gansen P. (1983). *In* "De la Molécule au Mammifère" (R. Thomas, J. P. Boon, and C. Vandecasserie, eds.), p. 243. Revue de l'Université de Bruxelles.
Van Gansen, P., Devos, L., and Ozoran, Y. (1977). *Biol. Cell.* **29,** 42a.
Van Gansen, P., Devos, L., Ozoran, Y., and Roxburgh, C. (1979). *Biol. Cell.* **34,** 255.
Van Gansen, P., Siebertz, B., Capone, B., and Malherbe, L. (1984). *Biol. Cell* **52,** 161.
Van Nest, G., Raman, R. K., and Rutter, W. J. (1983). *Dev. Biol.* **98,** 295.
Van Straaten, F., Müller, R., Curran, T., Van Beveren, C., and Verma, I. M. (1983). *Proc. Natl. Acad. Sci. U.S.A.* **80,** 3183.
Varmus, H. E. (1984). *Ann. Rev. Genet.* **18,** 553.
Vennström, B., and Bishop, J. M. (1982). *Cell (Cambridge, Mass.)* **28,** 135.
Vennström, B., Sheiness, D., Zabielski, J., and Bishop, J. M. (1982). *J. Virol.* **42,** 773.

Venolia, L., Gartlet, S. M., Wassman, E. R., Yen, P., Mohandas, T., and Shapiro, L. J. (1982). *Proc. Natl. Acad. Sci. U.S.A.* **79**, 2352.

Verma, A. K., Rice, H. M., and Boutwell, R. K. (1977). *Biochem. Biophys. Res. Commun.* **79**, 1160.

Verma, A. K., Shapas, B. G., Rice, H. M., and Boutwell, R. K. (1979). *Cancer Res.* **39**, 419.

Vogel, K. G., Kendall, V. F., and Sapien R. E. (1981). *J. Cell. Physiol.* **107**, 271.

Vollmers, H. P., and Birschmeier, W. (1983). *Proc. Natl. Acad. Sci. U.S.A.* **80**, 6863.

Vollmers, H. P., Imhof, B. A., Braun,S., Waller, C. A., Schirrmacher, V., and Birchmeier, W. (1984). *FEBS Lett.* **172**, 17.

Wachtel, S. S., Ohno, S., Koo, G. C., and Boyse, E. A. (1975). *Nature (London)* **257**, 235.

Waddington, C. H., Needham, J., and Brachet, J. (1936). *Proc. R. Soc. London, Ser. B* **120**, 173.

Wadsworth, S. C., Madhavan, K., and Bilodeau-Wentworth, D. (1985). *Nucl. Ac. Res.* **13**, 2153.

Waechter, D. E., and Baserga, R. (1982). *Proc. Natl. Acad. Sci. U.S.A.* **79**, 1106.

Wahli, W., Dawid, I. B., Ryffel, G. V., and Weber, R. (1981). *Science* **212**, 298.

Wahrmann, J. P., Winand, R., and Luzzati, D. (1973). *Nature (London), New Biol.* **245**, 112.

Wakelam, M. J. O. (1985). *Biochem. J.* **228**, 1.

Walford, R. L., Jawaid, S. Q., and Naeim, F. (1981). *Age* **4**, 67.

Walther, B. T., Pictet, R. L., David, J. D., and Rutter, W. J. (1974). *J. Biol. Chem.* **249**, 1953.

Walton, J. (1982). *Mech. Ageing Dev.* **19**, 217.

Wang, E. (1985). *J. Cell Biol.* **100**, 1466.

Wang, E., and Gundersen, D. (1984). *Exp. Cell Res.* **154**, 191.

Wang, J. Y. J., and Baltimore, D. (1983). *Mol. Cell. Biol.* **3**, 773.

Wang, S. Y., and Gudas, L. J. (1983). *Proc. Natl. Acad. Sci. U.S.A.* **80**, 5880.

Warren, S. T., Yotti, L. P., Moskal, J. R., Chang, C. C., and Trosko, J. E. (1981). *Exp. Cell Res.* **131**, 427.

Wartiovaara, J., Lehtonen, E., Nordlung, S., and Saxén, L. (1972). *Nature (London)* **238**, 407.

Waterfield, M. D., Scrace, G. T., Whittle, N., Stroobant, P., Johnsson, A., Wasteson, Å, Westermark, B., Heldin, C. H., Huang, J. S., and Deuel, T. F. (1983). *Nature (London)* **304**, 35.

Weatherall, D. J., and Clegg, J. B. (1982). *Cell (Cambridge, Mass.)* **29**, 7.

Weber, A., Gregori, C., and Schapira, F. (1976). *Biochim. Biophys. Acta* **444**, 810.

Weber, R. (1963). *In* "Lysosomes, Ciba Foundation Symposium" (A. V. S. De Renck and M. P. Cameron, eds.), p. 282. Little, Brown, Boston, Massachusetts.

Weber, R. (1967). *Compr. Biochem.* **28**, 145.

Webster, G. C., and Webster, S. L. (1983). *Mech. Ageing Dev.* **22**, 121.

Weeks, C. E., Herrmann, A. L., Nelson, F. R., and Slaga, T. J. (1982). *Proc. Natl. Acad. Sci. U.S.A.* **79**, 6028.

Weill, C. L., and Greene, D. P. (1984). *Nature (London)* **308**, 452.

Weinberg, R. A. (1980). *Cell (Cambridge, Mass.)* **22**, 643.

Weinberg, R. A. (1981). *Biochim. Biophys. Acta* **651**, 25.

Weinberg, R. A. (1982a). *Trends Biochem. Sci.* **7**, 135.

Weinberg, R. A. (1982b). *Cell (Cambridge, Mass.)* **30**, 3.

Weinberg, R. A. (1983). *J. Cell Biol.* **97**, 1661.

Weinberg, R. A. (1984). *Trends Biochem. Sci.* **9**, 131.

Weinstein, I. B. (1981). *J. Supramol. Struct. Cell Biochem.* **17**, 99.

Weinstein, I. B. (1983). *Nature (London)* **302**, 750.

Weinstein, I. B., Gattoni-Celli, S., Kirschmeier, P., Hsiao, W., Horowitz, A., and Jeffrey, A. (1984a). *Fed. Proc., Fed. Am. Soc. Exp. Biol.* **43**, 2287.

Weinstein, I. B., Gattoni-Celli, S., Kirschmeier, P., Lambert, M., Hsiao, W., Backer, J., and Jeffrey, A. (1984b). *J. Cell. Physiol. Suppl.* **3**, 127.

Weisbrod, S. (1982). *Nature (London)* **297,** 289.

Weisbrod, S., and Weintraub, H. (1979). *Proc. Natl. Acad. Sci. U.S.A.* **76,** 630.

Weiss, M. C., and Chaplain, M. (1971). *Proc. Natl. Acad. Sci. U.S.A.* **68,** 3026.

Weiss, R., Teich, N., Varmus, H., and Coffin, J., eds. (1982). "RNA Tumor Viruses." Cold Spring Harbor Lab., Cold Spring Harbor, New York.

Wells, R. S. (1982). *Cancer Res.* **42,** 2736.

Wertz, P. W., and Mueller, G. C. (1980). *Cancer Res.* **40,** 776.

Westin, E. H., Wong-Staal, F., Gelmann, E. P., Dalla Favera, R., Papas, T. S., Lautenberger, J. A., Eva, A., Reddy, E. P., Tronick, S. R., Aaronson, S. A., and Gallo, R. C. (1982). *Proc. Natl. Acad. Sci. U.S.A.* **79,** 2490.

Whatley, S. A., and Hill, B. T. (1979). *Cell Biol. Int. Rep.* **3,** 671.

White, F. A., and Bunn, C. L. (1985). *Mech. Ageing Dev.* **30,** 153.

Whiteley, B., Cassel, D., Zhuang, Y. X., and Glaser, L. (1984). *J. Cell Biol.* **99,** 1162.

Wiederanders, B., and Oelke, B. (1984). *Mech. Ageing Dev.* **24,** 265.

Williams, D. A., Lemischka, I. R., Nathan, D. G., and Mulligan, R. C. (1984). *Nature* **310,** 476.

Williams, G. T., and Johnstone, A. P. (1983). *Biosci. Rep.* **3,** 815.

Wilson, V. L., and Jones, P. A. (1983). *Science* **220,** 1055.

Winqvist, R., Saksela, K., and Alitalo, K. (1984). *EMBO J.* **3,** 2947.

Wittig, S., Hensse, S., Keitel, C., Elsner, C., and Wittig, B. (1983). *Dev. Biol.* **96,** 507.

Woerdeman, M. W. (1955). *In* "Biological Specificity and Growth" (E. Butler, ed.), p. 33. Princeton Univ. Press, Princeton, New Jersey.

Wojtyk, R. I., and Goldstein, S. (1980). *J. Cell. Physiol.* **103,** 299.

Wolf, S. F., and Migeon, B. R. (1982). *Nature (London)* **295,** 667.

Wolpert, L. (1971). *Curr. Top. Dev. Biol.* **6,** 183.

Wong-Staal, F., Dalla-Favera, R., Franchini, G., Gelmann, E. P., and Gallo, R. C. (1981). *Science* **213,** 216.

Wright, W. E. (1984a). *Exp. Cell Res.* **151,** 55.

Wright, W. E. (1984b). *J. Cell Biol.* **98,** 427.

Wright, W. E., and Aronoff, J. (1983). *J. Cell Biol.* **96,** 1571.

Wright, W. E., and Hayflick, L. (1972). *Exp. Cell Res.* **74,** 190.

Wright, W. E., and Hayflick, L. (1975). *Exp. Cell Res.* **96,** 113.

Wyllie, A. H., Kerr, J. F. R., and Currie, A. R. (1980). *Int. Rev. Cytol.* **68,** 251.

Xu, Y.-H., Ishii, S., Clark, A. J. L., Sullivan, M., Wilson, R. K., Ma, D. P., Roe, B. A., Merlino, G. T., and Pastan, I. (1984). *Nature (London)* **309,** 806.

Yaffe, D. (1969). *Curr. Top. Dev. Biol.* **4,** 37.

Yagi, M., and Koshland, E. (1981). *Proc. Natl. Acad. Sci. U.S.A.* **78,** 4907.

Yamada, T. (1950). *Biol. Bull. (Woods Hole, Mass.)* **98,** 98.

Yamada, T. (1967). *Compr. Biochem.* **28,** 113.

Yamada, T. and McDevitt, D. S. (1984). *Differentiation* **27,** 1.

Yamasaki, H., Fibach, E., Nudel, U., Weinstein, I. B., Rifkind, R. A., and Marks, P. A. (1977). *Proc. Natl. Acad. Sci. U.S.A.* **74,** 3451.

Yamasaki, H., Drevon, C., and Martel, M. (1982). *Carcinogenesis* **7,** 359.

Yamasaki, H., Martez, N., Fusco, A., and Ostertag, W. (1984). *Proc. Natl. Acad. Sci. U.S.A.* **81,** 2075.

Yancey, S. B., Edens, J., Trosko, J. E., Chang, C. C., and Revel, J. P. (1982). *Exp. Cell Res.* **139,** 329.

Yanishevsky, R. M., and Stein, G. H. (1980). *Exp. Cell Res.* **126,** 469.

Yotti, L. P., Change, C. C., and Trosko, J. E. (1979). *Science* **206,** 1089.

Young, P. R., and Tilghman, S. M. (1984). *Mol. Cell. Biol.* **4,** 898.

Yuasa, Y., Srivastava, S. K., Dunn, C. Y., Rhim, J. S., Reddy, E. P., and Aaronson, S. A. (1983). *Nature (London)* **303,** 775.

Yuasa, Y., Gol, R. A., Chang, A., Chiu, I. M., Reddy, E. P., Tronick, S. R., and Aaronson, S. A. (1984). *Proc. Natl. Acad. Sci. U.S.A.* **81,** 3670.

Yunis, J. J. (1983). *Science* **221,** 227.

Yuspa, S. H., Ben, T., Hennings, H., and Lichti, U. (1982). *Cancer Res.* **42,** 2344.

Zagris, N., and Eyal-Giladi, H. (1982). *Dev. Biol.* **91,** 208.

Zeelon, P., Gershon, H., and Gershon, D. (1973). *Biochemistry* **12,** 1743.

Ziegler, M. L. (1978). *Somatic Cell Genet.* **4,** 477.

Zimmerman, J., Brumbaugh, J., Biehl, J., and Holtzer, H. (1974). *Exp. Cell Res.* **83,** 159.

Zorn, G. A., Smith, B., and Hayflick, L. (1982). *J. Cell Biol.* **95,** 449a

Zwilling, E. (1964). *In* "Cellular Injury, Ciba Foundation Symposium" (A. V. S. De Reuck and J. Knight, eds.), p. 352. Little, Brown, Boston, Massachusetts.

Chapter 4

FINAL REMARKS

All the problems dealt with in this book have been examined in such detail that any further discussion would be both useless and repetitious. The aim of "Biochemical Cytology" had been to show that the biochemical and morphological approaches should always be combined if marked progress is to be made. This aim has now been fulfilled. Papers now published by cell biologists, in which electrophoretic analyses, electron micrographs, and photographs of sections stained using sophisticated cytochemical methods coexist, are becoming the rule rather than the exception. There is no doubt that in recent years, because of the development of biochemical and, later, molecular cytology, tremendous progress has been made in our understanding of the basic mechanisms of life in the cell. Yet, as pointed out repeatedly in this book, there is much more that we do not yet know; exciting discoveries still lie ahead for cell biologists. Like those before them, they will savor the wonderful flavor of discovery—the unique feeling of being the first to discover something and the only one to know about it. They will enjoy this feeling even if their new discovery adds only a small stone to the giant edifice being built by the whole scientific community.

In conclusion, we would like to present a few general ideas—they are not original, but they have perhaps not been expressed as clearly as one might have wished in this book. For instance, one of the major paradoxes of life, the fact that living organisms display *both unity and diversity,* was probably not made as apparent as it should have been. The reason for this is that the human mind seeks unity and is always looking for general, if not universal, laws; exceptions and contradictions are always unpleasant. We have often discussed "cells" without specifying the strains to which they belong. The currently available strains of cultured cells are innumerable, and they may differ slightly or profoundly from each other. Even in a well-known and very popular strain (3T3 fibroblasts, for instance), no cell is identical to its neighbors; cells may look alike, but each one retains its individuality. Variants appear due to somatic mutations. Even during early cleavage of a fertilized egg, a blastomere that contains specific cytoplasmic determinants (once called a *germinal localization*) is different from its neighbors.

Diversity in the plant and animal kingdoms is so obvious that it was portrayed by our distant ancestors when they were painting scenes of wildlife on the walls of their prehistoric caves. They knew very well the differences and similarities between a mammoth, a deer, and a horse.

The study of diversity in nature is the task of zoologists, botanists, and ecologists. Today, when increasing pollution is a menace for wildlife, ecology is attracting many students in the fields of botany and zoology. In Western Europe, this science is becoming more and more confused with politics. The "greens" fight for the preservation of nature, and for peace and against atomic plants and weapons. As a rule, ecologists who work on the "ground" have little sympathy for molecular and cell biologists who, in their laboratories, use radioisotopes and add (even a little) to the pollution of nature. However, without teamwork in well-equipped laboratories and without radioisotopes, very little progress would have been made in our understanding of cellular life.

Unity of living organisms results from the fact that they all obey the universal laws of molecular biology, a multidisciplinary science that attracts many students of biology, chemistry, physics, and medicine. Nucleic acids and proteins are at the core of all living things—from bacteria to humans. In fact, we are unable to imagine a form of life, on some distant planet, in which nucleic acids, proteins, and ATP are absent. Molecular biologists have solved the great mystery of heredity, working out the complex technology that has domesticated bacteria for the benefit of humanity. They are the priests of a new religion in which the name of God is spelled DNA; this religion has its dogmas that today, like many religious dogmas, are undergoing change. A few molecular biologists believe that we will know everything when human DNA has been completely sequenced. Yet, one may wonder whether knowledge of this sequence will ever explain how a human egg gives rise to a man (or woman) and how our brain works.

Ecologists and molecular biologists stand at two opposite poles among biologists, whereas cell biologists and embryologists, who are always confronted with the paradoxical coexistence of unity and diversity in cells, are in the middle. All animals and plants are composed of cells that have the same basic organization. Their nucleus (except in mammalian red blood cells) harbors DNA in chromatin, and this is where all types of RNA are synthesized. The nucleus is always limited by a double membrane, which is provided with nuclear pore complexes. In the cytoplasm, the ribosomes synthesize proteins and the mitochondria produce energy. The lysosomes specialize in intracellular digestion, while the Golgi bodies specialize in intracellular transport and distribution of the various proteins. The plasma membrane allows subtle exchanges between cells and with the outside medium. This is true for all cells; this amazing unity results from innumerable ages of evolution. However, cell differentiation, in even the lowest organisms leads to diversity. In a differentiated tissue, although all cells look alike—with the same morphology, the same physiological functions, and the same specific marker proteins—none of these cells is completely identical to its neighbor. Each has its own fate, as one may have gathered from what has been said about cell aging and death.

Nowhere is the acquisition of increasing diversity more visible than during embryonic development. This diversity results from genetic recombination at

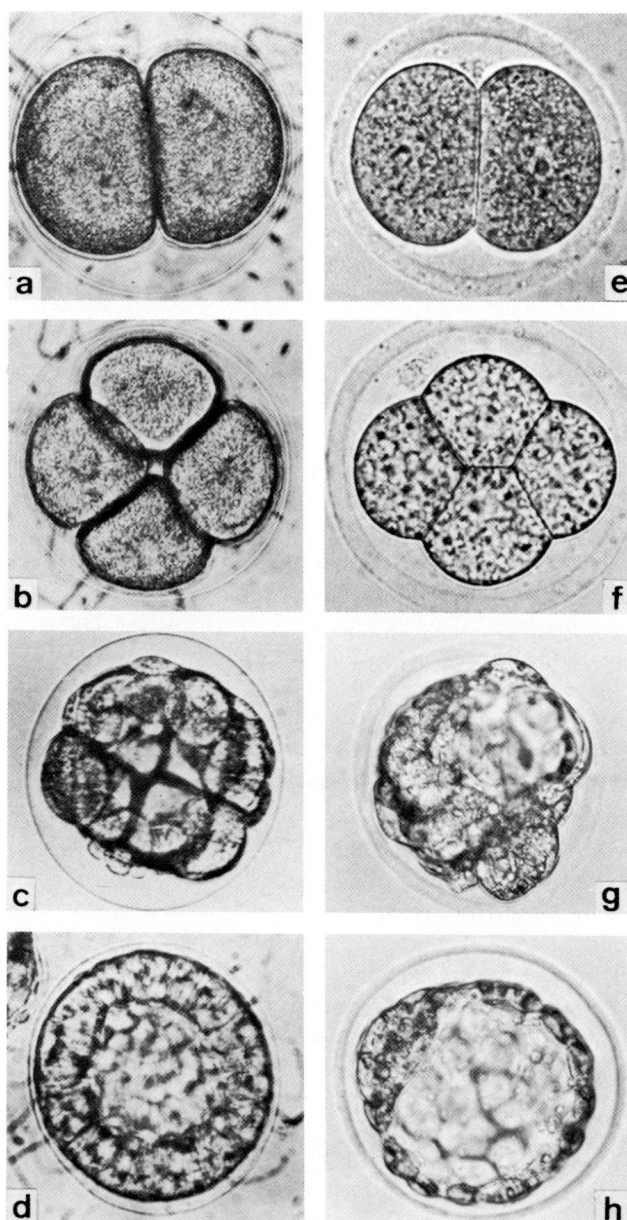

FIG. 1. Comparison between the cleavage stages of the sea urchin *Paracentrotus lividus* (a–d) and the mouse (e–h). (a,e) 2-cell stage; (b,f) 4-cell stage; (c,g) 32-cell stage; (g) nascent blastocyst of a mouse; (d) sea urchin blastula; (h) full-grown blastocyst of a mouse. [(a–d) Giudice, 1973; (e,h) courtesy of Dr. H. Alexandre.]

meiosis and from the association of maternal and paternal genes at fertilization. Each zygote possesses its own genetic program, which will be carried on until death. The early stages of development display a remarkable unity among animal species. During oogenesis, the oocyte accumulates RNA, proteins, glycogen, fats, etc. The large differences in size among eggs of various species depend mainly on the amount of yolk phosphoproteins that has been taken up by endocytosis. The basic structure of all oocytes, however, is the same. All have a large germinal vesicle containing extended lampbrush chromosomes (like the vegetative nucleus of the primitive alga *Acetabularia;* this is another striking indication of biological unity); their cytoplasm does not differ markedly from that of small somatic cells, except for the accumulation of variable amounts of yolk granules or platelets. The mechanisms of maturation and fertilization at both the morphological and biochemical levels are essentially the same in all species. Cleavage always leads to extensive and rapid DNA replication and cellularization, although there are some differences among species related to the yolk content of the eggs. Figure 1 compares the cleavage patterns of two eggs that are approximately the same size. Who, except an expert, would guess that one of them will give rise to a sea urchin and the other to a mouse? Later, cell-to-cell interactions, morphogenetic movements, inductions, and finally, embryonic differentiation will transform the similar-appearing blastulae into two completely different organisms. Our knowledge of the molecular mechanisms that form the basis of embryonic development has increased greatly. Nevertheless, the gap between the genetic program inscribed in the fertilized egg and the production of an adult remains gigantic. For this reason, embryogenesis and cell differentiation will remain fascinating fields of investigation for many years.

Of course, unity among living beings is due to the very fact that they are living. Very little has been said about life in this book, which deals with cells and embryos. What is the reason for this? Since the crystallization of tobacco mosaic virus (which was once believed to be a living macromolecule), the concepts of life and living have become more and more obscure for biologists. They prefer to leave the solution of the problem to philosophers and theoretical physicists, who certainly can provide many answers. In fact, it is always embarrassing to be asked whether an isolated nucleus, a chromosome, or a mitochondrion is alive. This question cannot be answered simply "yes" or "no," since, in the case of the cell and its organelles, life is often more a quantitative than an all-or-none concept. This is also true for organisms: is not a baby or a child more "lively" than an old person?

We have seen that anucleate fragments of unicellular organisms, eggs and fibroblasts (cytoplasts), are undoubtedly alive; however, their life spans and biochemical potentialities may greatly differ. An anucleate fragment of *Acetabularia* is more alive than an enucleate ameba, not only because it has a much longer life span, but because it is capable of extensive, prolonged morphogenesis; it is the seat of important DNA, RNA, and protein synthesis, due to the partial autonomy of the chloroplasts toward the nucleus. In contrast, an anucleate fragment of an ameba rounds up and loses locomotion within a few minutes; RNA synthesis stops immediately, and protein synthesis terminates after a few days. Still mysterious is the rapid resumption of locomotion if an ameba nucleus (even from a foreign species) is grafted into the anucleate cytoplasm. After a few days of enucleation, the anucleate half can no longer be reactivated by grafting a nucleus. Irreversible changes, symptoms of premature aging, have taken place in the anucleate half. Yet this half is still alive, since several more days will elapse before it "dies," i.e., before it disintegrates and vanishes. Cytoplasts from fibroblasts have an even shorter life span than anucleate fragments of A. proteus. However, they can be saved if they are fused quickly enough with karyoplasts. The resulting "reconstituted" cell may divide; this is, of course, the best proof that it is really alive. Even an anucleate half of Acetabularia is less "alive" than a whole alga; it has lost the main attribute of all living organisms—reproduction. The same story could be told for anucleate fragments of sea urchin and amphibian eggs. They survive for some time, but their response to parthenogenetic agents is ridiculous compared to that of intact unfertilized eggs. In summary, nonnucleated cytoplasm always remains "alive" for some time, but its biological activities are always inferior to those of whole cells.

Similar considerations hold for the cell nucleus. Nuclear transfer experiments have shown that nuclei isolated from amebas or amphibian embryos are "killed" if they remain in contact with the external medium for too long a time; improvement of the isolation medium (by the addition of polyamines, for instance) prolongs the "life" of somatic nuclei. In such experiments, the test for distinguishing "living" from "dead" nuclei is a stringent one: the capacity to undergo repeated mitotic divisions after injection into enucleated amebas or eggs. However, considerable biochemical work has shown that nuclei isolated from homogenates continue the already initiated DNA synthesis and synthesis of all types of RNA for some time (1–2 hr). It is very doubtful that all of these somatic nuclei would divide if they were reintroduced into enucleated cells (cytoplasts), but, as far as is known, no comparative study on the loss of the capacity of an isolated nucleus to divide after transplantation into cytoplasm and the loss of in vitro nucleic acid synthesis has been done. Such a study would probably show that isolated nuclei lose their ability to divide earlier than their capacity to synthesize DNA and RNA. If this is the case, nuclei that can still produce nucleic acids, but are no longer able to divide, might be considered less alive than those that have retained the capacity to enter into mitosis after reintroduction into an enucleated cell.

This discussion shows that, as correctly pointed out by B. Ephrussi in 1953, the cell as a whole should be considered a functional as well as a morphological unit; it is only for experimental purposes that we can dissect it. However, if we had not dissected it by various means (enucleation, homogenization followed by differential centrifugation, localized uv or laser irradiation, etc.), molecular cytology would not exist.

The more we delve into these problems, the more mysterious life remains. All we can do is to contribute our work to a partial solution of the biological riddles. What we must also do is to admire, respect, and love life—we must protect it and not destroy it. We biologists must work more than anyone else for peace.

REFERENCE

Giudice, G. (1973). "Developmental Biology of the Sea Urchin Embryo," p. 11. Academic Press, New York.

INDEX*

A

Acetabularia
  chloroplasts, **1**,136–138, 140, 142
  enucleated, **2**,23–50
    circadian rhythms, **2**,30–31
    energy production, **2**,28, 30
    enzyme synthesis, **2**,32
    nucleic acid synthesis, **2**,32–33, 35–39,
      43, 45–46
    protein synthesis, **2**,31–32
  life cycle, **2**,23–24
  nucleocytoplasmic interactions in, **2**,23–50
    interspecific grafts, **2**,24–26
    morphogenetic substances, **2**,24–27, 31–
      50
  vegetative nucleus, **2**,39–45
Acetylation, of histone, **1**,210, 285
Acetylcholinesterase, **2**,234–235
Acetyl-CoA-carboxylase, **1**,93
N-Acetylglucosamine phosphomannose, **1**,114
Achromatic apparatus, *see* Mitotic apparatus
Acid phosphatase
  in anucleate cytoplasm, **2**,19, 20, 38, 39
  detection, **1**,109, 110
  lysosomal, **1**,108, 109
  in senescent cells, **2**,423
Acquired immune deficiency syndrome
    (AIDS), **2**,389
Acridine orange, **1**,11, 15
  for nucleic acid binding, **1**,11, 13
Acrosin, **1**,116–117
Acrosome
  description, **1**,116, 118
  properties, **1**,117–118
Acrosome reaction, **2**,163
  induction, **2**,170
Actin
  in anaphase chromosome movement, **1**,328
  in cancer cells, **2**, 371–372
  capping, **1**,32, 33
  cross-linkage, **1**,45
  depolymerization, **1**, 36
  detection, **1**,14, 17

endocytosis and, **1**,45, 47
  F-, **1**,81, 84, 117, 320
    depolymerizing factor, **1**,84
  G-, **1**,81; **2**,117
    in acrosome reaction, **2**,163
  genes, **2**,310
  in germinal vesicle, **2**,117
  in mitotic apparatus, **1**,320
  in nuclear sap, **1**,245–246
  patching, **1**,32
  polymerization, **1**,81, 84
  protein binding, **1**,32, 81
  in senescent cells, **2**,421, 422
  in sol–gel transformation, **1**,68
  transcription, **2**,250
α-Actin
  gene, **2**,250
  as myogenesis marker, **2**,309
  RNA, **2**,250
β-Actin, **1**,81
δ-Actin, **1**,81
Actin-binding protein, **1**,81–84
Actin-capping protein, **1**,84
α-Actinin, **1**,81, 82
  actin-binding, **1**,81
  in cytokinesis, **1**,328
β-Actinin, **1**,83
Actin/myosin ratio, **1**,83
Actinomycin D
  cleavage inhibition, **2**,212
  protein synthesis, effects, **2**, 233
  tritiated, **2**,5
Actomyosin
  in cell locomotion, **1**,67–68
  in cytokinesis, **1**,328–329
  in microvilli contraction, **1**,46
Acumentin, **1**,84
Adenosine diphosphate, coupling mechanism,
    1,124–125
Adenosine ribosyltransferase, **2**,330–331
Adenosine triphosphatase, **1**,326
  mitotic activity peaks, **1**,334
Adenosine triphosphate
  coupling mechanism, **1**,124–125

*Boldface numbers indicate volume numbers.

477

DATE DUE

AUG 7 87			
AUG 8 1988			

DEMCO 38-297